3

AF770855

BIOCHEMISTRY RESEARCH TRENDS

BIOACTIVE HETEROCYCLES

SYNTHESIS AND BIOLOGICAL EVALUATION

BIOCHEMISTRY RESEARCH TRENDS

Additional books in this series can be found on Nova's website under the Series tab.

Additional e-books in this series can be found on Nova's website under the e-books tab.

BIOCHEMISTRY RESEARCH TRENDS

BIOACTIVE HETEROCYCLES

SYNTHESIS AND BIOLOGICAL EVALUATION

K. L. AMETA

R. P. PAWAR

AND

A. J. DOMB

EDITORS

Nova Biomedical

New York

Copyright © 2013 by Nova Science Publishers, Inc.

All rights reserved. No part of this book may be reproduced, stored in a retrieval system or transmitted in any form or by any means: electronic, electrostatic, magnetic, tape, mechanical photocopying, recording or otherwise without the written permission of the Publisher.

For permission to use material from this book please contact us:
Telephone 631-231-7269; Fax 631-231-8175
Web Site: http://www.novapublishers.com

NOTICE TO THE READER

The Publisher has taken reasonable care in the preparation of this book, but makes no expressed or implied warranty of any kind and assumes no responsibility for any errors or omissions. No liability is assumed for incidental or consequential damages in connection with or arising out of information contained in this book. The Publisher shall not be liable for any special, consequential, or exemplary damages resulting, in whole or in part, from the readers' use of, or reliance upon, this material. Any parts of this book based on government reports are so indicated and copyright is claimed for those parts to the extent applicable to compilations of such works.

Independent verification should be sought for any data, advice or recommendations contained in this book. In addition, no responsibility is assumed by the publisher for any injury and/or damage to persons or property arising from any methods, products, instructions, ideas or otherwise contained in this publication.

This publication is designed to provide accurate and authoritative information with regard to the subject matter covered herein. It is sold with the clear understanding that the Publisher is not engaged in rendering legal or any other professional services. If legal or any other expert assistance is required, the services of a competent person should be sought. FROM A DECLARATION OF PARTICIPANTS JOINTLY ADOPTED BY A COMMITTEE OF THE AMERICAN BAR ASSOCIATION AND A COMMITTEE OF PUBLISHERS.

Additional color graphics may be available in the e-book version of this book.

Library of Congress Cataloging-in-Publication Data

ISBN: 978-1-62257-636-4

Library of Congress Number: 2012944244

Published by Nova Science Publishers, Inc. † New York

Contents

Preface

The majority of pharmaceuticals and biologically active agrochemicals are heterocycles while countless additives and modifiers used in industrial applications ranging from cosmetics, reprography, information storage and plastics are heterocyclic in nature. One striking structural feature inherent to heterocycles, which continues to be exploited to great advantage by the drug industry, lies in their ability to manifest substituents around a core scaffold in defined three-dimensional representations. For more than a century, heterocycles have constituted one of the largest areas of research in organic chemistry.

They have been immensely conducive to the development of society from a biological and industrial point of view as well as to the understanding of life processes and the efforts to improve the quality of life. Amongst approximately 20 million chemical compounds identified by the end of the second millennium, more than two-thirds are fully or partially aromatic and approximately half of them are heterocyclic. The presence of heterocycles in all kinds of organic compounds of interest in electronics, biology, optics, pharmacology, material sciences and so on is widely known and recognized. However, specifically, sulfur and nitrogen-containing heterocyclic compounds have maintained the interest of researchers through decades of historical development of organic synthesis.

Nevertheless, heterocycles with other heteroatom such as oxygen, phosphorus and selenium also appear on the scene. Many natural drugs such as papaverine, theobromine, quinine, emetine, theophylline, atropine, procaine, codeine, reserpine and morphine are heterocycles. Almost all the compounds we know as synthetic drugs such as diazepam, chlorpromazine, isoniazid, metronidazole, azidothymidine, barbiturates, antipyrine, captopril and methotrexate fall under the same name. Some dyes (e.g. mauveine), luminophores, (e.g. acridine orange), pesticides (e.g. diazinon) and herbicides (e.g. paraquat) are also heterocyclic in nature. Thus, the purpose behind writing this book is to provide a succinct summary of methods for the synthesis and biological activities of various different sized bioactive heterocycles apart from highlighting anticancer, antifungal, antibacterial, antiHIV, anti-inflammatory, antioxidant and many more biological activities of heterocyclic compound.

The Editors

About the Editors

Dr. K. L. Ameta received his doctorate degree in Organic Chemistry from M. L. Sukhadia University, Udaipur, India in the year 2002. Presently, he is working as Assistant Professor of Chemistry, Faculty of Arts, Science and Commerce, Mody Institute of Technology and Science (Deemed University) Lakshmangarh, Rajasthan, India. He has vast experience of teaching both graduate and postgraduate level students. His research area involves synthesis, characterization and biological evaluation of different sized bioactive heterocyclic systems. In addition, to this he has keen interests in photocatalysis. He has published a number of research articles in various journal of national and international repute.

Dr. R. P. Pawar is serving as a Head and Associate Professor in Deogiri College, Aurangabad, Maharashtra, India. He completed his doctorate degree in Organic Chemistry from Swami Ramanand Teerth Marathwada University, Nanded, India in the year 1998. He did his postdoctoral training at Department of Medicine, The Hebrew University, Jerusalem, Israel. He has immense experience of graduate and postgraduate level teaching. His research area involves Synthetic Organic Chemistry, Synthesis and Biological evaluation of novel bioactive heterocyclic systems, Green Chemistry Approches. He has published a number of research articles in National and International reputed journals.

Dr. A. J. Domb is a Professor for Medicinal Chemistry and Biopolymers at the Faculty of Medicine of the Hebrew University, Jerusalem, Israel. He earned Bachelors degrees in Chemistry, Pharmaceutics and Law studies and Ph. D. degree in Chemistry from Hebrew University. He did his postdoctoral training at MIT and Harvard Univ. USA and was R&D manager at Nova Pharm. Co. Baltimore US during 1988-1992. Since 1992 he is a faculty member at the Hebrew university with interests in biopolymers, medicinal chemistry and forensic sciences.

Contributors

A. G. Ibragimov: Institute of Petrochemistry and Catalysis, Russian Academy of Sciences, 141 Prospekt Oktyabrya, Ufa, 450075, Russian Federation.

Abraham J. Domb: School of Pharmacy, Department of Medicine, The Hebrew University of Jerusalem, Jerusalem 91120, Israel

Ambadas B. Rode: Department of Chemistry, Kongju National University, 182 Shinkwan-dong, Kongju, Chungnam 314-701, Republic of Korea.

Anita R. Maguire: Department of Chemistry & School of Pharmacy, Analytical and Biological Chemistry Research Facility, University College Cork, Cork, Ireland.

Anjan R. Sawale: Department of Chemistry, Deogiri College, Aurangabad-431005, MS, India.

Chabubai. S. Patil: Department of Chemistry, Deogiri College, Aurangabad-431005, MS, India.

E. B. Rakhimova: Institute of Petrochemistry and Catalysis, Russian Academy of Sciences, 141 Prospekt Oktyabrya, Ufa, 450075, Russian Federation.

Elena Soriano: Laboratorio de Radicales Libres y Química Computacional (IQOG, CSIC), C/ Juan de la Cierva 3, 28006-Madrid, Spain.

Essam M. Hussein: Department of Chemistry, Faculty of Science, Assiut University, Assiut 71516, Egypt.

Florea Dumitrascu: Center of Organic Chemistry "C. D. Nenitzescu", Romanian Academy, Spl. Independentei 202B, Bucharest 060023, Romania.

J. L. Marco: Laboratorio de Radicales Libres y Química Computacional (IQOG, CSIC), C/ Juan de la Cierva 3, 28006-Madrid, Spain.

K. L. Ameta: Department of Chemistry, MITS University, P-Lakshmangarh, Rajasthan, India-332311.

M. Carmo Carreiras: Research Institute for Medicines and Pharmaceutical Sciences (iMed.UL), Faculty of Pharmacy, University of Lisbon. Av. Prof. Gama Pinto, 1649-003 Lisbon, Portugal.

Marcel Mirel Popa: Center of Organic Chemistry "C. D. Nenitzescu", Romanian Academy, Spl. Independentei 202B, Bucharest 060023, Romania.

Marie Kissane: Department of Chemistry, Analytical and Biological Chemistry Research Facility, University College Cork, Cork, Ireland (current address Chemical Product R&D, Lilly Research Laboratories, Eli Lilly and Company, Indianapolis, IN 46285, USA.

Maya V. Katariya: Department of Chemistry, Deogiri College, Aurangabad (MS) India 431005.

Mino R. Caira: Department of Chemistry, University of Cape Town, Rondebosch 7701, South Africa.

Namrata D. Erande: Department of Chemistry, Deogiri College, Aurangabad-431005, MS, India.

N. F. Galimzyanova: Institute of Biology, Ufa Scientific Centre, Russian Academy of Sciences, Ufa, 450075, Russian Federation.

Okram Mukherjee Singh: Department of Chemistry, Manipur University, Canchipur, Imphal-795 003, India.

Prakash T. Parvatkar: Department of Chemistry, Goa University, Taleigao Plateau, Goa, 403 206, India.

Rajendra P. Marathe: Department of Pharmacy, Govement College of Pharmacy, Aurgangabad-431 005, India.

Rajendra P. Pawar: Department of Chemistry, Deogiri College, Aurangabad-431005, MS, India.

Rajesh B. Navale: Department of Pharmacy, Govement College of Pharmacy, Aurgangabad-431 005, India.

Rajita D. Ingle: Department of Chemistry, Deogiri College, Aurangabad-431005, MS, India.

Ramesh C. kumar: Institute of Biotechnology and pharmaceuticals research, National Health Research Institute. Zhunan, Taiwan.

Rameshwar R. More: Department of Chemistry, Deogiri College, Aurangabad (MS) India 431005.

Rupali L. Magar: Department of Chemistry, Deogiri College, Aurangabad-431005, MS, India.

Sandeep B. Mane: Institute of Chemistry, Academia sinica, Taipei, Taiwan

Sandeep V. Shinde: Department of Chemistry, Indira Gandhi College, Cidco, Nanded, India.

Sanjay B. Ubale: Department of Chemistry, Deogiri College, Aurangabad, India.

Sanjay K Vyawahare: Department of Physics, Deogiri College, Aurangabad-431005, MS, India.

Santosh G. Tilve: Department of Chemistry, Goa University, Taleigao Plateau, Goa, 403 206, India.

Saroj R. Bembalkar: Department of Chemistry, Deogiri College, Aurangabad-431005, MS, India.

Satish A. Dake: Department of Chemistry, Deogiri College, Station Road, Aurgangabad-431 005, MS, India.

Satish U. Deshmukh: Department of Chemistry, Deogiri College, Aurangabad, India.

Sunil K. Wasmatkar: Department of Chemistry, Deogiri College, Aurangabad, India.

Sunil R. Mirgane: Department of Chemistry, J. E. S. College, Jalna-431203, India.

Sunil U. Tekale: Department of Chemistry, Shri Muktanand College, Gangapur-431109, MS, India.

Sushma S. Kauthale: Department of Chemistry, Deogiri College, Aurangabad (MS) India 431005.

Swapnil R. Sarda: Department of Chemistry, J. E. S. College, Jalna-431203, MS, India

Thokchom Prasanta Singh: Department of Chemistry, Manipur University, Canchipur, Imphal-795 003, India.

Vijaykumar V. Paike: Department of Chemistry, National Tsing Hua University, Hsinchu, Taiwan.

Vinod K. Ahirrao: Department of Chemistry, Deogiri College, Aurangabad-431005, MS, India.

Vinod V. Thorat: Department of Chemistry, Deogiri College, Aurangabad-431005, MS, India.

Vivekanand B. Jadhav: Department of Chemistry, Shri Muktanand College, Gangapur-431109, MS, India.

V. R. Akhmetova: Institute of Petrochemistry and Catalysis, Russian Academy of Sciences, 141 Prospekt Oktyabrya, Ufa, 450075, Russian Federation.

Wamanrao N. Jadhav: Organic Chemistry Synthesis Laboratory, Dnyanopasak College, Parbhani-431401, India.

Yeshwant B. Vibhute: Department of Chemistry, Yeshwant Mahavidyalaya, Nanded-431602, MS, India.

In: Bioactive Heterocycles
Editors: K. L. Ameta, R. P. Pawar and A. J. Domb
ISBN: 978-1-62257-636-4
© 2013 Nova Science Publishers, Inc.

Chapter 1

Multipotent 1,8-Naphthyridines, as New Tacrine Analogues, for the Treatment of Alzheimer's Disease

M. Carmo Carreiras[1,]*, Elena Soriano[2] and J. L. Marco[3]
[1]Research Institute for Medicines and Pharmaceutical Sciences (iMed.UL), Faculty of Pharmacy, University of Lisbon, Lisbon, Portugal
[2]SEPCO (IQOG, CSIC), Madrid, Spain
[3]Laboratorio de Química Médica (IQOG, CSIC), Madrid, Spain

Abstract

The synthesis and biological evaluation of 1,8-naphthyridines, new tacrine analogues, as acetylcholinesterase and butyrylcholinesterase inhibitors, as well as voltage-dependent Ca^{2+} channels modulators is reviewed. These compounds are hybrids from tacrine and 1,4-dihydropyridines, well-known Ca^{2+}-antagonists. They were synthesized from conveniently substituted and functionalized 2-amino-3-cyano substituted heterocycles *via* Friedlander-type reaction with diverse cycloalkanones. These compounds are moderate and selective AChE inhibitors, as well as Ca^{2+} channels antagonists. As the regulation of Ca^{2+} entry to cells has been described to play a key role in the cell apoptosis process, some of them were studied as neuroprotective agents against different toxic agents.

Keywords: Acetylcholinesterase; Butyrylcholinesterase; Friedländer reaction; Tacrine; 1,8-Naphthyridines; 1,4-Dihydropyridines; Voltage-dependent calcium channels; Neurodegenerative diseases; Alzheimer's disease

1. Introduction

Alzheimer's disease (AD) is a neurodegenerative, irreversible disorder clinically characterized by abnormal memory loss along with deterioration of other cognitive abilities as well as motor capacities [1]. Although the etiology of AD is still poorly understood, amyloid β (Aβ) deposits [2], τ-protein aggregation [3], oxidative stress [4] or low levels of acetylcholine (ACh) are of paramount importance in the progress of the disease [5-6]. Based on the cholinergic hypothesis, cholinesterase inhibitors (ChEI) such as rivastigmine, donepezil, and galanthamine [7,8] are currently used for the treatment of AD. ChEI, by inhibiting acetylcholinesterase (AChE) and butyrylcholinesterase (BuChE), prevent the hydrolysis of ACh at the synaptic cleft [9], enhancing the central cholinergic deficit, thus alleviating its symptomatology [10].

A renewed interest in AChEI [11, 12] emerged when beneficial effects, which were not directly linked to cholinergic neurotransmission, were demonstrated for some of them [13-17]. In this regard, galanthamine has been described to protect human neuroblastoma cells and cortical neurons against Aβ-induced neurotoxicity [18-19]. Furthermore, donepezil was confirmed to reduce τ-phosphorylation by inhibition of glycogen synthase kinase 3β [20-22]. In addition, AChE seems to accelerate Aβ fibril aggregation in the brain [23], an activity located in the peripheral anionic site (PAS) of AChE [24, 25]. *In vivo* studies of some AChEI-antiaggregating drugs showed reversion of the cognitive impairment due to scopolamine in the MWM (Morris Water Maze) test, as well as reduction of plaque load in the brains of human amyloid precursor protein (APP) transgenic mice [26].

The multifactorial profile of AD has prompted the search for multipotent drugs able to interact simultaneously in two or more targets [27]. The multitarget approach [28] has driven to hybrids of tacrine-donepezil [29, 30], tacrine-melatonin [31, 32], tacrine/8-hydroxyquinolein [33], coumarins that inhibit both MAO and the AChE enzyme [34, 35], and highly potent neuroprotective rasagiline-rivastigmine hybrids [36].

Since 1996 the group led by Marco-Contelles and colleagues has been investigating new multipotent drugs able to target the cell Ca^{2+} signal [37, 38] along with a good inhibitory activity of ChE enzymes [39-41].

Though Ca^{2+} overload is the main factor that triggers processes leading to cell death, the Ca^{2+} dyshomeostasis has been a poorly investigated target in order to develop new drugs for the treatment of AD [42]. The Ca^{2+} hypothesis of ageing [43], supported by the fact that sustained changes in the regulation of $[Ca^{2+}]_i$ are due to age-dependent alterations in the cellular mechanisms of Ca^{2+} homeostasis, has shown that calcium dysfunction can increase Aβ formation and tau hyperphosphorylation [44, 45]. More recently, some authors reported that the P86L-mutated channel CALHM1 generates both mitochondrial ($[Ca^{2+}]_m$) and cytosolic ($[Ca^{2+}]_c$) Ca^{2+} signal with a transient pattern, but Ca^{2+} entry through P86L-CALHM1 generates a plateau of sustained and prolonged in time $[Ca^{2+}]_m$, which could be relevant in the context of mitochondrial Ca^{2+} overload, neuronal death, and the association between P86L-CALHM1 and the risk of AD [46, 47]. Calcium entry through L-type Ca^{2+} channels causes both calcium overload and mitochondrial disruption, which leads to activation of the apoptotic cascade and cell death [48]. Hence, blocking the entrance of Ca^{2+} through this specific subtype of Ca^{2+} channel could be valuable to prevent cell death. This is the case of *nifedipine* (Figure 1), a neuronal L-type Ca^{2+} channel blocker that protects neurons from death evoked by focal stroke [49].

With these precedents in mind and following the Multi-Target Directed Ligand (MTDL) strategy for drug design, a series of compounds were designed based on the hybridization of *tacrine*, the first AChEI approved for the treatment of AD in USA [50], and 1,4-dihydropyridines (1,4-DHP) (Figure 1), well known Ca^{2+} channel antagonists of the VDCC L-type [51,52]. Thus, these molecules, *pyranotacrines* or *pyridotacrines*, are tacrine-1,4-DHP hybrids bearing the 1,8-naphthyridine heterocyclic ring system (Figure 1).

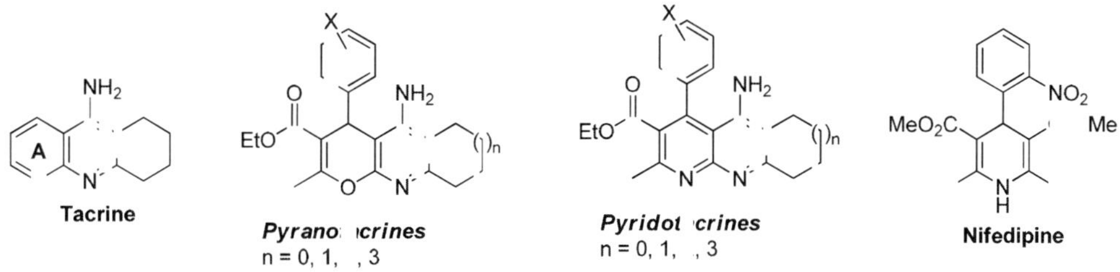

Figure 1. New tacrine analogues based on the hybridization of tacrine and nifedipine.

The expected compounds were obtained by the Friedländer-type reaction using an appropriate 2-amino-3-nitrile intermediate and a cycloalkanone [53].

Friedländer reaction represents one of the best methods for the synthesis of quinolines. This type of reaction is a base- or acid- promoted condensation of an aromatic 2-amino-substituted carbonyl compound (aldehyde, ketone, or equivalent) with an appropriately substituted carbonyl derivative containing a reactive methylene group, followed by cyclodehydration. Friedländer annulations are generally carried out either by refluxing an aqueous or alcoholic solution of the reactants, in the presence of a base or acid, or by heating a mixture of the reactants at 150-220 °C in the absence of a catalyst [53]. A modified Friedländer quinoline synthesis starts from anthranilamides *via* imine formation and base-mediated intramolecular cyclization. 2-Aminonitriles such as anthranilonitrile produce fused 4-aminoquinolines. Typical experimental conditions include the mixing of 2-aminonitrile derivatives with the selected ketone, in presence of a Lewis acid ($AlCl_3$ is one of the most extended catalysts), using 1,2-dichloroethane as solvent.

2. 4H-Pyrano [2,3-*b*]Pyridines (Pyranotacrines)

2.1. Synthesis

The scope and limitations of Friedländer-type reactions [53,54] in heterocyclic compounds showing a β-enaminonitrile-type moiety [55] started to be investigated. Based on our experience in the synthesis and reactivity of polyfunctionalized 4*H*-pyrans [56], we realized that the reactivity of 2-amino-3-cyano-4*H*-pyrans *1* with ketones had not been explored, although these heterocycles have been largely studied. In fact, the β-enaminonitrile moiety is a structural and functional subunit that has been involved in the synthesis of a variety of biologically active heterocyclic compounds such as pyridopyrimidines [57], quinolines [58], quinazolines [59], pyridines [60], indolizines [61], etc. For this reason, the preparation of tacrine analogues *via* Friedlander-type reactions of cyclic ketones with heterocyclic compounds related to 1,4-DHP would afford compounds with a high biological interest.

Thus, following the standard methodology ($AlCl_3$, $ClCH_2CH_2Cl$, reflux), Friedlander reaction of 4*H*-pyrans *1* gave *pyranotacrines 2-18* with low to excellent yields [55,62,63]. The synthesis of these compounds was achieved in three steps from commercial starting materials (Scheme 1). This approach was highly atom efficient and able to be scaled up. Yields were slightly lowered when cyclopentanone was used as the cyclic ketone (Table 1).

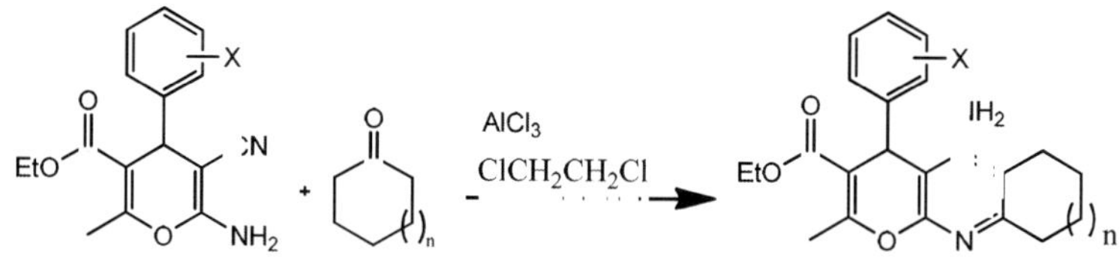

1 **2-18**

(X = *p*-OMe, *m*-OMe, *o*-OMe, *p*-Cl, *p*-CN, *m*-NO_2, *p*-Me, etc)
(n = 0, 1, 2)

Scheme 1. Synthesis of ***pyranotacrines*** *via* Friedländer-type reaction.

2.2. Pharmacology

Inhibition of AChE from *Electrophorus electricus* was carried out according to Rappaport [64]. All compounds able to inhibit AChE with an IC_{50} around 1 μM or higher were also evaluated against eqBuChE. Pyranotacrines *2-18* were highly selective but moderate AChEI, being compound *18* the best inhibitor, with an IC_{50} of 0.87 μM. Inhibitory activity of AChE was more effective when compounds exhibited a cyclohexane, instead of cyclopentane or cycloheptane, as cycloalkane fused-ring. Compound *4* showed the best selectivity, i.e., more than 1900-fold better AChEI than butyrylcholinesterase inhibitor (BuChEI). No requirements in the electronic effects of the aromatic substituents for the inhibition were found, but compounds possessing a *para*-substitution at the aromatic ring manifested better inhibitory activities.

Results emerging from inhibitor *18* docking studies suggested that the catalytic active site (CAS) of AChE is not flexible enough as to easily accommodate these compounds, thus the hypothesis that they could preferentially bind at the PAS of AChE was proposed [63,65]. In kinetic experiments compound *18* behaved as a non-competitive AChE inhibitor, suggesting that it was unable to pass through the gorge to reach the catalytic binding site. Hence, compound *18* might exert its inhibitory activity by interacting at the PAS of AChE, as it has already been noted for propidium [66].

The ability of pyranotacrines to block VDCC has also been investigated, using bovine chromaffin cells (BCC), a very interesting model since several subtypes of Ca^{2+} channels (L, N, P/Q, R) [67] are present. Chromaffins are neuroendocrine cells located mainly in the adrenal medulla of suprarenal glands [68]. Chromaffin cells generate action potentials and present voltage-dependent Ca^{2+}, Na^{+}, and K^{+} channels, and both neuronal nicotinic and muscarinic receptors [69-71].

Most compounds (Table 1) were more effective on blocking DMPP than K^{+}-induced $^{45}Ca^{2+}$ uptake (Table 1). Compounds *13*, *14*, and *18* were the most potent blockers of DMPP-stimulated $^{45}Ca^{2+}$ entry. The OMe group at the phenyl ring seems to enhance the blocking activity, preferentially at *meta position*. Since compounds *9*, *12*, *13*, and *14* scarcely blocked the K^{+}-evoked $^{45}Ca^{2+}$ entry through Ca^{2+} channels, compared to their ability to block DMPP-induced $^{45}Ca^{2+}$ entry, thus a possibility exists that these compounds are acting directly on nicotinic receptors (nAChR). Although nicotinic antagonists have been studied for several serious diseases [72], they have also been described to impair cognitive functions [73,74].

A more moderate blockade of DMPP-induced $^{45}Ca^{2+}$, in addition to an effect on high K^{+}-stimulated Ca^{2+} channels opening, would indicate a direct interaction on VDCC. Most compounds blocked $^{45}Ca^{2+}$ entry induced by high K^{+} above 20%. This result was interesting taking into account the fact that from all VDCC present in BCC, around 20% are L-type. Compounds *2*, *3*, *10*, and *13* blocked the K^{+} signal by 30%, and compounds *8*, *17*, and *18* inhibited K^{+}-induced $^{45}Ca^{2+}$ entry by 50% (Table 1). Moreover, the cyclohexane moiety seemed to promote higher blockade of VDCC. All compounds capable of blocking K^{+}-induced $^{45}Ca^{2+}$ entry by 30% or more, and compound *4*, were disposed to $^{45}Ca^{2+}$ uptake dose–response experiments. Only compounds *2*, *3*, *8*, and *18* showed a sufficient potency to describe their IC_{50} to block the $^{45}Ca^{2+}$ uptake by BCC.

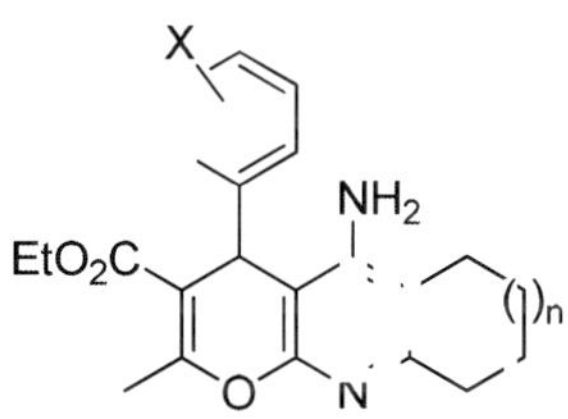

Pyranotacrines 2, *8*, and *18* showed IC_{50} of 8.2, 5.2, and 9.2, respectively, while *pyranotacrine 3*, with a cyclopentane-fused ring, showed an IC_{50} of 6.7 μM [62].

In contrast to the previously described for nicotinic antagonists, Ca^{2+} channels blockers have shown neuroprotective profile both *in vitro* and *in vivo* models of neuronal death [48]. For a better description of their L-type VDCC blocking activity, isolated organ experiments using K^{+}-pre-contracted rat aortas were performed. This vascular bed only presents L-type VDCC (Table 1).

Compound *3* was the most potent in relaxing K^{+}-pre-contracted isolated rat aortas, with an IC_{50} of 0.3 μM. Only six compounds showed an IC_{50} below 10 μM. I was observed that methoxy groups exerted better activities when placed at positions *para* and *ortho* than at *meta*, while compounds with electron-withdrawing groups showed little blockade (Table 1) [62].

To sum up, the *pyranotacrines* exhibited a moderate AChE inhibition, in addition to a decent Ca^{2+} entry blockade. Among all the compounds, attention was focused on *18* because it showed the highest K^{+}-stimulated Ca^{2+} channels blockade in BCC, along with its ability to relax isolated aortas with an IC_{50} of 1.6 μM. Moreover, it revealed the best inhibitory activity for AChE (0.87 μM).

Table 1. *Pyranotacrines* 2-18: IC_{50} values for the inhibition of *Ee*AChE and eqBuChE, blockade of $^{45}Ca^{2+}$ uptake induced by 70 mM K^+ or DMPP, and IC_{50} values for the relaxation of 35 mM K^+precontracted isolated rat aorta

Comp.	*X*	*n*	*Cholinesterases Inhibition IC_{50} (μM)*			*Blockade of $^{45}Ca^{2+}$. (% control)*		*Isolated organ relax. IC_{50} (μM)*	*Ref*
			AChE	*BuChE*	*BuChE/ AChE*	*K^+ 70 mM*	*DMPP 0.1 mM*	*K^+ 35 mM*	
2*	H	1	> 30	-	-	34	72	0.03	[62, 65]
3	H	0	3.9	300	77	35	77	13	[55, 62]
4	H	1	1.6	> 3000	> 1923	23	0	0.3	[55, 62]
5	*p*-Cl	1	1.9	-	-	20	44	-	[55, 62, 65]
6	*p*-CN	1	3.5	-	-	18	71	16	[55, 62, 65]
7	*m*-NO_2	1	1.9	-	-	13	68	21	[55, 62, 65]
8	*p*-CH_3	1	1.8	-	-	48	70	9.2	[55, 62, 65]
9	*o*-OCH_3	0	24	-	-	25	81	-	[62, 63]
10	*o*-OCH_3	1	7.7	-	-	28	50	8.4	[62, 63]
11	*o*-OCH_3	2	21	-	-	23	67	7.1	[62, 63]
12	*m*-OCH_3	0	> 100	-	-	22	81	3.4	[62, 63]
13	*m*-OCH_3	1	> 30	-	-	31	88	18	[62, 63]
14	*m*-OCH_3	2	13	-	-	18	85	17	[62, 63]
15	3,4-diOCH_3	0	27	-	-			11	[63]
16	3,4-diOCH_3	1	4.6	-	-	22	40	1.6	[63]
17	3,4-diOCH_3	2	33	-	-	45	45		[63]
18	*p*-OCH_3	1	0.87	6.2	7.1	50	90		[55, 62, 65]
Tacrine			0.18	0.036	0.2	17	55		[65]
Diltiazem						36	92		[62]

3. Furo and Thieno[2,3-*b*]pyridines (Furo and Thienotacrines)

3.1. Synthesis

Carreiras and co-workers described the pharmacological properties for related furo and thieno[2,3-*b*]quinolines *19-21* (Figure 2) [75-77], designed as new tacrine analogues by modifying the benzene ring of tacrine into a five-membered aromatic ring system *22-27* (Figure 2).

Compound *19* is a potential agent for the treatment of various memory dysfunctions, as it was demonstrated by its ability to restore cholinergic deficient memory [75]. The AChEI *20* shows an IC_{50} between 5 and 0.1 μM, [76], while drug *21* proved to be a powerful high affinity choline uptake enhancer [77].

Furotacrines 22-25 were prepared by Friedländer-type reaction with readily available 2-amino-4,5-diaryl-3-cyano substituted furans [78, 79] in low to good yields [62] (Scheme 2).

Thienotacrines 26 and *27* were prepared from 2-amino-3-cyano-4,5-diphenylthiophene, a new compound [80], prepared as reported for analogue derivatives [81,82], but affording low yields, possibly due to the poisoning of the catalyst by sulfur, since this atom coordinates the catalyst and promoter $AlCl_3$ (Scheme 2).

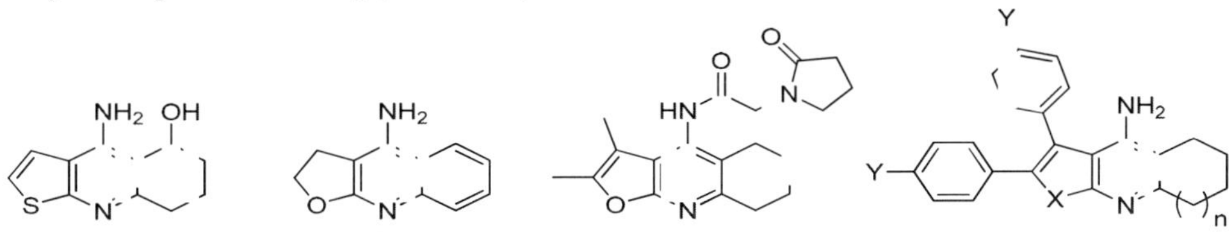

19 20 21 22 27

Figure 2. Some furo [2,3-*b*] quinolines with anti-neurodegenerative pharmacological properties and the general structure of *furotacrines* 22-25 (X = O) and *thienotacrines* 26, 27 (X = S).

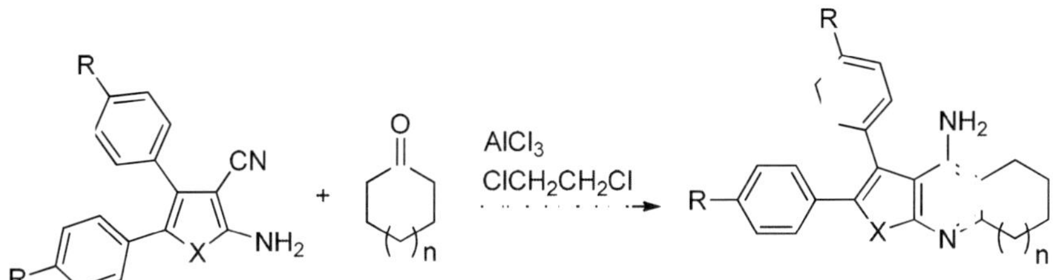

X = O, S

22-27

Scheme 2. Synthesis of *furo-* and *thienotacrines via* Friedländer-type reaction.

3.2. Pharmacology

The inhibitory activity against both *Ee*AChE and eqBuChE showed that *furotacrines* were submicromolar AChEI, while *thienotacrines* lost any ChE inhibitory activity (Table 2).

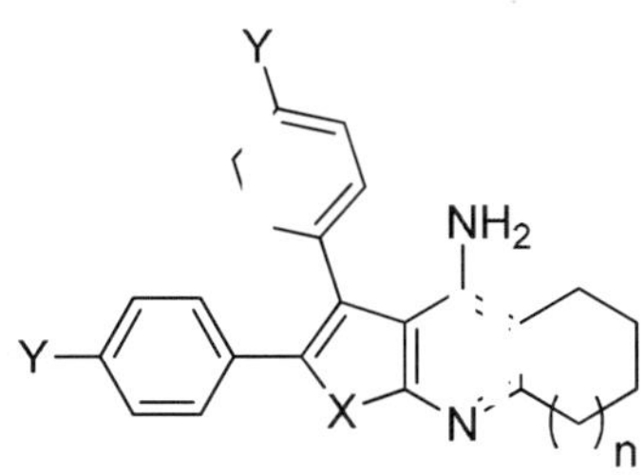

A series of experiments were carried out in order to determine the type of antagonism in innibitors *22-27*. All *furotacrines* were competitive inhibitors of AChE. Appropriate molecular modeling confirmed that the electron lone pairs of sulphur in *thienotacrines* were preventing a good interaction with the CAS of AChE [83]. It seems that substitution in C4' at the phenyl ring is well accepted in the AChE binding pocket, since no significant differences of IC_{50} were found among all the *furotacrines 22-25*. Unlike AChE inhibition values, only non-substituted tacrines were able to inhibit BuChE in the submicromolar range. A possibility exists that fitting into the hydrophobic pocket is very sensitive, and a tiny size enlargement would prevent the correct adjustment. This proposal would be confirmed by 2-fold lower activity of the cycloheptane-fused derivative *23* (Table 2) [80].

In the context of searching for multitarget compounds, *furo-* and *thienotacrines* were submitted to $^{45}Ca^{2+}$ uptake experiments in BCC in order to test their ability to regulate Ca^{2+} entry, as previously described (Table 2). *Thienotacrines* were unable to modulate $^{45}Ca^{2+}$ entry. Methyl and methoxy-substituted *furotacrines 24* and *25* showed some blocking activity of $^{45}Ca^{2+}$ uptake. (Table 2).

Furotacrines 22 and *23* were equipotent inhibitors of AChE and BuChE with no activity on cell Ca^{2+} signal. This dual effect on both ChE enzymes could be a pitfall if selective AChE inhibition was pursued. Dual ChE inhibitors are being considered for the beneficial effect of BuChE inhibition in AD [84,85]. Drugs *24* and *25* are selective AChEI that reveal moderate blocking activity on VDCC.

Table 2. *Furo-* and *thienotacrines* 22-27: IC_{50} values for the inhibition of *Ee*AChE and eqBuChE, and blockade of $^{45}Ca^{2+}$ uptake induced by 70 mM K^+ or DMPP [62, 80]

Comp.	*Y*	*X*	*n*	*Cholinesterases Inhibition IC_{50} (μM)*			*Blockade of $^{45}Ca^{2+}$. (% control)*	
				AChE	*BuChE*	*BuChE/ AChE*	*K^+70 mM*	*DMPP 0.1 mM*
22	H	O	1	0.44	0.37	0.8	3	26
23	H	O	2	0.32	0.77	2.4	10	36
24	Me	O	1	0.38	> 100	> 263	21	46
25	MeO	O	1	0.46	> 100	> 217	21	35
26	H	S	1	> 100	> 100	-	0	0
27	H	S	2	> 100	> 100	-	2	0
Tacrine				0.13	0.044	0.3	17	55
Diltiazem							36	92

4. 4-Aryl-1,8-naphthyrydines (Pyridotacrines)

4.1. Synthesis

Based on our previous results on *pyranotacrines*, the synthesis and biological evaluation of *pyridotacrines*, as 1,4-DHP-tacrine hybrids, was an obvious purpose, thus it was also investigated.

The key intermediates, bearing the 2-amino-3-nitrile moiety, were prepared from 4*H*-pyrans by reaction with ammonium acetate in acetic acid [63].

Friedländer reactions [55] were carried out to afford the expected *pyridotacrines* with the required cycloalkane fused ring (Scheme 3). *Pyridotacrines 28-50* were obtained in moderate to good yields, except when cyclopentanone was used, since the longer times needed to accomplish reactions led to decomposition of the intermediates (Table 3) [65].

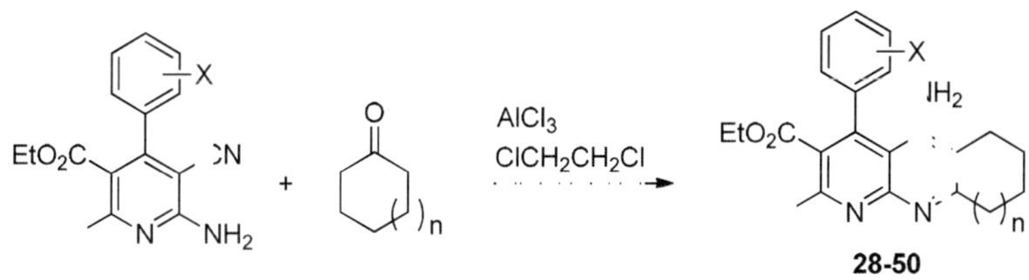

Scheme 3. Synthesis of *pyridotacrines via* Friedländer-type reaction.

4.2. Pharmacology

The inhibitory activity of *pyridotacrines* was assessed against *Ee*AChE. In general, compounds *28-50* with similar cycloalkane-fused ring, were better AChEI than *pyranotacrines 2-18*. Inhibitors *30*, *35*, and *39* showed IC_{50} values of 0.8, 0.9, and 0.7 μM, respectively.

Like *pyranotacrines 2-18*, *pyridotacrines 28-50* bearing a cyclohexane fused ring revealed better IC_{50} data. Electron-withdrawing groups seemed to improve the AChE inhibitory activity of cyclohexane fused ring derivatives. Compound *39*, displaying a fluorine atom at the phenyl *para* position, was the best inhibitor. Among the compounds with a cycloheptane fused-ring, only *47* showed an IC_{50} below 3 μM. BuChE inhibition proved to be less effective in compounds that showed a good AChE inhibition. Only compound *30* could be considered a dual ChEI, presenting selectivity to AChE (Table 3).

In order to analyze binding interactions and the necessary structural hits to improve AChE inhibition, molecular modeling was carried out with lead compound *30*, which has not any substituent at the phenyl ring, and shows an IC_{50} of 0.8 μM. Like *pyranotacrine 18*, the binding modes did not result in a good fitting at the CAS of AChE, though the stacking interactions were roughly preserved along the simulation [63]. AChE inhibition kinetic experiments indicated compound *30* was a non-competitive inhibitor [65], suggesting that, as hypothesized for *pyranotacrine 18*, these inhibitors might exert their inhibitory activity preferentially by interacting at the PAS of AChE.

The ability of *pyridotacrines 28-50* to regulate Ca^{2+} entry in BCC was evaluated with $^{45}Ca^{2+}$ uptake experiments. Compounds were assayed at the concentration of 3 μM in presence of DMPP, 70 mM K^+ or 70 mM K^+ and *nifedipine* (Figure 1) 3 μM as depolarizing stimuli (Table 3).

Like *pyranotacrines 2-18*, *pyridotacrines 28-50* were better blockers of DMPP-induced $^{45}Ca^{2+}$ entry than in the experiments evoked by high K^+. This blocking activity ranged from 50 to 100%. It is important to note that the *pyridotacrines* which bear electron-donating groups presented higher blockade of $^{45}Ca^{2+}$, i.e. 3,4-dimethoxyphenyl substituted derivatives *49* and *50* completely blocked the DMPP-induced Ca^{2+} entry. Among the best blockers, compounds *28*, *32*, *44*, *46*, and *50* scarcely blocked the K^+-evoked $^{45}Ca^{2+}$ entry through Ca^{2+} channels (Table 3).

As far as high K^+-stimulated $^{45}Ca^{2+}$ entry, most compounds were able to block about 20%. As 20% of Ca^{2+} channels in BCC are L-type, we wondered whether the *pyridotacrines* were acting selectively on this specific subtype. To answer this question, $^{45}Ca^{2+}$ uptake experiments were carried out in presence of *nifedipine* (Figure 1) 3 μM, a 1,4-DHP that selectively blocks L-type VDCC.

Thus, any compound capable of keeping the $^{45}Ca^{2+}$-entry blockade in presence of *nifedipine* (Figure 1), would be interacting with non-L type VDCC. Conversely, a compound which looses Ca^{2+} antagonism in presence of nifedipine would be a selective ligand of L-type Ca^{2+} channels. Comparing data of K^{+}-stimulated $^{45}Ca^{2+}$ entry blockade with and without *nifedipine* (Figure 1), it can be concluded that compounds *29*, *30*, *33*, *34*, *38*, *39*, and *49* were selective ligands of L-type VDCC.

To sum up, most of the *pyridotacrines 28-50* showed multitarget pharmacological profile, due to their activities on both AChE and cell Ca^{2+} regulation, two targets implicated in neurodegeneration. Compounds *30*, *35*, *39*, and *47* may be highlighted from all studied. Compound *30* showed good inhibitory activity against AChE in the submicromolar range with certain selectivity, and a moderate blocking activity on L-type VDCC in a micromolar level. Compound *39* also displayed good AChE inhibitory activity with some selectivity, and a remarkable blockade of L-type VDCC. Inhibitor *35* can be considered a multitarget drug as well, but presumably acting on non-L-type VDCC. It is well known that the L-type Ca^{2+} channel is the most related to neuroprotection and neurodegeneration processes [48], but further pharmacological studies with compound *35* could determine the real contribution of each Ca^{2+} channel type to this physiological events. Finally, compound *47* would deserve a further pharmacological characterization, for its moderate but high selective AChE inhibitory activity and the global activity on VDCC.

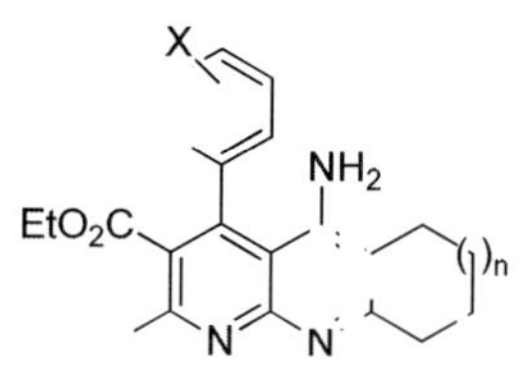

In correlation with the study of *pyridotacrine 30* pharmacological properties, we focused on improving its AChE inhibitory activity.

Docking studies pointed out that *pyridotacrine 30* aromatic substituent at C4 might hinder the movement onward through the gorge to reach the CAS of AChE [63]. Hence, we designed a series of compounds which lacked an aromatic substituent at C4. Presumably, this structural modification would favor the accommodation of the new compounds into the AChE catalytic binding site, thus potentiating the inhibitory activity against the ChE enzymes.

Friedländer-type reaction of the appropriate alkyl 6-amino-5-cyano-2-methylnicotinate (*I*) with the usual cycloalkanones provided compounds *51-58* bearing the expected structure of 1,8-naphthyridine (Scheme 4) [86], in 57-90% yields.

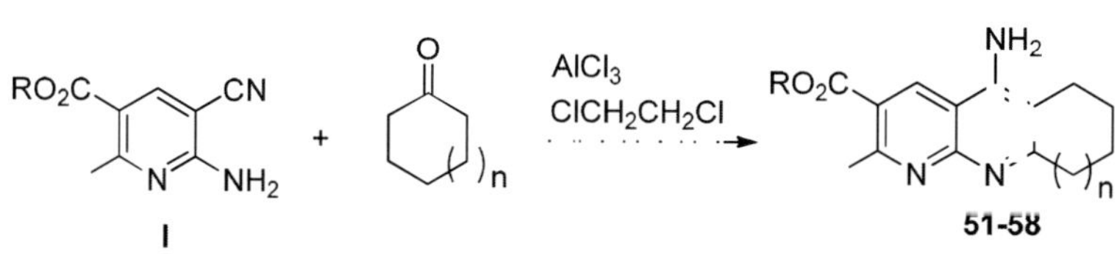

Scheme 4. Synthesis of *pyridotacrines* 51-58 *via* Friedländer-type reaction.

Table 3. *Pyridotacrines* 28-50: IC_{50} values for the inhibition of *Ee*AChE and eqBuChE, blockade of $^{45}Ca^{2+}$ uptake induced by 70 mM K^+, 70 mM K^+ + nifedipine 3 μM, or DMPP, and IC_{50} values for the relaxation of 35 mM K^+-pre-contracted isolated rat aorta

Comp.	X	n	Cholinesterases Inhibition IC_{50} (μM)			Blockade of $^{45}Ca^{2+}$. (% control)			Ref
			AChE	BuChE	BuChE/ AChE	K^+70 mM	70 mM K^+ + nife 3 μM	DMPP 0.1 mM	
28	H	1	1.2	12	10	10	0	82	[63]
29	H	0	12	-	-	24	0	52	[63]
30	H	1	0.8	5.0	6.2	20	0	52	[63]
31	H	2	21	-	-	12	7	70	[63]
32	H	3	17	-	-	10	0	82	[63]
33	*p*-Cl	1	1.3	52	40	34	0	83	[63]
34	*p*-Cl	2	19	-	-	27	1	63	[63]
35	*m*-NO_2	1	0.9	21	23	18	20	61	[63]
36	*m*-NO_2	2	14	-	-	13	2	74	[63]
37	*p*-CH_3	1	-	-	-	-	-	-	
38	*p*-CH_3	2	13	-	-	24	0	68	[63]
39	*p*-F	1	0.7	12	17	38	3	63	[63]
40	*p*-F	2	5	-	-	17	0	71	[63]
41	*o*-OCH_3	1	-	-	-	-	-	-	
42	*o*-OCH_3	2	22	-	-	17	0	89	[63]
43	*m*-OCH_3	1	1.2	31	25	15	5	50	[63]
44	*m*-OCH_3	2	> 30	-	-	9	7	62	[63]
45	*p*-OCH_3	0	14	-	-	15	25	71	[63]
46	*p*-OCH_3	1	9.0	-	-	10	0	76	[63, 65]
47	*p*-OCH_3	2	1.30	81	62	39	39	82	[63]
48	*p*-OCH_3	3	17	-	-	20	10	73	[63]
49	3,4-diOCH_3	1	1.3	> 100	> 77	26	0	99	[63]
50	3,4-diOCH_3	3	6.4	-	-	9	9	97	[63]
Tacrine			0.18	0.036	0.2	17	10	55	[63]
Diltiazem			-	-	-	36	-	92	[62]

Pyridotacrines 51-58 were evaluated as *Ee*AChE and eqBuChE inhibitors (Table 4) [87]. Compared to the *Ee*AChE inhibitory activities of *pyridotacrines 28-50*, compounds *51-58* were at least 10-fold more potent than those bearing an aryl substituent at C4, confirming our hypothesis that aryl substituents at C4 were barely accommodated into the catalytic binding site of AChE [86].

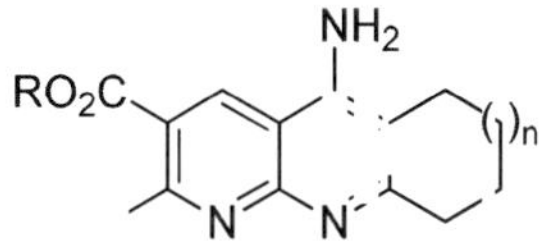

In agreement with our previous stud.es, t.ıe cyclohexane moiety was the best for the inhibition of *Ee*AChE. Replacement of the ethyl by a methyl ester slightly modified the *Ee*AChE inhibitory activity, either improving or lowering it (Table 4). The best *pyridotacrine* was cyclohexane-fused derivative *56* showing an IC_{50} of 60 nM, whose type of inhibition was elucidated from the analysis of Lineweaver-Burk reciprocal plots, suggesting a mixed-type inhibition [86], with a Ki value of 24 nM.

Table 4. IC_{50} values for the inhibition of *Ee*AChE and eqBuChE by *pyridotacrines* 51-58 [86]

Compound	*R*	*n*	*Cholinesterases Inhibition. IC_{50} (µM)*		
			AChE	*BuChE*	*BuChE/AChE*
51	Me	0	0.12	5	41.7
52	Me	1	0.093	3.4	36.6
53	Me	2	0.36	0.49	1.36
54	Me	3	0.25	4.07	16.3
55	Et	0	0.35	39	111
56	Et	1	0.060	12	200
57	Et	2	0.31	2.1	6.77
58	Et	3	0.40	5.3	13.2
Tacrine			0.027	0.0052	0.193

Inhibition of eqBuChE by *pyridotacrines 51-58* afforded IC_{50} values between 0.49 and 36 µM (Table 4). Compounds bearing the cycloalkane and methyl esters were better BuChEI than ethyl ester derivatives. Thus, cycloheptane-fused *pyridotacrines 53* and *57* seem to be the best inhibitors, with IC_{50} of 0.49 and 2.1 µM, respectively. In spite of the smaller potency of these *pyridotacrines* compared to *tacrine*, it is worth noting that compounds *51-58* presented better activities than *pyridotacrines 28-50* (Table 3). The poor inhibitory activity of BuChE found in previously described compounds could be a consequence of discarding either methyl ester or cycloheptane-fused derivatives in the chemical design of our tacrine-1,4-DHP hybrid compounds.

To explain the AChE inhibitory activity, Docking analysis of the family were carried out, using *56* as the best compound of the series [88]. Thus, *pyridotacrine 56* could be well-accommodated between Trp86 and Tyr337 *Ee*AChE, generating a π-π stacking interaction in the catalytic active site. This conformation would be facilitated by a hydrogen bond between protonated aromatic nitrogen and the carbonyl group of His447.

Although the overall anticholinesterasic activity was improved, the capacity to regulate intracellular [Ca^{2+}] was dramatically decreased, as reported by experiments of Ca^{2+} entry induced by K^+ depolarization [86], monitored in Fluo-4/AM-loaded SH-SY5Y human neuroblastoma cells incubated with *pyridotacrines 51-58* at 1 µM. Most compounds had no effect on Ca^{2+} uptake. Only inhibitor *54* significantly blocked Ca^{2+} entry by 20% [86]. The absence of Ca^{2+} channel regulatory activity of these new compounds in comparison to previous analogues was expected, since the 1,4-DHP-like moiety, which is believed to confer such activity, is absent due to the aryl substituent (Figure 1), which has been replaced by hydrogen in compounds *51-58*.

Conclusion

Here, we have summarized our research line in the last sixteen years, aimed at discovering new drugs for the potential treatment of neurodegenerative disorders *via* the multitarget directed ligand approach. AD, as well as other neurodegenerative disorders, appears to be multifactorial. Thus, the design of single-targeted potent drug is supposed to be worthless [28]. This approach has been focused on the ChE enzymes, and the regulation of cytosolic Ca^{2+} concentration since, this ion is implicated in cell apoptosis/survival processes.

We believe that the efforts undertaken in designing a series of multitarget compounds have been rewarded. Therefore, in a first proposal, we looked for the juxtaposition of two *drugable* sub-structures. In our case, both starting hits had different pharmacological properties, and many of the synthesized compounds kept both pharmacological activities, being simultaneously AChEI and Ca^{2+}-modulators.

Regarding the SARs, we found some general common trends along all the studied chemical families. Hence, cyclohexane-fused rings showed better inhibitory activities for AChE; the presence of electron-donating groups at the phenyl ring afforded better blockade of Ca^{2+} entry, if nicotinic agonist-induced or K^{+}-stimulated. However, some structural and electronic requirements were specific for each family. *Pyranotacrines* with substitution in *para* at the phenyl ring were better AChE inhibitors and VDCC blockers. *Para*-substitution in *furotacrines* afforded good VDCC blockade, but hindered the AChE inhibition. Interestingly, electron-withdrawing groups exerted good AChE inhibition in cyclohexane-fused *pyridotacrines*.

Finally, we found out the aromatic ring at C4 in *pyridotacrines* impairs the binding to ChE enzymes, but it is essential for blocking activity on VDCC. Therefore, the presence of a phenyl ring at C4 becomes a key point in the chemical design; either we search for highly potent or less potent ChEI, but with an additional activity on cell Ca^{2+}. Those compounds behaving as potential multitarget drugs lost certain pharmacological activity at the expense of interacting with both AChE and cell Ca^{2+} signal, but some of them exceed tacrine and 1,4-DHP in neuroprotection experiments. Moreover, the study of cell Ca^{2+} signal regulation demonstrated the important role of this ion in neurodegeneration. We believe that it is not the blockade, but the fine regulation of the $[Ca^{2+}]_c$, the driving force promoting the neuroprotection profile against Ca^{2+} overload-derived neuronal death, as the well-known SERCA depleting toxicant thapsigargin, and one of our compounds, namely *30*, elicited neuroprotection by slightly increasing $[Ca^{2+}]_c$.

References

[1] Goedert, M.; Spillantini, M. G. *Science 2006, 314*, 777.

[2] León, R.; García, A. G. G.; Marco-Contelles, J. *Med. Res. Rev.* doi: 10.1002/med.20248.

[3] Gong, C. X.; Iqbal, K. *Curr. Med. Chem. 2008, 15*, 2321.

[4] Pratico, D. Ann. N Y Acad. Sci. 2008, 1147, 70.

[5] Cummings, J. L. *Rev. Neurol. Dis. 2004, 1*, 60.

[6] Rafii, M. S.; Aisen, P. S. *BMC Med. 2009, 7*, 7.

[7] van Marum, R. J. Fundam. Clin. Pharmacol. 2008, 22, 265.

[8] Fillit, H. M.; Doody, R. S.; Binaso, K.; Crooks, G. M.; Ferris, S. H.; Farlow, M. R.; Leifer, B.; Mills, C.; Minkoff, N.; Orland, B.; Reichman, W. E.; Salloway, S. *Am. J. Geriatr. Pharmacother. 2006, 4 Suppl A*, S9-S24; quiz S25-S28.

[9] Racchi, M.; Mazzucchelli, M.; Porrello, E.; Lanni, C.; Govoni, S. *Pharmacol. Res. 2004, 50*, 441.

[10] Galdeano, C.; Viayna, E.; Arroyo, P.; Bidon-Chanal, A.; Blas, J.R.; Muñoz-Torrero, D.; Luque, F.J. *Curr. Pharm. Des. 2010, 16*, 2818.
[11] O'Neill, M. F. *Drug Discov. Today 2005, 10*, 1333.
[12] Bullock, R.; Touchon, J.; Bergman, H.; Gambina, G.; He, Y.; Rapatz, G.; Nagel, J.; Lane, R. *Curr. Med. Res. Opin. 2005, 21*, 1317.
[13] Li, W.; Mak, M.; Jiang, H.; Wang, Q.; Pang, Y.; Chen, K.; Han, Y. *Neurotherapeutics 2009, 6*, 187.
[14] Lorrio, S.; Sobrado, M.; Arias, E.; Roda, J. M.; García, A. G.; López, M. G. *J. Pharmacol. Exp. Ther. 2007, 322*, 591.
[15] Martorana, A.; Esposito, Z.; Koch, G. *CNS Neurosci. Ther. 2010, 16*, 235.
[16] Musial, A.; Bajda, M.; Malawska, B. *Curr. Med. Chem. 2007, 14*, 2654.
[17] Egea, J.; de los Ríos, C. *Curr. Top. Med. Chem. 2011, 11*, 2807.
[18] Arias, E.; Ales, E.; Gabilán, N. H.; Cano-Abad, M. F.; Villarroya, M.; García, A. G.; López, M. G. *Neuropharmacology 2004, 46*, 103.
[19] Melo, J. B.; Sousa, C.; Garção, P.; Oliveira, C. R.; Agostinho, P. *Eur. J. Neurosci. 2009, 29*, 455.
[20] Noh, M. Y.; Koh, S. H.; Kim, Y.; Kim, H. Y.; Cho, G. W.; Kim, S. H. *J. Neurochem. 2009, 108*, 1116.
[21] Shanks, M.; Kivipelto, M.; Bullock, R.; Lane, R. *Curr. Med. Res. Opin. 2009, 25*, 2439.
[22] Muñoz-Torrero, D. *Curr. Med. Chem. 2008, 15*, 2433.
[23] Inestrosa, N. C.; Álvarez, A.; Pérez, C. A.; Moreno, R. D.; Vicente, M.; Linker, C.; Casanueva, O. I.; Soto, C.; Garrido, J. *Neuron 1996, 16*, 881.
[24] Castro, A.; Martínez, A. *Mini Rev. Med. Chem. 2001, 1*, 267
[25] Bartolini, M.; Bertucci, C.; Cavrini, V.; Andrisano, V. *Biochem. Pharmacol. 2003, 65*, 407.
[26] García-Palomero, E.; Muñoz, P.; Usán, P.; García, P.; Delgado, E.; de Austria, C.; Valenzuela, R.; Rubio, L.; Medina, M.; Martínez, A. *Neurodegener. Dis. 2008, 5*, 153.
[27] Espinoza-Fonseca, L. M. *Bioorg. Med. Chem. 2006, 14*, 896.
[28] Cavalli, A.; Bolognesi, M. L.; Minarini, A.; Rosini, M.; Tumiatti, V.; Recanatini, M.; Melchiorre, C. *J. Med. Chem. 2008, 51*, 347.
[29] Alonso, D.; Dorronsoro, I.; Rubio, L.; Muñoz, P.; García-Palomero, E.; del Monte, M.; Bidon-Chanal, A.; Orozco, M.; Luque, F. J.; Castro, A.; Medina, M.; Martínez, A. *Bioorg. Med. Chem. 2005, 13*, 6588.
[30] Camps, P.; Formosa, X.; Galdeano, C.; Gómez, E.; Muñoz-Torrero, D.; Scarpellini, M.; Viayna, E.; Badía, A.; Clos, M. V.; Camins, A.; Pallas, M.; Bartolini, M.; Mancini, F.; Andrisano, V.; Estelrich, J.; Lizondo, M.; Bidon-Chanal, A.; Luque, F. J. *J. Med. Chem. 2008, 51*, 3588.
[31] Rodríguez-Franco, M. I.; Fernández-Bachiller, M. I.; Pérez, C.; Hernández-Ledesma, B.; Bartolomé, B. *J. Med. Chem. 2006, 49*, 459.
[32] Fernández-Bachiller, M. I.; Pérez, C.; Campillo, N. E.; Páez, J. A.; González-Muñoz, G. C.; Usán, P.; García-Palomero, E.; López, M. G.; Villarroya, M.; García, A. G.; Martínez, A.; Rodríuez-Franco, M. I. *ChemMedChem 2009, 4*, 828.
[33] Fernández-Bachiller, M. I.; Pérez, C.; González-Muñoz, G. C.; Conde, S.; López, M. G.; Villarroya, M.; García A. G.; Rodríguez-Franco, M. I. *J. Med. Chem. 2010, 53*, 4927.

[34] Bolea, I.; Juárez-Jiménez, J.; de los Ríos, C.; Chioua, M.; Pouplana, R.; Luque, F. J.; Unzeta, M.; Marco-Contelles, J.; Samadi, A. *J. Med. Chem. 2011*, *54*, 8251.

[35] Samadi, A.; Chioua, M.; Bolea, I.; de los Ríos, C.; Iriepa, I.; Moraleda, I.; Bastida, A.; Esteban, G.; Unzeta, M.; Gálvez , E.; Marco-Contelles, J. *Eur. J. Med. Chem. 2011*, *46*, 4665.

[36] Weinstock, M.; Kirschbaum-Slager, N.; Lazarovici, P.; Béjar, C.; Youdim, M. B.; Shoham, S. *Ann. NY Acad. Sci. 2001*, *939*, 148.

[37] León, R.; de Los Ríos, C.; Marco-Contelles, J.; López, M. G.; García, A. G.; Villarroya, M. *Eur. J. Med. Chem. 2008*, *43*, 668.

[38] León, R.; Marco-Contelles, J. *Curr. Med. Chem. 2011*, *11*, 552.

[39] León, R.; de los Ríos, C.; Marco-Contelles, J.; Huertas, O.; Barril, X.; Luque, F. J.; López, M. G.; García, A. G.; Villarroya, M. *Bioorg. Med. Chem. 2008*, *16*, 7759.

[40] Marco-Contelles, J.; León, R.; de los Ríos, C.; Samadi, A.; Bartolini, M.; Andrisano, V.; Huertas, O.; Barril, X.; Luque, F. J.; Rodríguez-Franco, M. I.; López, B.; López, M. G.; García, A. G.; Carreiras M. C.; Villarroya, M. *J. Med. Chem. 2009*, *52*, 2724.

[41] Samadi, A.; Marco-Contelles, J.; Soriano, E.; Álvarez-Pérez, M.; Chioua, M.; Romero, A.; González-Lafuente, L.; Gandía, L.; Roda, J. M.; López, M. G.; Villarroya, M.; García, A. G.; de los Ríos, C. *Bioorg. Med. Chem. 2010*, *18*, 5861.

[42] Verkhratsky, A.; Toescu, E. C. *Trends Neurosci. 1998*, *21*, 2.

[43] Khachaturian, Z. S. *Ann. NY Acad. Sci. 1994*, *747*, 1.

[44] Fraser, S. P.; Suh, Y. H.; Djamgoz, M. B. *Trends Neurosci. 1997*, *20*, 67.

[45] Kruman, I.; Guo, Q.; Mattson, M. P. *J. Neurosci. Res. 1998*, *51*, 293.

[46] Dreses-Werringloer, U.; Lambert, J. C.; Vingtdeux, V.; Zhao, H.; Vais, H.; Siebert, A.; Jain, A.; Koppel, J.; Rovelet-Lecrux, A.; Hannequin, D.; Pasquier, F.; Galimberti, D.; Scarpini, E.; Mann, D.; Lendon, C.; Campion, D.; Amouyel, P.; Davies, P.; Foskett, J. K.; Campagne, F.; Marambaud, P. A. *Cell 2008*, *133*, 1149.

[47] Moreno-Ortega, A. J.; Ruiz-Nuño, A.; García, A. G.; Cano-Abad, M. F. *Biochem. Biophys. Res. Commun. 2010*, *391*, 722.

[48] Cano-Abad, M. F.; Villarroya, M.; García, A. G.; Gabilán, N. H.; López, M. G. *J. Biol. Chem. 2001*, *276*, 39695.

[49] Lorrio, S.; Sobrado, M.; Arias, E.; Roda, J. M.; García, A. G.; López, M. G. *J Pharmacol Exp Ther 2007*, *322*, 591-599.

[50] Summers, W. K.; Majovski, L. V.; Marsh, G. M.; Tachiki, K.; Kling, A. *N. Engl. J. Med. 1986*, *315*, 1241.

[51] Mori, Y.; Mikala, G.; Varadi, G.; Kobayashi, T.; Koch, S.; Wakamori, M.; Schwartz, A. *Jpn. J. Pharmacol. 1996*, *72*, 83.

[52] Walgren, J. L.; Mitchell, M. D.; Thompson, D. C. *Crit. Rev. Toxicol. 2005*, *35*, 325-361.

[53] Marco-Contelles, J.; Pérez-Mayoral, E.; Samadi, A.; Carreiras M. C.; Soriano, E. *Chem. Rev. 2009*, *109*, 2652.

[54] Marco, J. L.; Carreiras, M. C. *The Friedländer Reaction*, LAP Lambert Academic Publishing AG & Co. KG, Saarbrücken, Germany, 2010.

[55] Martínez-Grau, A.; Marco, J. L. *Bioorg. Med. Chem. Lett. 1997*, *7*, 3165.

[56] Martín, N.; Martínez-Grau, A.; Seoane, C.; Marco, J. L.; Albert, A.; Cano, F. H. *J. Heterocyclic Chem. 1996*, *33*, 27.

[57] Quintela, J. M.; Peinador, C.; Botana, L.; Estévez, M.; Riguera, R. *Bioorg. Med. Chem. 1997, 5*, 1543.
[58] Veronese, A. C.; Callegari, R.; Salah, S. A. A. *Tetrahedron Lett. 1990, 31*, 3485.
[59] Seijas, J. A.; Vázquez-Tato, M. P.; Martínez, M. M. *Tetrahedron Lett. 2000, 41*, 2215.
[60] Shirai, K.; Fukunishi, K.; Yanagisawa, A.; Takahashi, H.; Matsuoka, M. *J. Heterocyclic Chem. 2000, 37*, 1151.
[61] Marchalin, T.; Cvopova, K.; Pham-Huu, D. P.; Chudik, M.; Kozisek, J.; Svoboda, I.; Daich, A. *Tetrahedron Lett. 2001, 42*, 5663.
[62] de los Ríos, C.; Marco, J. L.; Carreiras, M. D.; Chinchón, P. M.; García, A. G.; Villarroya, M. *Bioorg. Med. Chem. 2002, 10*, 2077.
[63] Marco, J. L.; de los Ríos, C.; García, A. G.; Villarroya, M.; Carreiras, M. C.; Martins, C.; Eleuterio, A.; Morreale, A.; Orozco, M.; Luque, F. J. *Bioorg. Med. Chem. 2004, 12*, 2199.
[64] Rappaport, F.; Fischl, J.; Pinto, N. *Clin. Chim. Acta. 1959, 4*, 227.
[65] Marco, J. L.; de los Ríos, C.; Carreiras, M. C.; Baños, J. E.; Badía, A.; Vivas, N. M. *Bioorg. Med. Chem. 2001, 9*, 727.
[66] Radic, Z.; Reiner, E.; Taylor, P. *Mol. Pharmacol. 1991, 39*, 98.
[67] Albillos, A.; García, A. G.; Olivera, B.; Gandía, L. *Pflug. Arch. 1996, 432*, 1030.
[68] Coupland, R. E.; Hopwood, D. *J. Anat. 1966, 100*, 227.
[69] García, A. G.; Garcia-De-Diego, A. M.; Gandía, L.; Borges, R.; García-Sancho, J. *Physiol. Rev. 2006, 86*, 1093.
[70] Patterson, P. H. *Cell 1990, 62*, 1035.
[71] Villarroya, M.; López, M. G.; Cano-Abad, M. F.; García, A. G. *Methods Mol. Biol. 2006, 312*, 135.
[72] Peng, Y.; Zhang, Q.; Snyder, G. L.; Zhu, H.; Yao, W.; Tomesch, J.; Papke, R. L.; O'Callaghan, J. P.; Welsh, W. J.; Wennogle, L. P. *Bioorg. Med. Chem. Lett. 2010, 20*, 4825.
[73] Cannady, R.; Weir, R.; Wee, B.; Gotschlich, E.; Kolia, N.; Lau, E.; Brotherton, J.; Levin, E. D. *Biochem. Pharmacol. 2009, 78*, 788.
[74] Arthur, D.; Levin, E. D. *Psychopharmacology 2002, 160*, 140.
[75] Shutske, G. M.; Kapples, K. J. *US Patent* 4753950, 1988.
[76] Desai, M. C. *Eur. Patent* 311303, 1989.
[77] Chaki, H.; Yamabe, H.; Sugano, M.; Morita, S.; Bessho, T.; Tabata, R.; Saito, K. I.; Egawa, M.; Tobe, A.; Morinaka, Y. *Bioorg. Med. Chem. Lett. 1995, 5*, 1495.
[78] Gewald, K. *Chem. Ber. 1966, 99*, 1002.
[79] Feng, X.; Lancelot, J. C.; Prunier, H.; Rault, S. *J. Heterocyclic. Chem. 1996, 33*, 2007.
[80] Marco, J. L.; de Los Ríos, C.; Carreiras, M. C.; Baños, J. E.; Badía, A.; Vivas, N. M. *Arch. Pharm. 2002, 335*, 347.
[81] Taylor, E. C.; Berger, J. G. *J. Org. Chem. 1967, 32*, 2376.
[82] Gewald, K.; Schinke, E.; Bottcher, H. *Chem. Ber. 1966, 99*, 94.
[83] Marco, J. L.; de los Ríos, C.; Carreiras, M. C.; Baños, J. E.; Badia, A.; Vivas, N. M.; Morreale, A.; Barrero, M.; Iriepa, I.; Gálvez, E. XII Congress of Spanish Society of Therapeutic Chemistry, Seville, Spain, 2001.
[84] Del Monte-Millán, M.; García-Palomero, E.; Valenzuela, R.; Usán, P.; de Austria, C.; Muñoz-Ruiz, P.; Rubio, L.; Dorronsoro, I.; Martínez, A.; Medina, M. *J. Mol. Neurosci. 2006, 30*, 85.

[85] Darvesh, S.; Hopkins, D. A.; Geula, C. *Nat. Rev. Neurosci. 2003, 4*, 131.
[86] De Los Ríos, C.; Egea, J.; Marco-Contelles, J.; León, R.; Samadi, A.; Iriepa, I.; Moraleda, I.; Gálvez, E.; García, A. G.; López, M. G.; Villarroya, M.; Romero, A. *J. Med. Chem. 2010, 53*, 5129.
[87] Ellman, G. L.; Courtney, K. D.; Andres, V., Jr.; Feather-Stone, R. M. *Biochem. Pharmacol. 1961, 7*, 88.
[88] Soriano, E.; Samadi, A.; Chioua, M.; de los Ríos, C.; Marco-Contelles, J. *Bioorg. Med. Chem. Lett. 2010, 20*, 2950.

In: Bioactive Heterocycles
Editors: K. L. Ameta, R. P. Pawar and A. J. Domb
ISBN: 978-1-62257-636-4
© 2013 Nova Science Publishers, Inc.

Chapter 2

Pyrrolo [1,2-a] Quinolines: Synthesis and Biological Activity

Marcel Mirel Popa[1], Mino R. Caira[2] and Florea Dumitrascu[1]
[1]Center of Organic Chemistry "C. D. Nenitzescu",
Romanian Academy, Bucharest, Romania
[2]Department of Chemistry, University of Cape Town,
Rondebosch, South Africa

Abstract

Pyrrolo[1,2-a]quinolines have raised interest as attractive targets for bioactivity evaluation and much attention has been focused on their chemistry. Herein is presented an insight into the synthetic tools and biological activity of these fascinating compounds.

1. Introduction

Remarkable compounds, due to their potential applications, pyrrolo[1,2-a]quinolines have been a source of ongoing interest for many years, prompting reports of numerous studies associated with their chemistry including their synthetic routes. Pyrrolo[1,2-a]quinoline, also known in the literature as benzopyrrocoline or benzo[e]indolizine, is an aromatic heterocycle formally obtained by the condensation of a quinoline with a pyrrole ring.

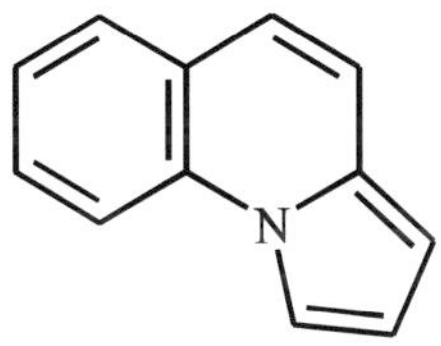

The first pyrrolo[1,2-a]quinoline framework was formally obtained by Einhorn and Sherman [1] in 1895. Unfortunately, they misassigned its structure as α-quinolylpropyl alcohol. Five years later, Koenigs [2] reinvestigated their work and reported that the correct structure of this compound is actually 1,2,3,3a,4,5-hexahydropyrrolo[1,2-a]quinolin-1-one.

The synthetic methods towards pyrrolo[1,2-a]quinolines attracted some interest [3] and can be classified according to the nature of the starting compounds: synthesis starting from quinoline and its derivatives, synthesis starting from substituted pyrroles, and synthesis starting from acyclic or other compounds. The interest in obtaining pyrrolo[1,2-a]quinoline-based compounds increased significantly over time, mainly due to their applications in many fields including those exploiting their bioactivity, discussed herein.

2. Synthetic Routes to Pyrrolo[1,2-a]Quinolines

Numerous synthetic routes to pyrrolo[1,2-a]quinolines are known, spanning more than one hundred years of their chemistry in wich the interest in such compounds remained essentially constant, with renewed interest evident in recent years owing to application-oriented studies.

2.1. Synthesis Starting from Quinoline and Its Derivatives

From Quinolines and Acetylenic Esters

The reaction between nitrogen-containing aromatic N-heterocyles and acetylenic esters, the most promising being dimethyl acetylenedicarboxylate (DMAD), was successfully used to obtain benzoindolizines. This reaction was first applied in the case of quinoline by Diels and Alder [4] in 1934 (Scheme 1A). Acheson et al. [5, 6] also employed non-symmetrical dipolarophiles for obtaining the pyrrolo[1,2-a]quinoline framework. Recently, [7-10] the intermediate proposed earlier by Acheson [11] obtained by the reaction of quinolines with DMAD could be trapped by suitable electrophiles or proton donors to afford the corresponding pyrrolo[1,2-a]quinolines **2** (Scheme 1B).

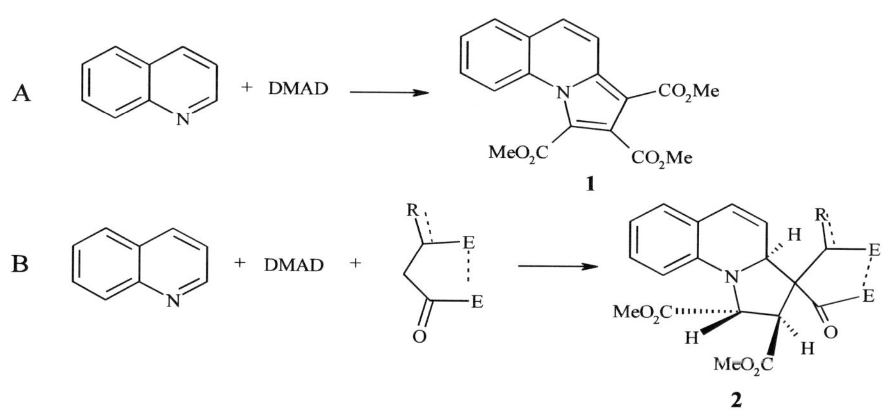

Scheme 1.

From Quinolines and Cyclopropenones

Lown and Matsumoto [12] studied the reaction between diphenylcyclopropenone and various N-heterocyclic aromatic compounds. After several years, Eicher [13] and later Weidner [14, 15] reinvestigated the reactions of diphenylcyclopropenone with N-heterocyclic compounds (Scheme 2). The obtained 3-hydroxy-1,2-diphenylpyrrolo[1,2-a]quinoline **3** could undergo further functionalization.

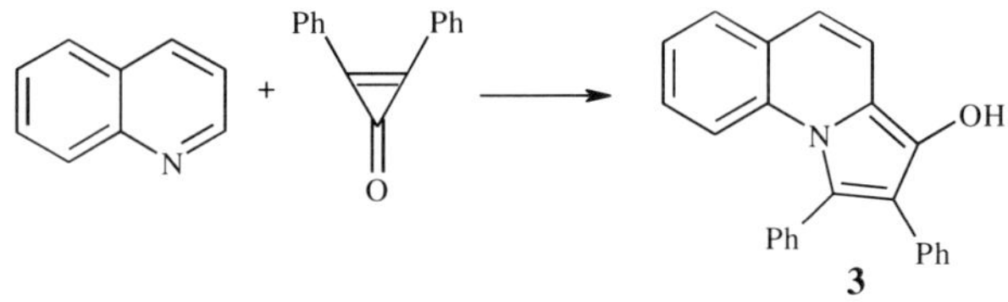

Scheme 2.

The pyrrolo [1,2-a]quinoline framework was also obtained via the nucleophilic addition of the corresponding heterocyclic bases to cyclopropenes [16-18] or cyclopropanes [19] in the presence of suitable catalysts (Scheme 3).

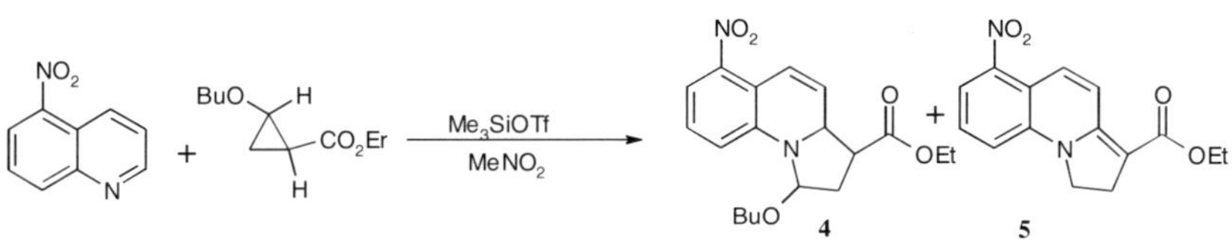

Scheme 3.

1,3-Dipolar Cycloaddition of Mesoionic Oxazoloquinolinone

The 1,3-dipolar cycloaddition of mesoionic 1,3-oxazole-5-ones with acetylenic dipolarophiles was efficiently applied in quinoline synthesis [20-23]. The mesoionic oxazolo[3,2-a]quinoline-2-ones **6a,b** could be generated *in situ* by the action of *N,N'*-dicyclohexylcarbodiimide (DCC) or acetic anhydride on 1(*H*)-quinolin-2-one acetic acids. In the presence of DMAD, **6a,b** will afford the corresponding pyrrolo[1,2-a]quinolines **7a,b** (Scheme 4).

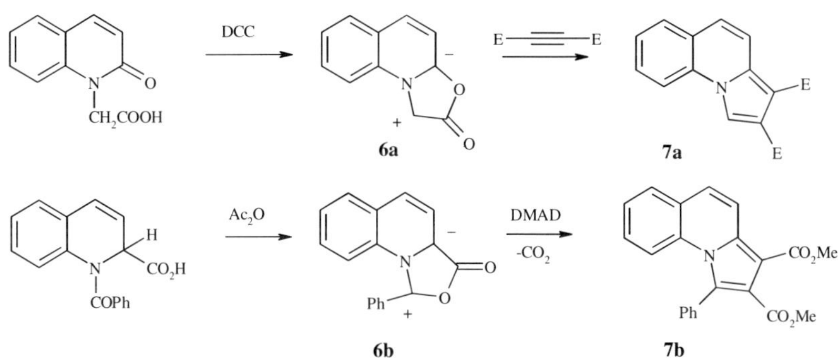

Scheme 4.

From Reissert Compounds of Quinoline

The reaction of Reissert compounds of six-membered N-heteroaromatic compounds is a useful tool for the synthesis of pyrroloazines with a bridgehead nitrogen atom. The quinoline Reissert compounds **8** were also used in the synthesis of pyrrolo[1,2-a]quinoline [21-25]. Anderson et al. [21, 22] managed to obtain pyrrolo[1,2-a]quinoline derivatives **9** and **10** (Scheme 5) in moderate-to-good yields.

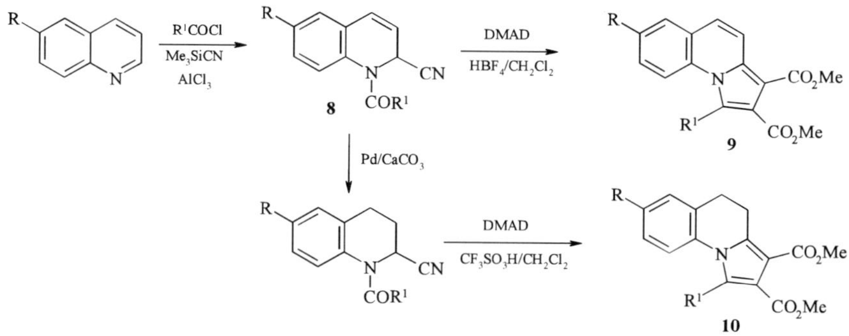

Scheme 5.

The Chichibabin Reaction

The Chichibabin reaction in the synthesis of indolizines was patented in 1927 by A. Chichibabin himself [26]. First attempts to extend the Chichibabin reaction for quinolines were reported by Wibberley et al. [27, 28] as unsuccessful. However, they managed to obtain in their studies the pyrrolo[1,2-a]quinoline scaffold [29]. Later, a series of pyrrolo[1,2-a]quinolines was obtained by the Chichibabin method starting from substituted 2-methylquinolines [30-32], but the yields were lower than 30% (Scheme 6).

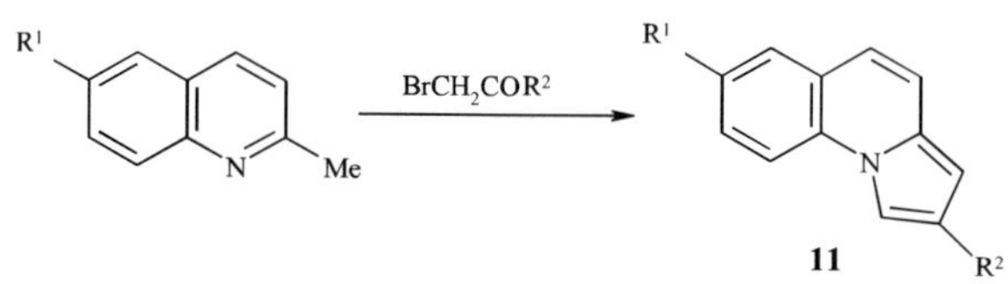

Scheme 6.

Starting from 2-substituted quinolines

Starting from 2-substituted quinolines, a series of synthetic strategies were developed to obtain pyrrolo[1,2-a]quinolines [33-62]. Under suitable conditions, the cyclization of the substituent at position 2 could be achieved (Scheme 7).

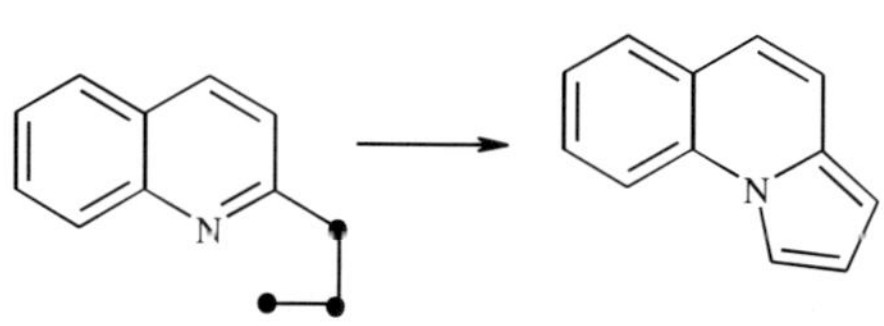

Scheme 7.

As an interesting example, starting from the monosubstituted malononitrile **12**, Tayyari et al. [48] observed that its passage through a chromatographic column (silica gel) serendipitously yielded the pyrrolo[1,2-a]quinoline **13** (Scheme 8).

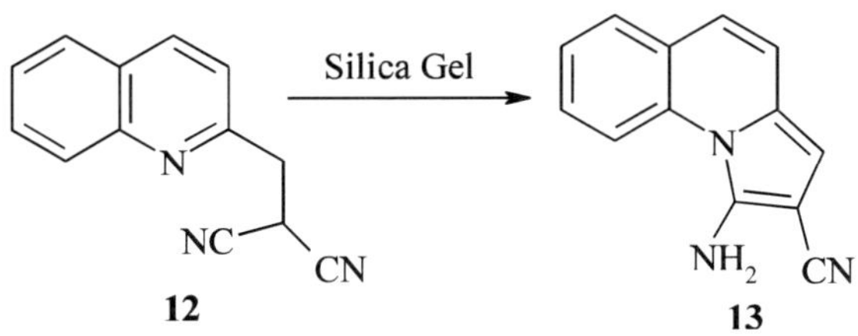

Scheme 8.

The Cu or Pd catalyzed cycloisomerization reaction [53-59] of 2-quinolyl substituted propargylic acetates **14** was employed [53-55, 57, 59] in the synthesis of pyrrolo [1,2-a]quinoline derivatives **15** (Scheme 9).

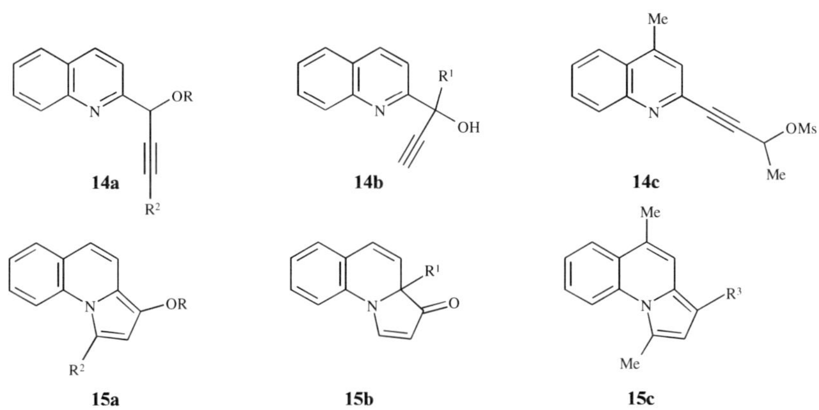

Scheme 9.

Starting from N-Substituted Quinolines

Pyrrolo[1,2-a]quinolines could also be obtained by starting from *N*-substituted quinolines [63-70]. Basavaiah et al. [69] reported the one-pot synthesis of benzoindolizines, including pyrrolo[1,2-a]quinolines, by the cyclization of the N-alkylation products of quinoline with differently substituted Baylis-Hillman bromo derivatives (Scheme 10).

Scheme 10.

1,3-Dipolar Cycloaddition of Quinolinium N-Ylides

The 1,3-dipolar cycloaddition between quinolinium *N*-ylides and activated olefinic or acetylenic dipolarophiles [71-103] is the most versatile tool for obtaining substituted pyrrolo[1,2-a]quinolines, owing to features such as readily available starting materials and simple procedures. A general representation of the 1,3-dipolar cycloaddition reaction is shown in Scheme 11. Under specific conditions, the ylide **17** is generated and further reacts with olefinic or acetylenic dipolarophiles to give compounds of type **18**, which under suitable reaction conditions afford the aromatic pyrrolo[1,2-a]quinolines **19**.

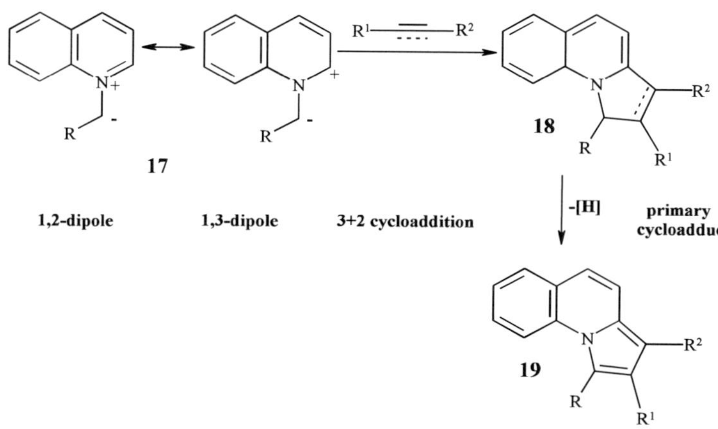

Scheme 11.

Efficient one-pot three component routes to pyrrolo[1,2-a]quinolines synthesis **20** were reported. One of them, involves the simple mixing of quinoline, bromoacetophenones and acetylenic dipolarophiles in an epoxide,which plays the role of reaction medium and HBr scavenger [97-100] (Scheme 12a). Another one-pot multicomponent strategy [103] involves the use of reactive alkenes generated *in situ*. The pyrrolo[1,2-a]quinolines **22** were obtained by the oxidation in the air of hydrogenated intermediates. Compounds of type **23** were also obtained in a similar manner by reacting quinoline with three molecules of bromoacetyl derivatives under the assistance of piperidine (Scheme 12b).

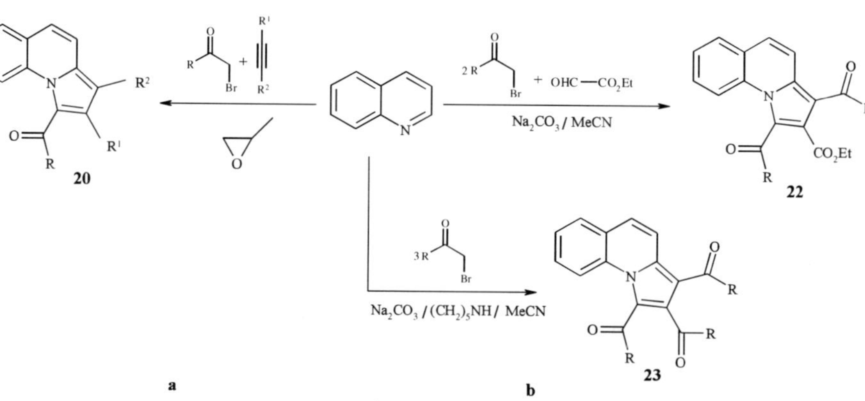

Scheme 12.

2.2. Synthesis Starting from Pyrroles

Synthesis Starting from Pyrroles and Corresponding Hydrogenated Analogues

The construction of aromatic polycyclic skeletons, including pyrrolo[1,2-a]quinolines, from various substituted pyrroles was intensively studied [104-128]. The corresponding hydrogenated pyrrolo[1,2-a]quinolines were synthesized from hydrogenated pyrrole frameworks, such as pyrrolines or pyrrolizines [129-145].

One of the early studies in the synthesis of pyrrolo[1,2-a]quinolines from substituted pyrroles is due to Garcia et al. [104, 105]. Thus, from the 1-phenylpyrrole **24**, the substituted pyrrolo[1,2-a]quinoline **25** was obtained (Scheme 13).

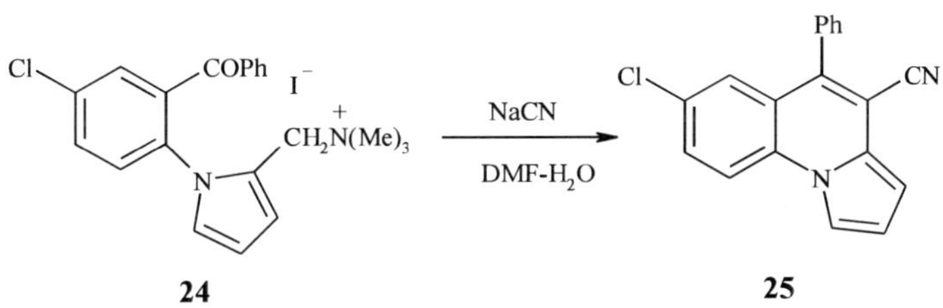

Scheme 13.

In recent studies, there has been reported the synthesis of iodo-pyrrolo[1,2-a]quinolines **26** by iodine-promoted cyclization of the corresponding alkyne-substituted N-arylpyrroles [128]. These compounds were further functionalized by replacing iodine with other groups.

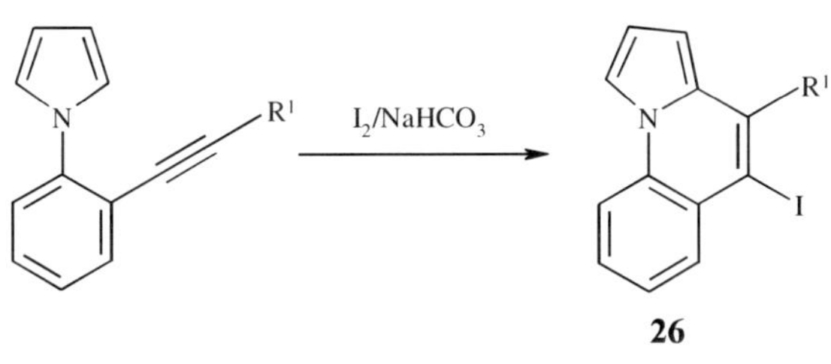

Scheme 14.

Synthesis by the Tert-Amino Effect

A synthetic strategy starting from pyrroles is associated with the tert-amino effect which generally describes the condensation reactions of *ortho* substituted N,N-dialkylanilines towards various heterocycles. Thus, starting from suitable *ortho* substituted phenyl-pyrrolidines (cycloalkyl anilines), substituted pyrrolo[1,2-a]quinolines were reported [146-162]. Of particular interest is the work of Reinhoudt et al. [126-134] who obtained the pyrrolo[1,2-a]quinoline framework 27 according to Scheme 15..

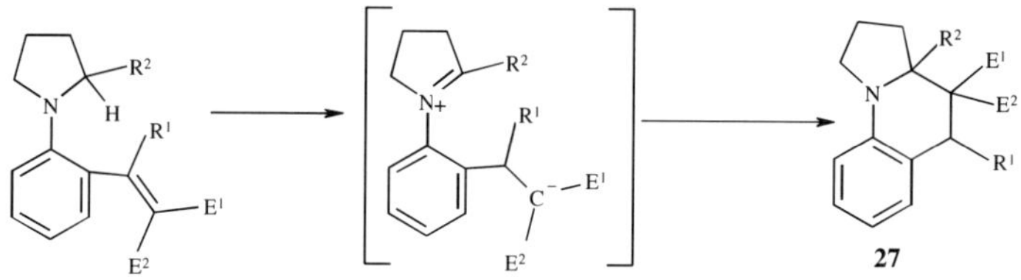

27

Scheme 15.

2.3. Other Methods for Obtaining Pyrrolo[1,2-a]Quinolines

Starting from quinolinium N-oxide, its reaction with suitable reagents led to the formation of the pyrrolo[1,2-a]quinoline framework in moderate yields [163, 164]. Other methods for the synthesis of pyrrolo[1,2-a]quinoline includes cyclomerization or 1,3-dipolar cycloadditions of xylilenes obtained by the pyrolysis of suitable starting heterocycles [165-168], the Claisen rearrangement of allyl 4-quinolyl ethers followed by intramolecular cyclization [169-171], by a rearrangement under suitable conditions of some spiro derivatives of quinoline [172], or by a thermal rearrangement of some izoxazoline-5-spiro and isoxazolidine derivatives [173, 174]. More recently, the $AlCl_3$-catalyzed transformation of 1-(2-isothiocyanoaryl)-2-(2-furyl)ethane in obtaining the pyrrolo[1,2-a]quinoline framework was reported [175]. 1,2,3-Triazolo[1,2-a]quinoline *N*-ylides in the presence of methyl propiolate or DMAD gave pyrrolo[1,2-a]quinolines [176, 177]. The key intermediate in this reaction is the betaine, resulting from the Michael addition of the N-ylide to the acetylenic dipolarophile.

The pyrrolo[1,2-a]quinoline framework could also be obtained by double cyclization reactions starting from suitable reagents [178-184]. Recently [183, 184], it was demonstrated that gold or gold/silver catalyzed reactions of 1,4-aminoalkynes with alkynes resulted in the formation of pyrrolo[1,2-a]quinoline framework (Scheme 16).

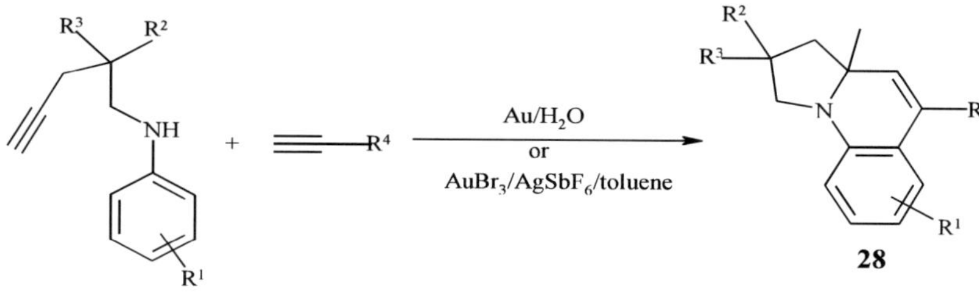

28

Scheme 16.

Thermal and photochemical reactions of *o*-azidophenylethylfurans gave pyrrolo[1,2-a]quinolines [185, 186]. An intramolecular cyclization of substituted phenacyl azides [187] and azidoquinones [188] led to the formation of pyrrolo[1,2-a]quinolines.

An interesting rearrangement reaction [189] starting from epoxyazides (Scheme 17) affords the hydrogenated pyrrolo[1,2-a]quinoline framework **29** in high yields.

Other synthesizing, involving quinoline and alkenyldiazo derivatives [190], α,β-unsaturated carboxylic acids [191] or by methylenecyclopropanes [3+2] cycloaddition to nitrones [192], were also reported recently.

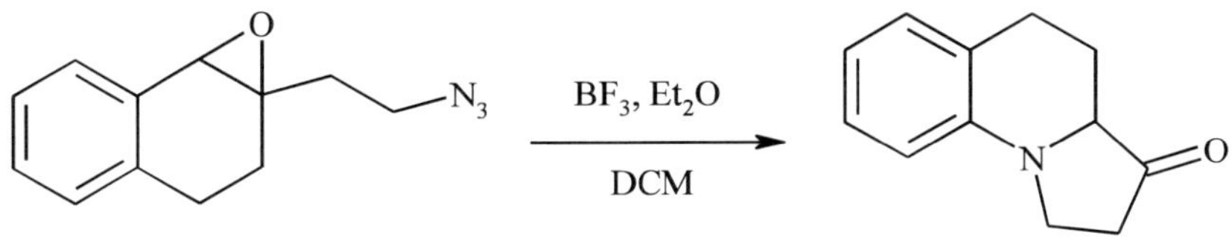

29

Scheme 17.

2.4. Bioactive Compounds Based on Pyrrolo[1,2-a]Quinolines

Natural Alkaloids with the Pyrrolo[1,2-a]Quinoline Core

Natural alkaloids are of great interest owing to their interesting biological activities with a potential for medicinal purposes. Among the vast number of natural alkaloids, Gephyrotoxin **30** is one that contains the pyrrolo[1,2-a]quinoline framework. Gephyrotoxin was the first isolated [193] from skin secretions of an amphibian, *Dendrobates Histrionicus,* together with other alkaloids and its activity was studied as a muscarinic antagonist [194, 195]. Gephyrotoxin was the subject of formal and total synthesis [196-223].

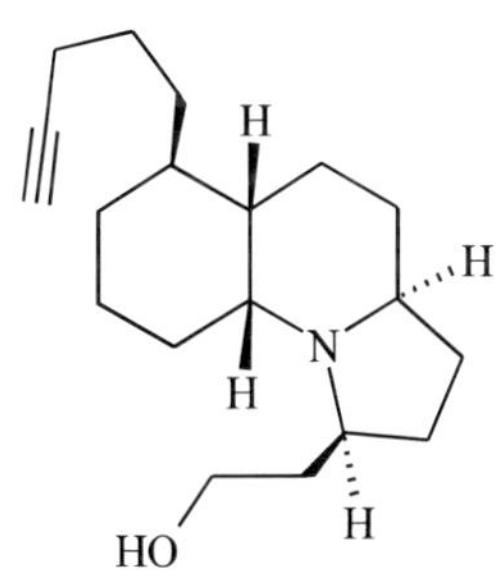

30

Gephyrotoxin was remarkable for its low toxicity and potential neurological activity in neuromuscular transmission or its interaction with specific receptors (non-competitive blockers of nicotinic channels) [220-223].

Bioactive Pyrrolo[1,2-a]Quinoline Derivatives

Pyrrolo[1,2-a]quinoline derivatives were also studied with the aim of evaluating their potential biological activity. The 4,5-dihydro compounds **31** were more active than the fully aromatic analogues against P388 lymphocytic leukemia [22].

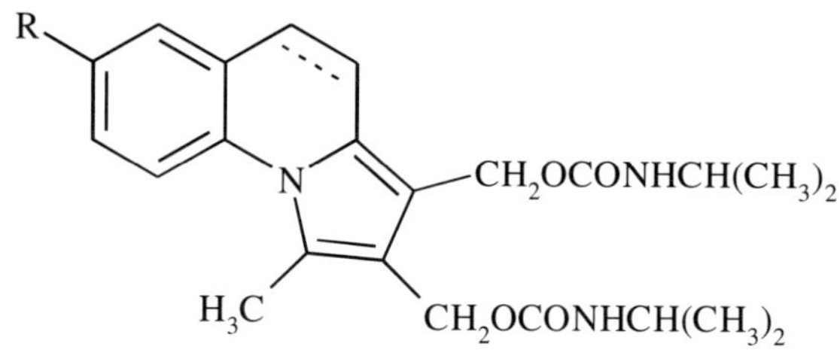

R = H; OCH_3

31

The 1-glyoxylilamide indolizines were found to possess cytotoxic activity [224] against cancer cells such as the human uterine sarcoma cell line MES-SADX5 and human myeloid leukemia cell line HL60/TX1000. The authors of the patent give the pyrrolo[1,2-a]quinoline **32** as an example among the indolizines described therein.

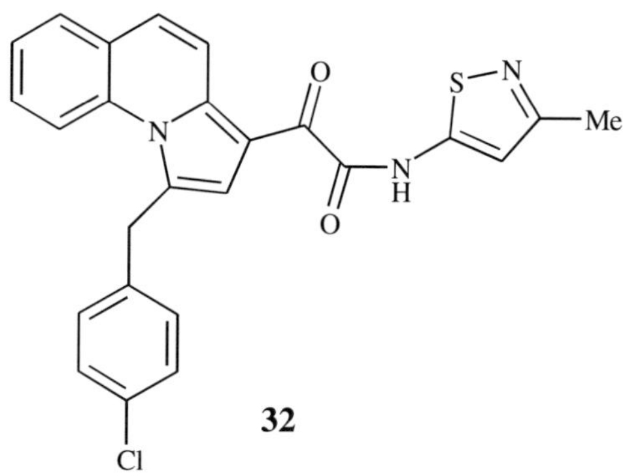

32

A pyrrolo[1,2-a]quinoline framework **33** with a cyclopropylcarbonyl group synthesized by 1,3-dipolar cycloaddition reaction of corresponding N-ylides was evaluated [225] as an EGFR kinase inhibitor in order to asses its antiproliferative activity on Hep-G2, the human hepatocellular liver carcinoma.

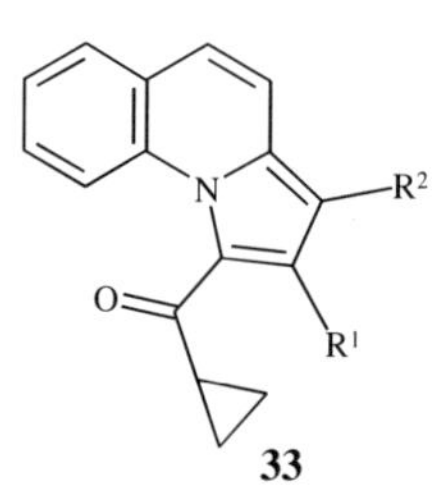

33

The tryciclic compounds **34** were found to possess significant activity in DNA-gyrase and topoisomerase-II inhibition [226], rendering them as potential candidates for anti-cancer therapy.

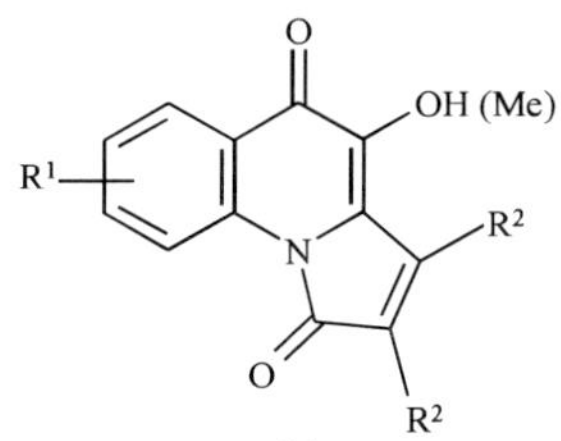

34

The compound **35** was chosen for anticancer studies [183] in order to assess its activity against the HeLa (cervical epithelioid carcinoma) cell line (IC_{50}=34.7 μm) and against the normal lung fibroblast (CCD-19 Lu) cell line (IC_{50}=82.3 μm). The higher activity of these compounds could open the way to further studies on its analogs.

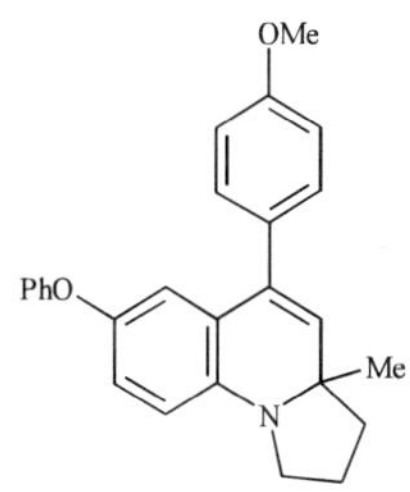

35

Hypoxia-inducible (Hif-α) [227] transcriptome is a factor that influences tumor growth by increasing resistance to apoptosis.

Thus, compound **36** was one of the compounds studied [227] having as target the factor-A (Hif-α) with the aim of developing novel cancer therapeutics. Compound **36** was found to be active on specific cell lines of breast cancer, melanoma and renal cancer.

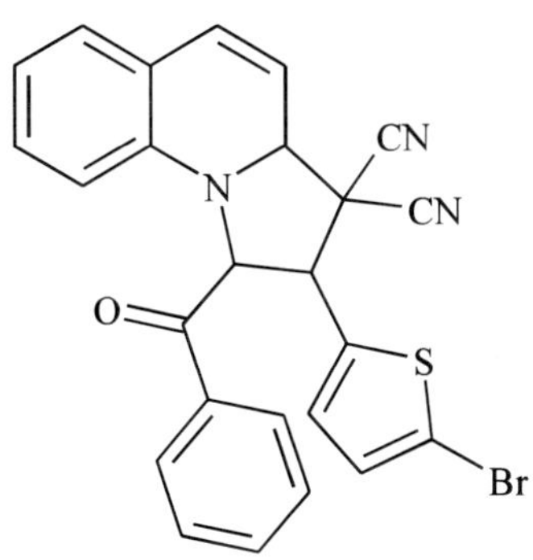

36

Pyrrolo[1,2-a]quinoline derivatives **37-40** were also patented as apoptosis inducers and anti-cancer agents [228-231].

Their SAR studies were performed in order to optimize the target molecules. Thus, the presence of CN groups enhances the activity, whereas the presence of an electron-withdrawing group at the *para* position of the benzoyl moieties (the substituents Ar in **37** and **38**) enhances the potency and selectivity to the target.

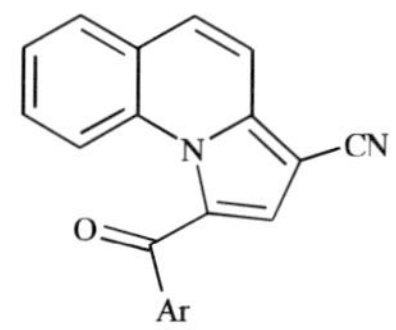

37

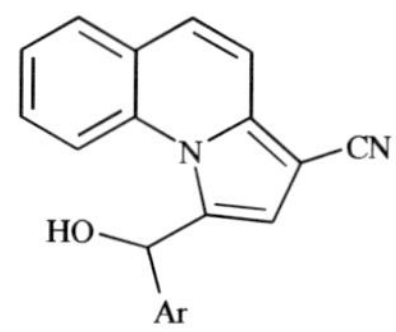

38

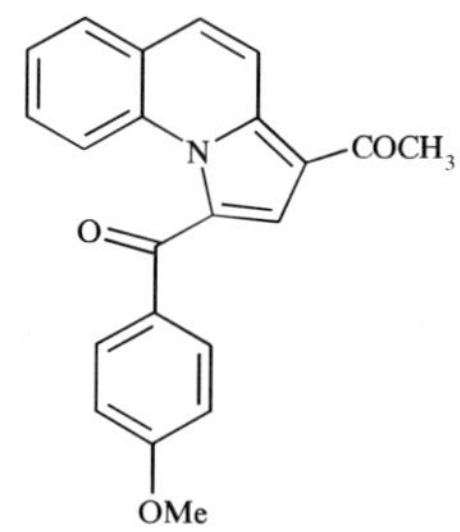

39

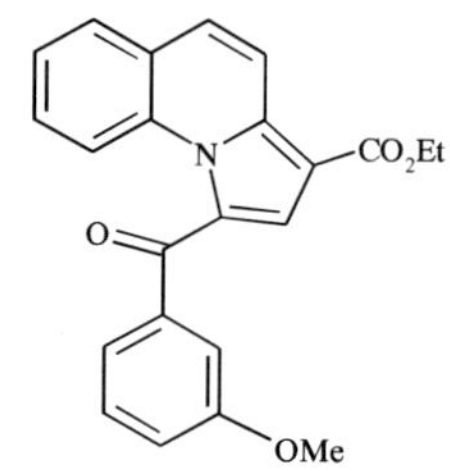

40

Various hydrogenated pyrrolo[1,2-a]quinolines **41** were patented as cancer cell migration inhibitors for different cell lines such as pancreatic cancer [232, 233].

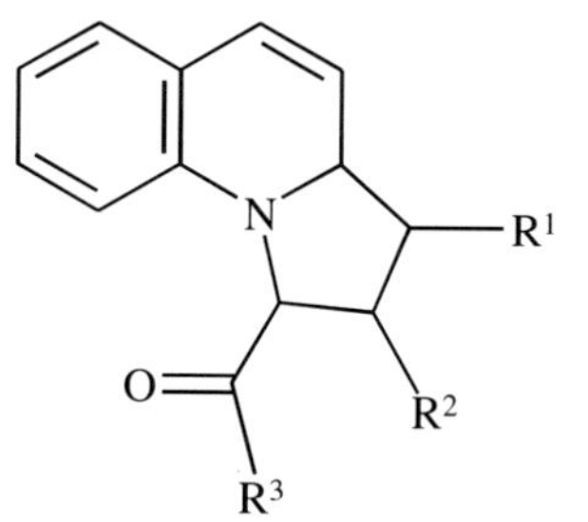

41

Pyrrolo[1,2-a]quinolines **42** were studied [125] for their interaction with the NK1 receptor in order to obtain responses such as transmission of pain and stress signals, smooth muscle contraction, the induction of neurogenic inflammation, endothelium-dependent vasodilation and angiogenesis.

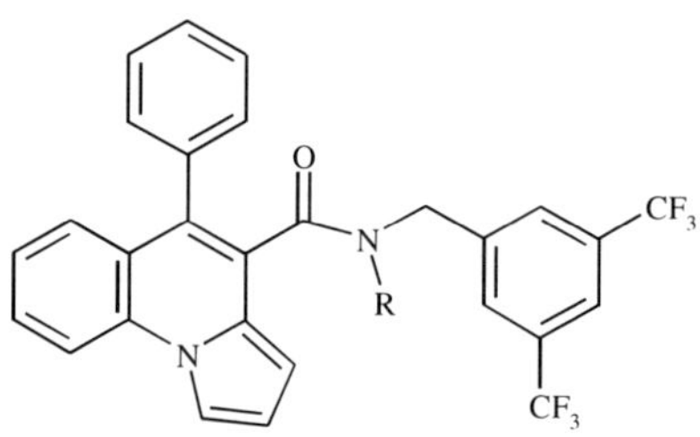

R = H, Me

42

Compounds of type **43** with the 4-quinolonic skeleton were tested against *Aspergillius negri*, *Candida Albicans* and *Pseudomona Aeruginosa*, *S. Aureus*, and *E. Coli*, but the results were unsatisfactory regarding their potential against these fungi and bacteria [133].

Another study [137] showed that in the case of compound **43** with $R_1 = R_4 = H$, $R_3 = $ 4-methylpiperazinyl and $R_2 = F$, some minor activity against *S. Aureus* and *E. Coli* were observed.

Compound **43a** with $R_1 = R_4 = H$, $R_3 = $ 4-methylpiperazinyl and $R_2 = NO_2$, was reported [234] to possess the highest potential of such derivatives, with good activity against gram-positive bacteria such as *E. Coli*, *S. Typhi*, *Sh. Disenteriae*, *Kl. Pneumoniae,* on strains of *Ps. aeruginosa*, *Proteus vulgaris* and *S. Aureus.* These compounds showed a higher activity than Pefloxacin or Tarivid (Ofloxacin).

In the case of *in vivo* studies, **43a** showed chemotherapeutic properties in induced septicemia with *Typhoid bacilli* on mice, with a 65% survival rate at the maximum dose as compared with dioxacin and pefloxacin (potent antibiotics from the class of quinolone drugs) which had a 100% survival rate at half dosage.

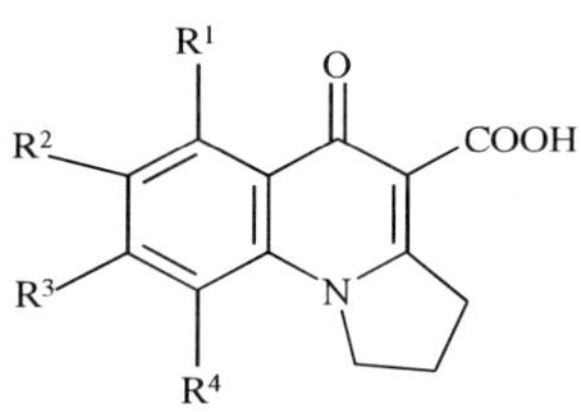

43

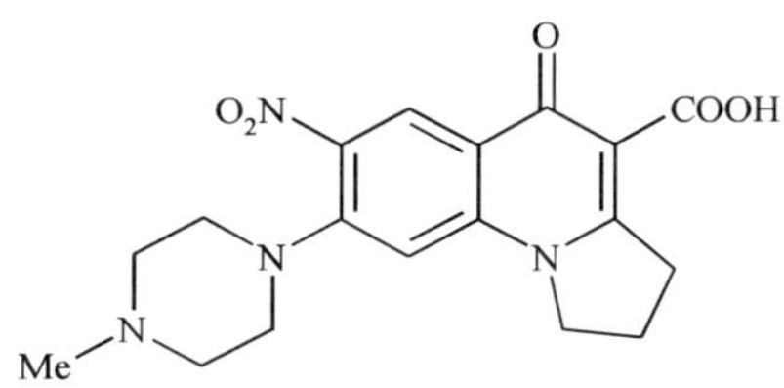

43a

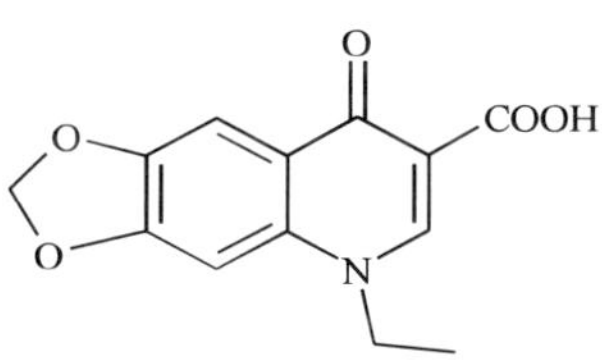

Dioxacin

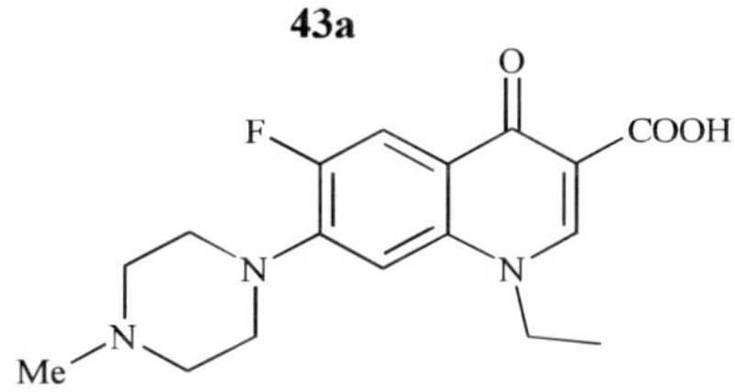

Pefloxacin

Trypanosomiasis, sleeping sickness, is a fatal disease caused by the protozoan parasites *Trypanosoma brucei gambiense*. Compound **44** was found to be an effective anti-trypanosomal agent [235]. The corresponding carboxylic acid of **44** was studied and found to be ineffective against *P. aeruginosa*, *E. coli* and *Proteus vulgaris* [236].

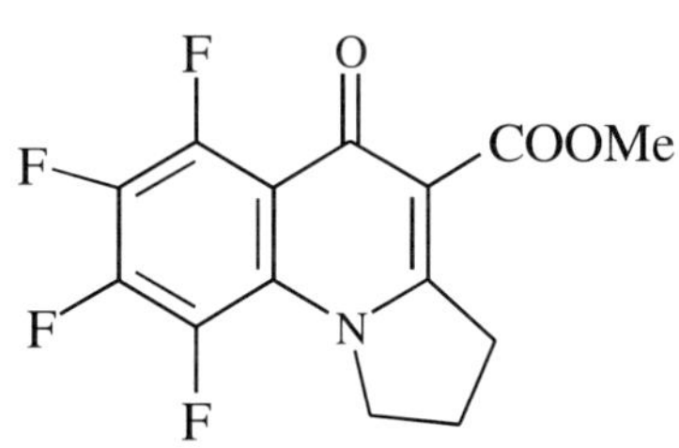

44

Pyrrolo[1,2-a]quinoline of formula **45** is claimed [237] to be useful in antifungal therapy against *A. flavus*, *A. fumigatus*, *A. niger*, *A. terreus* in combination with another antifungal agent. Substituted 5,7-dimethyl-2-phenylpyrrolo[1,2-a]quinolines were also reported with activity against *S. typhi* and *A. flavus* [238].

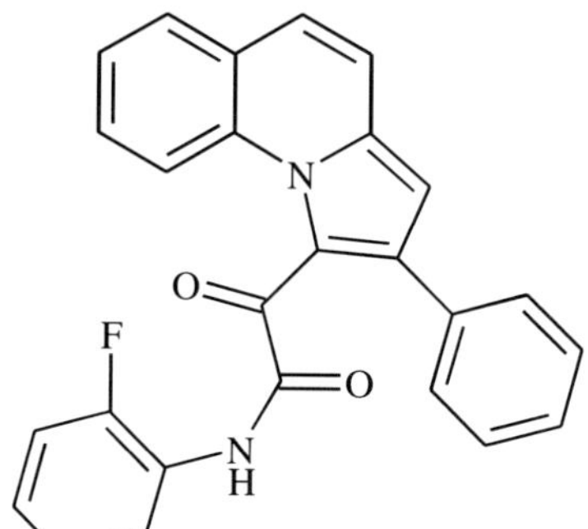

45

Pyrrolo[1,2-a]quinolines **45** were studied [31] as antiallergic compounds and found to possess very low activity, whereas compounds of structure **46** were found with potential antioxidant activity [239].

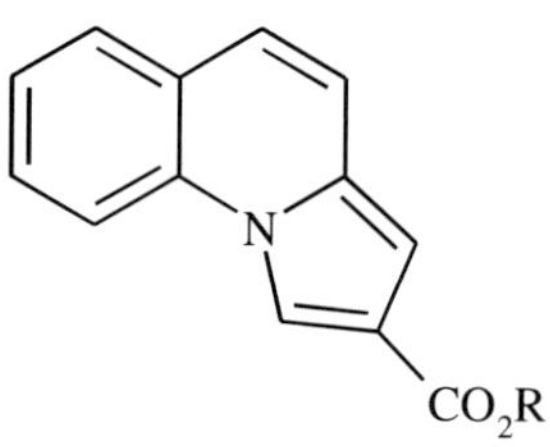

R = H; Et

45

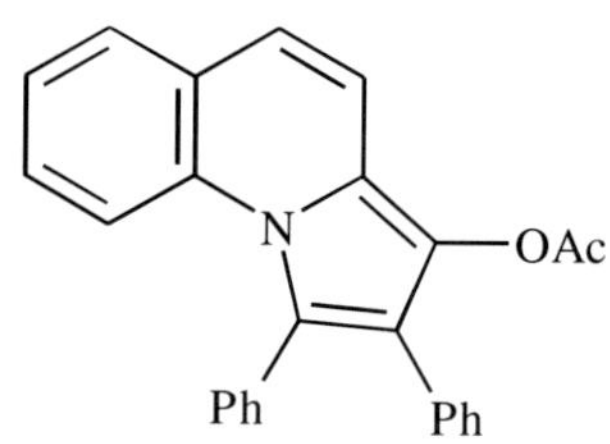

46

Compounds **47** were patented [240] as central nervous system (CNS) agents. Such hexahydropyrrolo[1,2-a]quinolines, or some of their stereoisomers in particular cases, were found to possess activities as anticonvulsants, diuretics and antitussives, or have potential in glaucoma treatment.

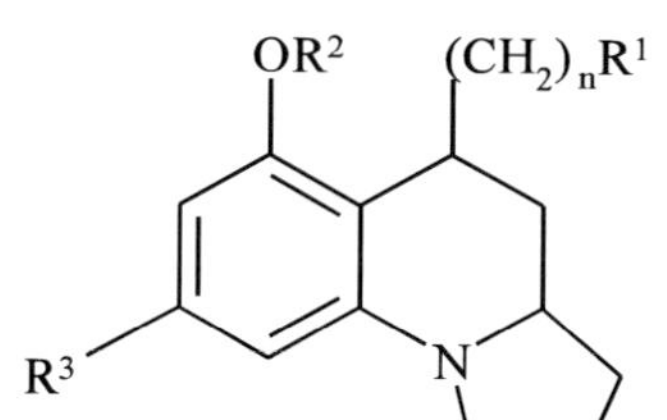

47

It is noteworthy that R^1, R^2 and R^3 cover a large variety of subsituents, giving a high structural diversity to the title compounds.

Conclusion

The interest in the chemistry of pyrrolo[1,2-a]quinolines has been maintained for over a century. It is important to note that the studies of their bioavailability are rather scarce by comparison with those of their indolizine and pyrrolo[2,1-a]isoquinoline analogs. One factor may be the low occurrence of the pyrrolo[1,2-a]quinoline framework among the natural alkaloids. Thus, the biological properties of pyrrolo[1,2-a]quinolines may warrant further investigation in the future.

Acknowledgments

MRC is grateful to the University of Cape Town and the National Research Foundation for financial support.

References

[1] Einhorn A.; Sherman, P. *Justus Liebigs Ann. Chem.* 1895, 287, 26-49.
[2] Koenigs, W. Ber. 1900, 33, 218-228.
[3] El-Sayed, H. E.-A.; El-Sayed, I. I. *Adv. Heterocycl. Chem.* 2003, 84, 83-97.
[4] Diels, O.; Alder, K. *Justus Liebigs Ann. Chem.* 1934, 510, 87-128.
[5] Acheson, R. M.; Verlander, M. S. *J. Chem. Soc. C* 1969, 2311-2316.
[6] Acheson, R. M.; Robinson, D. A. *J. Chem. Soc. C* 1969, 1633-1638.
[7] Yavari, I.; Hossaini, Z.; Sabbaghan, M. *Tetrahedron Lett.* 2006, 47, 6037-6040.
[8] Nair, V.; Devipriya, S.; Suresh, E. *Synthesis* 2008, 551-554.
[9] Yavary, I.; Mirzaei, A.; Moradi, L.; Hosseini, N. *Tetrahedron Lett.* 2008, 49, 2355-2358.
[10] Yavari, I; Piltan, M.; Moradi, L. *Tetrahedron* 2009, 65, 2067-2071.
[11] Acheson, R., M.; Procter, G. *J. Chem. Soc. Perkin Trans.* 1 1979, 2171-2179.
[12] Lown, J. W.; Matsumoto, K. *Can. J. Chem.* 1971, 49, 1165-1175.
[13] Eicher, T.; Rhode, R. *Synthesis* 1985, 619-625.
[14] Weidner, C. H.; Michaels, F. M.; Beltman, D. J.; Montgomery, C. J.; Wadsworth, D. H.; Briggs, B. T.; Picone, M. L. *J. Org. Chem.* 1991, 56, 5594-5598.
[15] Weidner, C. H.; Wadsworth, D. H.; Bender, S. L.; Beltman, D. J. *J. Org. Chem.* 1989, 54, 3660-3664.
[16] Hauck, G.; Durr, H. Angew. Chem. 1979, 91, 1010; *Angew. Chem. Int.* Ed. 1979, 18, 945.
[17] Durr, H.; Gross, H.; Zils, K.-D.; Hauck, G.; Klauck, G.; Hermann, H. *Chem. Ber.* 1983, 116, 3915-3925.
[18] Ahmed, S. A. *J. Phys. Org. Chem.* 2006, 19, 402-414.
[19] Morra, N. A.; Morales, C. L.; Bajtos, B.; Wang, X.; Jang, H.; Wang, J.; Yu, M.; Pagenkopf, B. L. *Adv. Synth. Catal.* 2006, 348, 2385-2390.
[20] Potts, K. T.; Yao, S. J. Org. Chem. 1979, 44, 977-979.
[21] Anderson, W. K.; De Ruiter, J.; Heider, A. R. *J. Org. Chem.* 1985, 50, 722-724.

[22] Anderson, W. K.; Heider, A. R.; Raju, N.; Yucht J. A. *J. Med. Chem.* 1988, 31, 2097-2102.
[23] McEwen, W. E.; Mineo, I. C.; Shen, Y. H.; Han, G. I. *Tetrahedron Lett.* 1968, 49, 5157-5161.
[24] McEwen, W. E.; Mineo, I. C.; Shen, Y. H.*; J. Am. Chem. Soc.* 1971, 93, 4479-4484.
[25] Uff, B. C.; Budhram, R. S.; Consterdine, M. F.; Hicks, J. K.; Slingsby, B. P.; Pemblington, J. A. *J. Chem. Soc. Perkin Trans. 1* 1977, 2018-2022.
[26] Chichibabin, A. DRP Patent 464481 1927.
[27] Melton, T.; Taylor, J.; Wibberly, D. *J. Chem. Commun.* 1965, 151-152.
[28] Hurst, J.; Melton, T.; Wibberly, D. J. *J. Chem. Soc.* 1965, 2948-2955.
[29] Irwin, W. J.; Wibberly, D. J. *J. Chem. Soc. Perkin* 1 1974, 250-252.
[30] Kuo, H.-S.; Yoshina, S.; Tung, Y.-C. *J. Heterocycl. Chem.* 1979, 393-395.
[31] Ager, I. R.; Barnes, A. C.; Danswan, G. W.; Hairsine, P. W.; Kay, D. P.; Kennwell, P. D.; Matharu, S. S.; Miller, P.; Robson, P.; Rowlands, D. A.; Tully, W. R.; Westwood, R. *J. Med. Chem.* 1988, 31, 1098-1115.
[32] Chai, W.; Kwok, A.; Wong, V.; Carruthers, N. I.; Wu, J. *Synlett* 2003, 2086-2088.
[33] Osawa, A.; Abe, Y.; Igeta, H. *Bull. Chem. Soc. Jpn.* 1980, 53, 3273-3275.
[34] Deady, L. W.; Werden, D. M. *J. Org. Chem.* 1987, 52, 3930-3932.
[35] Greenhill, J. V.; Loghmani-Khouzani, H.; Maitland, D. J. *J. Chem. Soc.* Perkin Trans. 1 1991, 2831-2840.
[36] Andrews, H.; Skidmore, S.; Suschitzky, H. *J. Chem. Soc.* 1962, 2370-2374.
[37] Matoba, K.; Itoh, K.; Kondo, K.; Yamazaki T.; Nagata, M. *Chem. Pharm. Bull.* 1981, 29, 2442-2450.
[38] Pohjala, E. K. *Acta Chim. Scand.* B 1980, 34, 79-83.
[39] Bode, M. L.; Kaye, R. T. *J. Chem. Soc. Perkin Trans* 1 1993, 1809-1813.
[40] Kaloko, J.; Hayford, A. *Org. Lett.* 2005, 7, 4305–4308.
[41] Leonard, N. J.; Boyer, J. H. *J. Am. Chem. Soc.* 1950, 72, 2980-2985.
[42] Roberts, E. M.; Gates, M.; Boekelheide, V. *J. Org. Chem.* 1955, 20, 1443-1447.
[43] Baumgarten, H., E.; Beckerbrauer, R.; Debrunner, M. R. *J. Org. Chem.* 1961, 26, 1539-1546.
[44] Nagata, M. *Yakugaku Zasshi* 1966, 86, 608-613.
[45] Cannon, J. G.; Lazaris, S. A.; Wunderlich, T. A. *J. Heterocycl. Chem.* 1967, 4, 259-261.
[46] Kuo, H.-S.; Ma, Y.-T.; Tung, Y.-C. Hua Hsueh 1978, 77-79; *Chem. Abstr.* 1980, 92, 94216z.
[47] Kocian, O.; Ferles, M. *Coll. Czech. Chem. Commun.* 1978, 43, 1413-1430.
[48] Tayyari, F.; Wood, D. E.; Fanwick, P.; Sammelson, R. E. *Synthesis* 2008, 279-285.
[49] Fujita, R.; Hoshino, M.; Tomishava, H. *Chem. Pharm. Bull.* 2006, 54, 334-337.
[50] Jukic, L.; Bratusek, U.; Skof, M.; Svete, J.; Stanovnik, B. *Khim. Geterosikl. Soedin.* 1996, 1510-1514.
[51] Sun, S.; Wang, M.; Deng, H.; Liu, Q. *Synthesis* 2008, 4, 573-583.
[52] Zhu, L.; Vimolratana, M.; Brown, S. P.; Medina, J. C. *Tetrahedron Lett.* 2008, 49 1768-1770.
[53] Yan, B.; Zhou, Y.; Chen, J.; Liu, Y. *J. Org. Chem.* 2007, 72, 7783-7786.
[54] Liu, Y.; Song, Z.; Yan, B. *Org. Lett.* 2007, 9, 409-412.
[55] Smith, C. R.; Bunnelle, E. M.; Rhodes, A.; Sarpong, R. *Org. Lett.* 2007, 6, 1169-1171.

[56] Bunnelle, E. M.; Smith, C. R.; Lee, S. K.; Singaram, S. W.; Rhodes, A. J.; Sarpong, R. *Tetrahedron* 2008, 64, 7008-7014.
[57] Chernyak, D.; Gadamsetty, S., B.; Gevorgyan, V. *Org. Lett.* 2008, 10, 2307-2310.
[58] Chernyak, D.; Skontos, C.; Gevorgyan, V. *Org.Lett.* 2010, 12, 3242-3245.
[59] Kim, H.; Lee, K.; Kim, S.; Lee P. H. *Chem. Commun.* 2010, 46, 6341-6343.
[60] Bai, Y; Zeng; Ma, J. J.; Gorityala, B. K.; Liu X.-W. *J. Comb. Chem.* 2010, 12, 696-699.
[61] Tukulula, M.; Klein, R.; Kaye, P. T. *Synthetic Commun.* 2010, 40, 2018–2028.
[62] Rueping, M.; Hubener, L. *Synlett* 2011, 1243-1246.
[63] Yoon, U. C.; Quillen, S, L.; Mariano, P. S. *Tetrahedron Lett.* 1982, 23, 919-922
[64] Yoon, U. C.; Quillen, S. L.; Mariano, P. S.; Swanson, R.; Stavinoha, J. L.; Bay, E. *J. Am. Chem. Soc.* 1983, 105, 1204-1218.
[65] Kato, H.; Chiba, T.; Kimura, H. *Chem. Pharm. Bull.* 1977, 25, 203-208.
[66] Milot, G.; Meyers, A. I. *J. Org. Chem.* 1993, 58, 6538-6540.
[67] Kumari, S.; Parkash, Su.; Goel, V. K. *Indian J. Chem. Sect.* B 1996, 35B, 1329-1330.
[68] Osornio, Y. M.; Miranda, L. D.; Cruz-Almanza, R.; Muchowski, J. M. *Tetrahedron Lett.* 2004, 45, 2855-2858.
[69] Basavaiah, D.; Devendar, B.; Lenin, D. V.; Satyanarayana T. *Synlett* 2009, 411-416 .
[70] Khadem,. S.; Udachin, K. A.; Arya, P. *Synlett* 2010, 199-202.
[71] Henrick, C. A.; Ritchie, E.; Taylor W. C. Aust. J. Chem. 1967, 20, 2467-2477.
[72] Tewari, R. S.; Dubey, A. K.; Misra, N. K. *J. Chem. Eng. Data* 1982, 27, 101-103
[73] Tewari, R. S.; Dixit, P. D.; Dubey, A. K. *J. Chem. Eng. Data* 1983, 28, 283-285.
[74] Stetinova, J.; Jurasek, A.; Kovac, J.; Dandarova, M.; Safar, P. *Coll. Chem. Czech. Commun.* 1986, 51, 412-418.
[75] Dumitrascu, F.; Mitan, C. I.; Draghici, C.; Caproiu, M. T.; Raileanu D. *Tetrahedron Lett.* 2001, 42, 8379-8382.
[76] Dumitrascu, F.; Mitan, C. I.; Draghici, C.; Caproiu, M. T. *Rev. Roum. Chim.* 2002, 47, 881-884.
[77] Georgescu, E.; Georgescu, F.: Danila, M. D.; Filip, P. I.: Draghici C.; Caproiu M. T. Arkivoc 2002, (ii), 30-45.
[78] Kheder, N. A.; Darwish, E. S.; Dawood, K. M. *Heterocycles* 2009, 78, 177-188.
[79] Shestopalov, A. M.; Chunikhin, K. S.; Rodionovskaya, L. A. *Chem. Heterocycl. Compd.* 2002, 38, 310-313.
[80] Shestopalov, A. M.; Litvinov, V. P.; Sharanin, Yu. A.; Khoroshilov, G. E. *Dokl. Acad. Nauk.* 1990, 312, 1156.
[81] Fröhlich, J.; Kröhnke, F. *Chem. Ber.* 1971, 104, 1621-1628.
[82] Kiel, W.; Kröhnke, F. *Chem. Ber.* 1972, 105, 3709-3715.
[83] Zhang, L.; Liang, F.; Sun, L.; Hu, Y.; Hu, H. *Synthesis* 2000, 12, 1733-1737.
[84] Zhang, X.; Cao, W.; Wei, X.; Hu, H. *Synthetic Commun.* 1997, 27, 1395-1403.
[85] Wu, K.; Chen, Q.-Y. *Synthesis* 2003, 35-40.
[86] Komatsu, M.; Kasano, Y.; Yamaoka, S. Minakata, S. *Synthesis* 2003, 1398-1402.
[87] Anary-Abbasinejad, M.; Charkhati, K.; Hassanabadi, A. *J. Chem. Res.* 2009, 95-97.
[88] Georgescu, E.; Georgescu, F.; Roibu, C.; Draghici C.; Caproiu M. T. *Rev. Roum. Chim.* 2001, 46, 357-362.
[89] Wei, X.; Hu, Y.; Li, T.; Hu, H. *J. Chem. Soc. Perkin* 1 1993, 2487-2489.
[90] Khlebnikov A. F.; Kostikov, R. R. *Chem. Heterocycl. Compd.* 1987, 23, 708-709.
[91] Komatsu, M.; Minakata, S.; Oderaotoshi, Y. Arkivoc 2006, (vii), 370-389.

[92] Ma, C.; Ren, Y.; Zhang, Q.; Ding, K.; Zhao, J.; Zhang, D. *Chem. Lett.* 2007, 36, 1152-1153.
[93] Liu, Y.; Zhang, Y.; Shen,Y.-M.; Hu, H.-W.; Xu, J.-H. *Org. Biomol. Chem.* 2010, 8, 2449–2456.
[94] Dawood, K. M.; Ragab, E. A.; Mohamed, S. N. Z. *Naturforsch.* 2009, 64b, 434-438.
[95] Dawood, K. M.; Ragab, E.A.; Khedr, N. A. *J. Chin. Chem. Soc.* 2009, 56, 1180-1185.
[96] Kianmehr, E.; Estiri, H.; Bahreman, A. *J. Heterocycl. Chem.* 2009, 46, 1203-1206.
[97] Georgescu, E.; Dumitrascu, F.; Georgescu, F.; Draghici, C.; Popa, M. M. *Rev. Roum. Chim.* 2010, 55, 217-221.
[98] Georgescu, E.; Georgescu, F.; Draghici, C.; Caproiu, M. T.; Dumitrascu. F. *Rev. Roum. Chim.* 2010, 55, 1027-1031.
[99] Caira, M. R.; Georgescu, E.; Georgescu, F.; Popa, M. M.; Dumitrascu, F. *Arkivoc* 2009, (xii), 242-253.
[100] Georgescu, E.; Caira, M. R.; Georgescu, F.; Draghici, B.; Popa, M. M.; Dumitrascu, F. *Synlett* 2009, 1795-1799.
[101] Glushchenko, T. P.; Aksenov A. V.; Goncharov, V. I. *Chem. Heterocycl. Compd.* 2009, 45, 351-356.
[102] Hazra, A;. Mondal, S.; Maity, A.; Naskar, S.; Saha, P.; Paira, R.; Sahu, K. B.; Paira, P.; Ghosh, S.; Sinha, C.; Samanta, A.; Banerjee, S.; Mondal, N. B. *Eur. J. Med. Chem.* 2011, *46*, 2132-2140.
[103] Mao, Zhenjun.; Li, X.; Lin , X.; Lu, P.; Wang, Y. *Tetrahedron* 2012, 68, 85-91.
[104] Garcia, E. E.; Riley, J. G.; Fryer, R. I. *J. Org. Chem.* 1968, 33, 1359-1363.
[105] Garcia, E. E. *Org. Prep. Proced. Int.* 1974, 6, 11-13.
[106] Matsuo, T.; Mihara, S. *Bull. Chem. Soc. Jpn.* 1975, 48, 3660-3664.
[107] Vomero, S.; Chimenti, F.; Giuliano, R.; Artico, M. *Farmaco* 1976, 31, 681-690.
[108] Cardinaud, I.; Gueiffier, A.; Debouzy, J. C.; Milhavet, J. C.; Chapat, J. P. *Heterocycles* 1993, 36, 2513-2522.
[109] Abdellah, E.-C.; Lahboub, B. J. *Mar. Chim. Heterocycl.* 2006, 5, 62-64.
[110] Cheesman, G. W. H.; Eccleshall, S. A.; Thorton, T. *J. Heterocycl. Chem.* 1985, 22, 809-811.
[111] Bouyazza, L.; Lancelot, J.-C.; Rault, S.; Robba, M. *J. Heterocycl. Chem.* 1991, 28, 77-80.
[112] Hulcoop, D.G.; Lautens, M. *Org. Lett.* 2007, 9, 1761-1764.
[113] Cadogan, J. I. G.; Hewage, C. M.; McNab, H.; MacPherson, A. D.; Nicolson, I. S.; Reed, D.; Sadler, I. H. *Org. Biomol. Chem.* 2006, 4, 2446-2451.
[114] McNab, H.; Reed, D.; Tipping, I. D.; Tyas, R. G. Arkivoc 2007, (xi), 85-95.
[115] Kobayashi, K.; Takanohashi, A.; Hashimoto, K.; Morikawa, O.; Konishi, H. *Tetrahedron* 2006, 62, 10379-10382.
[116] Kobayashi, K.; Nakahashi, R.; Takanohashi, A.; Kitamura, T.; Morikawa, O.; Konishi, H. *Chem. Lett.* 2002, 624-625.
[117] Furstner, A.: Mamane, V. *J. Org. Chem.* 2002, 67, 6264-6267.
[118] Seung H. C., Sukbok C. *Angew. Chem. Int. Ed.* 2008, 47, 2836-2839.
[119] Kobayashi, K.; Takanohashi, A.; Himei, Y.; Sano, T.; Fukamachi, S.; Morikawa, O.; Konishi, H. *Heterocycles* 2007, 71, 2717-2720.
[120] Kobayashi, K.; Himei, Y.; Izumi, Y.; Fukamachi, S.; Morikawa, O.; Konishi, H. *Heterocycles* 2007, 71, 691-697.

[121] Kanno, K.; Liu, Y.; Iesato, A.; Nakajima, K.; Takahashi, T. *Org. Lett.* 2005, 7, 5453–5456.
[122] Mamane, V.; Hannen, P.; Furstner, A. *Chem. Eur. J.* 2004, 10, 4556-4575.
[123] Mamane, V.; Fort, Y. *Tetrahedron Lett.* 2006, 47, 2337-2340.
[124] Fuerstner, A.; Mamane, V.; Seidel, G.; Laurich, D. *Org. Synth.* 2006, 83, 103-110.
[125] Cappeli, A.; Giuliani, G.; Anzini, M.; Riitano, D.; Giorgi, G.; Vomero, S. *Bioorg. Med. Chem.* 2008, 16, 6850-6959.
[126] Hwang, S., J.; Cho, S., H.; Chang, S. *J. Am. Chem. Soc.* 2008, 130, 16158-16159.
[127] Chai, D. I.; Lautens, M. *J. Org. Chem.* **2009**, *74*, 3054–3061.
[128] Verma, A. K.; Shukla, S. P.; Singh, J. Rustagi, V. *J. Org. Chem.* 2011, 76, 5670–5684.
[129] Pranab, H.; Gandhi K. K.; Jayanta, K. R. *Tetrahedron Lett.* 2003, 44, 7433-7435.
[130] Alkhathlan H. Z. *J. Chem. Res.* (S) 1992, 260-261.
[131] Gulyakevich, O. V.; Kurman, P. V.; Mikhal'chuk, A. L. *Russ. J. Gen. Chem.* 2004, 74, 801-802.
[132] Michael, J. P.; de Koning, C. B.; Stanbury, T. V. *Tetrahedron Lett.* 1996, 37, 9403-9406.
[133] Michael, J. P.; de Koning, C. B.; Hosken, G. D.; Stanbury, T. V. *Tetrahedron* 2001, 57, 9635-9648.
[134] Kelly, R. B.; Alward, S. J. *Can. J. Chem.* 1978, 56, 320-325.
[135] Kumar, P.; Dinesh, C. U.; Pandey, B. *Tetrahedron Lett.* 1994, 35, 9229-9232.
[136] Liu, Y.; Yu, C.,Y.; Wang, M. X. *Arkivoc* 2003, (ii), 146-154.
[137] Chu, D. T. W.; Claiborne, A. K. *J. Heterocycl. Chem.* 1987, 24, 1537-1539.
[138] Santos, L. S.; Pilli, R. A. *Synthesis* 2002, 87-93.
[139] Rubinov, D., B.; Rubinova, I. L. Russ. *J. Org. Chem.* 2004, 40, 1187-1189.
[140] Zhang, W.; Huang, L.; Wang, J. *Synthesis* 2006, 12, 2053-2063.
[141] Miyata, O.; Shirai, A.; Yoshino, S.; Takeda, Y.; Sugiura, M.; Naito, T. *Synlett* 2006, 893-896.
[142] Michael, J. P.; B. de Koning, C.; Stanbury, T. V. *Tetrahedron Lett.* 1996, 37, 9403-9406.
[143] Alkhathlan, H., Z.; Al-Farhan, K. A. *Heterocycles* 1998, 48, 641-655.
[144] Allan, M. G.; Parsons, A. F.; Pons, J. F. *Synlett* 2002, 1431-1434.
[145] Li, X.; Li, C.; Zhang, W.; Lu, X.; Han, S.; Hong, R. Org. Lett. 2010, 12, 1696–1699.
[146] Kelderman, E.; Noorlander-Bunt, H. G.; Van Eerden, J.; Verboom, W.; Reinhoudt, D. N. *Recl. Trav. Chim. Pays-Bas* 1991, 110, 115-123.
[147] Van Hulst, N. F.; Heesink, G. J. T.; Bölger, B.; Kelderman, E.; Verboom, W.; Reinhoudt, D. N. *Proc. SPIE* 1361 1991, 581-588.
[148] Nijhuis, W. H. N.; Verboom, W.; Abu El-Fadl, A.; Harkema, S.; Reinhoudt, D. N. J. *Org. Chem.* 1989, 54, 199-209.
[149] Groenen, L. C.; Verboom, W.; Nijhuisa, W. H. N.; Reinhoudt, D. N.; Van Hummel, G. J.; Feil, D. *Tetrahedron* 1988, 44, 4637-4644.
[150] Kelderman, E.; Verboom, W.; Engbersen, J. F. J.; Harkema, S.; Heesink, G. J. T.; Lehmusvaara, E.; Van Hulst, N. F.; Reinhoudt, D. N.; Derhaeg, L.; Persoons, A. *Chem. Mat.* 1992, 4, 626-631.
[151] Verboom, W.; Reinhoudt, D. N.; Visser, R.; Harkemalc, S. *J. Org. Chem.* 1984, 49, 269-276.

[152] Nijhuis, W. H. N.; Verboom, W.; Abu El-Fadl, A.; Van Hummel, G. J.; Reinhoudt, D. N. *J. Org. Chem.* 1989, 54, 209-216.
[153] Nijhuis, W. H. N.; Werboom, W.; Reinhoudt, D. N. *J. Am. Chem. Soc.* 1987, 109, 3136-3138.
[154] Kelderman, E.; Verboom, W.; Engbersen, J. F. J.; Reinhoudt, D. N.; Harkema, S. *Acta Cryst.* 1996, C52, 458-460.
[155] Gerlach, U. *Tetrahedron Lett.* 1995, 36, 5159-5162.
[156] Tverdokhlebov, A. V.; Gorulya, A. P.; Tolmachev, A. A.; Kostyuk, A. N.;. Chernega, A. N.; Rusanov, E. B. *Synthesis* 2005, 2161-2170.
[157] Kaval, N.; Dehaen, W.; Mátyus P.; Van der Eycken, E. *Green Chem.* 2004, 6, 125-127.
[158] Glukhareva, T. V.; D'yachenko, E. V.; Morzherin Yu. Yu. *Chem. Heterocycl. Compd.* 2002, 38, 1426-1427.
[159] Kaval, N.; Halasz-Dajka, B.; Vo-Thanh, G.; Dehaen, W.; Van der Eycken, J.; Matyus, P.; Loupy, A.; Van der Eycken, E. *Tetrahedron* 2005, 61, 9052-9057.
[160] D'yachenko, E. V.; Gluchareva, T. V.; Morzherin, Yu. Yu. *Chem. Heterocycl. Compd.* 2003, 39, 1532-1533.
[161] D'yachenko, E. V.; Gluchareva, T. V.; Dyudya, L. V.; Eltsov, O. V.; Morzherin, Yu. *Molecules* 2005, 10, 1101-1108.
[162] Lobodin, V. V.; Ovcharenko, V. V.; Pihlaja, K.; Morzherin, Yu. Yu.; Lebedev, A. T.; *Rapid Commun. Mass. Spectrom.* 2004; 18, 724-728.
[163] Watanabe, M.; Kodera, M.; Kinoshita, T.; Furukawa, S. *Chem. Pharm. Bull.* 1975, 23, 2958-2604.
[164] Kobayashi, Y.; Kumadaki, I.; Fujino, S. *Heterocycles* 1977, 7, 871-876.
[165] Hodgets, I.; Noyce, S. J.; Storr, R. C. *Tetrahedron Lett.* 1984, 47, 5435-5438.
[166] Bowen, R. D.; Davies, D. E.; Fishwick, C. W. G.; Glasbey, T. O.; Noyce, S. J.; Storr R. C. *Tetrahedron Lett.* 1982, 23, 4501-4504.
[167] Baydar, A. E.; Boyd, G. V.; Lindley, P. F.; Watson, F.; Manley, P. W.; Storr, R. C. *J. Chem. Soc. Perkin Trans.* 1 1981, 2878-2883.
[168] Wojciechowski, K. *Tetrahedron* 1993, 49, 7277-7286.
[169] Makisumi, Y. *Tetrahedron Lett.* 1964, 13, 699-704.
[170] Makisumi, Y. *J. Org. Chem.* 1965, 30, 1989-1994.
[171] Newhouse, B. J.; Bordner, J.; Augeri, D. J.; Litts, C. S.; Kleinman, E. F. *J. Org. Chem.* 1992, 57, 6991-6995.
[172] Zeiler, A. G.; Paudler, W. W. *J. Heterocycl. Chem.* 1974, 11, 1091.
[173] Occhiato, E. G.; Guarna, A.; Brandi, A.; Goti, A.; De Sarlo, F. *J. Org. Chem.* 1992, 57, 4206-4211.
[174] Brandi, A.; Durust, Y.; Cordero, F. M.; De Sarlo, F. *J. Org. Chem.* 1992, 57, 5666-5670.
[175] Butin, A, V.; Tsiunchik, F. A.; Abaev, V. T.; Bosikova, K. V. Arkivoc 2009, (iv), 79-87.
[176] Abarca, B.; Ballesteros, R.; Houari, N. *Tetrahedron* 1997, 53, 12765-12770.
[177] Abarca, B.; Ballesteros, R.; Houari, N.; Samadi, A. *Tetrahedron* 1998, 54, 3913-3918.
[178] Barbieri, W.; Bernardi, L.; Colo, V.; Patelli, B. *Chim. Ind.* 1972, 54, 786-789.
[179] Ito, Y.; Miyata, S.; Nakatsuka, M.; Saegusa, T. *J. Am. Chem. Soc.* 1981, 103, 5250-5251.
[180] Deady, L. V.; Werden, D. M. *Synth. Commun.* 1987, 17, 319-328.

[181] Alcaide, B.; Almendros, P.; Alonso, J. M.; Aly, M. F. *Chem. Eur. J.* 2003, 9, 3415-3426.
[182] Bunce, R. A.; Schammerhorn, J. E.; Slaughter, L. M. *J. Heterocycl. Chem.* 2006, 43, 1505-1511.
[183] Liu, X. Y.; Che, C. M. *Angew. Chem. Int. Ed.* 2008, 47, 3805 –3810.
[184] Zhou, Y.; Feng, E.; Liu, G.; Ye, D.; Li, J.; Jiang, H.; Liu, H. *J. Org. Chem.* 2009, 74, 7344–7348.
[185] Yakushijin, K.; Tsuruta, T.; Furukawa, H. *Heterocycles* 1979, 12, 1021-1026.
[186] Yakushijin, K.; Tsuruta, T.; Furukawa, H. *Chem. Pharm. Bull.* 1982, 30, 140-151.
[187] Kim, S. H.; Jeong, J. U.; Choi, D. O.; Lee, K-J. *Bull. Korean Chem. Soc.* 1993, 14, 11-12.
[188] Naruta, Y.; Nagai, N.; Arita, Y.; Maruyama, K. *J. Org. Chem.* 1987, 52, 3956-3967.
[189] Lang, S.; Kennedy, A. R.; Murphy, J. A.; Payne, A. H. *Org. Lett.* 2003, 5, 3655-3658.
[190] Barluenga, J.; Lonzi, G.; Riesgo, L.; Lopez, L. A.; Tomas, M. *J. Am. Chem. Soc.* 2010, 132, 13200-13202.
[191] Yang, Y.; Xie, C.; Xie, Y.; Zhang, Y. Org. *Lett.* 2012, 14, 957–959.
[192] Tran, T. Q.; Diev, V. V.; Starova, G. L.; Gurzhiy, V. V.; Molchanov, A. P. *Eur. J. Org. Chem.* 2012, 2054-2061.
[193] 193. Daly, J. W.; Witkop, B.; Tokuyama, T.; Nishikawa, T.; Karle, I. L. *Helv. Chim. Acta* 1977, 60, 1128-1140.
[194] Daly, J., W.; Brown, G., B.; Mensah-Dwumah, M.; Myers, C., W. *Toxicon* 1978, 16, 163-188.
[195] Mensah-Dwumah, M.; Daly, J. W. *Toxicon* 1978, 16, 189-194.
[196] Habermehl, G. G.; Thurau, O. *Naturwissenschaften* 1980, 67, 193-194.
[197] Overman, L. E.; Fukaya, C. *J. Am. Chem. Soc.* 1980, 102, 1454-1456.
[198] Crans, D. C.; Snyder, J. P. *J. Am. Chem. Soc.* 1980, 102, 7154-7156.
[199] Overman, L. E.; Freerks, R. L. *J. Org. Chem.* 1981, 46, 2833-2835.
[200] Hart, D. J. *J. Org. Chem.* 1981, 46, 3576-3578.
[201] 201. Fujimoto, R.; Kishi, Y.; Blount, J. F. *J. Am. Chem. Soc.* 1980, 102, 7154-7156.
[202] Fujimoto, R.; Kishi, Y. *Tetrahedron Lett.* 1981, 22, 4197-4198.
[203] Hart, D. J.; Tsai, Y.-M. *J. Am. Chem. Soc.* 1982, 104, 1430-1432.
[204] Hart, D. J.; Tsai, Y.-M. *J. Org. Chem.* 1982, 47, 4403-4409.
[205] Ito, Y.; Nakajo, E.; Nakatsuka, M.; Saegusa, T. *Tetrahedron Lett.* 1983, 24, 2881-2884.
[206] Hart, D. J.; Kanai, K. *J. Am. Chem. Soc.* 1983, 105, 1255-1263.
[207] Overman, L. E.; Lesuisse, D.; Hashimoto, M. *J. Am. Chem. Soc.* 1983, 105, 5373-5379.
[208] Ibuka, T.; Chu, G.-N.; Yoneda, F. *J. Chem. Soc. Chem. Commun.* 1984, 597-599.
[209] Ibuka, T.; Chu, G. N.; Yoneda, F. *Tetrahedron Lett.* 1984, 25, 3247-3250.
[210] Ibuka, T.; Chu, G. N. *Chem. Pharm. Bull.* 1986, 34, 2380-2390.
[211] Edwards, M. W.; Bax, A. *J. Am. Chem. Soc.* 1986, 108, 918-923.
[212] Pearson, W. H.; Poon, Y.-F. *Tetrahedron Lett.* 1989, 30, 6661-6664.
[213] Holmes, A. B.; Hughes, A. B.; Smith, A. L.; Williams, S. F. *J. Chem. Soc. Perkin* I 1992, 1089-1099.
[214] Pearson, W. H.; Fang, W-K. *J. Org. Chem.* 2000, 65, 7158-7174.
[215] Pearson, W. H.; Hutta, D. A.; Fang, W-K. *J. Org. Chem.* 2000, 65, 8326-8332.
[216] Wei, L.-L.; Hsung, R. P.; Sklenicka, H. M.; Gerasyuto, A. I. *Angew. Chem. Int. Ed.* 2001, 40, 1516-1518.

[217] Comins, L. D.; Kuethe, J. T.; Miller, T. M.; Fevier, F. C.; Brooks, C. A. *J. Org. Chem.* 2005, 70, 5221-5234.
[218] Santarem, M.; Vanucci-Bacque, C.; Lhommet, G. *J. Org. Chem.* 2008, 73, 6466-6469.
[219] Miao, L.; Shu, H.; Noble, A. R.; Fournet, S. P.; Stevens, E. D. Trudell, M. L. *Arkivoc* 2010, (iv), 6-14.
[220] Souccar, C.; Varanda, W. A.; Aronstam, R. S.; Daly, J. W.; Albuquerque, E. X. *Mol. Pharmacol.* 1984, 25, 395-400.
[221] Aronstam, R. S.; Daly, J. W.; Spande, T. F.; Narayanan, T. K.; Albuquerque, E. X. *Neurochem. Res.* 1986, 11, 1227-1240.
[222] Daly, J. W.; Nishizawa, Y.; Edwards, J. A.; Waters, J. A.; Aronstam, R. S. *Neurochem. Res.* 1991, 16, 489-500.
[223] Daly, J. W. *J. Nat. Prod.* 1998, 162-172.
[224] Koya, K.; Sun, L.; Ono, M.; Ying, W.; Li, H. US 6861436 B2 2005.
[225] Shen, Y.-M.; Lv, P.-C.; Chen, W.; Liu, Pe.-G.; Zhang, M.-Z.; Zhu, H.-L. *Eur. J. Med. Chem.* 2010, 45, 3184-3190.
[226] Sui, Z.; Altom, J.; Nguyen, V. N.; Fernandez, J.; Bernstein, J. I.; Hiliard, J. J.; Barrett, J. F.; Podlogar, B. L.; Ohemeng, K. A. *Bioorg. Med. Chem.* 1998, 6, 735-742.
[227] Jones, D. T. Harris, A. L. *Mol. Cancer. Ther.* 2006, 5, 2193-2202.
[228] Cai, S. X.; Drewe, J. A.; Jiang, S.; Kasibhatla, S.; Kuemmerle, J. D.; Sirisoma, N. S.; Zhang, H. Z. WO2004055163, 2004.
[229] Cai, S. X.; Drewe, J. A.; Jiang, S.; Kasibhatla, S. Kummerle, J. D.; Sirisoma, N. S.; Zhang, H.-Z. US7135480 B2 2006.
[230] Kemnitzer, W.; Kuemmerle, J.; Jiang, S.; Zhang, H.-Z.; Sirisoma, N.; Kasibhatla, S.; Crogan-Grundy, C.; Tseng, B.; Drewe, J.; Cai, S. X. *Bioorg. Med. Chem. Lett.* 2008, 18, 6259–6264.
[231] Kemnitzer, W.; Kuemmerle, J.; Jiang, S.; Sirisoma, N.; Kasibhatla, S.; Crogan-Grundy, C.; Tseng, B.; Drewe, J.; Cai, S. X. *Bioorg. Med. Chem. Lett.* 2009, 19, 3481-3484.
[232] Edwards, D. L.; Berens, M. E.; *Beaudry*, C. US2006/0128742 A1 2006.
[233] Edwards, D. L.; Berens, M. E.; *Beaudry*, C. US7700615 B2 2010.
[234] Glushkov, R. G.; Marchenko, N. B.; Padeiskaya, A. N.; Shipilova, L. D. *Pharm. Chem. J.* 1990, 24, 460-465.
[235] Keiser, J.; Burri, C. *Trop. Med. Int. Health* 2001, 6, 369-389.
[236] Dannhardt, G.; Bauer. A. *Pharmazie* 1996, 51, 805-810.
[237] Birch, M.; Sibley, G. E. M.; Law, D. Oliver, J. D. WO/144473 A1 2009.
[238] Goel, S. R. *Indian J. Heterocycl. Chem.* 2006, 15, 401-402.
[239] Rise, F.; Wikstrom, H.; Ugland, S.; Ddijkstra, D.; Gundersen, L.-L.; De Boer, P.; Bast, A.; Haenen, G.; Antonsen, Ö. G.; Liao, Y.; Nasir, A. I. WO/021662 1996.
[240] Eggler, J. F.; Johnson, M. R.; Melvin, L. S. EP0090516 A1 1983.

In: Bioactive Heterocycles
Editors: K. L. Ameta, R. P. Pawar and A. J. Domb
ISBN: 978-1-62257-636-4
© 2013 Nova Science Publishers, Inc.

Chapter 3

Aziridines: Synthesis and Bioactivity

Vijaykumar V. Paike[1*], Sushma S. Kauthale[2], Ramesh C. Kumar[3], Maya V. Katariya[2], Rameshwar R. More[2], K. L. Ameta[4] and Sandeep B. Mane[5]

[1]Department of Chemistry, National Tsing Hua University, Hsinchu, Taiwan
[2]Department of Chemistry, Deogiri College, Aurangabad (MS) India
[3]Institute of Biotechnology and pharmaceuticals research, National Health Research Institute. Zhunan, Taiwan
[4]Department of Chemistry, MITS University, P-Lakshmangarh, Rajasthan, India
[5]Institute of Chemistry, Academia sinica, Taipei, Taiwan

Abstract

Three-membered nitrogen, containing heterocycle aziridines, has arracted the attention of chemists due to its association with a wide spectrum of biological activities. It is the central core of many natural products. The current chapter deals with the synthesis and bioactivities of the aziridine nucleus containing heterocyclic compounds.

1. Introduction

Aziridine is a versatile class of nitrogen containing three, small-membered heterocyclic compounds, first discovered in 1888 by Gabriel [1]. They are of great importance as versatile synthetic intermediates and natural products.

They are commonly used as reagents, chiral auxiliaries and monomers for polymer synthesis. Different scientists are involved in their synthesis and the transformations of these

[*] Email:vv.paike@yahoo.co.in.

compounds in various higher-membered heterocyclic compounds. Thus, aziridines are attractive organic molecules due to their great synthetic utilities.

The representative structures of these three membered N-heterocyclic rings are shown as:

(1) (2) (3)

The compound (1) is the aziridine. The partially reduced form of the aziridine is commonly referred as 1-H azirine (2) and 2-H azirine (3).

Aziridines play a significant role in organic synthesis and pharmaceuticals as important intermediates. Aziridines are the main precursors for the synthesis of various nitrogen-containing bioactive molecules, such as different heterocyclic compounds, alkaloids and amino acids. They behave as carbon electrophiles capable of reacting with different nucleophiles and their ability to undergo regioselective ring-opening reactions, mainly contributing to their synthetic utility.

Aziridines have been synthetic targets as well as useful intermediates and building blocks in the synthesis of biological active natural products. N-alkenyl aziridines do not behave like typical enamines on account of the inability of the nitrogen atom lone-pair of electrons to delocalize.

Synthetic chemists have been attracted by the increased strain and unique reactivity of the three-membered ring. In this chapter, various synthetic routes for the preparation of aziridines and the theurepetic potential of aziridines are briefly discussed.

2. Overview and General Features

Aziridines are less basic than acyclic amines because of the increased *s*-character of the nitrogen lone pair. The conjugate acid of aziridine **(1)** has a pKa value of 7.98 as compared to an acyclic secondary amine which has a pKa value of 10.98 for the conjugate acid **(4)**. N-substituted electron with-drawing groups (*i.e.* CO_2R, SO_2R and COR) helps to activate aziridine, whereas electron donating groups (*i.e.* H, Alkyl and Aryl) are deactivating the aziridines.

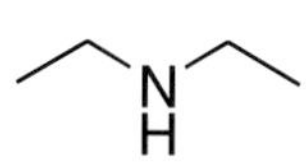

(4)

The high-ring strain (27 Kcal/mol^{-1}) renders them susceptible towards regioselective and stereoselective ring opening reactions which led to the hope that these properties might be translated into biological activities. The chemistry associated with the small ring of aziridines has expanded with considerable intensity in the past decade or two because of its theoretical, mechanistic, synthetic and biological applications.

The reactivity of aziridines is mainly due to the ring strain energy (SE) of 26.7kcal/mol for nonsubstituted aziridine. The reactivity of aziridines constitues an important component in synthetic processes and several natural products. Hence, aziridines, especially optically active aziridines, are of great importance in organic chemistry. Even though aziridines possess high ring strain energy, unactivated aziridines are generally not much more susceptible to a ring opening via a nucleophilic attack. These normally require the presence of Lewis acid for the purpose of opening the ring.

3. Synthesis of Aziridines

Amino alcohols are the most convenient precursors for the synthesis of aziridines. Numerous synthetic methods were developed by different researchers for the synthesis of aziridines.

These methods include the ring closing reaction of 1,2-amino alcohol and their derivatives, the ring opening of epoxides with metal salts, and the addition of different reagents to imines, etc. Studies show that normally a bulky or an electron-withdrawing substituent at the nitrogen atom decreases the yields for aziridine products. Many review articles giving information about the synthesis, reactions and biological potential of aziridines are available in literature [2]. Different synthetic methods are found in the literature and allow for access to various aziridines. The different protocols for the synthesis of aziridines may be categorized under the following headings:

1. From addition to alkenes
2. From α,β-unsaturated compounds
3. From imines
4. From ring closure of amino alcohols
5. By ring transformation of epoxides
6. Miscellaneous synthesis

3.1. From Addition to Alkenes

The synthesis of N-tosyl azides (7), from the cyclization between olefins (5) and tosyl amine (6) catalyzed by dirhodium (II) caprolactamate complex (8) was reported by A. J. Catino [3]. The synthesis was performed at room temperature. An important advantage of this method is that the reaction can be performed even in an atmosphere of air due to the stability of the complex.

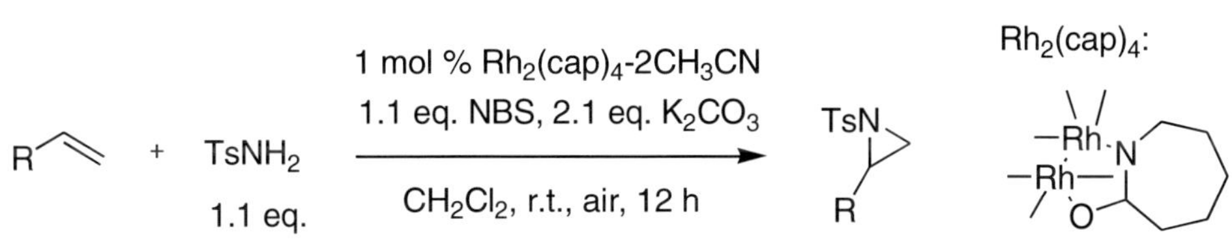

(5) (6) (7) (8)

An efficient synthesis of N-tosyl aziridines (10) from *N*-alkenylamides (9) using tert-butyl hypoiodite as the cyclizing agent was developed by S. Minakata and coworkers [4]. The reagent, being powerful, works under mild conditions.

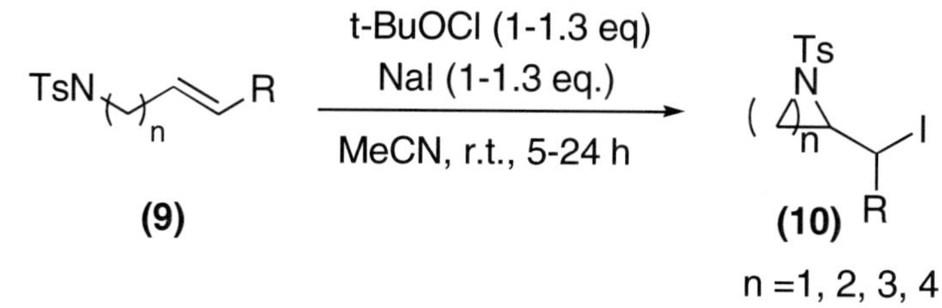

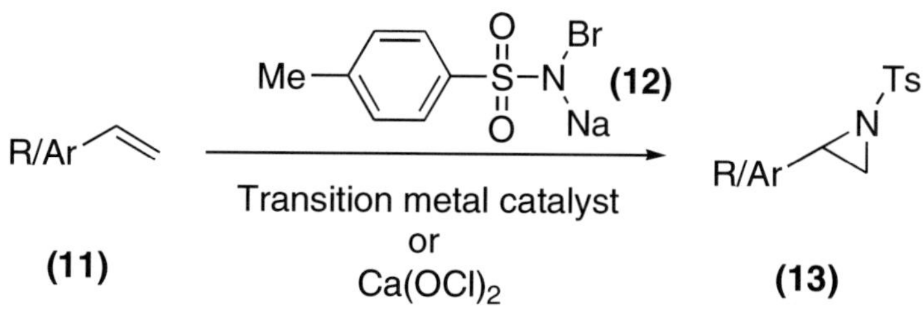

The direct aziridination of alkenes (11) with bromamine-T (12) as the source of nitrene catalyzed by either transition metal or calcium oxychloride catalyzed the reaction which facilitates the synthesis of N-tosylated aziridines (13) as reported by B. M. Chanda et al. [5]. It was found that, among the transition metal catalysts, the Cu-halides were the best catalysts for this transformation.

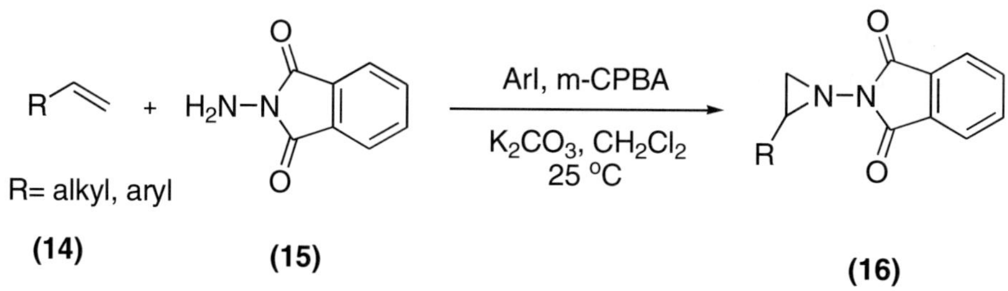

N-aminophthalimides (15) react with olefins (14) mediated by aryl iodides and m-CPBA in the preasence of a potassium carbonate base at room temperature conditions forming the corresponding N-aziridinyl-phthalimides (16) in excellent yields (upto 94%) [6].

An efficient synthesis of aziridines (19) can be achieved by the reaction of electron-rich alkenes (17) with $HN(OCH_3)_2$ in the presence of trimethylsilyl triflate followed by a reaction with a sodium hydroxide base and then a reduction with Li/ammonia [7].

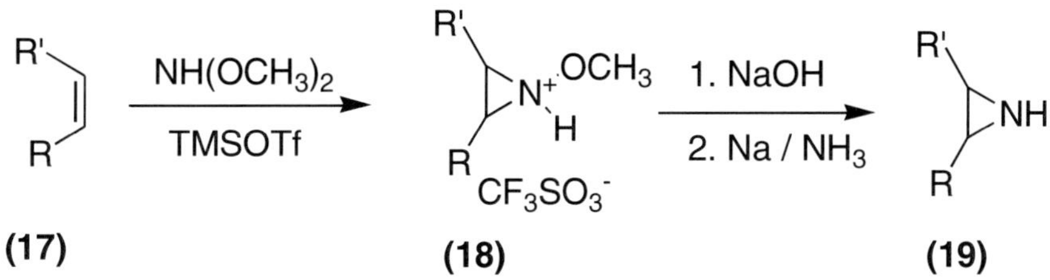

The *N*-enoyl oxazolidinones (20) were reacted with *N*-aminophthalimide using lead tetraacetate in presence of a chiral derivative of camphor as the ligand, resulting in the formation of corresponding aziridines (21) in high enantioselectivities [8]. The reactions were carried out under cooling conditions (0 °C).

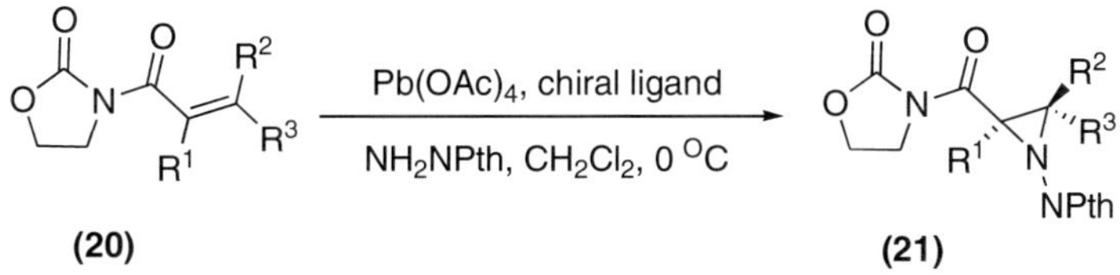

J. Li obtained the aziridines (25), (26) by aziridination of olefins (24) catalyzed by iodobenzene diacetate in the presence of a potassium carbonate base [9].

A good-to-excellent yield and high conversions were obtained by this method.

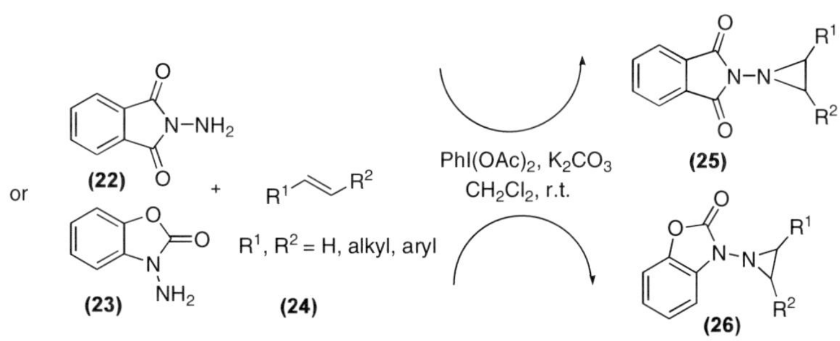

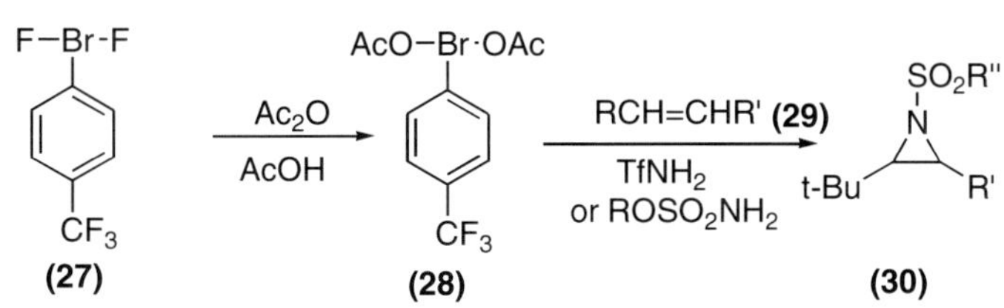

p-$CF_3C_6H_4BrF_2$ (27) reacts with the acetic anhydride in acetic acid where an exchange of the fluoro and acetoxy groups occurs to form the (diacetoxybromo)benzene (28) [10]. This reacts with olefins (29) directly in the presence of $TfNH_2$ to afford the N-sulfonyl aziridines (30) under mild reaction conditions. The aziridination of olefins (32) by an electrochemical process was documented by T. Sun et al. [11]. Both electron-rich and electron-poor olefins were successfully converted to aziridines (33) with high efficiency.

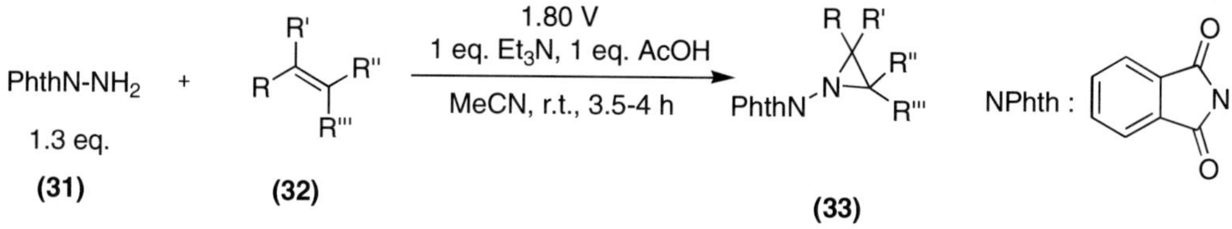

Quaternary salts of *Cinchona* alkaloids (37) were reported as phase-transfer catalysts for the synthesis of chiral N-arylaziridines (36) for reactions associated with N-acyl-N-arylhydroxylamines (34) with olefins (35). The conversions were reported in toluene at room temperature [12].

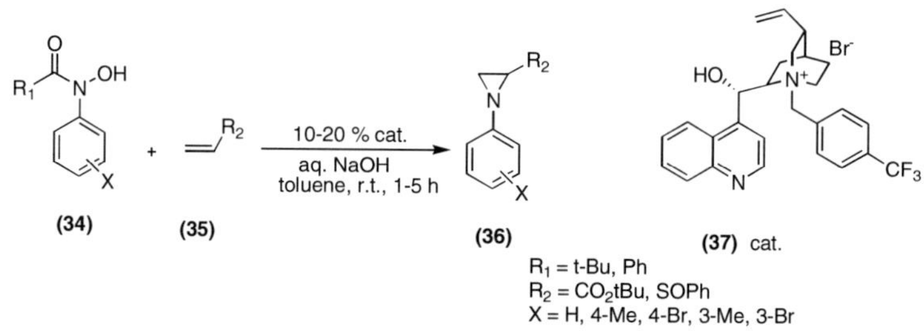

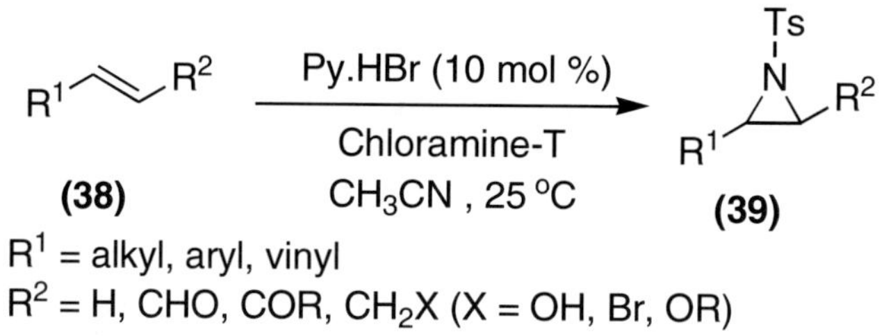

Pyridinium hydrobromide/perbromide was reported as an efficient catalyst for the aziridination of olefins (38) using chloramines-T (*N*-chloro-*N*-sodio-*p*-toluenesulfonamide) as the source of nitrogen in an acetonitrile solvent at room temperature in moderate-to-good yields [13].

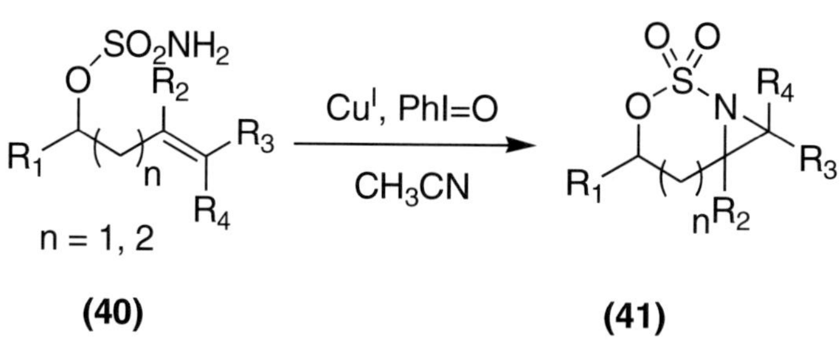

Established by F. Duran [14], unsaturated sulfamates (40) can undergo intramolecular cyclization in the presence of a copper-catalyzed reaction mediated by iodosylbenzene.

The aziridination of alkenes (42), using *N*, *N*-dichloro-*p*-toluenesulfonamide (43) as the source of nitrogen catalyzed by a palladium catalyst, was reported by J. Han [15]. The reaction was carried out in a DMF solvent and K_2CO_3 base. In this method, several alkenes, including the α,β-unsaturated alkenes, undergo cyclization to form the aziridines in excellent yield.

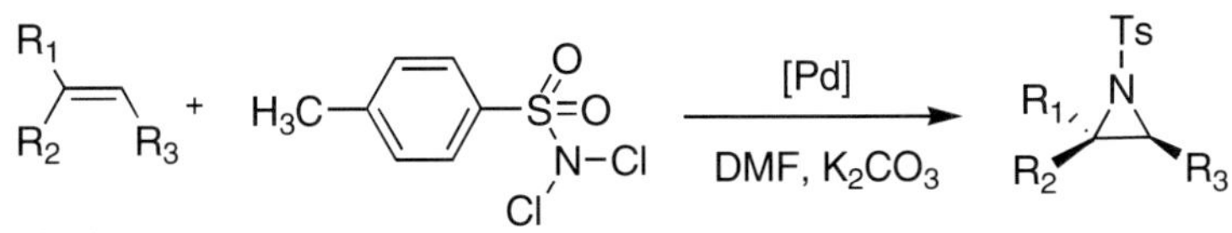

(42) **(43)** **(44)**

As reported by T. Ando and his coworkers [16], the reaction of alkenes (45) with anhydrous chloramine-T, serving as the source of nitrogen in the presence of copper catalysts and MS-5A, affords the corresponding aziridines (46) in moderate-to-good yields.

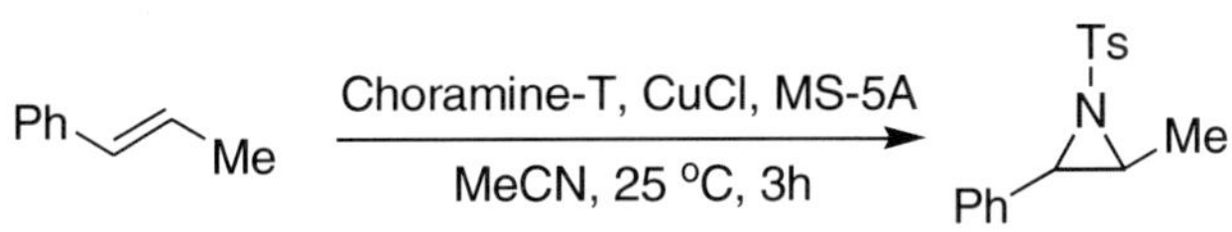

(45) **(46)**

A one-pot method for synthesizingng N-sulfonyl aziridines (49), using the copper-catalyzed and $PhI(OAc)_2$-mediated cyclization of olefins (47) and sulfonamides (48), was developed by H. L. Kwong [17]. Only a catalytic amount (5 mol%) of the chiral copper catalyst (50) was sufficient for completiing the reaction to accomplish an enantioselectivity up to 75%.

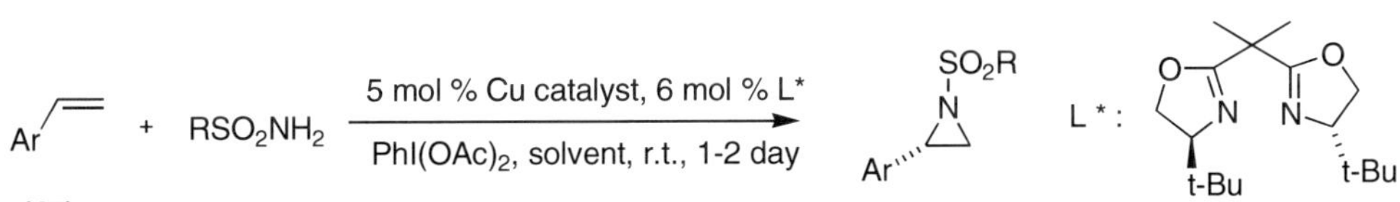

(47) **(48)** **(49)** **(50)**

The action of arenesulphonyloxycarbamates ($ArSO_3NHCO_2Et$) on α-nitroalkenes (51) in the presence of calcium oxide in dichloromethane results in the formation of α-nitroaziridines (52) in high yields and good purity under mild conditions [18].

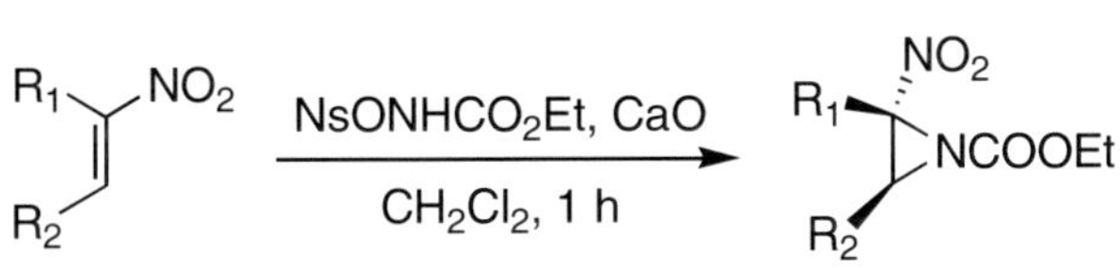

(51) **(52)**

Aziridinyl phosphonates (54) are the precursors of amino phosphoric acids. These ziridinyl phosphonates are obtained by the reaction of α, β-unsaturated phosphonate (53) with $NsONHCO_2Et$; involving the addition of an (ethoxycarbonyl) amino group onto the double bond [19].

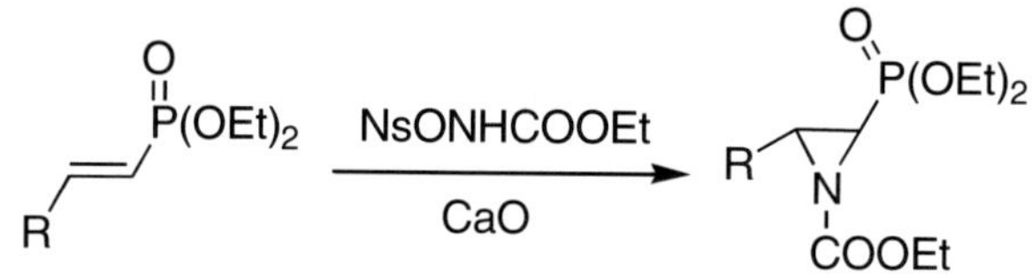

(53) (54)

N-tosyl aziridines (56) were synthesized by the cyclization of olefins (55) with chloramine-T (*N*-chloro-*N*-sodio-*p*-toluenesulfonamide) as the source of nitrogen in the presence of the *N*-bromoamides catalyst [20]. Both the electron-deficient and electron-rich alkenes reacted smoothly at an ambient temaperature in excellent yields to afford the corresponding aziridines.

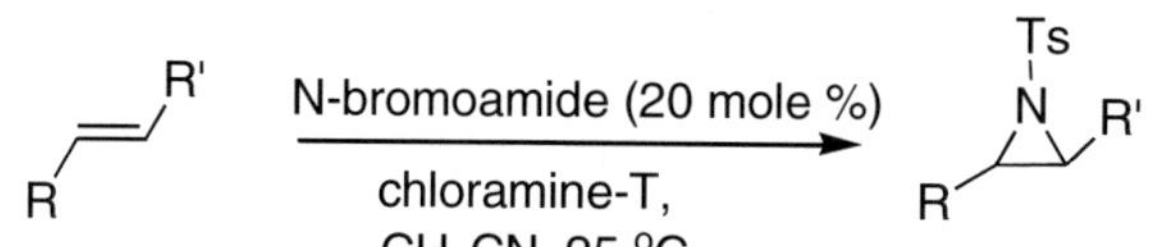

(55) (56)

R = aryl, H
R^1 = H, alkyl, aryl, acyl, CH_2OH, COOR, etc.

Olefins (57) react with aqueous chloramine-T in the presence of H_2O_2 and HBr to form the corresponding aziridines. This method provides an excellent transitional, metal-free green approach for the synthesis of aziridines (58) [21].

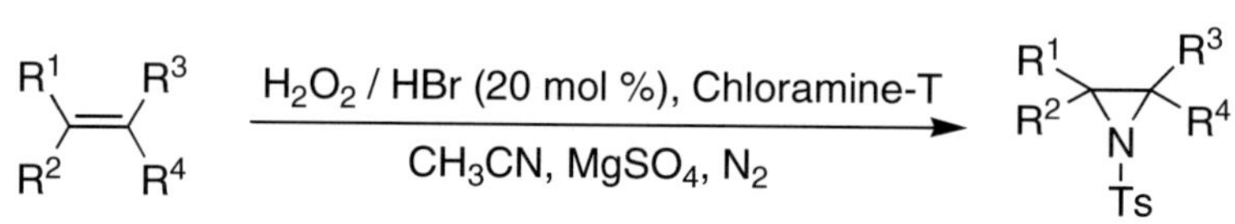

(57) (58)

N-iodo-*N*-potassio-*p*-toluenesulphonamide was employed as the nitrene precursor for the aziridination of alkenes (59) catalyzed by copper catalysts [22].

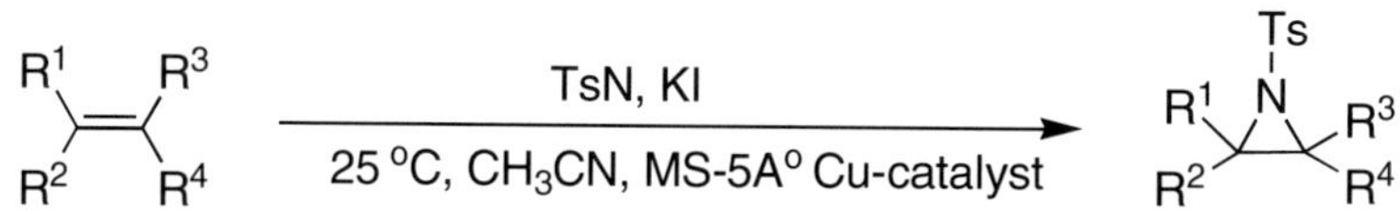

(59) (60)

The alkenes (61) undergo smooth cyclization with *p*-toluenesulfonamide (62) in the presence of $PhI(OAc)_2$ and I_2 under mild reaction conditions to afford the corresponding aziridines (63) [23]. This protocol provides an alternative to the expensive metal catalysts and the use of the unstable and explosive reagent PhI ═NTs.

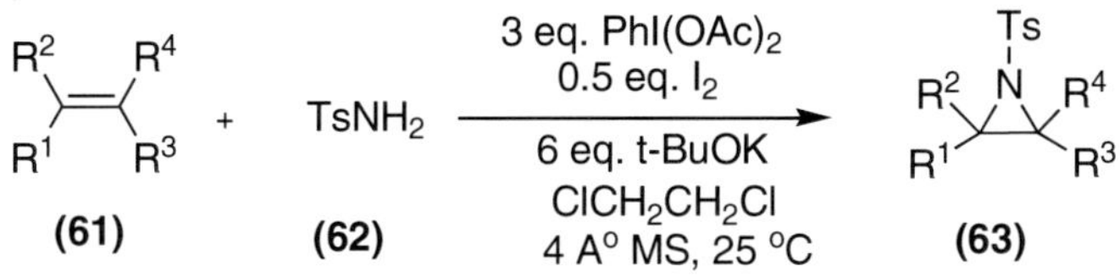

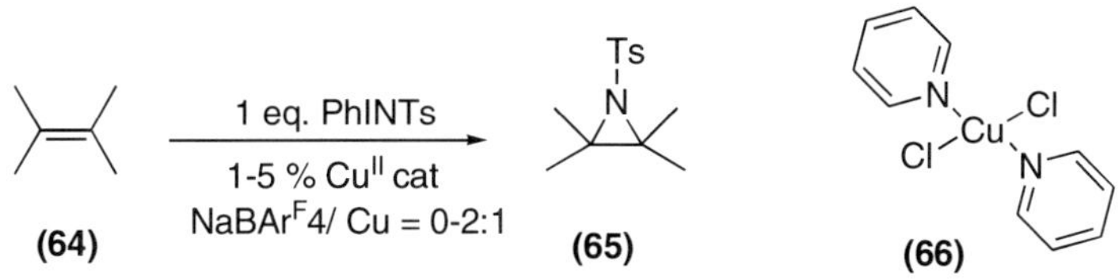

The direct aziridination of olefins (64), by treating with the PhINTs catalyzed by the dichlorocopper (II) complex (66), was developed by F. Mohr et al. [24]. It was observed that the aziridines (65) could be obtained in more than a 90% yield when the ratio of olefin/PhINTs was 1:1 and the catalyst loading was 1−5 mol%.

3.2. From α, β-Unsaturated Carbonyl Compounds

The aziridines (69) were synthesized by the treatment of α, β-unsaturated carbonyl compounds (67) with phosphoramides (68) in the presence of N-methyl morpholine and a sodium hydroxide base [25]. The reaction required prolonged time for completion.

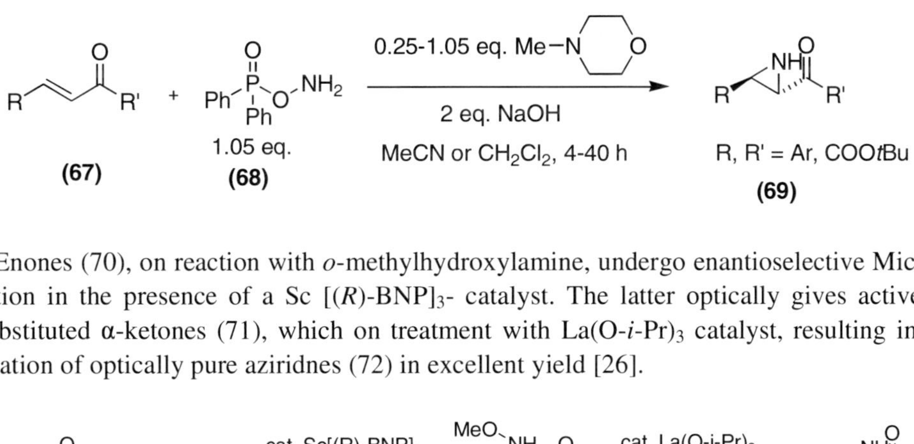

Enones (70), on reaction with *o*-methylhydroxylamine, undergo enantioselective Michael addition in the presence of a Sc [(*R*)-BNP]$_3$- catalyst. The latter optically gives active *N*-unsubstituted α-ketones (71), which on treatment with La(O-*i*-Pr)$_3$ catalyst, resulting in the formation of optically pure aziridnes (72) in excellent yield [26].

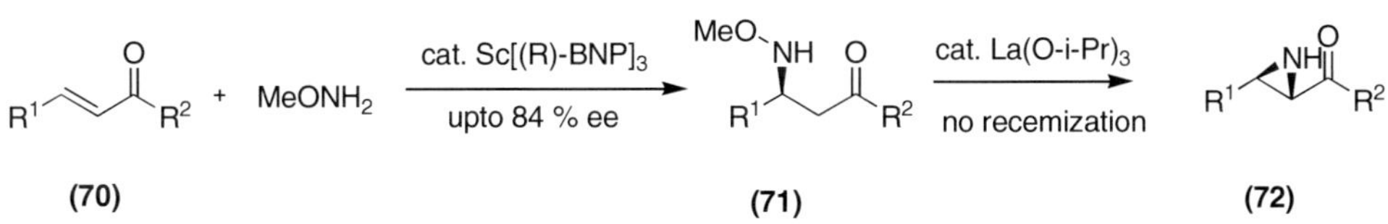

N-tosyl aziridines (75), (77) can be effectively synthesized by the copper catalyzed conversion of α,β-unsaturated esters and ketones (73), (76) in a moderate-to-good yield [27]. The resulting *N-p*-tosyl-aziridine-2-ketones were obtained in an excellent anti-stereoselectivity manner.

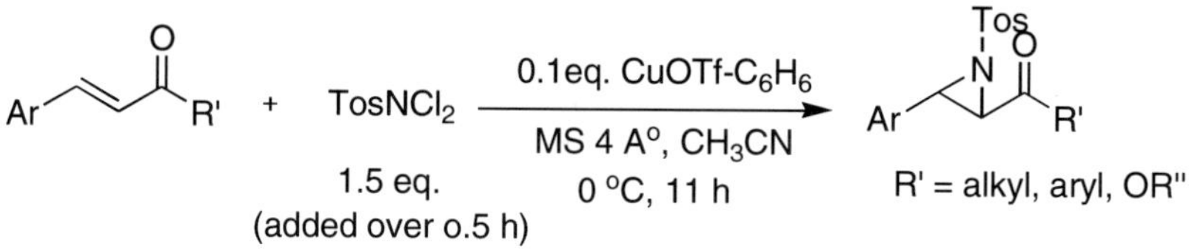

(73) (74) (75)

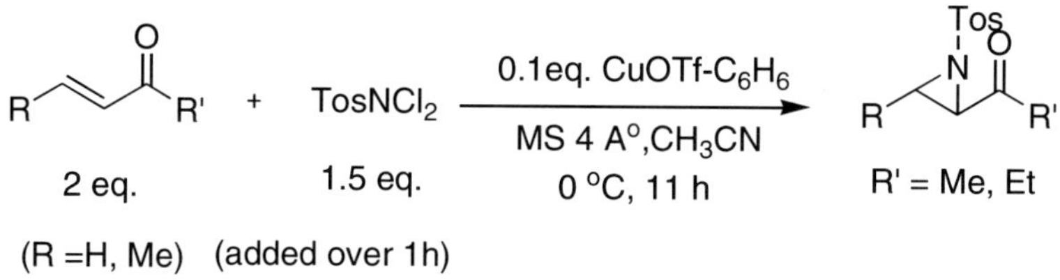

(76) (77)

3,3-pentamethylenediaziridine (78) reacts with butyl lithium to form the lithium salts which react rapidly in THF with α, β-unsaturated amides to afford the corresponding aziridines (80) in excellent yield [28].

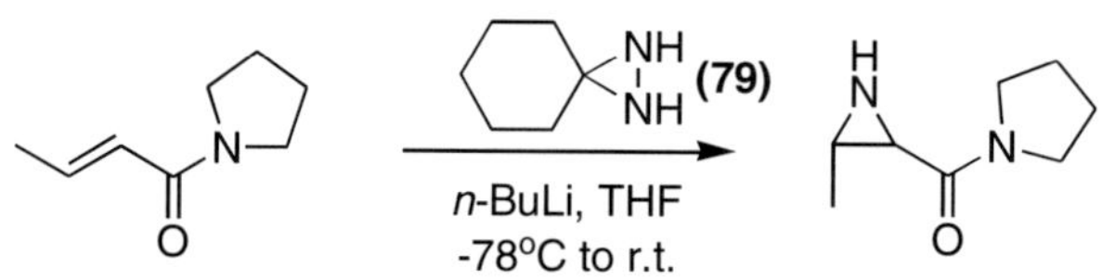

(78) (80)

A simple protocol for the aziridination of azide (81) with α, β-unstaurated carbonyl compounds (82) was standardized by J. M. Mahoney [29]. The reaction was catalyzed by triflic acid at 0°C. The reactions required a prolonged time for completion.

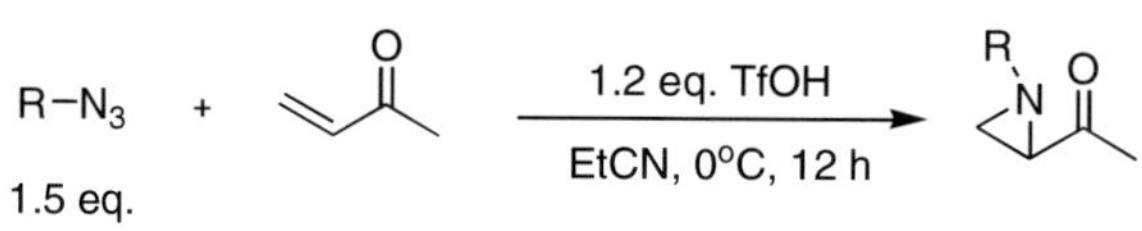

(81) (82) (83)

α-substituted and α, β-unsaturated aldehydes (84) can be easily converted into N-tosyl aziridines (87), catalyzed by diarylprolinol trimethylsilyl ether (86) [30]. The products were obtained in high yields and excellent enantioselectivities.

Bottom of Form:

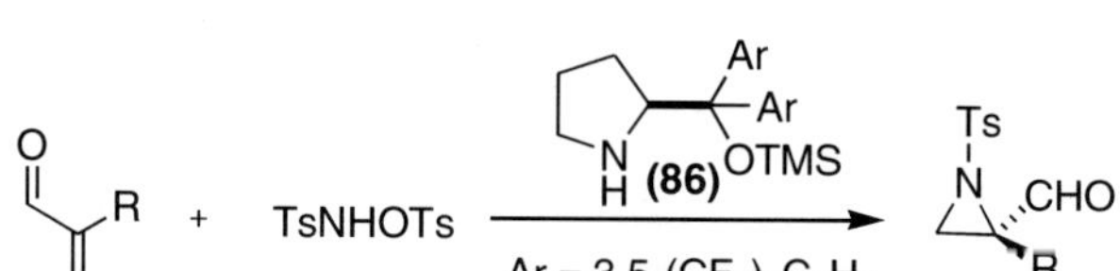

(84) (85) (87) 90 % ee

α, β, γ, δ-unsaturated carbonyl compounds (88) were converted into aziridines (89) by A. Armstrong using an amine-promoted reaction to afford the corresponding N-unsubstituted vinyl aziridines at room temperature conditions [31]. The yield of this reaction was moderate to excellent.

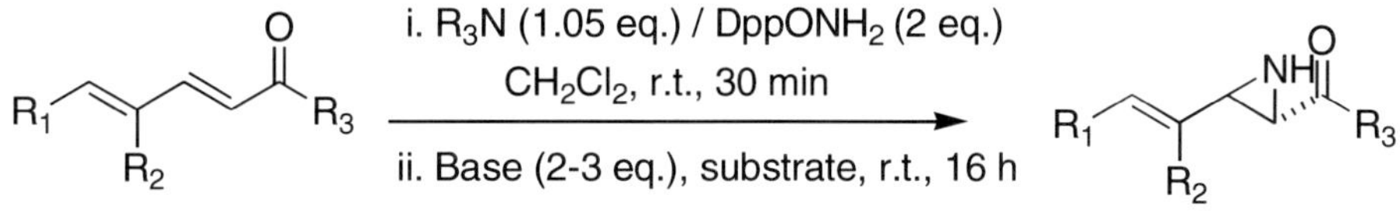

(88) **(89)**

The chiral aziridines (91) were synthesized by a dipolar cycloaddition of *N,O*-bis(trimethylsilyl)hydroxylamine and α, β-unsaturated malonates (90), catalyzed by copper triflate catalyst [32]. The reaction resulted in high enantioselectivity.

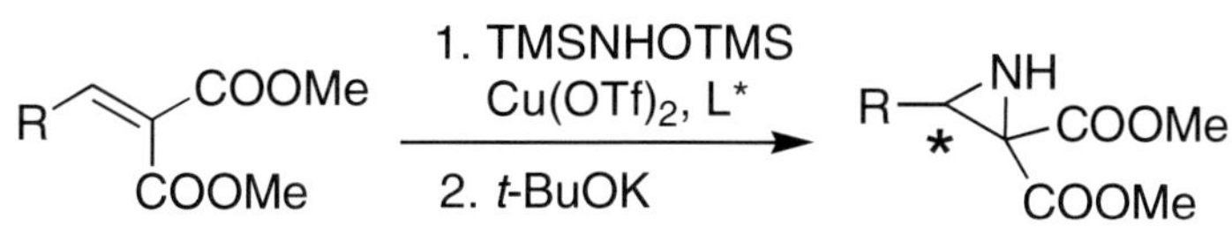

(90) **(91)**

3.3. From Imines

Imines also represent important precursors in aziridine synthesis. An intramolecular cyclization of the imines (92) into aziridines (93) can simply be carried out by the treating imines with sodium methoxide in methanol in good yields [33].

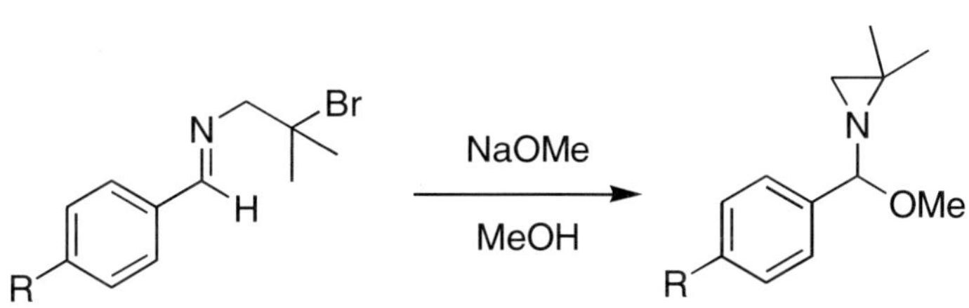

(92) **(93)**

M. R. Biscoe et al. reported the synthesis of N-protected aziridines (96) by the reaction of benzal dihalides (94), benzaldehyde imines (95) mediated by magnesium in ether [34]. For this reaction, two different pathways were discerned: one involving a nucleophilic attack by the alpha-halo Grignard species on the imine and one involving an electrophilic attack on imine by phenylcarbene resulting in the formation of an azomethine ylide.

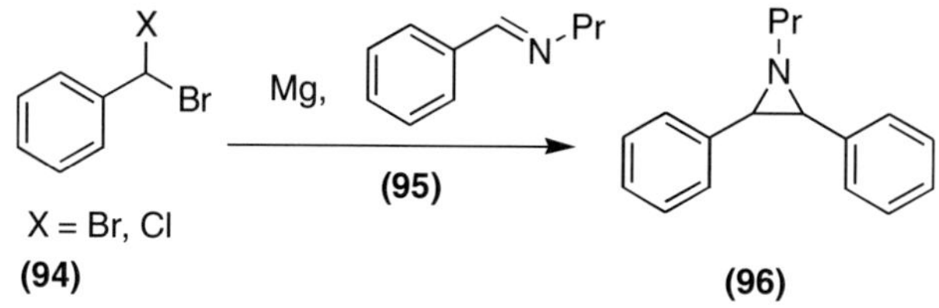

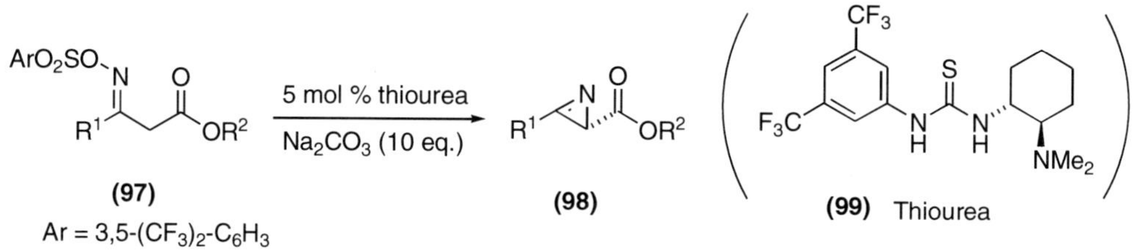

β-ketoxime sulfonates (97) reacted with a catalytic amount of thiourea (99) under basic conditions resulting in the formation of aziridine esters (98) in good yields with enantioselectivities upto 93% ee [35].

As reported by A. Mazumdar [36], Yb $(OTf)_3$ catalyzed the synthesis of aziridines (102) from phenyldiazomethanes (101) with α-imino esters (100) at -78°C to room temperature conditions. Dichloromethane was found to the appropriate solvent for the reaction.

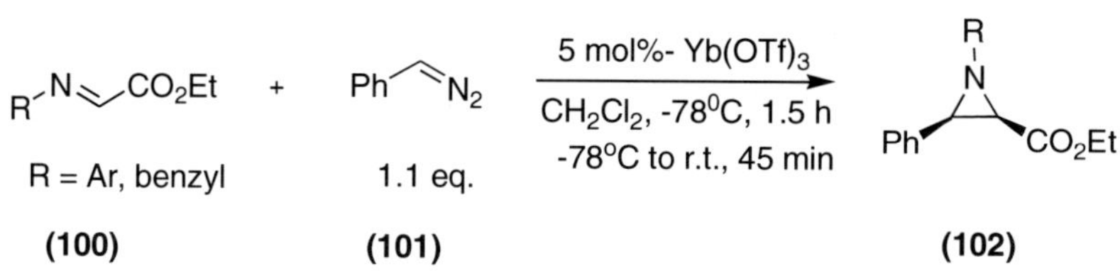

T. L. Troyer carried out the synthesis of cis-aziridines (107) by a triflic acid catalyzed reaction of trans-triazolines [37]. α-diazonium-β-amino ester (106) was found to be formed as an intermediate during the course of the reaction.

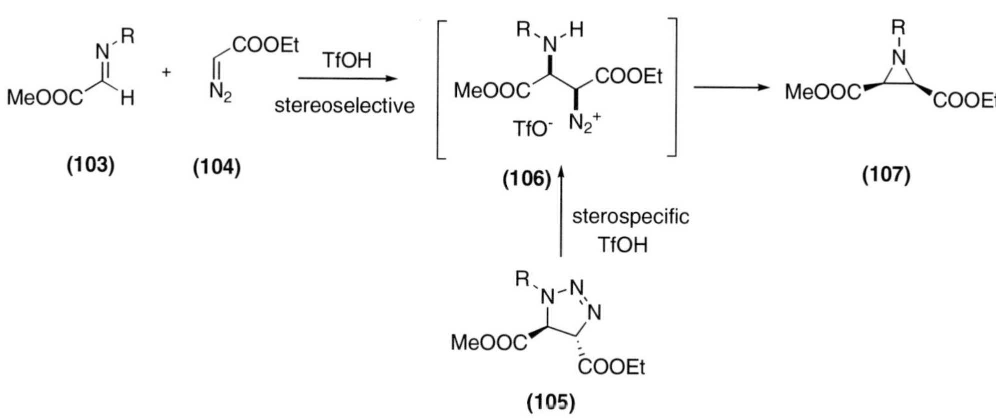

An enantioselective synthesis of aziridines (111) was reported by H. Ren by treating acids (108) with imines (109) in excellent yields [38].

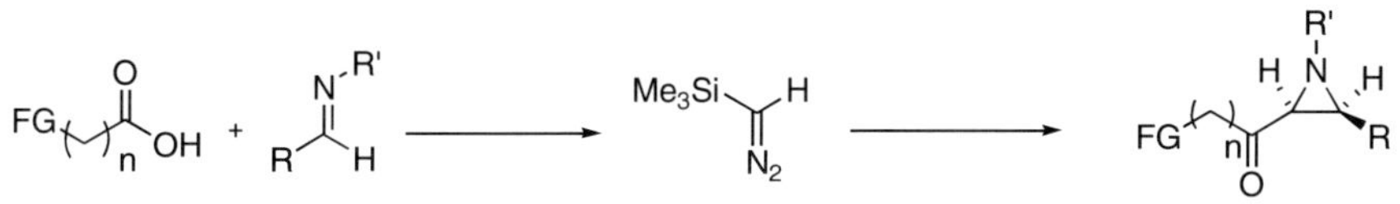

(108) (109) (110) (111)

FG = ester, alkene, alkyne, halide, imide, ketal

R. Fan reported the synthesis of difunctionalized N-sulfonyl aziridines (114) by the reaction of *N*-sulfonyl imines (112) with methylene compounds (113) using PhIO and in the presence of a catalytic amount of potassium iodide [39].

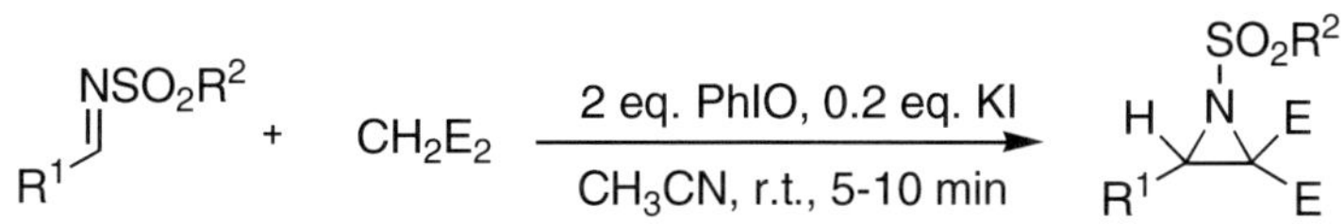

(112) (113) (114)

78-96 %

N-tert-butylsulfinyl aziridines (118) can be efficiently obtained by the organocerium reagent catalyzed (116) reaction of (α-chloro) sulfinylimine (117) in an excellent yield [40].

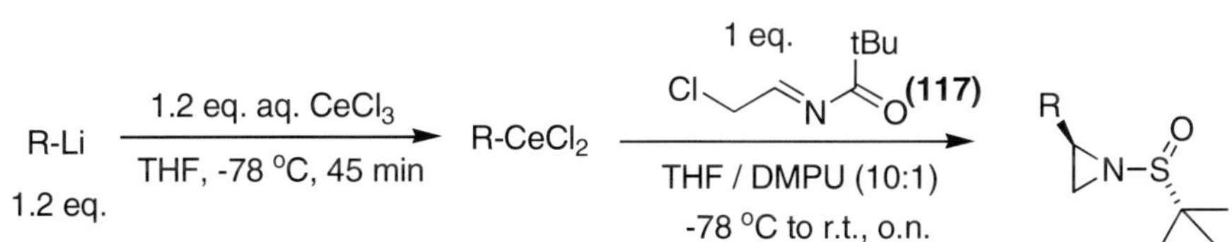

1.2 eq.

(115) (116) (118)

N-tosylaldimines (119) undergo cyclization when treated with 1-bromonitroalkanes (120) in the presence of sodium acetate in methanol under room temperature conditions resulting in the formation of diastereoselective N-tosyl aziridines (121) in an excellent yield [41].

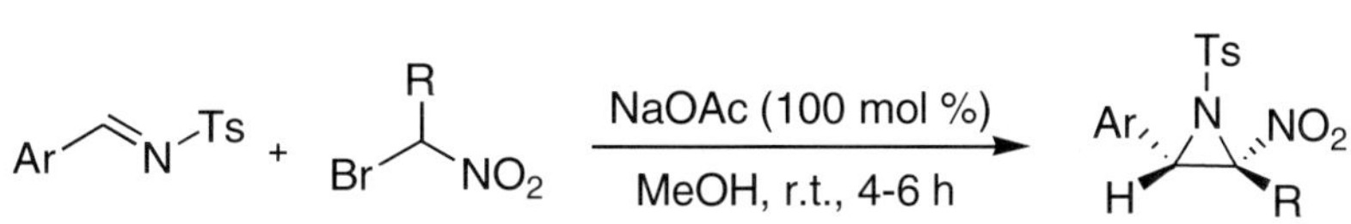

R = H, Me, Et

(81-98 % Z)

(119) (120) (121)

A magnesium-promoted defluorosilylation of trifluoromethyl imines (122) in the presence of trimethyl silyl chloride, followed by the reaction with dimethylsulfonium methylide, affords the corresponding difluoromethylaziridines (124) [42].

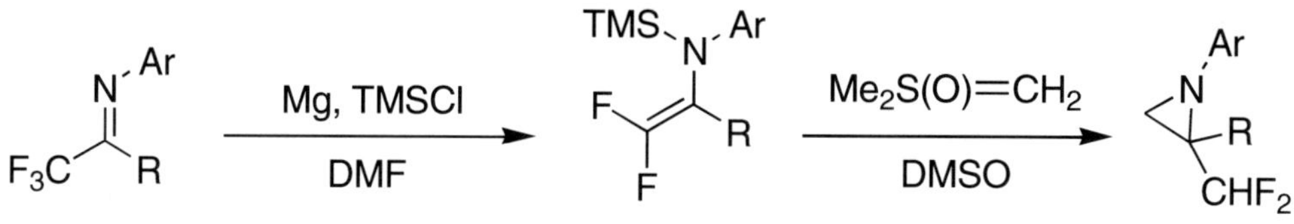

(122) (123) (124)

Montmorillonite K-10 was used as a solid acid catalyst for the synthesis of aziridines (127) under solvent free conditions from imines (125) with ethyl diazoacetate (126) at room temperature in high yields and excellent selectivity [43].

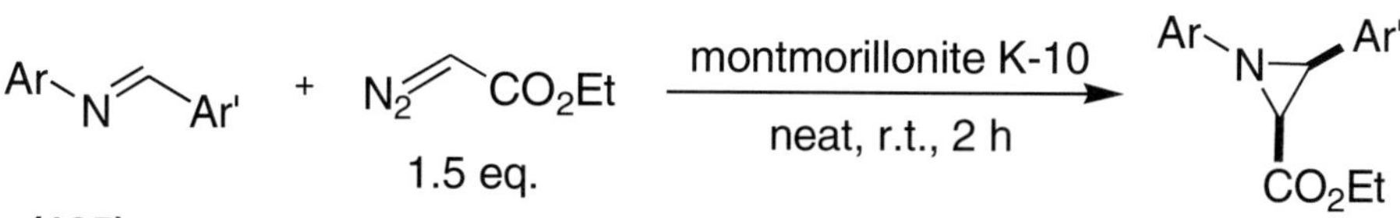

(125) (126) (127)

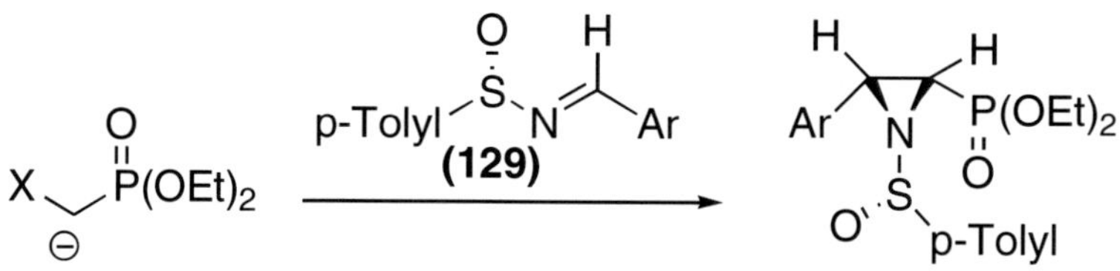

(128) (129) (130)

The chloromethylphosphonate anions (128) were treated with highly enantiopure sulfinimines (129) to afford the corresponding aziridine 2-phosphonates (130) [44].

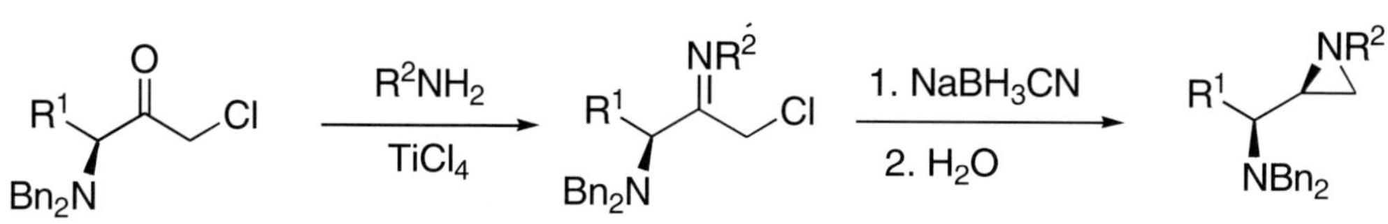

(131) (132) (133)

1-aminoalkyl chloromethyl ketones (131) were converted into chiral 1-aminoalkyl chloromethyl ketimines (132) by treating with various amines in the presence of titanium chloride that, upon reduction followed by intramolecular cyclization, affords the corresponding aminoalkyl aziridines (133) in highly enantiopure forms [45]. The asymmetric synthesis of trisubstituted aziridines from N-Boc imines (134) and α-diazo esters (135) or α-diazo-N-acyloxazolidinones was carried out by L. Huang [46]. The reaction was catalyzed by a (R)-VANOL (137) borate catalyst in excellent yields.

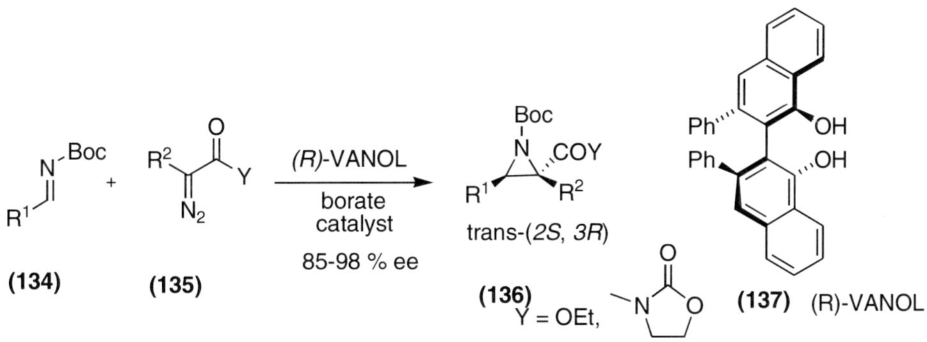

The synthesis of 2-(chloromethyl) aziridines (139) from chloro imines (138) in prsence of $NaBH_4$ in methanol was reported by S. Stankovic et al. [47].

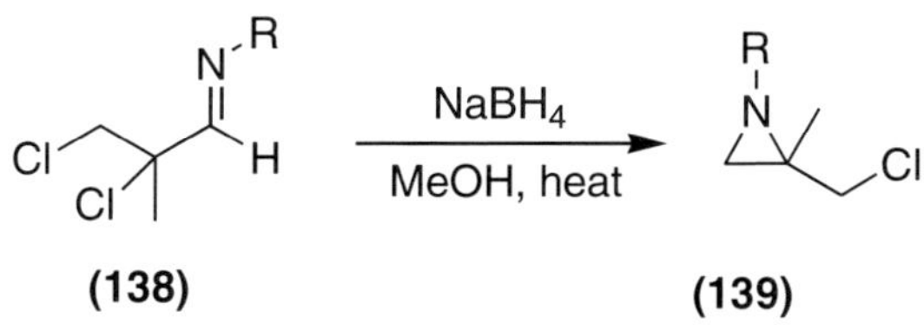

Vinyl aziridines (142) can be synthesized from the reaction of DPP or tosyl-protected imines (140) in the presence of *S*-allyl tetrahydrothiophenium bromide (141) [48].

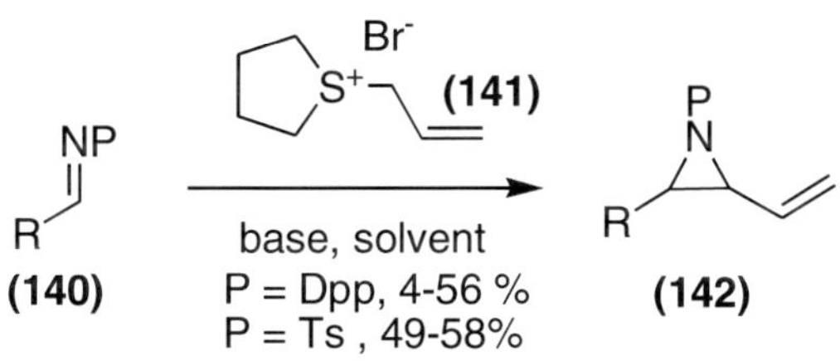

As reported by Chigboh et al. [49], the synthesis of aziridines (145) with diasteromeric excess up to 99% by the reaction of (S_S)-*tert*-butylphenylsulfinimine (143) with different ylides (144) in the presence of lithium ter. butoxide. The reactions proceeded to give products with high yields and a good trans:cis ratio, varying between 1:1 and 0:100.

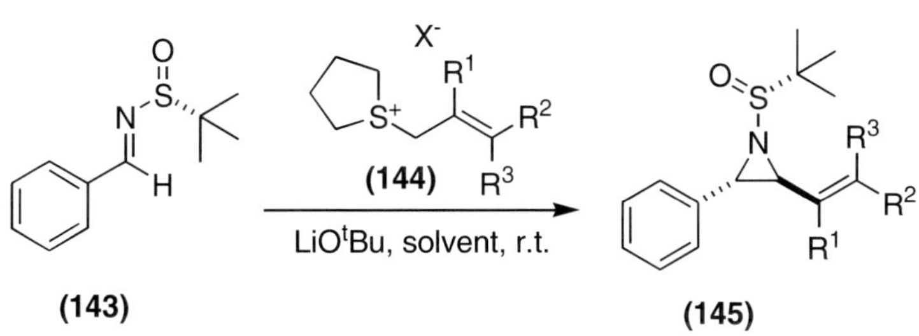

3.4. From Ring Closure of Haloamines and Amino Alcohols

Amino alcohols are organic compounds containing amino and hydroxyl functional groups. Similarly, amino halides/haloamines contain both the functional groups (i.e., halogen and amine).

These two compounds serve as good substrates for the synthesis of azirdines on account of the neighboring group participation shown by the two functional groups. Several methods of synthesizing aziridines from haloamines and aminoalcohols are documented in literature; of which, Wenker's synthesis is common.

An important modification to the Wenker synthesis of aziridines was reported by X. Li et al. Aziridines (148) were successfully obtained by the conversion of amino alcohols (146) with chlorosulfonic acid [50]. Firstly, the amino alcohols were converted into unstable sulfates (147), followed by a treatment with a sodium hydroxide base.

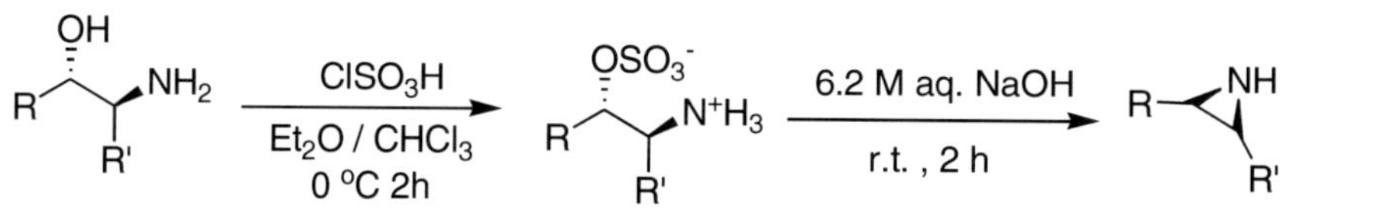

(146) (147) (148)

2-amino alcohols (149) were converted to *N*-tosyl aziridines (150) by two standard protocols developed by L. W. Bieber, et al. [51]. Studies have been revealed that the higher substituted amino alcohols afford better yields on using potassium carbonate in an acetonitrile solvent.

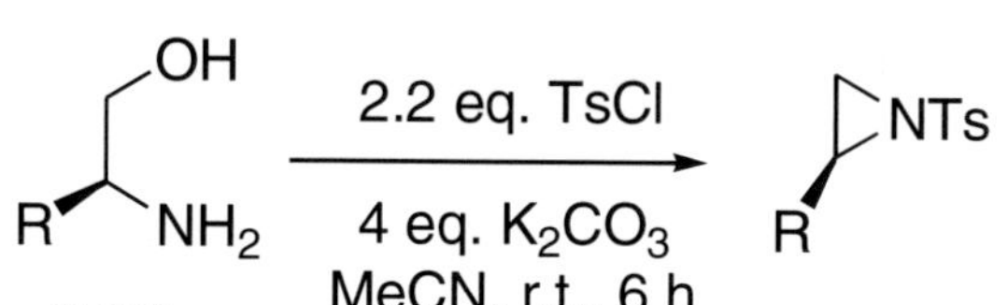

(149) (150)

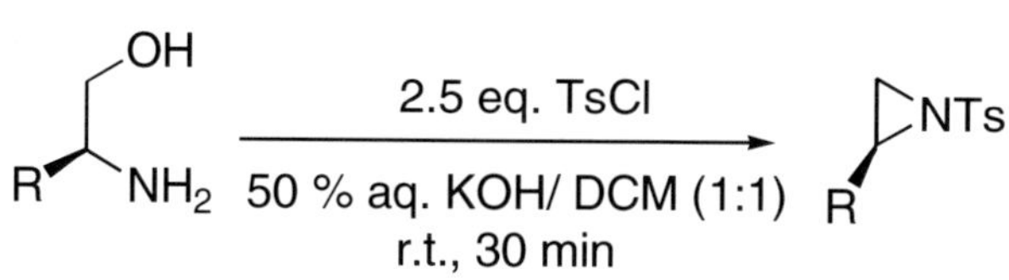

H. Xu reported the conversion of different chiral amino alcohols (151) into N-sulfonylaziridines (153) when the amino alcohols are treated with sulfonyl chlorides (152) under microwave irradiation [52].

The microwave-assisted protocol helps in reducing the reaction time from 24 to 0.5 hours, and yields of the reaction were improved to 84–93% with a minimization of by-product formation.

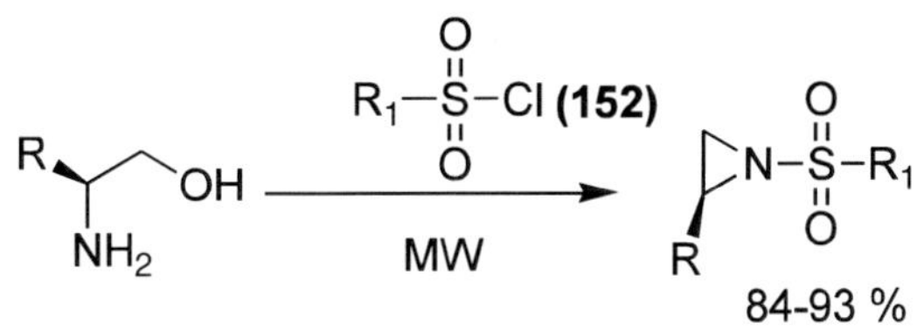

(151) (153)

3.5. By Ring Transformation of Epoxides

An epoxide is a cyclic ether with three ring atoms: two carbons and one oxygen. A variety of epoxides are converted into aziridines using different protocols. *N*-β-hydroxyethylaziridines are the biologically active compounds. Such aziridines (156) can be efficiently obtained by a ring opening reaction of epoxides (154) with β-chloroethylamine hydrochloride (155) under basic conditions in dioxane /water (1:1) at 90°C temperature [53].

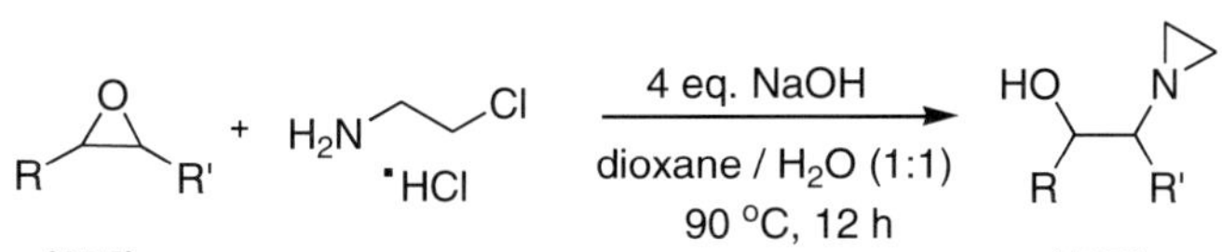

(154) (155) 2 eq. (156)

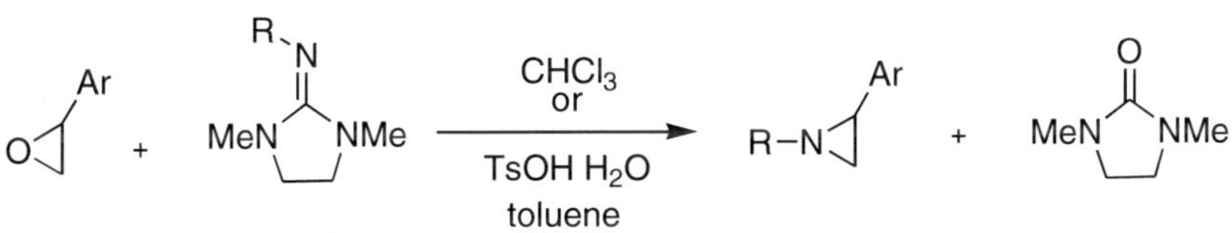

(157) (158) (159) (160)

The transformation of epoxides (157) into the aziridines (159) was reported using guanidine (158) as the nitrogen source [54].

The inversion of configuration at the chiral centers of epoxides was observed without any loss in terms of optical purity.

3.6. Miscellaneous Synthesis

N-acyl-2-hydroxymethyl aziridines (163) can be synthesized from the phthalimide-protected amino acylazides (162) and triphenyl phosphine followed by a reaction with glycidol under basic conditions [55].

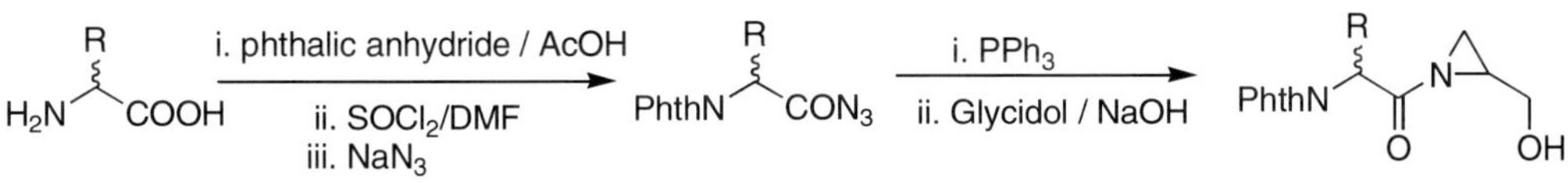

(161) (162) (163)

α-iodo- α, β-unsaturted cyclic ketones (164) can be transformed into N- substituted aziridines (165) under basic conditions in the presence of phenanthroline and xylene as the solvent [56].

Chiral aziridines could be obtained by treating the cyclic ketones with chiral amines.

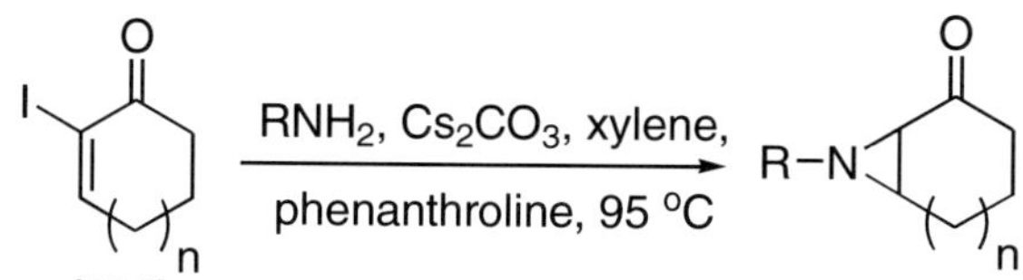

(164) **(165)**

R. L. Weller [57] reported that the anti steroisomer of ethyl -4, 5-dibromo-2-pentenoate (166) reacts with primary amines in the presence of DBU to afford the conjugated aziridines (167). A multicomponent reaction, involving the synthesis of aziridines (170) from the reaction between aryl /heteroaryl ketoximes (168), acetylene and aliphatic ketones (169) in the presence of bases, such as hydroxides of lithium or potassium, or cesium fluoride in a DMSO solvent, was accomplished by B. A. Trofimov et al [58].

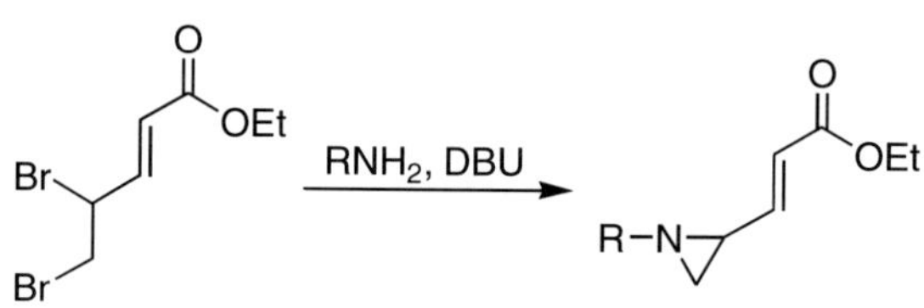

(166) **(167)**

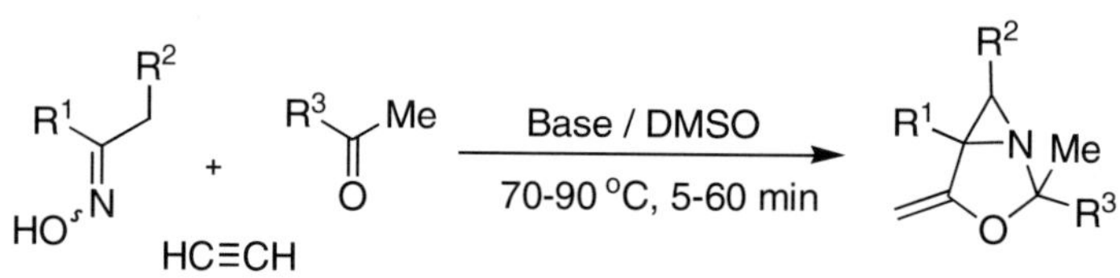

(168) **(169)** Base = KOH, LiOH/ CsF **(170)**

ω-unsaturated sulfonamides (171) undergo aziridination upon treatment with a tertiary butyl oxychloride, NaOH in an aqueous medium followed by a reaction with phenyltrimethylammonium tribromide (PTAB) in acetonitrile to form aziridines (173) [59]. This method provides an alternative to the copper-catalyzed synthesis of aziridines.

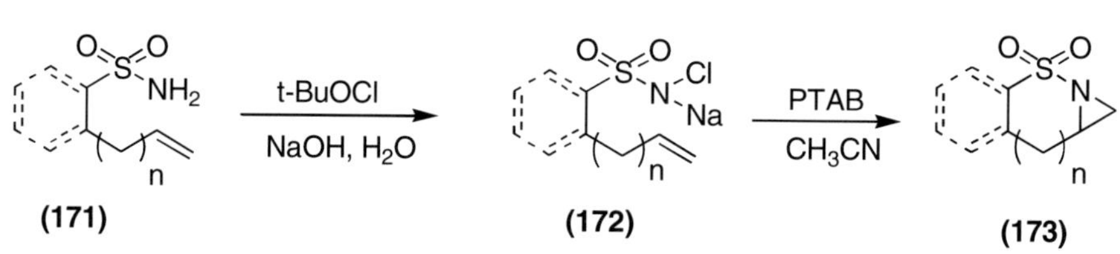

(171) **(172)** **(173)**

The reaction of hexahydro-1, 3, 5-triazines (174) with alkyldiazoacetates, in the presence of Lewis acid, results in the formation of aziridine-2-carboxylates (175) in excellent yields [60].

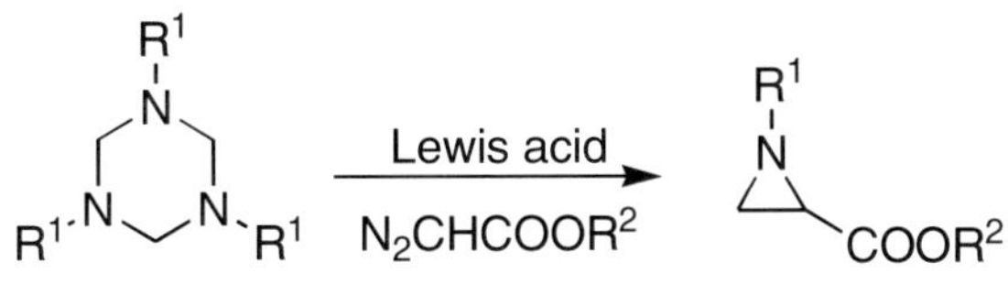

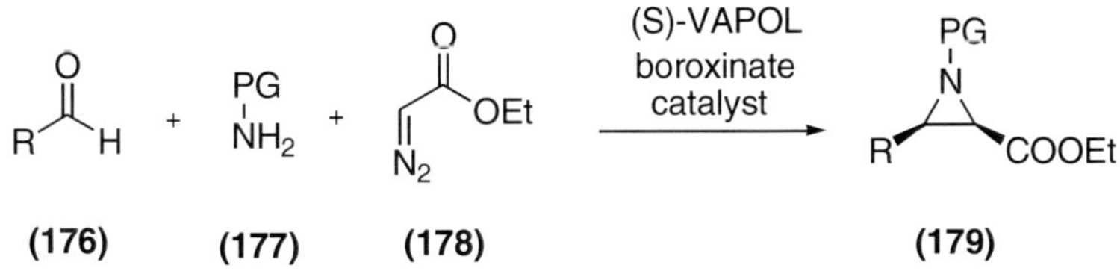

R = Aryl, heteroaryl, 1°, 2°, and 3° aliphatic

A. K. Gupta has recently developed a multicomponent method for the synthesis of aziridines (179) from various aldehydes (176), amines (177) and α-diazoesters (178), catalyzed by a (S)-VAPOL boroxinate catalyst in high diastereo- and enantioselectivity [61].

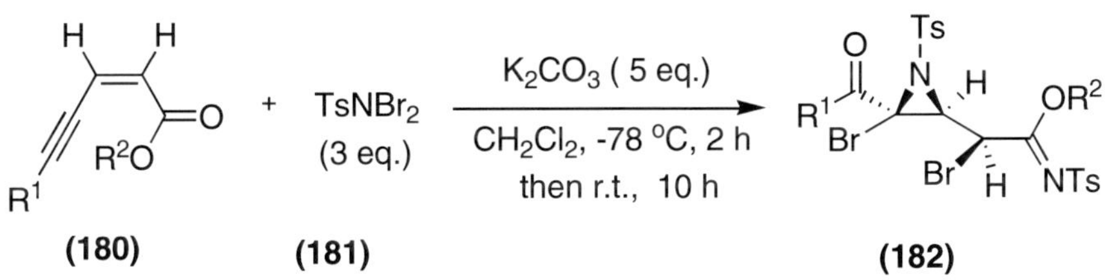

R. Shen and X. Huang [62] carried out the stereoselective synthesis of substituted aziridines (182) by the reaction of (*Z*)-alk-2-en-4-ynoates (180) and $TsNBr_2$ in presence of base potassium carbonate. The cyclization was reported to occur at a very low tempertature (i. e., –78°C).

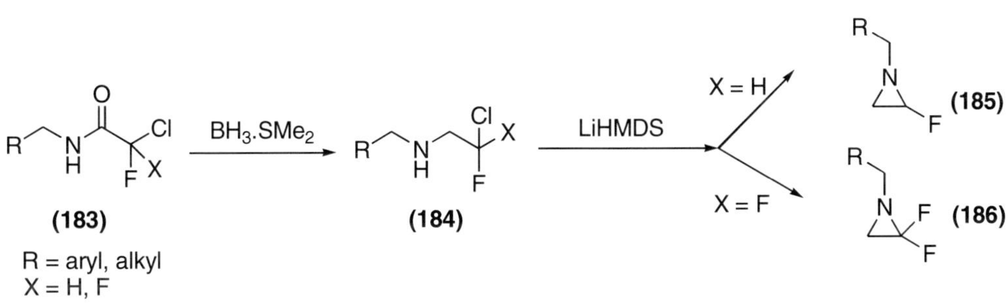

α-fluorinated amides (183), upon their reduction with borane, give the β-fluorinated β-chloroamines (184) which, upon treatment with LiHMDS, undergo cyclization to afford the corresponding 2-fluorinated aziridines (185), (186) [63].

The isoxazolyl olefins (187) on bromination results in the formation of dibromocompounds (188), cyclized under basic conditions at 0°C to room temperature to afford corresponding isoxazolyl aziridines (189) as exclusively cis diastereoisomers [64].

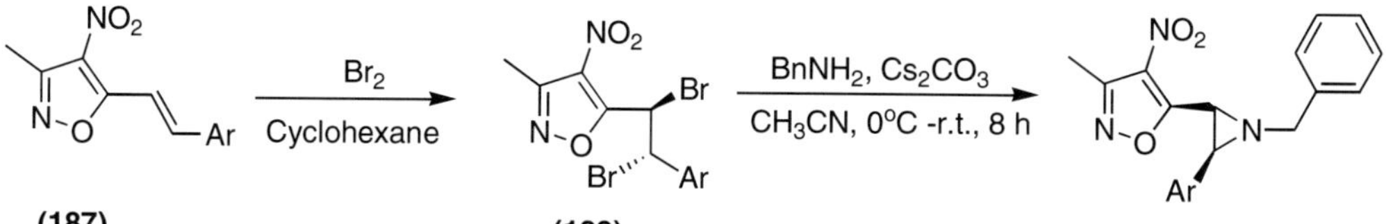

4. Biological Activity of Aziridines

A number of natural products possessing an aziridine ring have been shown to have potent biological activity, closely associated with the reactivity of the strained heterocyclic. For example mytomycin A, B, C, together with porfiromycin and mitiromycin, represents an important class of naturally occurring mitosanes isolated from soil extracts of streptomyces verticillatus. These mitosanes exhibit both antitumor and antibiotic activities, resulting from their ability to cross-link with DNA. Structural activity relationships have identified that the presence of aziridine was the cause for such antitumor activity. The aziridines that contain biologically active natural products are isolated from several marine sponges and found to have potent antitumor antibiotic activities. A wide spectrum of pharmacological activities is associated with this type of aziridine containing natural products and also aziridine derivatives, making them promising heterocycles. Fyaz Ismail has reviewed the therapeutic potential of some aziridines containing alkaloids [65].

Aziridines mainly exhibit the following biological activities:

1. Cytotoxic and anticancer activities
2. Antimicrobial activities
3. Protein inhibiting activities
4. Antibiotic and miscellaneous activities

4.1. Cytotoxic and Anticancer Activities

Compounds that can form covalent DNA cross-links often display interesting medicinal, carcinogenic, or toxic properties. Aziridines are activated spontaneously or by an enzymatic oxidative reaction. Following activation, alkylation can occur at a number of nucleophilic sites in DNA, RNA, protein and other molecules such as glutathione. The natural product azinomycin B contains electrophilic epoxide and aziridine groups that react with nucleophilic sites in duplex DNA to form interstrand cross-links. During the biological activity studies, strong cytotoxicity was observed in specific aziridines which were susceptible to fragmentation reactions by chemical oxidation. The latter shows that the products of fragmentation play an important role in the cytotoxic activity of aziridines [66].

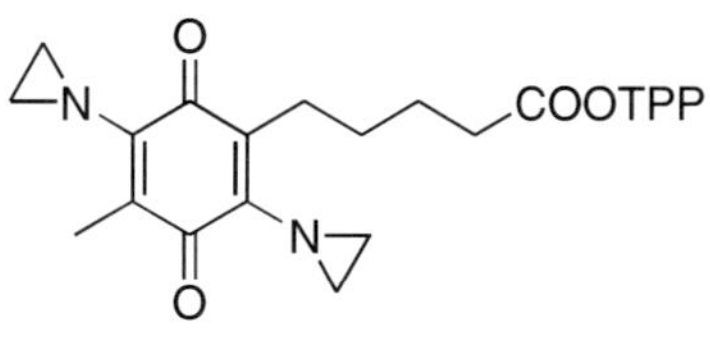

(190)

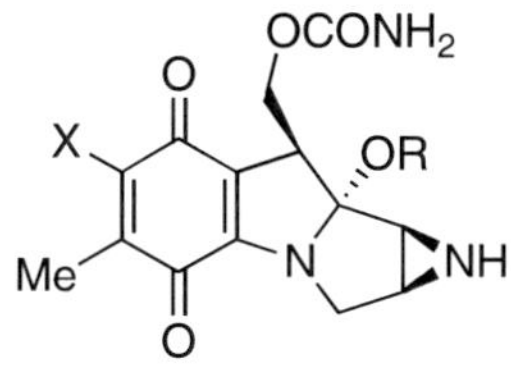

(191)

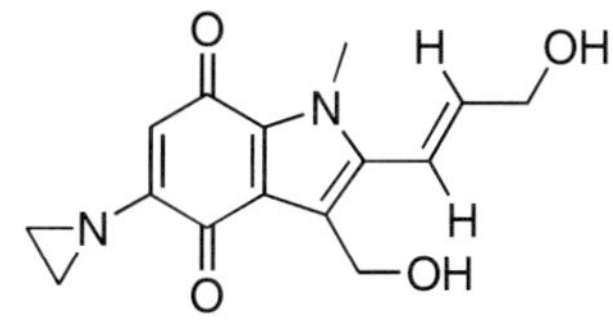

(192)

Reed et al. observed that diaziridinylquinone-activated ester (190) is suitable for conjugating with a variety of amine-containing, targeting ligands for evaluation as possible anticancer agents [67].

The mitomycins are the aziridine-containing natural products isolated from *Streptomyces caespitosus* or *Streptomyces lavendulae.* Mytomycin is has anticancer properties by its nature [68]. The compound Mitomycin (191) has antitumor antibiotic activity and is widely used as a chemotherapeutic agent. Apaziquone (192) is used for the treatment of bladder cancer [69].

Azinomycin (193) is an antitumor compound; studies prove that its action is due to the formation of interstrand cross-links within the major groove of DNA [70]. R. Vicik reported the cytotoxic activity of the aziridine-2, 3-dicarboxylate derivative (194) against macrophages [71]. The aziridine containing molecules (195) were evaluated for cytotoxicity and tubulin-binding polymerization by A. Regueiro-Ren [72].

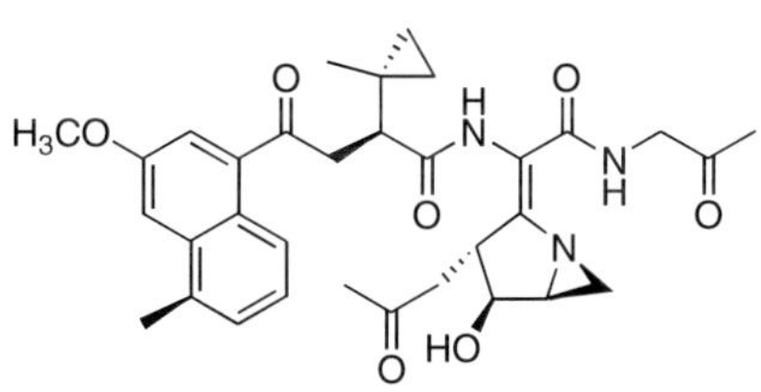

(193)

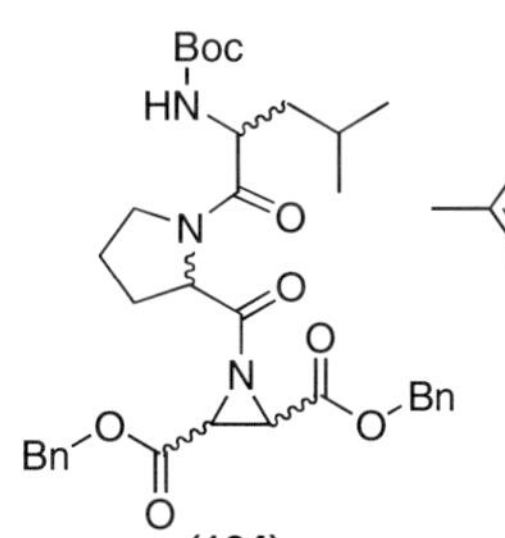

(194)

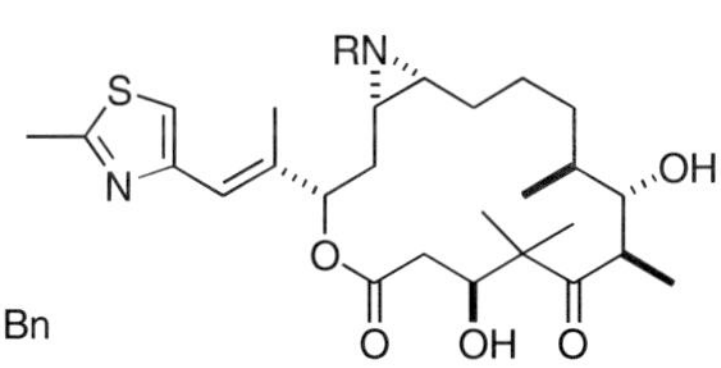

(195)

A study of cytotoxicity and anticancer activity of of triaziquone (TZQ) analogues (196) was carried out by C. H. Huang et al. The studies focused on three cell lines: breast cancer (BC-M1), oral cancer (OEC-M1), and larynx epidermal cancer [73]. The unique cytotoxic azacyclopropene, R-dysidazirine (197), was isolated from the marine sponge *Dysidea fragilis* [74].

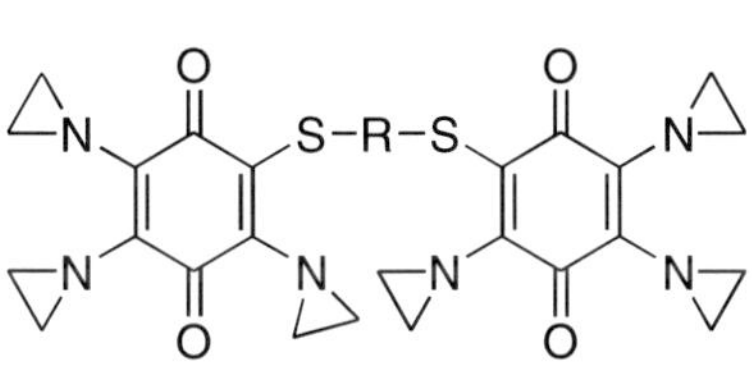

(196)

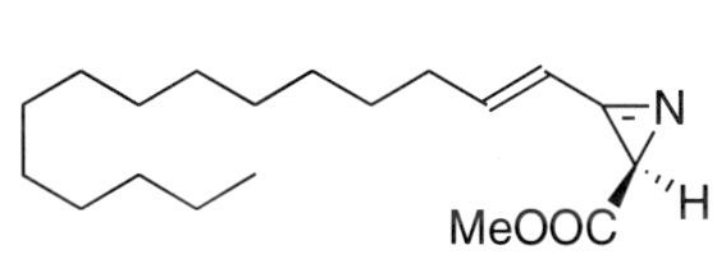

(197)

4.2. Antimicrobial Activities

Phosphonoaziridine compounds (198) have been found to posess promising antibacterial acitivites associated with them [75]. Ficellomycin (199) inhibits the *in vitro* growth of gram-positive bacteria and *Staphylococcus aureus* infections as studied in the case of mice [76].

The naturally occurring peptides madurastatin A1 (200) and madurastatin B1 (201), having an aziridine ring, were discovered from natural sources. These metabolites have showed antibacterial activity against *Micrococcus luteus* [77].

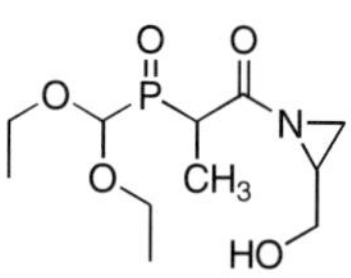

(198)

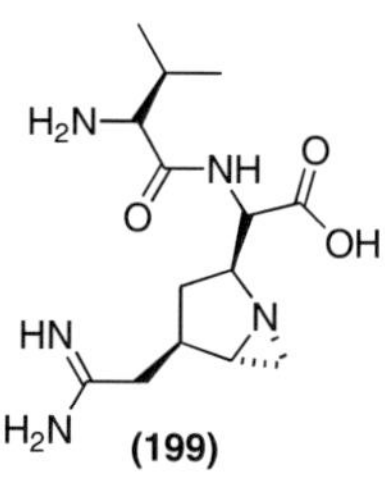

(199)

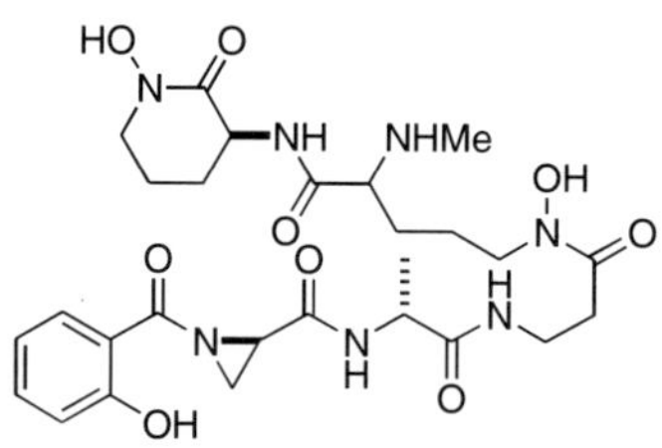

(200)

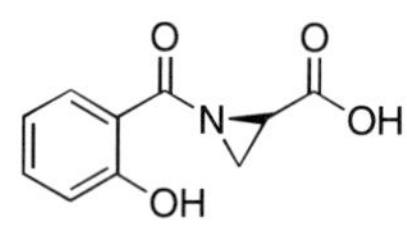

(201)

4.3. Protein Inhibitor Aziridines

A number of aziridines were shown to exhibit useful biological properties. The compound 2-(4-amino-4-carboxylbutyl)-aziridine-2-carboxylic acid (202) is a potent, irreversible inhibitor of the bacterial enzyme diaminopimelic acid epimerase [78]. The 2-(3-carboxypropyl)-aziridine-2-carboxylic acid (203) is an irreversible inhibitor of glutamic racemase [79]. These dicarboxylic aziridines (204) and related derivatives function as the inhibitors of cysteine proteases [80].

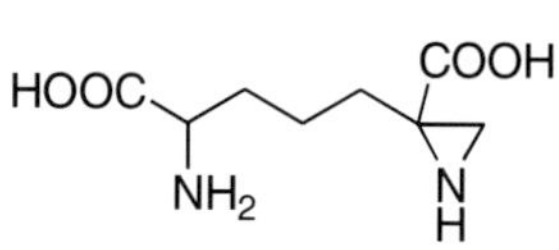

(202)

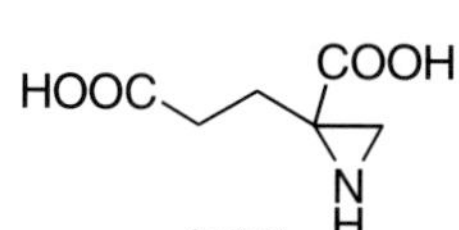

(203)

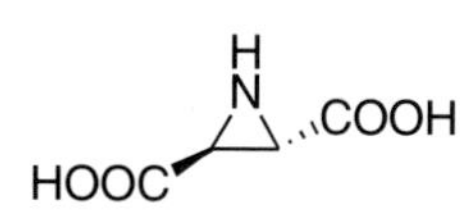

(204)

Miraziridine A (205) causes the inhibition of proteolytic activity of trypsin-like serine proteases [81]. These compounds (206) well represent substrates for studying enzyme mechanisms. Mainly, these are studied as potent protease inhibitors [82].

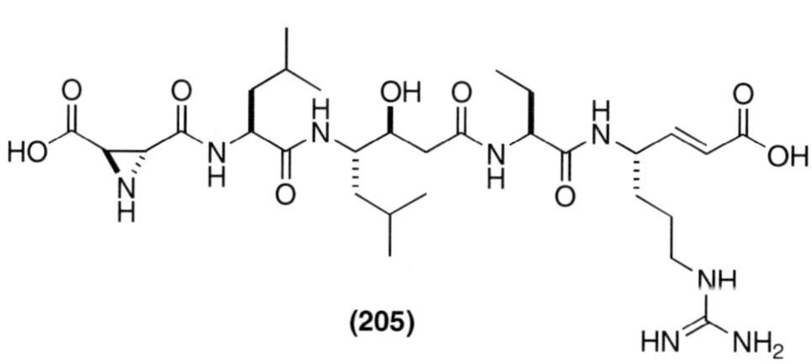

(205)

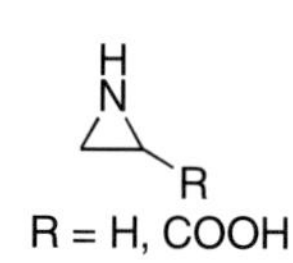

(206)

4.4. Antibiotic and Miscellaneous Activities of Aziridines

The methyl ester of 3-methyl-2H-azirine-2-carboxylic acid (207), was isolated from *Streptomyces aureus* which exhibited a broad spectrum of antibiotic activity *in vitro* against both gram-positive and gram-negative bacteria [83].

Its acid was reported to be most active against *Staphylococcus aureus*. Azicemicins (208) are the angucycline-type antibiotics synthesized by the actinomycete, *Kibdelosporangium* sp. [84]. These compounds also posseses antimicrobial activity [85]. The aziridine containing a quinone derivative (209) has potential for antineoplastic activity. It alkylates and cross-links to DNA during all the phases of the cell cycle, hence disrupting the DNA function, cell cycle arrest, and resulting in DNA damage [86].

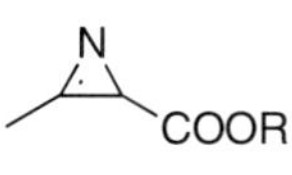

R = H, Me

(207)

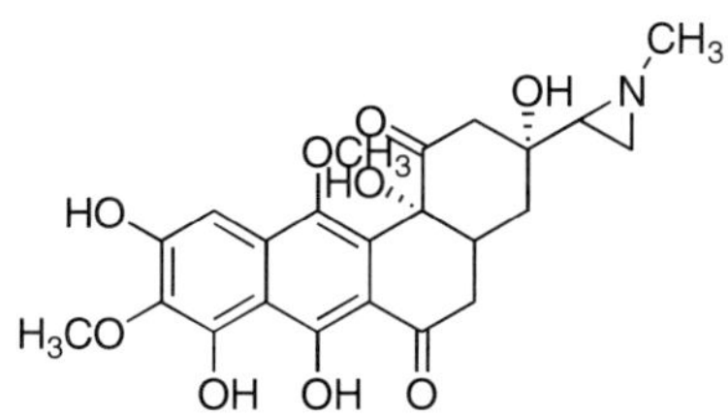

(208)

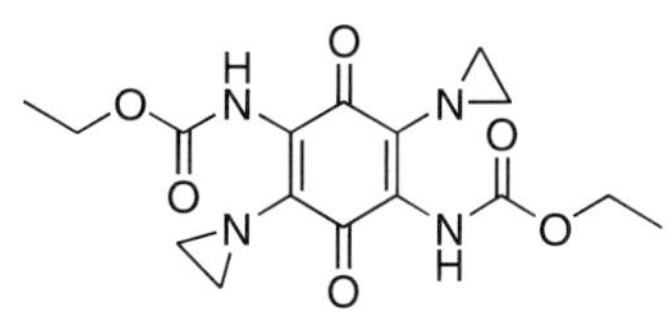

(209)

Conclusion

In summary, the present chapter mainly focus on the synthesis, as well as the biological and theurepetic potential, of 3-membered nitrogen containing aziridine heterocycles. We believe that the current chapter will certainly be fruitful for the researchers in the field of aziridine chemistry.

References

[1] Gabriel, S. Gabriel Ethylenimine Method. *Chemische Berichte,* 1888, *21*, 2664-2666.

[2] Padwa, A. and Murphree, S. S. Epoxides and aziridines - A mini review. *Arkivoc,* 2006, *iii*, 6-33 and references therein.

[3] Catino, A. J.; Nichols, J. M.; Forslund, R. E.; Doyle, M. P. Efficient aziridination of olefins catalyzed by mixed-valent dirhodium (II, III) caprolactamate. *Organic Letters*, 2005, *7*, 2787-2790.

[4] Minakata, S.; Morino, Y.; Oderaotoshi, Y. and Komatsu, M. Practical and convenient synthesis of *N*-heterocycles: Stereoselective cyclization of *N*-alkenylamides with *t*-BuOI under neutral conditions. *Organic Letters,* 2006, *8*, 3335-3337.

[5] Chanda, B. M.; Vyas, R. and Bedekar, A. V. Investigations in the transition metal catalyzed aziridination of olefins, amination, and other insertion reactions with bromamine-T as the source of nitrene. *Journal of Organic Chemistry*, 2001, *66,* 30–34.

[6] Aryl iodide mediated aziridination of alkenes. Li, J.; Chan, P. W. H. and Che, C-M. *Organic Letters*, 2005, *7*, 5801–5804.

[7] Vedejs, E. and Sano, H. Synthesis of N-methoxy and N-H aziridines from alkenes. *Tetrahedron Letters*, 1992, *33*, 3261–3264.

[8] Yang, K-S. and Chen, K. Enantioselective aziridination of alkenes with *N*-aminophthalimide in the presence of lead tetraacetate-mediated chiral ligand. *Organic Letters*, 2002, *4*, 1107-1109.

[9] Li, J.; Liang, J-L. ; Chan, P. W. H. and Che, C-M. Aziridination of alkenes with N-substituted hydrazines mediated by iodobenzene diacetate. *Tetrahedron Letters*, 2004,*45*, 2685–2688.

[10] Hoque, M. M.; Miyamoto, K.; Tada, N.; Shiro, M. and Ochiai, M. Synthesis and structure of hypervalent diacetoxybromobenzene and aziridination of olefins with imino-λ^3-bromane generated in situ under metal-free conditions. *Organic Letters,* 2011, *13*, 5428–5431.

[11] Siu, T. and Yudin, A. K. Practical olefin aziridination with a broad substrate scope. *Journal of American Chemical Society*, 2002, *124*, 530-531.

[12] Aires-de-Sousa, J.; Lobo; A. M. and Prabhakar, S. A new enantioselective synthesis of N-arylaziridines by phase-transfer catalysis. *Tetrahedron Letters*, 1996, *37*, 3183–3186.

[13] Ali, S. I.; Nikalje, M. D.; and Sudalai, A. Pyridinium hydrobromide perbromide: A versatile catalyst for aziridination of olefins using chloramine-T. *Organic Letters*, 1999, *1*, 705–707.

[14] Duran, F.; Leman L.; Ghini, A.; Burton, G.; Dauban, P. and Dodd, R. H. Intramolecular Ph=O mediated copper-catalyzed aziridination of unsaturated sulfamates: A new direct access to polysubstituted amines from simple homoallylic alcohols. *Organic Letters*, 2002, *4*, 2481–2483.

[15] Han, J.; Li, Y.; Zhi, S.; Pan, Y.; Timmons, C. and Li, G. Palladium-catalyzed aziridination of alkenes using *N, N*-dichloro-*p*-toluenesulfonamide as nitrogen source. *Tetrahedron Letters*, 2006, *47*, 7225–7228.

[16] Ando, T.; Minakata, S.; Ryu, I. and Komatsu, M. Nitrogen atom transfer to alkenes utilizing chloramine-T as a nitrogen source. *Tetrahedron Letters*, 1998, *39*, 309–312.

[17] Kwong, H-L.; Liu, D.; Chan, K-Y.; Lee, C-S.; Huang, K-H. and Che, C-M. Copper (I)-catalyzed asymmetric alkene aziridination mediated by PhI $(OAc)_2$: a facile one-pot procedure. *Tetrahedron Letters*, 2004, *45*, 3965-3968.

[18] Fioravanti, S.; Pellacani, L.; Stabile, S.; Tardella, P. A. and Ballini, R. (1997). Aziridination of conjugated nitroalkenes. *Tetrahedron Letters*, *38*, 3309–3310.

[19] Fazio, A.; Loreto, M. A. and Tardella, P. A. Aziridination of α, β-unsaturated phosphonic esters. *Tetrahedron Letters*, 2001, *42*, 2185–2187.

[20] Thakur, V. V. and Sudalai, A. *N*-Bromoamides as versatile catalysts for aziridination of olefins using chloramine-T. *Tetrahedron Letters*, 2003, *44*, 989–992.

[21] Jain, S. L.; Sharma, V. B. and Sain, B. An efficient transition metal-free aziridination of alkenes with chloramine-T using aqueous H_2O_2/HBr. *Tetrahedron Letters*, 2004, *45*, 8731-8732.

[22] Jain, S. L. and Sain, B. Aziridination of alkenes using *N*-iodo-*N*-potassio-*p*-toluenesulphonamide as a nitrene precursor. *Tetrahedron Letters,* 2003, *44*, 575–577.

[23] Fan, R.; Pu, D.; Gan, J. and Wang, B. PhI $(OAc)_2/I_2$ induced aziridination of alkenes with $TsNH_2$ under mild conditions. *Tetrahedron Letters*, 2008, *49*, 4925–4928.

[24] Mohr, F.; Binfield, S. A.; Fettinger, J. C. and Vedernikov, A. N. A practical, fast and high-yielding aziridination procedure using simple Cu (II) complexes containing N-donor pyridine-based ligands. *Journal of Organic Chemistry*, 2005, *70,* 4833–4839.

[25] Armstrong, A.; Baxter, C. A.; Lamont, S. G.; Pape, A. R.; Wincewicz, R. Amine-promoted, organocatalytic aziridination of enones. *Organic Letters*, 2007, *9*, 351-353.

[26] Sugihara, H.; Daikai, K.; Jin, X. L.; Furuno, H. and Inanaga, J. Catalytic conversion of conjugated enones into optically active α-keto aziridines using chiral rare earth metal complexes. *Tetrahedron Letters*, 2002, *43*, 2735–2739.

[27] Chen, D.; Timmons, C.; Guo, L.; Xu, X.; Li, G. One-pot stereoselective synthesis of *anti* 3-alkyl and 3-aryl-*N*-*p*-tosyl-aziridine-2-ketones and 3-aryl-*N*-*p*-tosyl-aziridine-2-carboxylates. *Synthesis*, 2004, 2479-2484.

[28] Hori, K.; Sugihar, H.; Ito, Y. N. and Katsuki, T. *Cis-selective* aziridination of *cis*- or *trans*-α,β-unsaturated amides using diaziridine. *Tetrahedron Letters*, 1999, *40*, 5207–5210.

[29] Mahoney, J. M.; Smith, C. R.; Johnston, J. N. Bronsted acid-promoted olefin aziridination and formal *anti*-aminohydroxylation. *Journal of American Chemical Society,* 2005, *127*, 1354-1355.

[30] Desmarchelier, A.; Pereira de Sant'Ana, D.; Terrasson, V.; Campagne, J. M.; Moreau, X.; Greck, C.; de Figueiredo, R. M. Organocatalyzed aziridination of α-branched enals: enantioselective synthesis of aziridines with a quaternary stereocenter. *European Journal of Organic Chemistry,* 2011, *20*, 4046-4052.

[31] Armstrong, A.; Pullin, R. D. C.; Jenner, C. R. and Scutt, J. N. Amine-promoted synthesis of vinyl aziridines. *Journal of Organic Chemistry*, 2010, *75*, 3499–3502.

[32] Cardillo, G.; Fabbroni, S.; Gentilucci, L.; Gianotti, M.; Percacciante, R. and. Tolomelli, A. Enantioselective synthesis of aziridine 2, 2-dicarboxylates. Part I: Copper (II)-bisoxazoline complex-catalysed Michael reaction on alkylidene malonates. *Tetrahedron Asymmetry*, 2002, *13*, 1407-1410.

[33] D'hooghe, M.; Hofkens, A. and Kimpe, N. D. A new route towards *N*-(α-methoxybenzyl) aziridines. *Tetrahedron Letters*, 2000, *44*, 1137–1139.

[34] Biscoe, M. R. and Fry, A. J. Dual reaction pathways in the magnesium-mediated synthesis of aziridines from benzal halides and imines. *Tetrahedron Letters*, 2001, *42*, 2759–2762.

[35] Sakamoto, S.; Inokuma, T. and Takemoto, Y. Organocatalytic asymmetric Neber reaction for the synthesis of 2*H*-azirine carboxylic esters. *Organic Letters*, 2011, *13*, 6374–6377.

[36] Mazumdar, A.; Xue, Z.; Mayer, M. F. A catalytic synthesis of aziridines without the usual byproducts. *Synlett*, 2007, 2025-2028.

[37] Troyer, T. L.; Muchalski, H.; Hong, K. B. and Johnston, J. N. Origins of selectivity in Brønsted acid-promoted diazoalkane–azomethine reactions (The Aza-Darzens aziridine synthesis). *Organic Letters*, 2011, *13*, 1790–1792.

[38] Ren, H. and Wulff, W. D. Trimethylsilyldiazomethane as a versatile stitching agent for the introduction of aziridines into functionalized organic molecules. *Organic Letters,* 2010, *12,* 4908–4911.

[39] Fan, R.; Wang, L.; Ye and Zhang, J. Facile iodine (III)-induced oxidative cycloaddition of *N*-sulfonyl imines with methylene compounds under neutral conditions. *Tetrahedron Letters*, 2009, *50*, 3857–3859.

[40] Hodgson, D. M.; Kloesges, J. and Evans, B. Terminal aziridines by addition of Grignard reagents or organoceriums to an (α-chloro) sulfinylimine. *Synthesis*, 2009, 1923-1932.

[41] Yadav, L. D. S.; Garima and Kapoor, R. The first diastereoselective nitroaziridination of *N*-tosylaldimines with 1-bromonitroalkanes. *Tetrahedron Letters*, 2009, *50*, 5420-5423.

[42] Mae, M.; Matsuura, M.; Amii, H. and Uneyama, K. A facile synthesis of difluoromethylaziridines. *Tetrahedron Letters*, 2002, *43*, 2069–2072.

[43] Borkin, D.; Carlson, A. and Torok, B. K-10-catalyzed highly diastereoselective synthesis of aziridines. *Synlett*, 2010, 745-748.

[44] Davis, F. A.; Wu, Y.; Yan, H.; McCoull, W. and Prasad, K. R. Asymmetric synthesis of aziridine 2-phosphonates from enantiopure sulfinimines (*N*-sulfinyl imines). Synthesis of α-amino phosphonates. *Journal of Organic Chemistry*, 2003, *68*, 2410–2419.

[45] Concellon, J. M.; Bernad, P. L.; Riego, E.; Garcia-Granda, S. and Forcen-Acebal, A. The first synthesis of enantiopure α-amino ketimines and amino aziridines. *Journal of Organic Chemistry*, 2001, *66*, 2764–2768.

[46] L. Huang and W. D. Wulff. Catalytic asymmetric synthesis of trisubstituted aziridines. *Journal of American Chemial Society*, 2011, *133*, 8892–8895.

[47] Stankovic, S.; D'hooghe, M.; Dewulf, J.; Bogaert, P.; Jolie, R. and Kimpe, N. D. Synthesis and reactivity of non-activated 2-(chloromethyl) aziridines. *Tetrahedron Letters,* 2011, *35*, 4529-4532.

[48] Arini, L. G.; Sinclair, A.; Szeto, P. and Stockman, R. A. Direct synthesis of aliphatic vinyl aziridines. *Tetrahedron Letters*, 2004, *45,* 1589-1591.

[49] Chigboh, K.; Morton, D.; Nadin, A. and Stockman, R. A. Synthesis of chiral non-racemic substituted vinyl aziridines. *Tetrahedron Letters*, 2008, *49*, 4768–4770.

[50] Li, X.; Chen, N. and Xu, J. An improved and mild Wenker synthesis of aziridines. *Synthesis*, 2010, 3423-3428.

[51] Bieber, L. W. and de Araujo, M. C. F. Short and efficient synthesis of optically active *N*-tosyl aziridines from 2-amino alcohols. *Molecules,* 2002, *7*, 902-906.

[52] Xu, H.; Tian, H.; Zheng' L.; Liu, Q.; Wang, L. and Zhang, S. One-pot and microwave-assisted synthesis of *N*-sulfonylaziridines. *Tetrahedron Letters*, 2011, *52*, 2873–2875.

[53] Kim, H. Y.; Talukdar, A. and Cushman, M. Regioselective synthesis of N-β-hydroxyethylaziridines by the ring-opening reaction of epoxides with aziridine generated in situ. *Organic Letters*, 2006, *8*, 1085-1087.

[54] Tsuchiya, Y.; Kumamoto T. and Ishikawa, T. Guanidines as a nitrogen source for direct conversion of epoxides to aziridines. *Journal of Organic Chemistry*, 2004, *69*, 8504–8505.

[55] Medjahed, W.; Zatla, A. T.; Mulengi, J. K.; Ahmed, F. Z. B. and Merzouk, H. The synthesis of *N*-acyl-2-hydroxymethyl aziridines of biological interest. *Tetrahedron Letters,* 2004, *45*, 1211-1213.

[56] Barros, M. T.; Maycock, C. D. and Ventura, M. R. A synthesis of aziridines from α-iodoenones. *Tetrahdron Letters*, 2002, *43*, 4329-4331.

[57] Weller, R. L. and Rajski, S. R. Aziridination of γ, δ dibromoethyl 2 pentenoate with primary amines: extension of the Gabriel–Cromwell reaction. *Tetrahedron Letters*, 2004, *45*, 5807-5810.

[58] Trofimov , B. A.; Schmidt, E. Y.; Mikhaleva, A. I.; Ushakov, I. A.; Protsuk, N. I.; Senotrusova, E.Y.; Kazheva, O. N. ; Aleksandrov, G. G. and Dyachenko, O. A. One-pot assembly of 4-methylene-3-oxa-1-azabicyclo [3.1.0] hexanes from alkyl aryl (hetaryl) ketoximes, acetylene, and aliphatic ketones: a new three-component reaction. *Tetrahedron Letters*, 2009, *50*, 3314–3317.

[59] Dauban, P. and Dodd, R. H. Intramolecular bromine-catalyzed aziridination: a new direct access to cyclic sulfonamides. *Tetrahedron Letters*, 2001, *42*, 1037–1040.

[60] Ha, H-J.; Kang, K-H.; Suh, J-M. and Ahn, Y-G. A new synthesis of aziridine-2-carboxylates by reaction of hexahydro-1, 3, 5-triazines with alkyldiazoacetates in the presence of tin (IV) chloride. *Tetrahedron Letters*, 1996, *37*, 7069–7070.

[61] Gupta, A. K.; Mukherjee, M. and Wulff,http://pubs.acs.org/ doi/abs/10.1021/ol202472z - cor1 W. D. Multicomponent catalytic asymmetric aziridination of aldehydes. *Organic Letters*, 2011, *13*, 5866–5869.

[62] Shen, R. and Huang, X. Unexpected multiple electrophilic addition reaction of (*z*)-alk-2-en-4-ynoates with *N, N*-dibromo-*p*-toluenesulfonamide ($TSNBr_2$): A highly diastereoselective synthesis of densely functionalized aziridines. *Organic Letters*, 2009, *11*, 5698–5701.

[63] Verniest, G.; Colpaert, F.; Hende, E. V. and Kimpe, N. D. (2007). Synthesis and reactivity of 1-substituted 2-fluoro- and 2, 2-difluoroaziridines. *Journal of Organic Chemistry*, *72*, 8569–8572.

[64] Adamo, M. F. A.; Bruschi, S.; Suresh, S. and Piras, L. Aziridination of 3-methyl-4-nitro-5-styrylisoxazoles. *Tetrahedron Letters*, 2008, *49*, 7406–7409.

[65] Fyaz, M. D. Ismail, Levitsky, D. O.; Dembitsky, V. M. Aziridine alkaloids as potential therapeutic agents. *European Journal of Medicinal Chemistry*, 2009, *44*, 3373–3387 and references therein.

[66] Hata, Y.; Watanabe, M.; Matasubara, T. and Touchi, A. Fragmentation reaction of ylide. 5. A new metabolic reaction of aziridine derivatives. *Journal of American Chemical Society*, 1976, *98*, 6033-6036.

[67] Reed, M. W. Wald, A. and Meyer R. B. Triplex-Directed Interstrand DNA cross-Linking by Diaziridinylquinone–Oligonucleotide Conjugates. *Journal of American Chemical Society*, 1998, *120*, 9729-9734.

[68] Wang, Z. and Jimenez L. S. A total synthesis of (±)-mitomycin K. Oxidation of the mitosene C9-9a double bond by (hexamethylphosphoramido) oxodiperoxomolybdenum (VI) (MoO_5 HMPA). *Tetrahedron Letters*, 1996, *37*, 6049-6052.

[69] Puri, R.; Palit, V. and Loadman, P. M. Phase I/II pilot study of intarvesical apaziquone (EO9) for superficial Bladder Cancer". *The Journal of Urology*, 2006, *176*, 1344-1348.

[70] B. Kelly G.T.; Liu, C.; Smith, R.; Coleman, R.S. and Watanabe, C. M. Cellular effects induced by the antitumor agent azinomycin. *Chemical Biology*, 2006, *13*, 485-92.

[71] Vicik, R. ; Hoerr, V. ; Glaser, M. ; Schultheis, M.; Hansell, E. ; McKerrow, J. H.; Holzgrabe, U. ; Caffrey, C. R.; Ponte-Sucre, A. ; Moll, H. ; Stich, A. ; Schirmeister, T. Aziridine-2, 3-dicarboxylate inhibitors targeting the major cysteine protease of Trypanosoma brucei as lead trypanocidal agents. *Bioorganic and Medicinal Chemistry Letters*, 2006, *16*, 2753-2757.

[72] Regueiro-Ren, A.; Borzilleri, R. M.; Zheng, X.; Kim, S-H. ; Johnson, J. A. ; Fairchild, C. R. ; Lee, F. Y. F. ; Long, B. H. and Vite, G. D. Synthesis and biological activity of novel epothilone aziridines. *Organic Letters*, 2001, *3*, 2693–2696.

[73] Huang, C. H.; Kuo, H-S. ; Liu, J-W. and Lin, Y-L. Synthesis and antitumor evaluation of novel bis-triaziquone derivatives. *Molecules,* 2009, *14*, 2306-2316.

[74] Molinski, T.F. and Ireland, C. M. *Journal of Organic Chemistry,* 1988, *53*, 2103–2105.

[75] Keniche, A.; Mezrai, A. and Mulengi, J. K. Synthesis of a novel class of phosphonoaziridines as interesting antibacterial agents. *Bentham. The Open Conference Proceedings Journal,* 2011, *2,* 29-35.

[76] Argoudelis, A. D.; Mizsak, S. A.; Baczynskyj, L. and Wnuk, R. J. The structure of feldamycin, *The Journal of Antibiotics,* 1976, *29*, 1117-1119.

[77] Harada, K. I. ; Tomita, K. ; Fujii, K. ; Masuda, K. ; Mikami, Y. ; Yazawa, K. and Komaki, H. Isolation and structural characterization of siderophore, madurastatins produced by a pathogenic *Actinomadura madurae. Journal of Antibiotics,* 2004, *57,* 125-135.

[78] Gerhart, F. Higgins, W. Tardif C. and Duceo J. B. (2-(4-Amino-4-carboxybutyl) aziridine-2-carboxylic acid. A potent irreversible inhibitor of diaminopimelic acid epimerase. Spontaneous formation from alpha-(halomethyl) diaminopimelic acids. *Journal of Medicinal Chemistry,* 1990, *33,* 2157-2162.

[79] Tanner, M. E. and Miao S. The synthesis and stability of aziridino-glutamate, an irreversible inhibitor of glutamate racemase. *Tetrahedron Letters*, 1994, *35*, 4073-4076.

[80] Schirmeister, T. and Peric, M. Aziridinyl peptides as inhibitors of cysteine proteases: effect of a free carboxylic acid function on inhibition. *Bioorganic and Medicinal Chemistry,* 2000, *8*, 1281-91.

[81] Schaschke, N. Miraziridine A: natures blueprint towards protease class-spanning inhibitors. *Bioorganic and Medicinal Chemistry Letters,* 2004, *14*, 855-857.

[82] Korn, A.; Rudolph-Boehner, S. and Moroder, L. On the synthesis of (2S)-aziridine-2-carboxylic acid containing peptides. *Tetrahedron*, 1994, *50*, 1717–1730. b) T. Schirmeister and M. Peric. Aziridinyl peptides as inhibitors of cysteine proteases: Effect of a free carboxylic acid function on inhibition. *Bioorganic and Medicinal Chemistry,* 2000, *8*, 1281–1291.

[83] Stapley, E. O.; Hendlin, D.; Jackson, M.; Miller, A.K.; Hernandez, S. and Martinez, M. Azirinomycin. I Microbial production and biological characteristics. *The Journal of Antibiotics,* 1971, *24*, 42–47. b). Miller, T.W.; Tristram, E.W. and Wolf, F. J. Azirinomycin II. Isolation and chemical characterization as 3-methyl-2(2H) azirinecarboxylic acid. *The Journal of Antibiotics,* 1971, *24*, 48–50.

[84] Ogasawara, Y. and Liu, H-W. Biosynthetic studies of aziridine formation in azicemicins. *Journal of American Chemical Society,* 2009, *131,* 18066–18068.

[85] Tsuchida, T.; Inuma, H.; Kinoshita, N.; Ikeda, T.; Sawa, T.; Hamada, M. and Takeuchi, T. Azicemicins A and B, a new antimicrobial agent produced by Amycolatopsis. I. Taxonomy, fermentation, isolation, characterization and biological activities. *Journal of Antibiotics,* 1995, *48*, 217-221.

[86] Fisher, G. R. and Gutierrez, P. L. The reductive metabolism of diaziquone (AZQ) in the S9 fraction of MCF-7 cells: Free radical formation and NAD(P)H: Quinone-acceptor oxidoreductase (DT-diaphorase) activity. *Free Radical Biology and Medicine,* 1991,*10*, 359-370.

In: Bioactive Heterocycles
Editors: K. L. Ameta, R. P. Pawar and A. J. Domb
ISBN: 978-1-62257-636-4
© 2013 Nova Science Publishers, Inc.

Chapter 4

Bioactivity and Synthesis of Substituted Imidazole Motifs

Satish A. Dake[1], Sandeep V. Shinde[2], Sanjay K. Vyawahare[1], Rajendra P. Marathe[3], Rajesh B. Navale[3], Anjan R. Sawale[1] and Rajendra P. Pawar[1]*

[1] Department of Chemistry, Deogiri College, Station Road,Aurgangabad, India
[2] Department of Chemistry, Indira Gandhi College, Cidco, Nanded, India
[3] Department of Pharmacy, Govement College of Pharmacy, Aurgangabad, India

Abstract

Substituted imidazole is the class of heterocyclic compounds. They are biologically important due to their wide range of medicinal applications. The present chapter focuses on the novel routes of their synthesis and various biological activities shown by them.

1. Introduction

The heterocyclic imidazole ring is one of the most important structures found in a large number of natural products and pharmacologically active compounds such as anti-ulcerative agent cimetidine, the proton pump inhibitor omeprazole and the benzodiazepine antagonist flumazenil. Trifenagrel, 2,4,5-triaryl-1*H*-imidazole reduces the platelet aggregation in several animal species and humans. 2,4,5-Triaryl-1*H*-imidazole compounds have gained remarkable importance due to their wide biological activities.

These compounds also play an important role in biochemical processes [1]. Imidazole derivatives show estrogen receptor and cytotoxic inhibitors of the cyclooxygenase [2], antifungal [3], antihelmitic [4], pesticidal or herbicidal properties. Substituted imidazole possesses potentially novel therapeutic activities [5] and absolute controlled processes [6].

*E-mail: rppawar@yahoo.com and satish_dake57@yahoo.com.

In addition, they also showed analgesic [7], fungicidal [8], anti-inflammatory [9] and antithrombotic activities [10].

Imidazoles are the core structures of many biological systems *viz.*histidine, histamine and biotin, an active component in several drug molecules (e.g. Losartan, Olmesartan, Eprosartan, and Trifenagrel), peptides etc. With an increasing environmental consciousness worldwide, the chemical industry is obliged to reexamine conventional methodologies to seek ways of developing and applying more efficient and environmental strategies that meet the challenges of green chemistry for future sustainable growth. The development of dry media solid supported reactions using microwave make a significant contribution to eco-friendly synthesis. To explore the biopotential of imidazoles and environmental role of dry media synthesis under microwaves, it was thought worthwhile to study the synthetic aspects of substituted imidazoles. Further the role of different inorganic solid supports under microwaves for target synthesis has been studied. In 1980, Breslow discovered the Diels-Alder reaction performed in water could be subjected to huge rate accelerations. This observation leads to increased interest for chemists regarding organic reactions in water or aqueous media. To date, several organic transformations have been carried out in water or aqueous media.

Science is attracted to compounds belonging to the imidazole family and the development of new synthetic methodologies and heterocyclic compounds containing nitrogen and their resulting pharmacological interests.

In this chapter, the synthesis and biological activity of substituted imidazole derivatives has been discussed. This novel methodology maintains both the atom economy and the environmentally friendly approach. An optimum reaction condition for the ecofriendly synthesis of substituted Benzimidazole derivatives has been also discussed.

2. Biological Importance of Substituted Imidazole Derivatives

2.1. Cyclooxygenase-2 (COX-2) Inhibitors Activity of imidazoles

The classical non-steroidal anti-inflammatory agents are non-selective COX inhibitors that reduce the formation of physiological prostaglandins produced by COX-1. The isoform of COX, which is expressed constitutively in most tissues, includes the kidneys and gastrointestinal tract [11]. These compounds include the substituted 4,5-diaryl-1*H*-imidazoles **(1),** [12], Fenflumizole **(2),(5)** [12, 13], Triflumizole **(6)** [13], Flumizole **(3)** [14]. The most active compound **11a** was found inside the COX-2 active site [15].

Generally, the COX-2 enzyme is not detected in normal tissues, but it is induced by pro-inflammatory cytokines IL-1 and TNF-α and results in enhanced synthesis of prostaglandins in neoplastic and inflamed tissues [16, 17]. Thus, selective COX-2 inhibitors have developed lesser gastrointestinal side effects compared to conventional non-steroidal anti-inflammatory drugs.

In 1990 the first compound, DUP-697 **(7),** [18] with a clear COX-2 specificity was developed and served for the development of new COX-2 inhibitors such as rofecoxib **(8),**

[19] celecoxib **(9),** [20] and 4-[4-chloro-5-(3-fluoro-4-methoxy-phenyl)imidazol-1-yl]benzene sulfonamide (cimicoxib) **(10)**.

Compound (**10**) was recognized as a highly selective COX-2 inhibitor showing high potential in all inflammation tests, together with excellent pharmacokinetics activity [21]. Recently, one more 1,5-diaryl-1*H*-imidazole, compound (**11**) has been shown to have strong inhibitory activity on COX-2-catalyzed PGE2 production [22]. Compound 4,5-diaryl-1*H*-imidazole (**4**) was also found to be a COX-2 inhibitor. It is 6750-fold more selective against COX-1; it is somewhat less active then celecoxib **(9)** (IC50 690 *vs* 28 nM) [23].

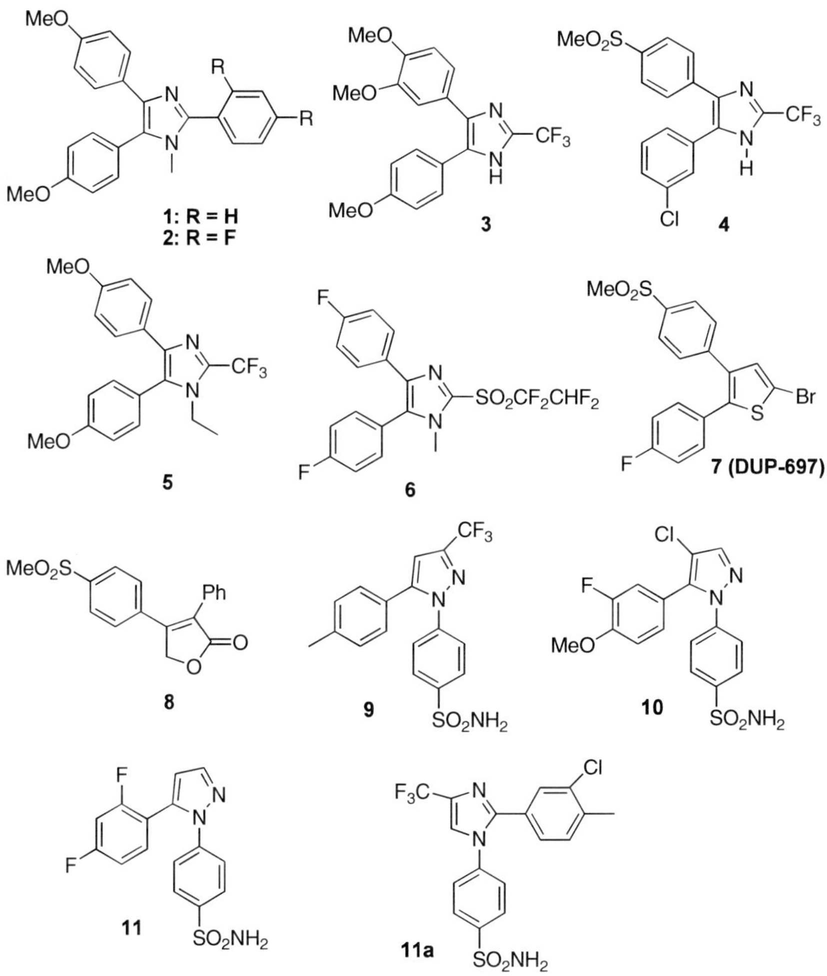

2.2. Inhibitors of p38 MAP Kinase Activity of imidazoles

The mitogen-activated protein (MAP) kinases are of ubiquitously distributed enzymes, able to mediate intracellular signal transduction and participate in the number of physiological as well as pathophysiological cellular processes including cell growth, differentiation and apoptosis [24, 25, 26]. The mammalian MAP kinase family member that includes ERK1,

ERK2, ERK3α, ERK3β, JNK1, JNK2, JNK3, p38α, p38β, p38γ, p38δ and ERK5 share sequence similarity and conserved structural domains [27, 28].

The extracellular-signal regulated kinases (ERKs) are activated by growth and mitogen factors *via* a Ras-dependent pathway [28, 29]. In difference, Jun N-terminal kinases (JNKs) and p38 kinases are activated in response to the pro-inflammatory cytokines interleukin-1 (IL-1) and tumor necrosis factor a (TNF-α) and by environmental stress including UV light, shock, toxins, heat and cellular injury [30]. The human p38aMAPkinase was originally identified as the molecular target of the pyridinylimidazole class of compounds to inhibit the biosynthesis of inflammatory cytokines such as IL-1 and TNF in lipopolysaccharide (LPS)-stimulated human monocytes. At present, the pyridinyl-imidazole drugs also inhibit p38b and that the p38 kinases, p38g and p38d, are insensitive to these drugs [31]. Nevertheless, during the last 10 years a number of novel chemotypes of p38 kinase inhibitors, including a variety of 4,5-diaryl and 2,4,5-triaryl-1*H*-imidazoles, have been discovered *via* high-throughput screening [32, 33]. In particular, substituted 4,5-diaryl-1*H*-imidazoles in which an aryl substituent is a pyridine group or a pyrimidine group, the second aryl substituent is a 4-fluorophenyl group formed an important class of inhibitors, strongly pursued by number of pharmaceutical companies and research institutes as anti-inflammatory drugs [33, 34, 35, 36].

The number of examples of substituted 5(4)-aryl-4(5)-(4-fluorophenyl)-1*H*-imidazoles are p38 MAP kinase inhibitors and are also known RK or HOG1,CSBP inhibitors.

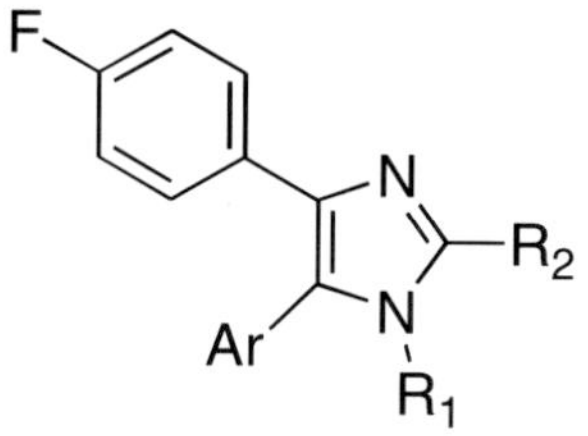

The compounds (**12**) (SB-210313) [34-36], 5(4)-Aryl-4(5)-(4-fluorophenyl)-1*H*-imidazole derivatives are p38 mitogen-activated protein kinase inhibitors (**13**) (SB-203580) [37, 38-41], (**14**) (SB-202190) [42], (**15**) (RPR-200765A), [35] and (**16**) (RWJ-67657) [43, 44], are examples of pyridinyl-imidazoles in which the 4-pyridinyl-substituent was considered as an essential element in the early p38 inhibitors [24]. Indeed, the strong hydrogen bonding between the p38 amide-NH of Met [45] and the pyridine nitrogen is a key determinant of the binding affinity common to all pyridinyl-imidazoles p38 MAP kinase inhibitors [24]. On the other hand, these substances also potently inhibit human hepatic cytochrome P-450 enzymes inhibitors known for drug-drug interactions or lead to hepatic changes such as P-450 enzyme induction [35, 46].

Potent *invitro* inhibitory activity of these substances for *cytochrome* P-450 isoenzymes is due to the fact that both the imidazole ring and pyridine ring are ligands for the heme iron of cytochrome P-450 [47]. Their consequent severe liver toxicity prevented the clinical development of these first-generation p38 inhibitors. Thus, significant modifications in their structure were introduced with the successful replacement of the 4-pyridyl moiety in related heterocycles. In particular, replacement of the pyridine ring by a pyrimidine moiety in a number of series of pyridinyl-imidazoles has been an active area in the search for improved p38 inhibitors, which has frequently resulted in analogues, such RPR-238677 [48] and RPR-

203494 [49], with minimal inhibition of cytochrome P-450. Compound Pyrimidinyl-imidazole SB-24235 has been reported to have entered phase I clinical trials [24].

	Ar	R_1	R_2
12 (SB-210313)	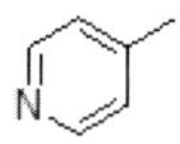	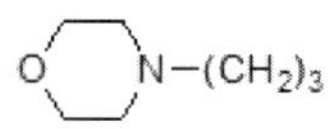	-H
13 (SB-203580)	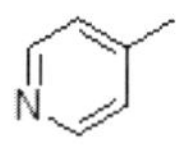	-H	4-$MeSOC_6H_4$
14 (SB-202190)	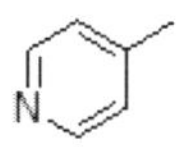	-H	4-HOC_6H_4
15 (RPR- 200765A)	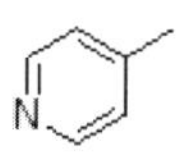	-H	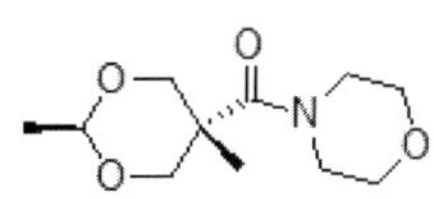
16 (RWJ-67657)	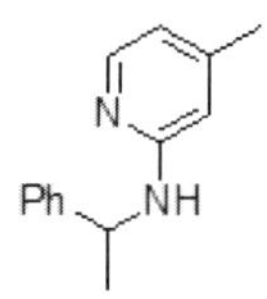	-Me	-SMe

It is important to note that p38 MAP kinase inhibitors imidazole derivatives, without a 4-fluorophenyl group, have also been described [26, 45].

2.3. Combretastatin A-4 (CA-4) Analogues with Antitumor Activity

Tubulin is a globular protein which makes up microtubules and is a major target for anti-cancer drug design [50]. Natural compounds including colchicine, combretastatin, podophyllotoxin and steganacine inhibit Tubulin polymerization by binding at a common site, the colchicines binding site [51]. Combretastatins are natural antimitotic agents isolated from the bark of the South African Combretumcaffrumtree [52-55], show antitumor activity for targeting the vasculature of tumors [56, 57]. In fact, these substances are able to cause pronounced shut down in blood flow to solid tumors, resulting in extensive tumor-cell necrosis, while leaving the blood flow in normal tissues relatively intact [58]. Among these natural products, combretastatin A-4 (CA-4) **(17)** possesses the most effective antitumor activity. It shows strong cytotoxicity against a variety of cell lines, including multidrug resistant cancer cell lines [59, 60]. Most prominently they have demonstrated potent cancer anti-vascular properties [61]. However, the low water solubility of CA-4 limits its efficacy *in vivo* and the water-soluble disodium phosphate pro-drug CA-4P (**18**) has already been entered in clinical trials [62]. In endothelial cells in culture, CA-4P causes rapid re-organization of the actin cytoskeleton, mediated by disruption of the tubulin cytoskeleton. An increase in vascular permeability is likely to be an important component of the mechanisms that lead to the shutdown of tumor blood flow and other vascular-disrupting agents (VDAs) structurally related to CA-4P [58]. The VDAs were previously known as vascular-targeting agents (VTAs).

Recently, it has been found that CA-4P selectively targets endothelial cells, but not smooth muscle cells and induces regression of unstable nascent tumor neovessels by disrupting the molecular engagement of endothelial cell-specific junctional molecular vascular endothelial-cadherin (VE-cadherin) *in vitro* and *in vivo* in mice [63]. Compound (**19**) a synthetic analogue of CA-4 with different substitution on ring B, is currently under clinical evaluation as a tumor vascular targeting agent [64, 66]. This compound is the pro-drug of AC-7739 (**20**).

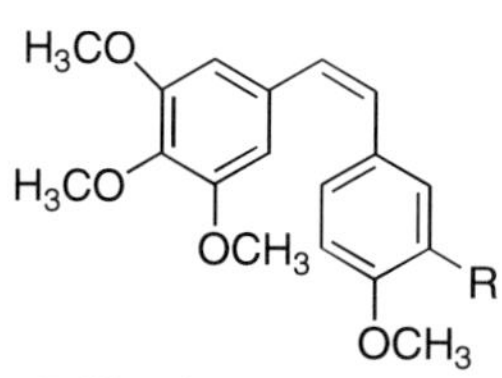

17 (CA-4) : R = -OH
18 (CA-4P) : R = -OH
19 (AVE-8062) : R = NHSerHCl
20 (AC-7739) : R = -NH_2HCl

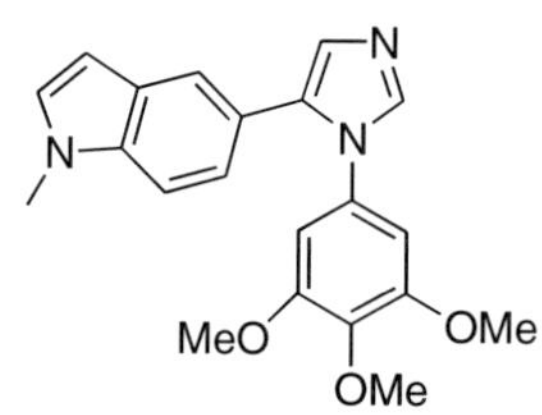

21

2.4. Modulators of P-glycoprotein (P-gp)-mediated Multidrug Resistance (MDR)

Modulators of P-glycoprotein (P-gp)-mediated a multidrug resistance (MDR), which is now recognized as one of the causes of cancer chemotherapy failure is due to an over-expression of the plasmamembrane P-gp molecule, a protein that functions as an ATP-dependent pump to efflux of diverse anticancer drugs from MDR cells [68]. The expression level of P-gp correlates with the degree of resistance [69]. Hundred of compounds have been selected by different approaches with the properties of inhibiting P gp [70].

In 2000, 2-aryl-4,5-(4-dimethylamino)phenyl-1*H*-imidazole (**22**) were well-known as potent non-toxic modulators of P-gp mediated MDR [71]. These compounds were optimized *via* structure-activity relationship (SAR) studies [72] and optimized imidazole OC-144-093 (**23**), was generated using combinatorial chemistry. It was shown to be able to reverse MDR to paclitaxel, doxorubicin and vinblastine in human lymphoma, uterine, breast, ovarian and colorectal carcinoma cell lines expressing P-gp [73].

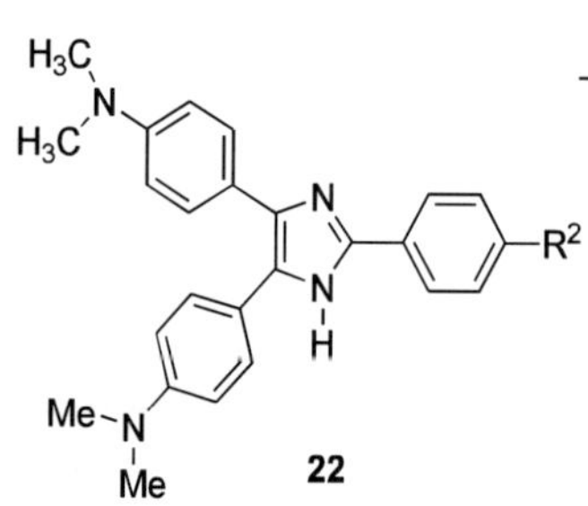

22

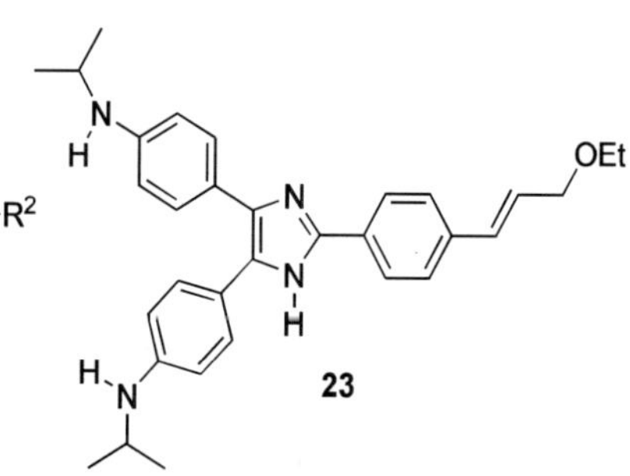

23

Chen *et.al* in 2004 reported three imidazole derivative compounds (**24, 25**) and (**26**) possessing a 3 to 4 fold stronger reversal of MDR activity than verapamil; a well-known positive MDR modulator [74]. These people also demonstrated that [62] (FG-020318) is a highly potent, efficacious MDR modulator, not only *in vitro,* but also *in vivo* [75].

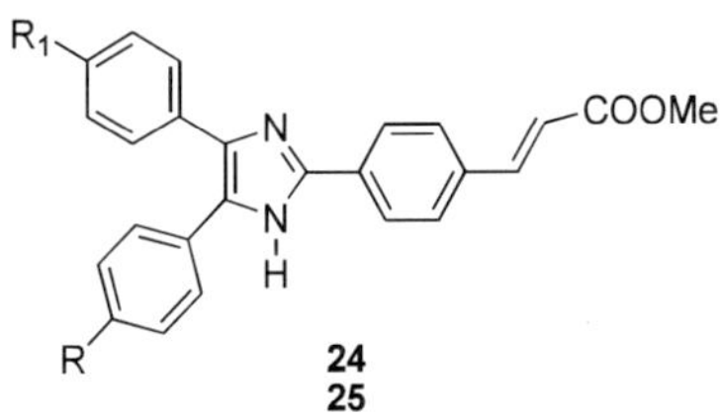

24
25

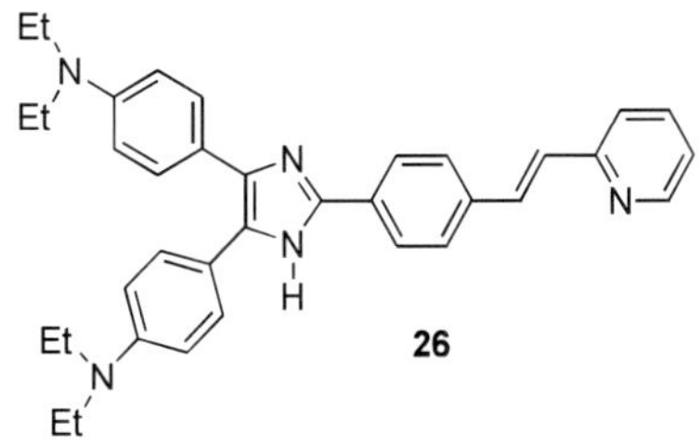

26

2.5. GSK3 Inhibitors Activities for Substituted imidazoles [76]

Synthesis, design and the GSK3 inhibitory activities of 7-hydroxy benzimidazole analogs are discussed. The solid-phase synthetic route was developed for the preparation of analogs consisting of novel ATP competitive scaffold. In this study, the structure-activity relationship (SAR) of 7-hydroxy benzimidazole analogs and their biological activities were reported. Some of the examples of GSK3 inhibitors are Indirubin (**27**), Staurosporine (**28**), Tideglusib (NP-12) (**29**), AR-AO014418 (**30**) as follows.

Examples of GSK3 inhibitors

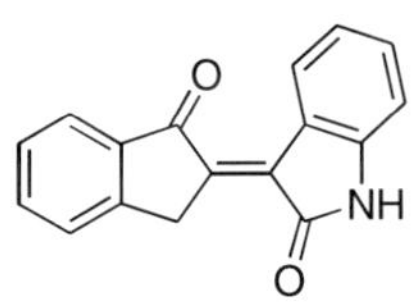

27: Indirubin

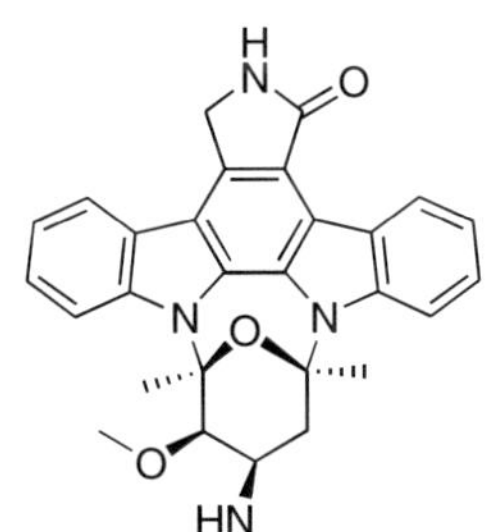

28: Staurosporine

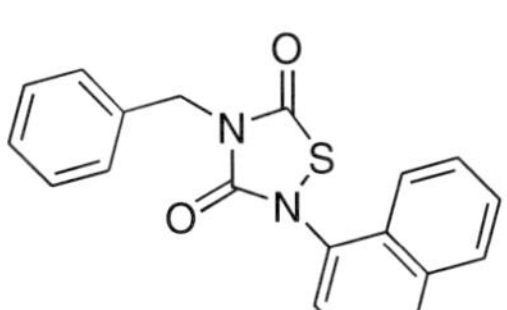

29: Tideglusib (NP-12)

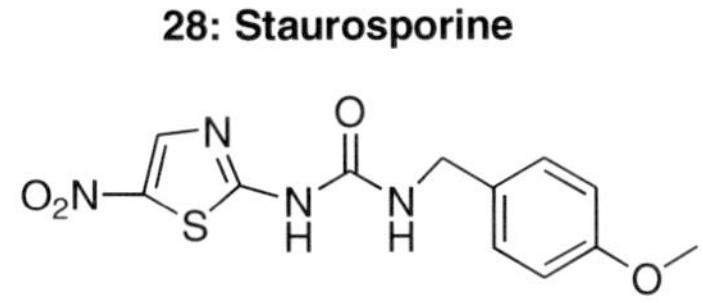

30: AR-AO014418

2.6. Anti-inflammatory Activity for Substituted imidazole Derivatives

Lombardino, J. G. *et al.* [77] discussed a number of trisubstituted imidazoles have been synthesized and found to be potent anti-inflammatory agents, when examined in the carrageenan rat paw edema test. An anti-inflammatory activity of these imidazoles is retained inadrenalectomized rats. One compound 4,5-bis(4-methoxyphenyl)-2-trifluoromethyl imidazole (eflumizole), is more potent than indomethacin in the rat pawedema test. In contrast to some related compounds, it does not cause photosensitization in mice.

2.7. Antimicrobial Activity of Substituted imidazoles i.e. Antibacterial & Antifungal Activity

Rogers, S. A. *et al.* [78] described the 2-aminoimidazole and menthylcarbamate classes of anti-biofilm agents **(31)** resulting in the development of a novel hybrid scaffold of biofilm modulators. These compounds were evaluated against bacterial strains for anti-biofilm and anti-microbial activity.

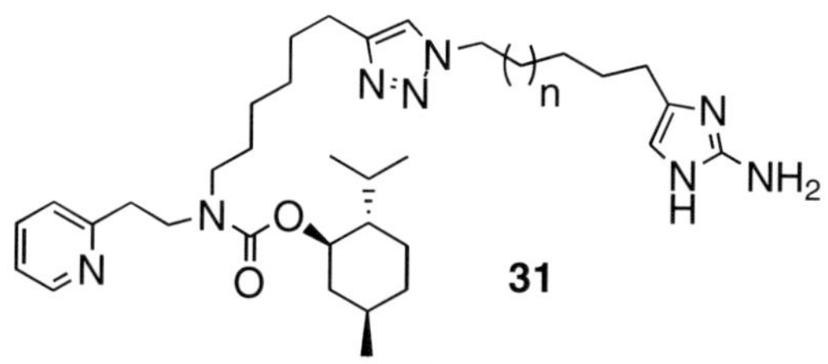

31

Soni, B. *et al.* [79] reported the series of 5-[2-(1,3-benzothiazol-2-yl-amino) ethyl]-4-(arylideneamino)-3-mercapto-(4*H*)-1,2,4-triazoles **(32)** as antimicrobial agents.

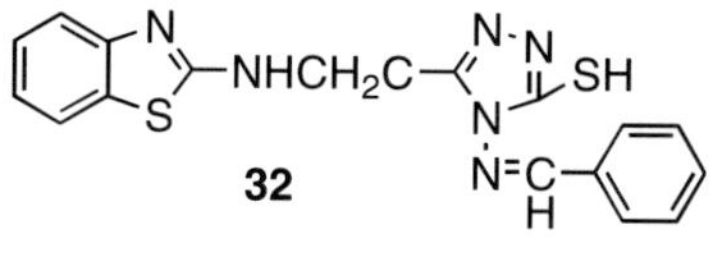

32

Ansari, K. F. *et al.* [80] showed good activity of 2-substituted-1-[{(5-substituted alkyl/aryl)-1,3,4-oxadiazol-2-yl} methyl]-1*H*-benzimidazole derivatives towards Gram-positive bacteria and negligible activity towards Gram-negative bacteria.

Khan, M. S. *et al.* [81] reported activity ofsubstituted2,4,5-trisubstituted imidazoles **(33)** against eight different human pathogenic bacteria and fungi. These compounds showed a variation in activity and were reported active against Gram-positive as well as Gram-negative bacteria.

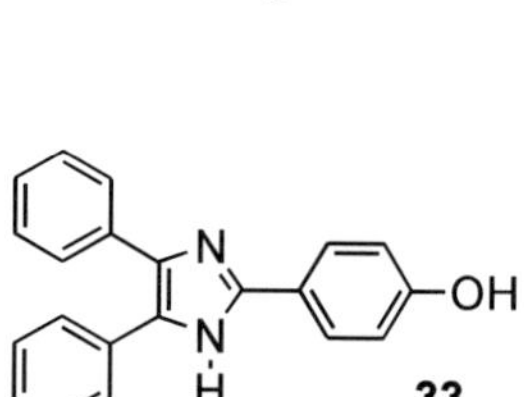

33

4-(4,5-Diphenyl-1H-imidazol-2-yl)-phenol.

Bandyopadhyay, P. *et al.* [82] in these study deals with *in vitro* time point quantitative antibacterial evaluation of newly synthesized substituted 1,2-disubstituted benzimidazole and 2-substituted benzothaizole derivatives against Gram-positive bacteria *Staphylococcus aureus, Bacillus cereus* and Gram-negative bacteria *Vibrio cholerae, Shigelladysenteriae and*

Escherichia coli. Bio-evaluation studies revealed that some of these compounds exhibited moderate to good antibacterial activity against all the tested bacterial stains. Some of them shown enhanced inhibitory activity as compared to drug ciprofloxacin against *V. cholerae, B. cereus, and S. dysenteriae.* Additionally, the compounds (**34**) displayed complete bactericidal activity within 24 hrs, where ciprofloxacin took 48 hrs to kill those bacteria completely.

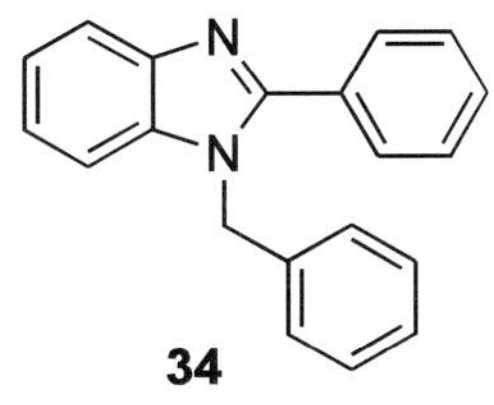

34

An antibacterial and antifungal activity of benzimidazole compounds **(35),** using a structure-activity relationship (SAR), study was reported by**Ozkay, Y.*et al.* [83]**.

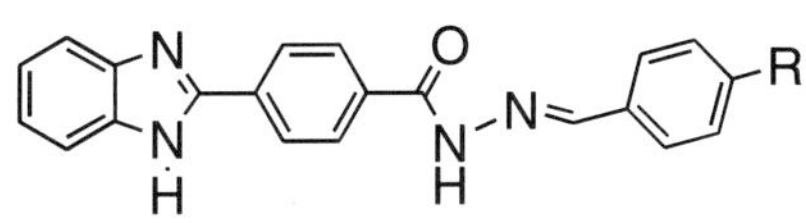

35

Canan, K. *et al* [84] described methyl 5(6)-fluoro-6(5)-substituted-1H-benzimidazole carbamate **(36, 37)** derivatives with antifungal activities against *Candida albicans.*

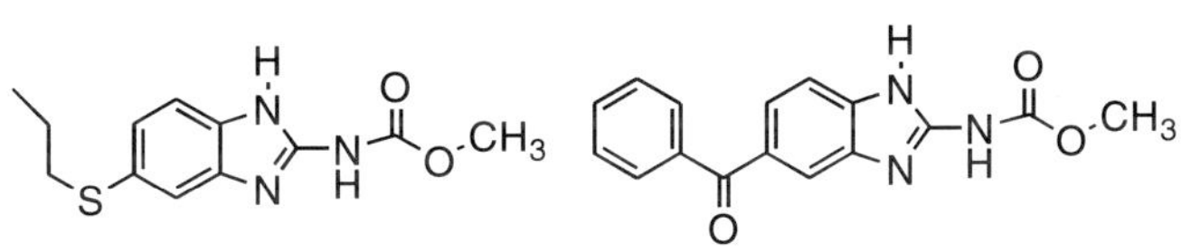

36 **37**

2.8. Other Biological Activity of Substituted imidazoles

Shen, H. C. *et al.* [85] reported benzimidazole pyrrolidinyl amides (**38**) containing a piperidinyl group which act as prolylcarboxypeptidase (PrCP) inhibitors and studied *in vivo* for its effect on weight loss and food intake in an eDIO mouse model.

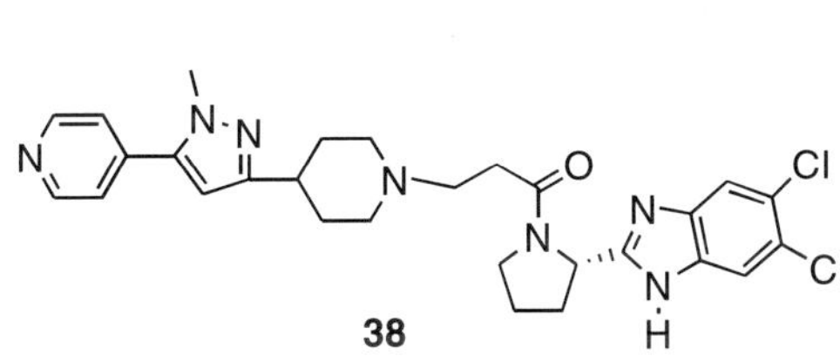

38

Kwak, Y. *et al.* [86] reported 2-[(2,6)-dimethylphenyl]benzimidazole **(39, 40, 41)** are strong potential analogs for mutagenicity following metabolic activation in either TA98 or TA100 *Salmonella typhimurium* strains.

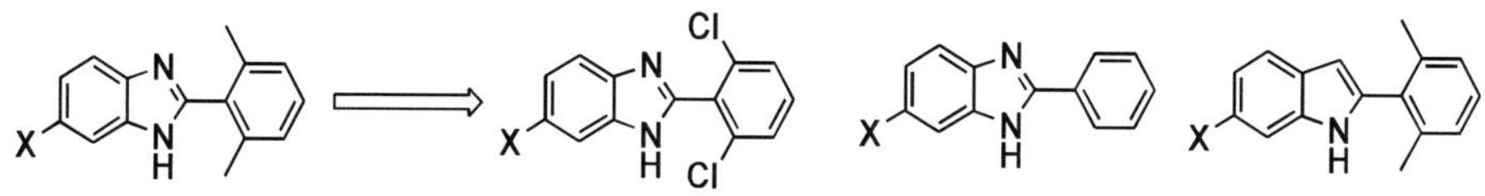

Ames positive **Ames negative** **Ames negative** **Ames negative**

39 **40** **41**

Econazole-derived nitroimidazoles **(42)** as anti-tubercular agents are reported by Lee, S. H. *et al.* [87]

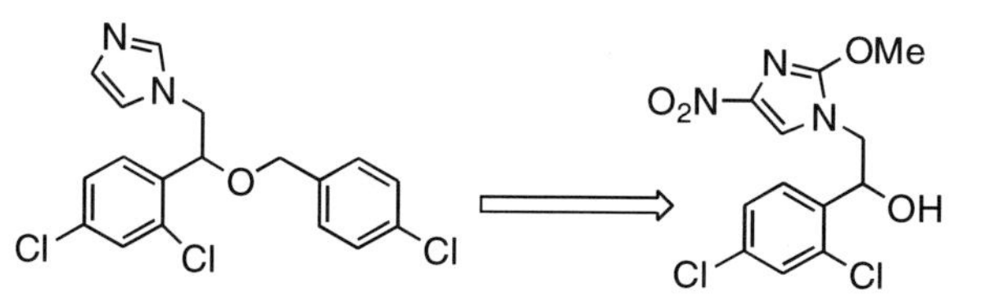

econazole
MIC 16μg/mL

42

MIC 0.5 μg/mL

Hans G. F. *et al.* [88] reported 4-{(*S*)-2-[2-(4-chloro-phenyl)-5,6-difluoro-benzo imidazole-1-yl]-2-cyclohexyl-acetylamino}-3-fluorobenzoic acid **(43)** as a highly potent and selective FXR agonist with excellent physicochemical, ADME properties and potent lipid lowering activity after oral administration to LDL receptor deficient mice.

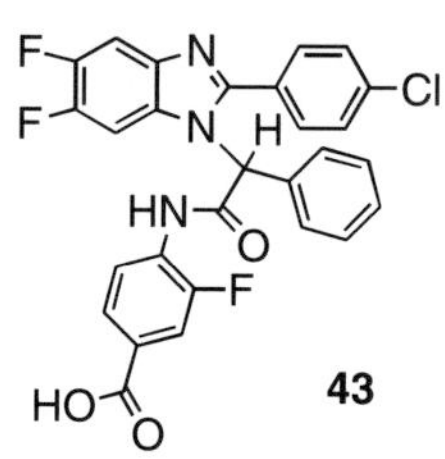

43

Wiglenda, T. *et al.* [89] described 1*H*-imidazoles as biologically active *in vitro*. The introduction of an ethyl chain at C_5 of the 1,2,4-tris(4-hydroxyphenyl)-1*H*-imidazole **(44)** caused hormonal activity in estrogen receptor positive MCF-7-2a cells.

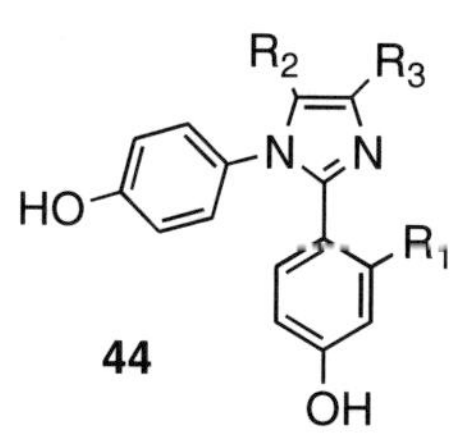

44

3. Methods of Synthesis for Substituted Imidazoles

Recently, several methods were reported in the literature for the synthesis of substituted 2,4,5-triaryl-1*H*-imidazoles using different catalyst such as,

Dake, S. A. *et al.* approach (2012) [90]: One-pot, three-component condensation of benzil/benzoin, substituted aromatic aldehydes, and ammonium acetate in an ethanol-water (1:1, *v/v*) solvent system using a sulfated tin oxide catalyst under reflux condition afforded corresponding 2,4,5-triaryl-1*H*-imidazoles with excellent yield.

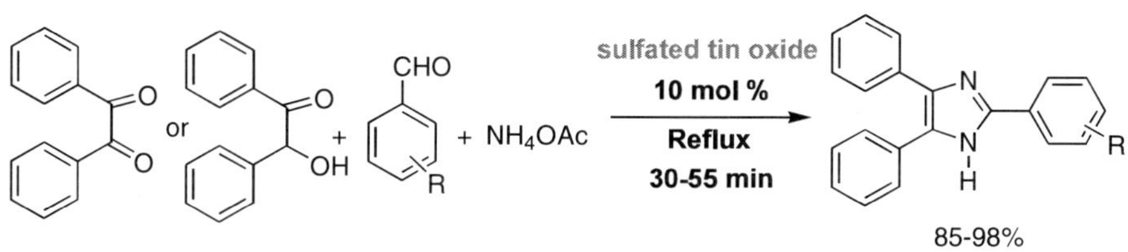

Padhy, A. K. *et al.* approach (2010) [91]: Hydroxy-2,4,5-trisubstituted imidazoles were synthesized starting from 1,2-diketones using zinc oxide nano particles. An unusual intermolecular hydrogen bonding through the association of water molecule has been reported.

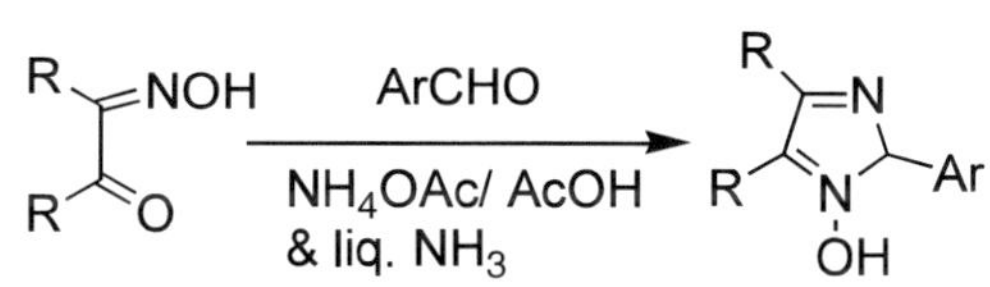

Kong, L. *et al.* approach (2010) [92]: The synthesis of 2,4,5-trisubstituted imidazoles has been reported using a continuous flow micro reactor system under pressure; aryl- alkyl- and hetero aryl substituted imidazoles has been under superheating conditions.

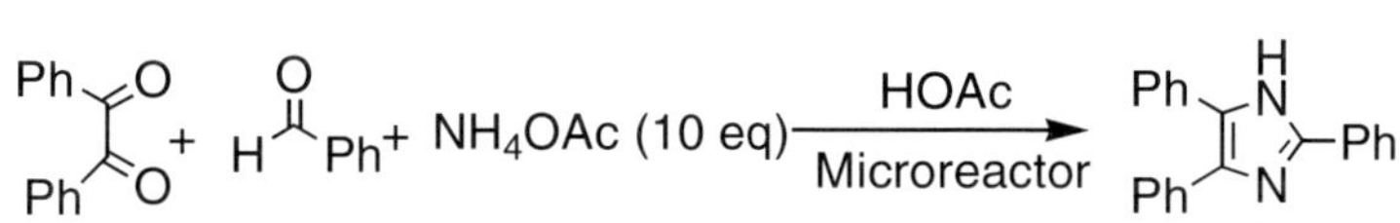

Hojati, S. F. *et al.* approach (2010) [93]: Synthesis of benzoxazoles, benzimidazoles, and oxazolo-[4,5-b]-pyridines were reported by the condensation of orthoesters with o-substituted anilines or 2-amino-3-hydroxypyridine in the presence of catalytic amount of 1,3-dibromo-5,5-dimethylhydantoin (DBDMH) under solvent-free conditions.

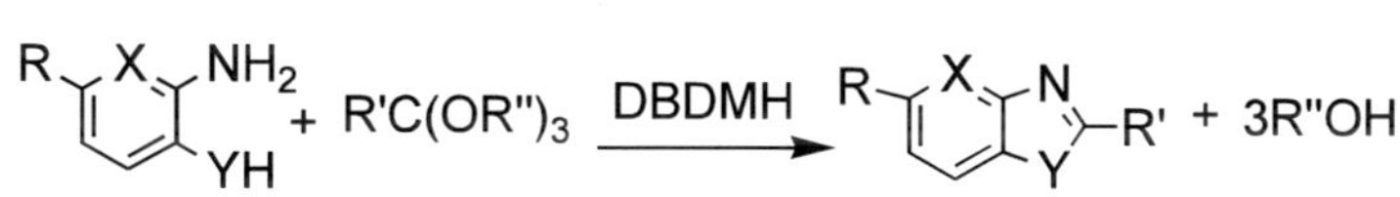

Safari, J. *et al.* approach (2010) [94a]: $(NH_4)_6Mo_7O_{24}\cdot 4H_2O$ was used as a catalyst for the synthesis of 2,4,5-trisubstituted imidazoles by a three-component, one-pot condensation of benzil, aryl aldehydes and ammonium acetate under solvent-free conditions using microwave irradiation.

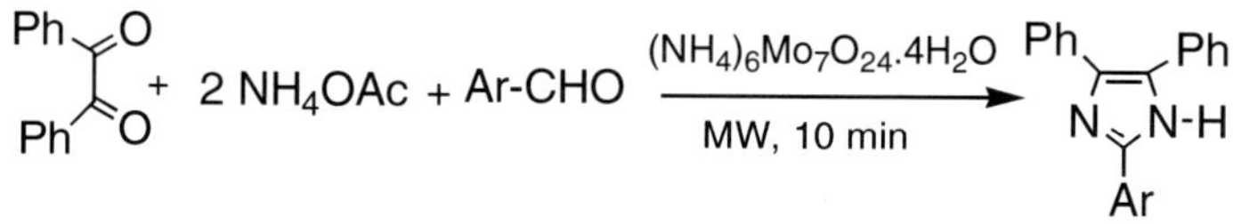

Nanocrystalline magnesium oxide **[94b]:** with high specific surface area has been used as a catalyst for the synthesis of biologically active 2,4,5-trisubstituted imidazoles, by condensation of 1,2-diketones and aryl aldehydes, under solvent-free and conventional heating conditions.

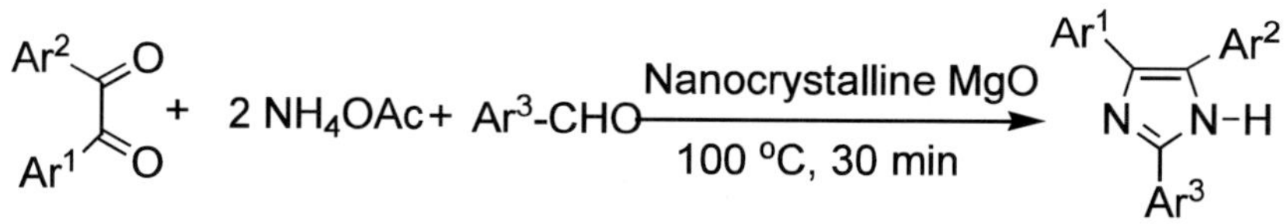

Nalage, S. V. *et al.* approach (2010) [95]: reported the synthesis of 2, 4, 6-triaryl-1*H*-imidazole in polyethylene glycol under microwave irradiation.

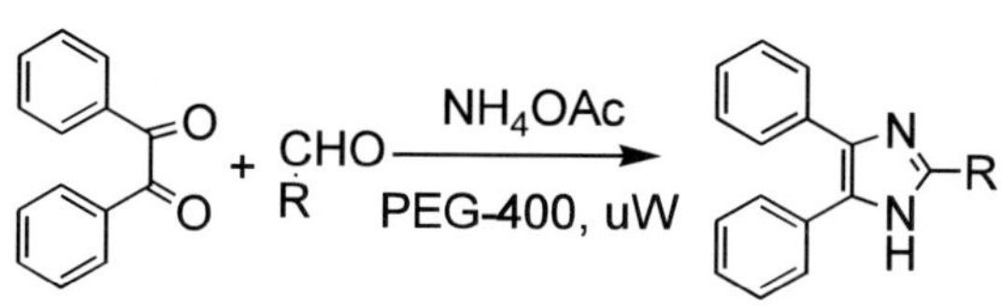

Parveena, A. *et al.* approach (2010) [96]: Synthesis of 2, 4, 5-triaryl imidazole was carried out at 75 ^{0}C using aromatic aldehyde, benzil or benzoin in the presence of a catalytic amount of ceric ammonium nitrate.

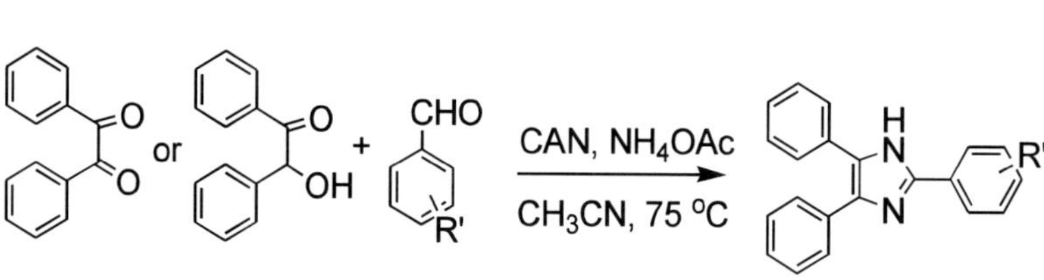

Tripathy, H. *et al.* approach (2010) [97]: Tri substituted imidazoles were synthesized using a synthetic microwave oven, which showed significant reduction in reaction time, increased yield and synthesis of a library of compounds. In a very short time the compounds were screened for anti-inflammatory activity by rat paw edema method and they showed good activity when given orally.

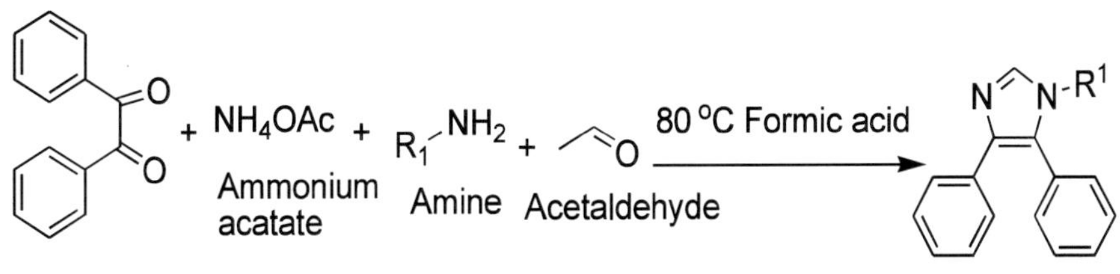

Joshi, R. S. *et al.* approach (2010) [98]: The synthesis of benzimidazole was carried out by the reaction of the substituted aldehyde with *o*-phenylenediamine in water under ultrasonic irradiation at ambient temperature using 5 mol% of TBAF.

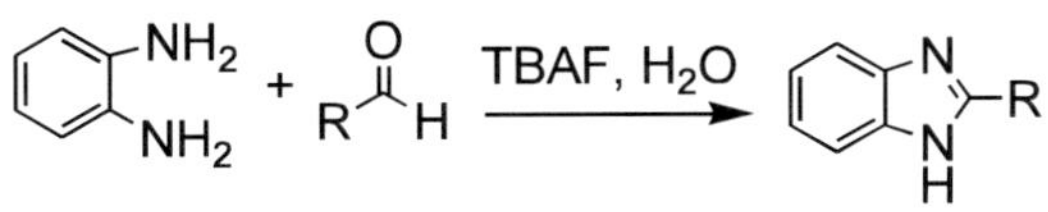

Hasaninejad, A. *et al.* approach (2010) [99]: Catalyst-free one-pot four component methodology for the synthesis of 1,2,4,5-substituted imidazoles under conventional heating and microwave irradiation using 1-butyl-3-methylimidazolium bromide, [Bmim]Br were reported.

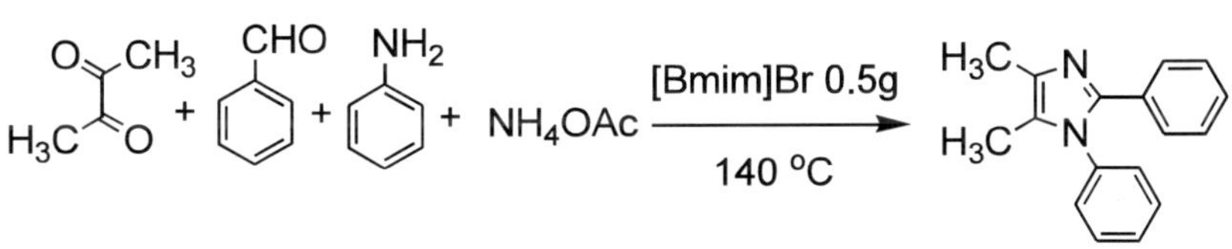

Samai, S. *et al.* approach (2009) [100]: Synthesis of 2,4,5-trisubstituted imidazoles by three component cyclocondensation of 1,2-dicarbonyl compound, aldehyde and ammonium acetate using *L*-proline as a catalyst in methanol at moderate temperature.

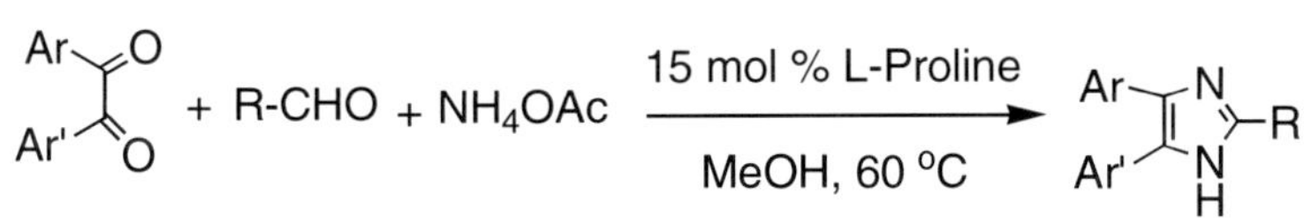

Bratulescu's approach (2009) [101]: Synthesis of 4,5-disubstituted imidazoles was carried out using 1,2-diketones and urotropine in the presence of ammonium acetate, under solvent free microwave irradiation.

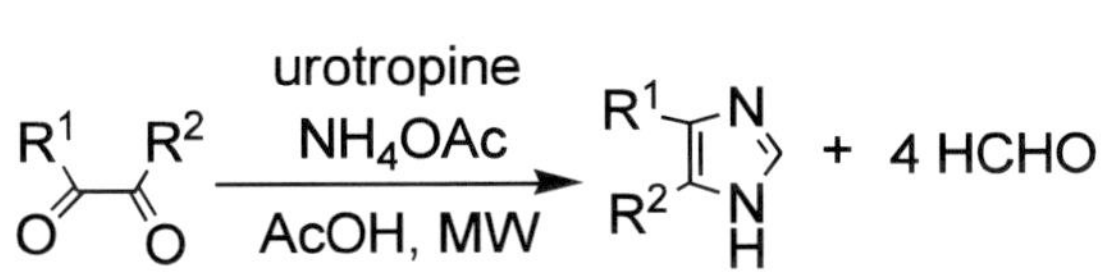

Pastor, I. *et al.* approach (2009) [102]: Different aspects of imidazole derivatives with one nitrogen atom bearing a substituted or unsubstituted phenyl moiety, such as their presence in bioactive compounds, their activity, as well as the main strategies for their preparation has been reported.

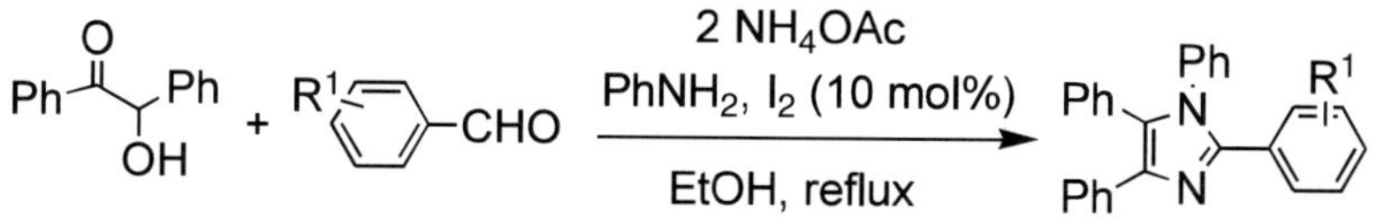

Rathod S. B.*et al.* approach (2009) [103]: CeO_2-MgO-ZrO_2 (CMZO) solid heterogeneous catalyst has been used for the synthesis of 2,4,5-triarylimidazole derivatives by the condensation of benzoin, aromatic aldehydes and ammonium acetate in EtOH:H_2O at 70 ^{O}C.

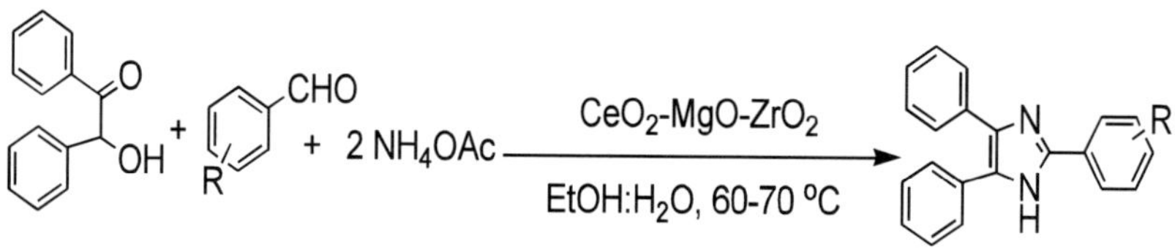

Wang, L. *et al.* approach (2009) [104]: A polymer-supported Lewis acid catalyst was used for the preparation of imidazole derivatives *via* a three-component condensation of benzil, aldehydes and ammonium acetate.

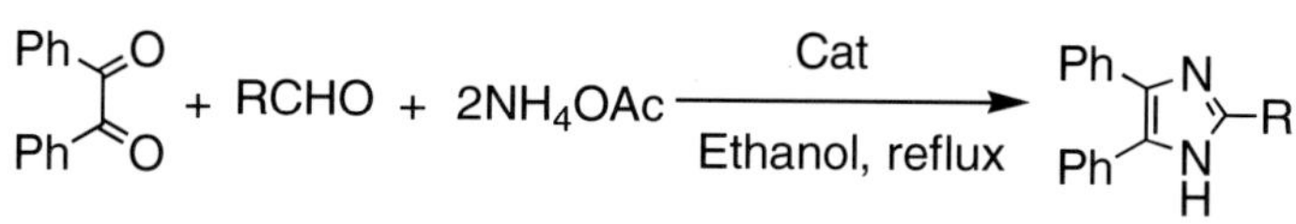

Rostamizadeh, S. *et al.* approach (2009) [105]: Substituted 2-aryl benzimidazoles were prepared in water solvent using $ZrOCl_2.nH_2O$ supported on montmorillonite K-10 as an efficient water tolerating Lewis acid.

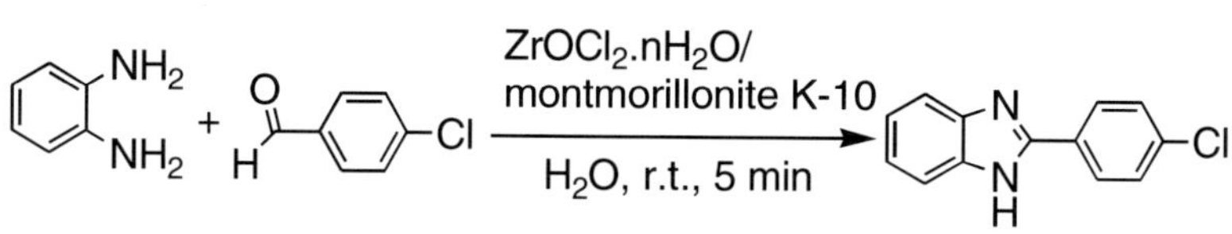

Gadekara, L. S. *al.* approach (2009) [106]: Natural scolecite has been found as a catalyst for one-pot synthesis of 2,4,5-triarylimidazole derivatives *via* three component reaction using benzil or benzoin, aldehydes and ammonium acetate.

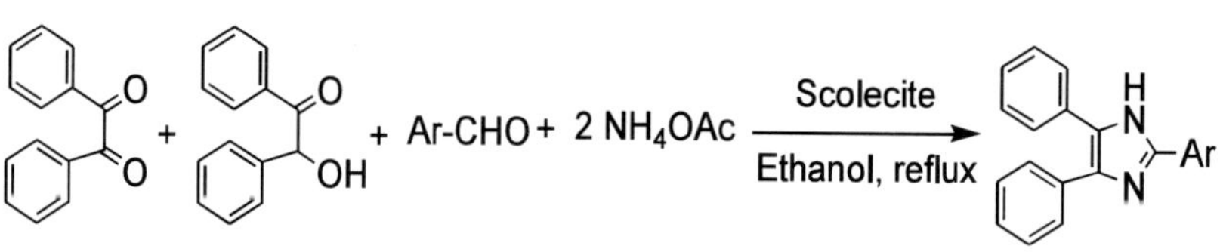

Shelke, K. F. *et al.* approach (2009) [107]: Boric acid (BO_3H_3) as a catalyst was reported for the synthesis of 2,4,5-triaryl-1*H*-imidazoles by a three component condensation of benzil/benzoin, an aldehydes and ammonium acetate in aqueous media under ultrasound at room temperature.

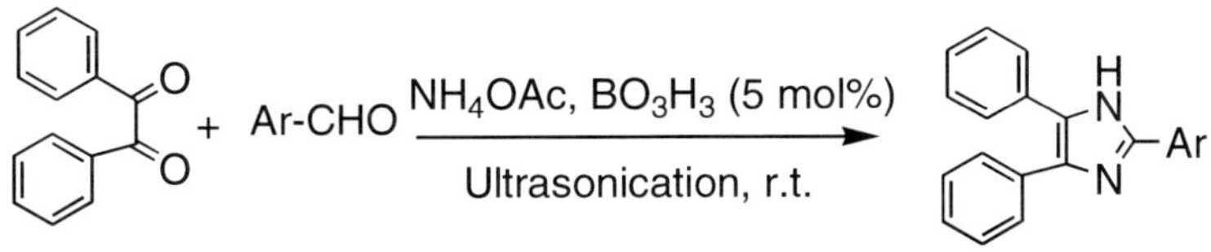

Shitole, N. V. *et al.* approach (2009) [108]: L-Proline has been found to be an organocatalyst used for one-pot synthesis of 2,4,5-triaryl substituted imidazole by the reaction of an aldehyde, benzil and ammonium acetate.

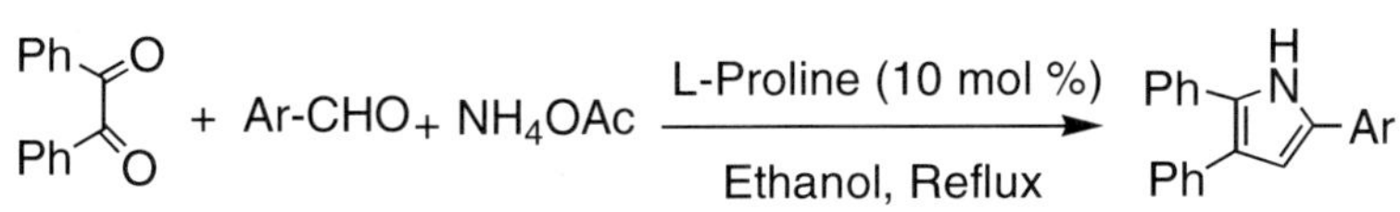

Ghorbani-Vaghei, R. *et al.* approach (2009) [109]: Poly(N, N-dibromo-N-ethyl-benzene-1, 3-disulfonamide) (PBBS) and N, N, N, N-tetrabromobenzene-1,3-disulfonamide (TBBDA) were used as a catalysts for the one-pot synthesis of 2-aryl-1-arylmethyl-1*H*-1,3-benzimidazoles.

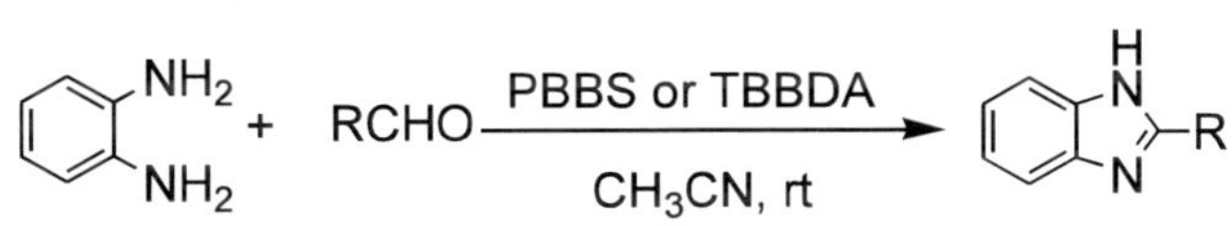

Rafiee, E. *et al.* approach (2009) [110]: A series of 12-phosphotungstic acid (PWA) supported on various porous carriers, such as silica, alumina, titania, clay, and carbon were prepared and their catalytic performance was reported in the synthesis of imidazoles under solvent-free condition.

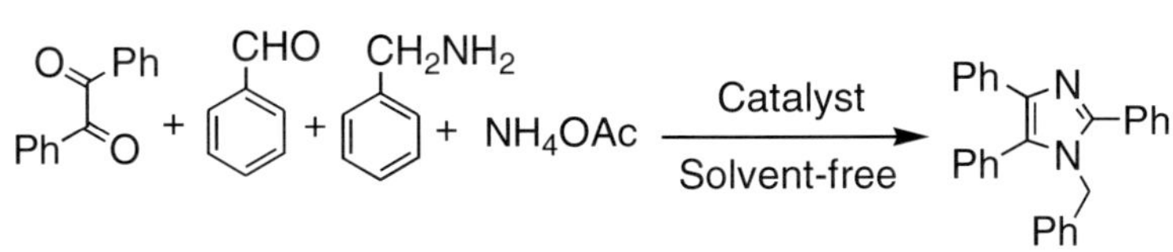

Karimi A. R. *et al.* approach (2009) [111]: Heteropolyacid $H_6P_2W_{18}O_{62} \cdot 24H_2O$ (WD) supported on silica (WD/SiO_2) has been used as a catalytic system for the synthesis of various 1,2,4,5-tetrasubstituted imidazoles by four component condensation of benzil, aldehydes, amines and ammonium acetate under solvent-free conditions.

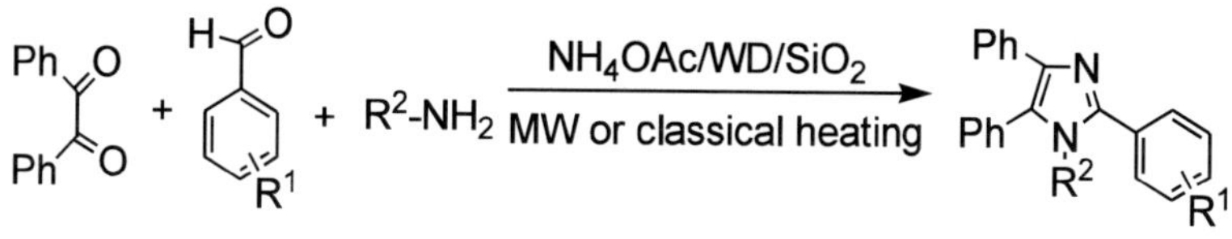

Sangshetti, J. N. *et al.* approach (2008) [112]: 2,4,5-triaryl-1*H*-imidazoles were synthesized from benzoin or benzil, ammonium acetate, aromatic aldehydes and $ZrOCl_2.8H_2O$ catalyst.

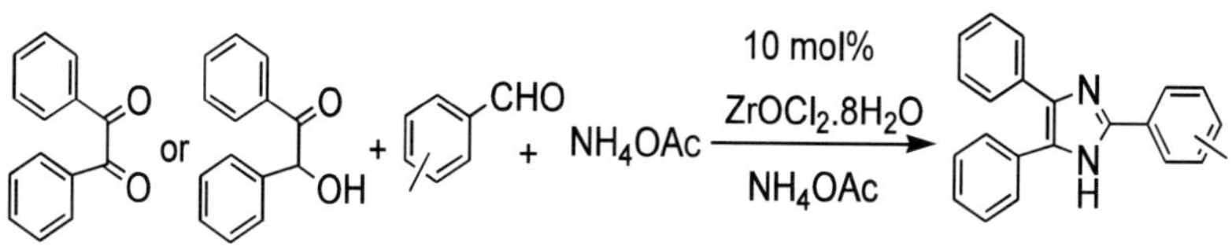

Khan, M. S. *et al.* approach (2008) [113]: Synthesized of 2,4,5-trisubstituted imidazoles from aryl aldehydes and 1,2-diketones or α-hydroxyketone using $[Hbim]BF_4$

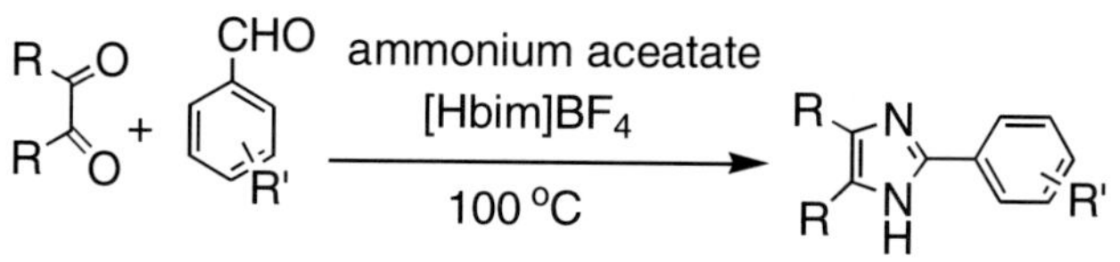

Khosropour's approach (2008) [114]: Aromatic, heteroaromatic, aliphatic compounds and unsaturated aldehydes react with benzil and ammonium acetate in the presence of room temperature ionic liquid to give the imidazoles.

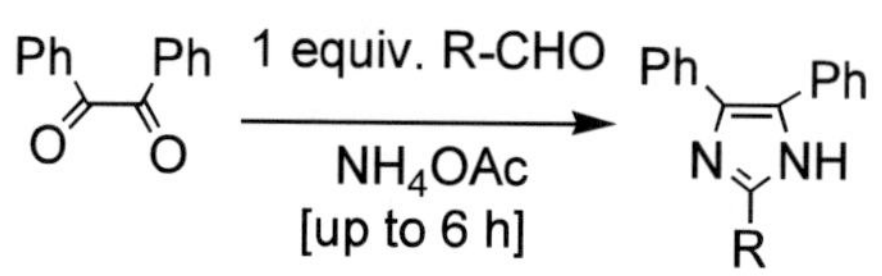

Shelke, K. *et al.* approach (2008) [115]: The condensation of benzil/benzoin, aldehydes and ammonium acetate under microwave irradiation in the presence of Glyoxylic acid (CHOCOOH) in a solvent-free condition has been reported for synthesis of 2,4,5-triarylimidazole derivatives.

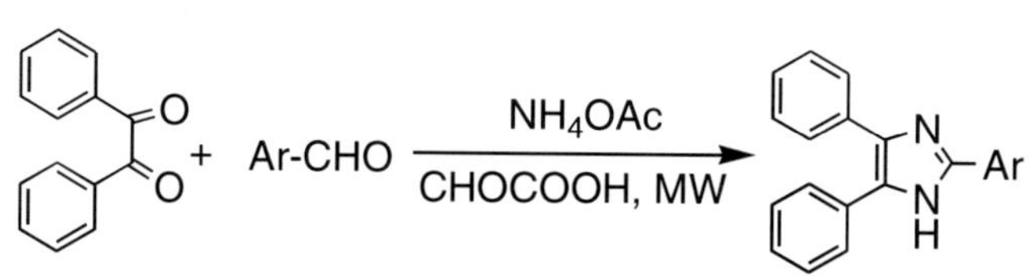

Sangshetti, J. N. *et al.* approach (2008) [116]: Ceric ammonium nitrate (CAN) was used as a catalyst for the synthesis of 2,4,5-triaryl-1*H*-imidazoles *via* condensation of benzoin/benzil, ammonium acetate and aromatic aldehydes.

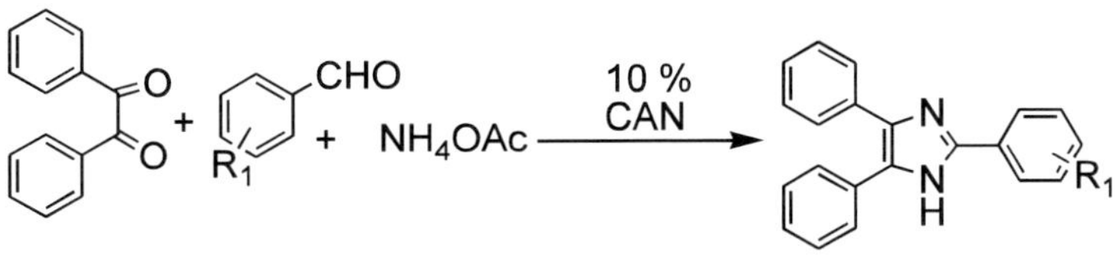

Mohammed, A. F. *et al.* approach (2007) [117]: Synthesis of 2,4,5-triarylimidazoles from benzoin or benzil, ammonium acetate and aromatic aldehydes using a sulphanilic acid catalyst were described.

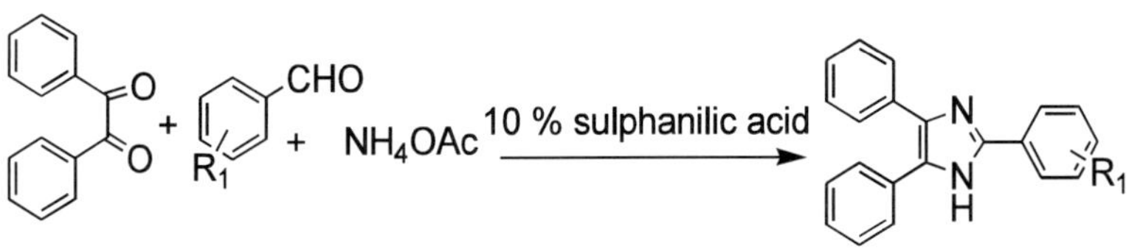

Parveen, A. M. *et al.* approach (2007) [118]: 2,4,5-triaryl substituted imidazoles were synthesized under solvent-free conditions by grinding 1,2-diketones, aromatic aldehydes and ammonium acetate in the presence of a molecular iodine catalyst.

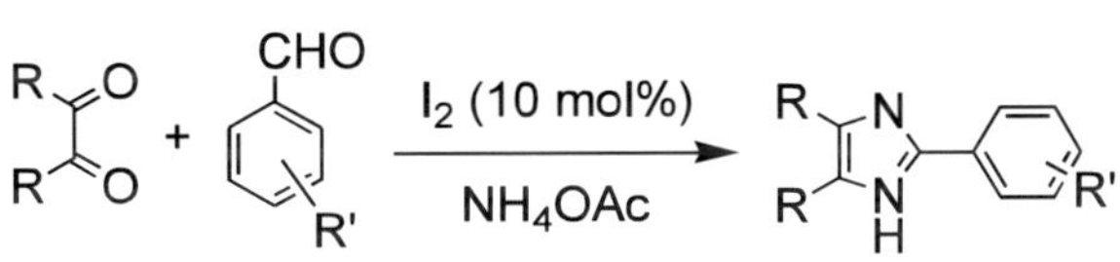

Kokare, N. D. *et al.* approach (2007) [119]: Oxalic acid was used as a catalyst for the synthesis of 2-aryl-1-arylmethyl-1*H*-benzimidazoles and 2-aryl-4,5-diphenyl-1*H*-imidazoles forming moderate to excellent isolated yields.

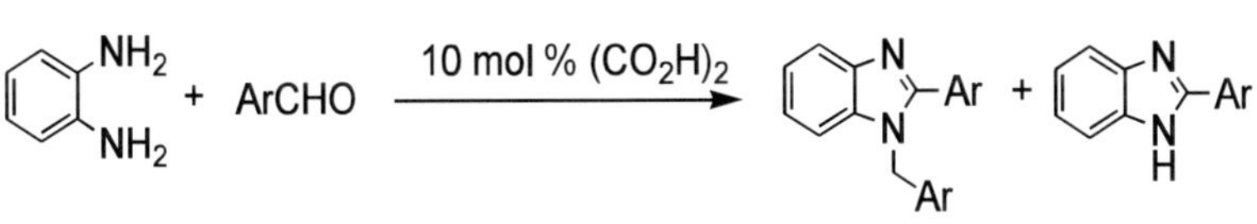

Khodaei, M. M. *et al.* approach (2007) [120]: Synthesis of 2,4,5-trisubstituted imidazoles from 1, 2-diketone or α-hydroxy ketone, aldehyde and ammonium heptamolybdatetetra-hydrate ionic liquid, tetrabutylammonium iodide (TBAI) in molten state using catalytic amounts of *p*-TSA has been reported.

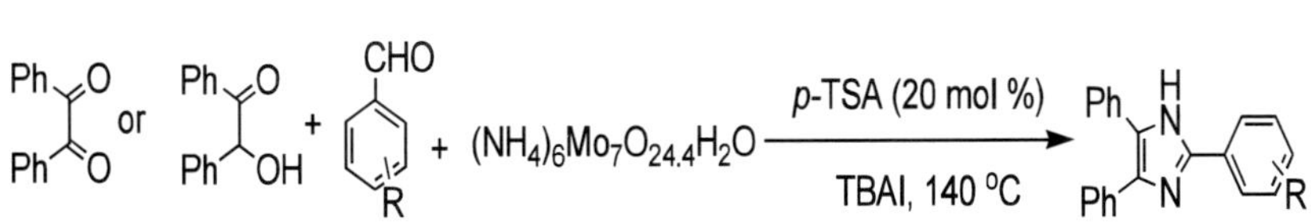

Xu, J. M. *et al.* approach (2006) [121]: Ionic liquid i.e.1-methyl-3-butylimidazolium hydroxide ([bmim]OH) has been used as a catalyst and reaction medium for the Markovnikov addition of N-heterocycles to vinyl esters.

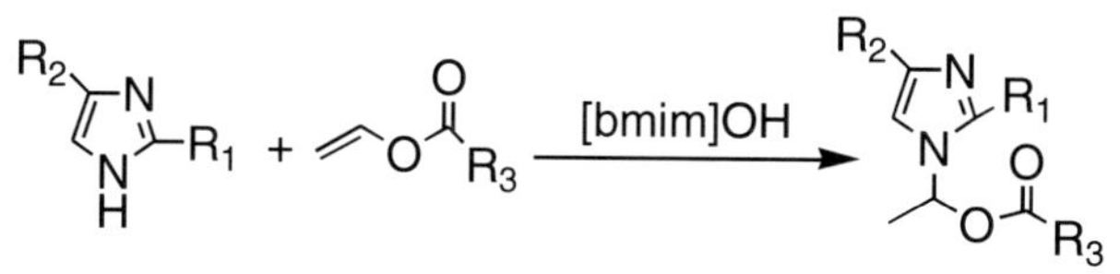

Pathan, M.Y. *et al.* approach (2006) [122]: Synthesis of 2-substituted 2-imidazolines under microwave irradiation was reported.

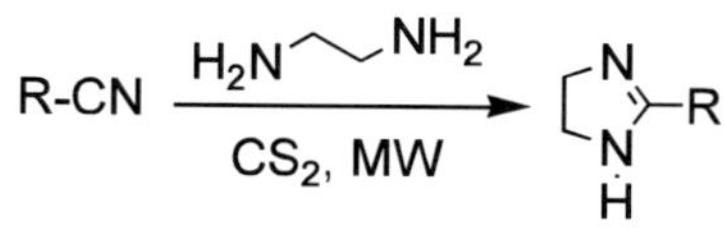

Nagawade, R. R. *et al.* approach (2006) [123]: Substituted benzimidazoles were synthesized under solvent-free conditions using *o*-phenylenediamine and aldehydes in the presence of $BF_3{\cdot}OEt_2$ as a catalyst.

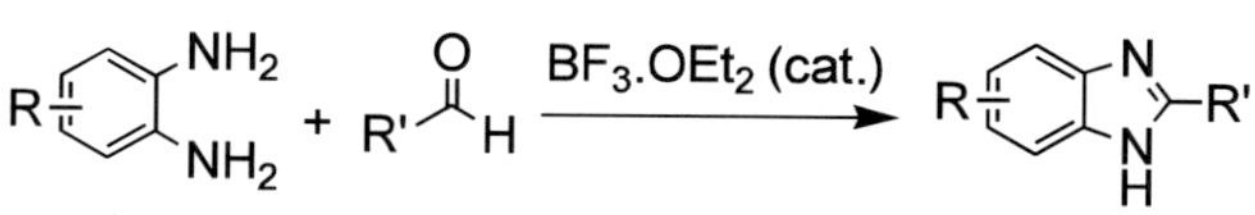

Bellina, F. *et al.* approach (2006) [124]: 1,2-diaryl-1*H*-imidazoles, were regioselectively synthesized by direct coupling of 1-aryl-1*H*-imidazoles with aryl iodides or bromides in DMF, CsF and a catalytic amount of $Pd(OAc)_2$ under ligandless conditions.

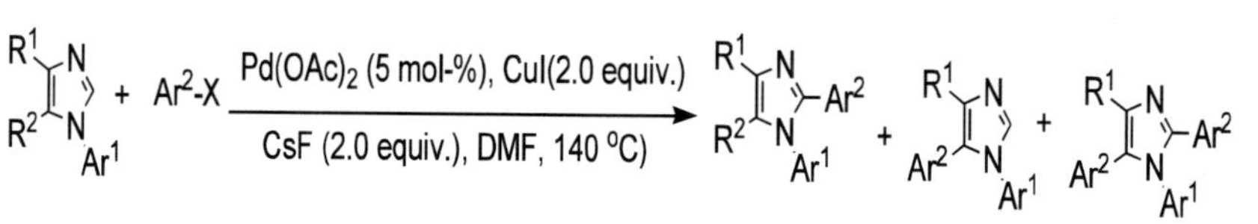

Siamaki A. R. *et al.* approach (2006) [125]: One step synthesis of imidazoles from imines and acid chlorides was reported in the presence of a palladium catalyzed multicomponent coupling reaction.

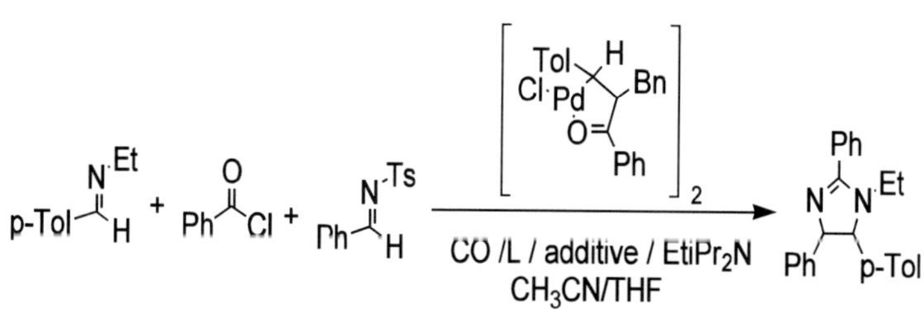

Bellina, F. *et al.* approach (2005) [126]: 1,5-diaryl-1*H*-imidazoles have been regioselectively synthesized by direct coupling of 1-aryl-1*H*-imidazoles with aryl iodides or bromides in DMF, CsF and a catalyst precursor consisting of a mixture of $Pd(OAc)_2$ and $AsPh_3$.

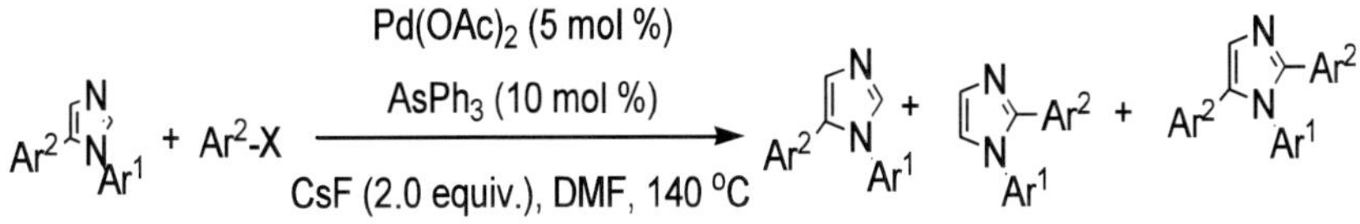

Kidwai, M. *et al.* approach (2005) [127]: The efficient synthesis of 2,4,5-trisubstituted and 1,2,4,5-tetrasubstituted-1*H*-imidazoles were reported under microwave irradiation.

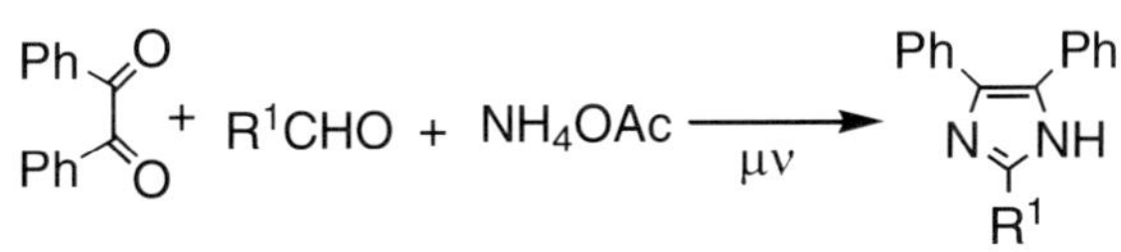

Sparks, R. B. *et al.* approach (2004) [128]: 2,4,5-Triaryl-imidazoles were synthesized directly from the keto-oxime *via* cyclization to the N-hydroxyimidazole and an unprecedented in situ thermal reduction of the NO bond upon microwave irradiation at 200 ^{O}C.

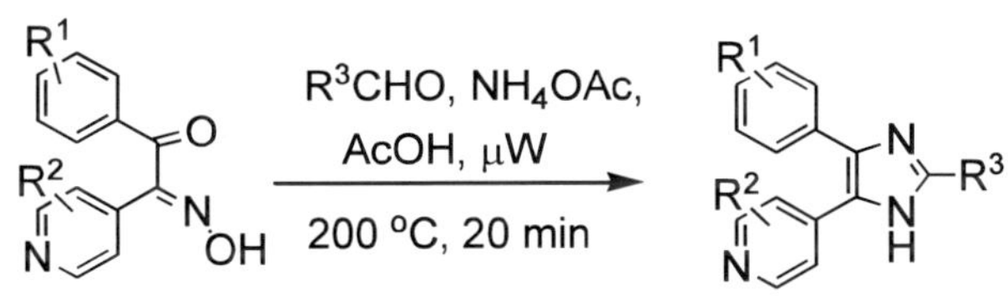

Golankiewicz, B. *et al.* approach (2004) [129]: Two new types of imidazole derivatives: N-(2-R1-5-R2-1*H*-imidazol-4-yl) thioureas and N-(2-R^1-5-R^2-1*H*-imidazol-4-yl) formamides were synthesized by hydrolytic degradation of 6-R1-8-R2-2-thioxo-2,3-dihydroimidazo[1,5-a]-1,3,5-triazin-4(1H)-ones and 6-R^1-8-R^2-imidazo[1,5-a]-1,3,5-triazin-4(3H)-ones.

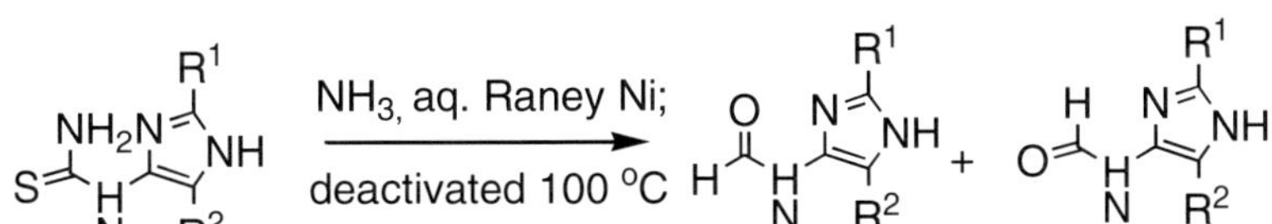

Frantz, D. E. *et al.* approach (2004) [130]: Synthesis of substituted imidazoles was reported by the thiazolium-catalyzed addition of an aldehyde to an acyl imine to generate the corresponding γ-ketoamide *insitu* followed by ring closure to the imidazole.

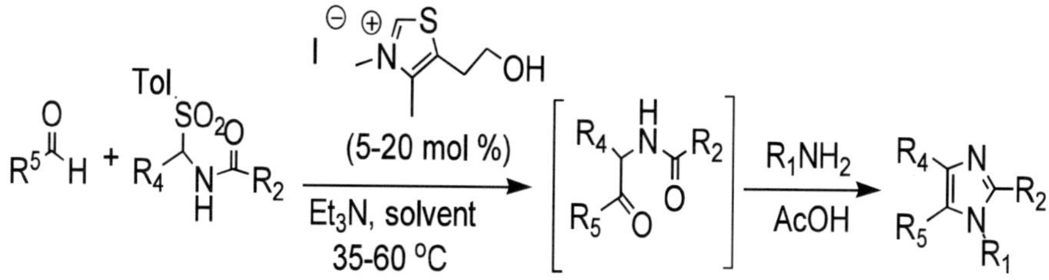

Wolkenberg, S. E. *et al.* approach (2004) [131]: Synthesis of 2,4,5-trisubstituted imidazoles from 1,2-diketones and aldehydes in the presence of NH_4OAc were described under microwave irradiation.

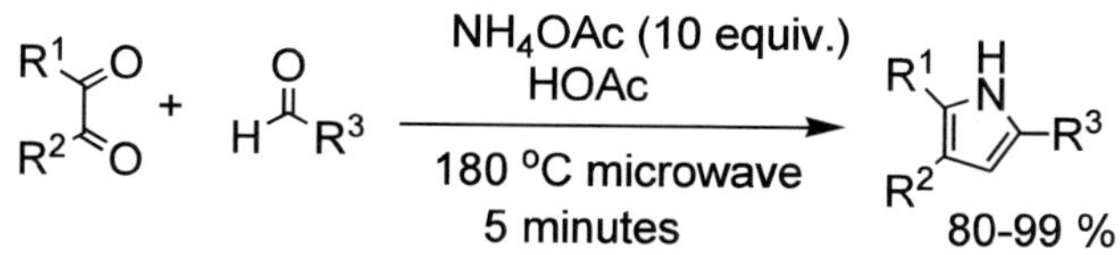

Rao, A. *et al.* approach (2004) [132]: Synthesis of 1*H*,3*H*-thiazolo[3,4-a]benzimidazoles, 2-aryl-1-benzylbenzimidazoles and 2,3-diaryl-1,3-thiazolidin-4-ones was carried out under microwave irradiation.

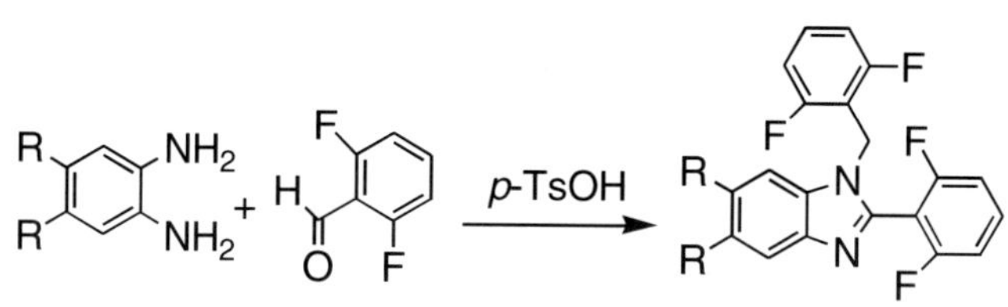

Proskurnina, M. V. *et al.* approach (2002) [133]: 2,4,5-triaryl-4,5-dihydroimidazoles were synthesized by the reaction of aromatic aldehydes with ammonium acetate.

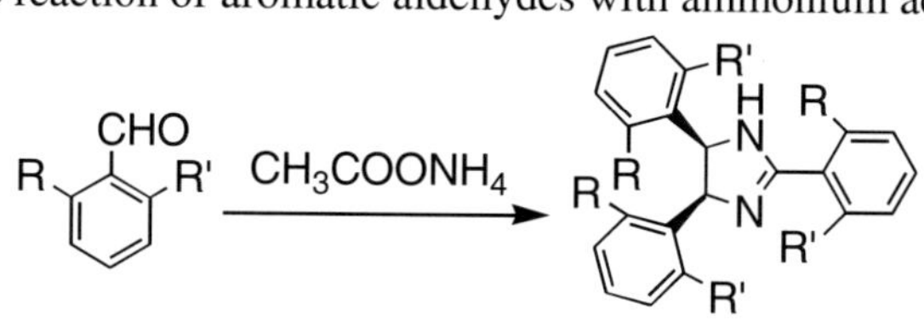

Shilcrat, S. C. *et al.* approach (1997) [134]: 1,2-disubstitued-1*H*-imidazole-5-carboxaldehydes were synthesized by the reaction of N-mono substituted amidines with 2-halo-3-alkoxy-2-propenals.

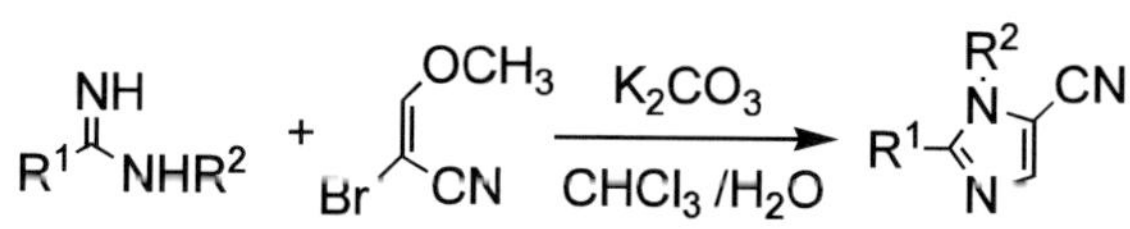

Conclusion

The substituted imidazoles have been investigated from the early 1980s, but the most considerable advances in both the biological and the syntheticmethodologies evaluation of these diazole derivatives have been made in the previous decade. A number of strategies and methodologies have been applied to conveniently achieve the synthesis of substituted imidazole derivatives for further research. Many designs have been undertaken in order to develop efficient, practical and scalable synthetic routes to some of these compounds and their analogues for biological and preclinical studies.

Thus, the reported methods are relatively ecofriendly and supportive to green chemistry. The notable merits are

Short reaction times,
Simple work-up procedure,
Excellent yield of products,
Cost efficiency and
Recovery & reusability of the catalyst.

References

[1] (a)Lombardino, J. G.; Wiseman, E. H. *J. Med. Chem.* 1974, *17*, 1182-1188. (b) Sundberg, R. J.; Martin, R. B. *Chem. Rev.* 1974, *74*, 471-517.
[2] Wiglenda, T.; Ott, I.; Kircher, B.; Schumacher, P.; Schuster, D.; Langer, T.; Gust, R. *J. Med. Chem.* 2005, *48(20),* 6516-6521.
[3] Gbadamassi, M.; Barascut, J.; Imbach, J. L.; Gayral, P. *Eur. J. Med. Chem.* 1988, *23*, 225-232.
[4] Hazelton, J.; Iddon, B.; Redhouse, A. D.; Suschitzky, H. *Tetrahedron.* 1995, 51, 5597-5608.
[5] Lee, J. C.; Laydon, J. T.; McDonnell, P. C.; Gallagher, T. T.; Kumar, S.; Green, D.; McKulty, D.; Blumenthal, M.; Heys, J. R.; Landvatter, S. W.; Strikler, J. E.; McLaughlin, M. M.; Siemen, I. R.; Fisher, S. M.; Livi, G. P.; White, J. R.; Adams, J. L. *Nature.* 1994, *372*, 739-746.
[6] Li, H. S.; Rampersaud, A. A.; Zimmerman, K. *J. Chin. Chem. Soc.* 1993, *40*, 273-282.
[7] Wolkenberg, S. E.; Wisnosk, D. D.; Leister, W. H.; Wang, Y.; Zhao, Z.; Lindsley, C. W. *Org. Lett.* 2004, *6*, 1453-1456.
[8] Pozherskii, A. F.; Soldatenkov, A. T.; Katritzky, A. R. Heterocycles in Life, Society; Wiley: New York, 1997, 179.
[9] Lombardino, J. G.; Wiseman, E. H. *J. Med. Chem.* 1974, *17*, 1182.
[10] Phillips, A. R.; White, H. L.; Rosen, S. *Eur. Pat. Appl. EP 58,* 890; *Chem. Abstr.*1982, *98*, 53894z.
[11] Ohsumi, K.; Hatanaka, T.; Fujita, K.; Nakagawa, R.; Fukuda, Y.; Nihei, Y.; Suga, Y.; Morinaga, Y.; Akiyama, Y.; Tsuji, T. *Bioorg. Med. Chem. Lett.* 1998, 8, 3153-3158.
[12] Lombardino, J. G.; Wiseman, E. H. *J. Med. Chem.* 1974, 17, 1182-1188.
[13] Corell, T.; Hasselmann, G. Acta Pharm. *Toxicol. (Copenh)* 1983, 53, 288-296.

[14] Wiseman, E. H.; McIlhenny, H. M.; Bettis, J.W. *J. Pharm. Sci.* 1975, 64, 1469-1475.

[15] Desiraju, G. R.; Gopalakrishnan, B.; Jetti, R. K. R.; Nagaraju, A.; Raveeendra, D.; Sarma, J. A. R. P.; Sobhia, M. E.; Thilagavathi, R. *J. Med. Chem.* 2002, 45, 4847-4857.

[16] Xie, W.; Robertson, D. L.; Simmons, D. L. *Drug Dev. Res.* 1992, 25, 249-265.

[17] For reviews on COX-2, its overexpression in several human cancers and the use of its selective inhibitors in the prevention and treatment of cancer, see: (a) Hussain, T.; Gupta, S.; Mukhtar, H. *Cancer Lett.*2003, *191,* 125-135; (b) Subbaramaiah, K.; Dannenberg, A. *J. Trends Pharmacol. Sci.* 2003, 24, 96-102; (c) Fu, S.-L.; Wu, Y.-L.; Zhang, Y. P.; Qiao, M. M.; Chen, Y. *World J. Gastroenterol.* 2004, *10,* 1971–1974; (d) Rodrigues, C. R.; Veloso, M. P.; Verli, H.; Fraga, C. A. M.; Miranda, A. L. P.; Barreiro, E. *J. Med. Chem. Rev.* 2004, *1,* 73-90; (e) Mazhar, D.; Ang, R.; Waxman, J. Br. *J. Cancer* 2006, *94,* 346–350; (f) Shiff, S. J.; Shivaprasad, P.; Santini, D. L. *Curr. Opin. Pharmacol.* 2003, *3,* 352-361; (g) Marnett, L. J.; DuBois, R. N. *Annu. Rev. Pharmacol. Toxicol.* 2002, *42,* 55-80; (h) Wo, J. W.; Song, J. Z.; Li, Q. *Med. Chem. Rev.* 2004, *1,* 291–295.

[18] Gans, K. R.; Gailbraith, W.; Roman, R. J.; Haber, S. B.; Kerr, J. S.; Schmidt, W. K.; Smith, C.; Hewes, W. E.; Ackerman, N. R. *J. Pharmacol. Exp. Ther.* 1990, *254,* 180–187.

[19] Prasit, P.; Riendeau, D. *Annu. Rep. Med. Chem.* 1997, *32,* 211–220.

[20. Penning, T. D.; Talley, J. J.; Bertenshaw, S. R.; Carter, J. S.; Collins, P. W.; Docter, S.; Graneto, M. J.; Lee, L. F.; Malecha, J. W.; Miyashiro, J. M.; Rogers, R. S.; Rogier, D. J.; Yu, S. S.; Anderson, G. D.; Burton, E. G.; Cogburn, J. N.; Gregory, S. A.; Koboldt, C. M.; Perkins, W. E.; Seibert, K.; Veenhuizen, A. W.; Zhang, Y. Y.; Isakson, P. C. *J. Med. Chem.* 1997, *40,* 1347-1365.

[21] Almansa, C.; Alfon, J.; de Arriba, A. F.; Cavalcanti, F. L.; Escamillo, K. I.; Gomez, L. A.; Miralles, A.; Soliva, R.; Bartrol, J.; Carceller, E.; Merlos, M.; Garca-Rafanell, J. *J. Med. Chem.* 2003, *46,* 3463-3475.

[22. Tuyen, T. N.; Sin, K. S.; Kim, H. P.; Park, H. Arch. Pharm. Res. 2005, *28,* 1013-1018.

[23] Barta, T. E.; Stealey, M. A.; Collins, P. W.; Weier, R. M. *Bioorg. Med. Chem. Lett.* 1998, *8,* 3443-3448.

[24] Bolos, J. *Mini-Rev. Med. Chem.* 2005, *5,* 857–868.

[25] Cobb, M. H.; Goldsmith, E. J. *J. Biol. Chem.* 1995, 270, 14843-14846.

[26] Davis, R. *J.Mol. Reprod. Dev.* 1995, *42,* 459-467.

[27] Wilson, K. P.; McCaffrey, P. G.; Hsiao, K.; Pazhanisamy, S.; Galullo, V.; Bemis, G. W.; Fitzgibbon, N. J.; Caron, P. R.; Murcko, M. A.; Su, M. S. S. Chem. Biol. 1997, *4,* 423-431.

[28] Chen, Z.; Gibson, T. B.; Robinson, F.; Silvestro, L.; Pearson, G.; Xu, B.E.; Wright, A.; Vanderbilt, C.; Cobb, M. H. Chem. Rev. 2001, *101,* 2449-2476.

[29] Bokemeyer, D.; Sorokin, A.; Dunn, M. J. Kidney Int. 1996, *49,* 1187–1198.

[30] (a) Han, J.; Lee, J. D.; Bibbs, L.; Ulevitch, R. J. *Science* 1994, *265,* 808-811; (b) Derijard, B.; Hibi, M.; Wu, I.H.; Barrett, T.; Su, B.; Deng, T.; Karin, M.; Davis, R. J. Cell 1994, 76, 1025–1037; (c) Ono, K.; Han, *J. Cell. Signal.* 2000, *12,* 1 13.

[31] Lee, J. C.; Laydon, J. T.; McDonnell, P. C.; Gallagher, T. F.; Kumar, S.; Green, D.; McNulty, D.; Blumenthal, M. J.; Keys, J. R.; Vatter, S. W. L.; Strickler, J. E.;

McLaughlin, M. M.; Siemens, I. R.; Fisher, S. M.; Livi, G. P.; White, J. R.; Adams, J. L.; Young, P. R. *Nature* 1994, *372,* 739-746.

[32] Diller, D. J.; Lin, T. H.; Metzger, A. *Curr. Top. Med. Chem.* 2005, *5,* 953–965

[33] (a) Newton, R.; Holden, N. S. *Drug Discov. Today: Dis. Mech.* 2006, *3,* 53-61; (b) Mayer, R. J.; Callahan, J. F. *Drug Discov. Today: Dis. Mech.* 2006, *3,* 49-54.

[34] Adams, J. L.; Boehm, J. C.; Gallagher, T. F.; Kassis, S.;Webb, E. F.; Hall, R.; Sorenson, M.; Garigipati, R.; Griswold, D. E.; Lee, J. C. *Bioorg. Med. Chem. Lett.* 2001, *11,* 2867-2870.

[35] Adams, J. L.; Boehm, J. C.; Kassis, S.; Gorycki, P. D.; Webb, E. F.; Hall, R.; Sorenson, M.; Lee, J. C.; Ayrton, A.; Griswold, D. E.; Gallagher, T. F. *Bioorg. Med. Chem. Lett.* 1998, *8,* 3111-3116.

[36] Boehm, J. C.; Smietana, J. M.; Sorenson, M. E.; Garigipati, R. S.; Gallagher, T. F.; Sheldrake, P. L.; Bradbeer, J.; Badger, A. M.; Laydon, J. T.; Lee, J. C.; Hillegass, L. M.; Griswold, D. E.; Breton, J. J.; Chabot-Fletcher, M. C.; Adams, J. L. *J. Med. Chem.* 1996, *39,* 3929-3937.

[37] Plummer, C.W.; Finke, P. E.; Mills, S. G.;Wang, J.; Tong, X.; Doss, G. A.; Fong, T. M.; Lao, J. Z.; Schaeffer, M.-T.; Chen, J.; Shen, C. P.; Stribling, D. S.; Shearman, L. P.; Strack, A. M.; Van der Ploeg, L. H. T. *Bioorg. Med. Chem. Lett.* 2005, *15,* 1441-1446.

[38] Cuenda, A.; Rouse, J.; Doza, Y. N.; Meier, R.; Cohen, P.; Gallagher, T. F.; Young, P. R.; Lee, J. C. FEBS *Lett.* 1995, *364*, 229-233.

[39] Badger, A. M.; Bradbeer, J. N.; Votta, B.; Lee, J. C.; Adams, J. L.; Griswold, D. E. *J. Pharmacol. Exp. Ther.* 1996, *279,* 453-461.

[40] Cuenda, A.; Cohen, P.; Buee-Scherrer, V.; Goedert, M. *EMBO J.* 1997, *16,* 295-305.

[41] Kumar, S.; McDonnell, P. C.; Gum, R. J.; Hand, A. T.; Lee, J. C.; Young, R. R. *Biochem. Biophys. Res. Commun.* 1997, *35,* 533-538.

[42] Karahashi, H.; Nagata, K.; Ishii, K.; Amano, F. *Biochim. Biophys. Acta* 2000, *1502,* 207-223.

[43] Wadsworth, S. A.; Cavender, D. E.; Beers, S. A.; Lalan, P.; Schafer, P. H.; Malloy, E. A.; Wu, W.; Fahmy, B.; Olini, G. C.; Davis, J. E.; Pellegrino-Gensey, J. L.; Wachter, M. P.; Siekierka, J. J. *J. Pharmacol. Exp. Ther.* 1999, *291,* 680-687.

[44] Fijen, J. W.; Zijlstra, J. G.; De Boer, P.; Spanjersberg, R.; Tervaert, J. W. C.; van Der Werf, T. S.; Ligtenberg, J. J. M.; Tulleken, J. E. Clin. Exp. Immunol. 2001, *124,* 16–20.

[45] Revesz, L.; Blum, E.; Di Padova, F. E.; Bihl, T.; Feifel, R.; Gram, H.; Hiestand, P.; Manning, U.; Rucklin, G. *Bioorg. Med. Chem. Lett.* 2004, *14,* 3595–3599.

[46] Howard, M. O.; Schwartz, L.W.; Newton, J. F.; Qualls, C.W., Jr.; Yodis, L. A.; Ventre, J. R. *Toxicol. Pathol.* 1991, *19,* 115-122.

[47] Testa, B.; Jenner, P. *Drug Metab. Rev.* 1981, *12,* 1-117.

[48] McKenna, J. M.; Halley, F.; Souness, J. E.; McLay, I. M.; Pickett, S. D.; Collis, A. J.; Page, K.; Ahmed, I. *J. Med. Chem.* 2002, *45,* 2173-2184.

[49] Collis, A. J.; Foster, M. L.; Halley, F.; Maslen, C.; McLay, I. M.; Page, K. M.; Redford, E. J.; Souness, J. E.; Wilsher, N. E. *Bioorg. Med. Chem. Lett.* 2001, *11,* 693-696.

[50] (a) Nam, N. H. *Curr. Med. Chem.* 2003, *10,* 1697–1722; (b) Hadfield, J. A.; Ducki, S.; Hirst, N.; McGown, A. T. *Prog. Cell Cycle Res.* 2003, *5,* 309-325.

[51] Tron, G. C.; Pirali, T.; Sorba, G.; Pagliai, F.; Busacca, S.; Genazzani, A. A. J. *Med. Chem.* 2006, *49,* 3033-3044.

[52] Cirla, A.; Mann, *J. Nat. Prod. Rep.* 2003, *20,* 558-564.

[53] Pettit, G. R.; Singh, S. B.; Niven, M. L.; Hamel, E.; Schmidt, J. M. *J. Nat. Prod.* 1987, *50,* 119-131.

[54] Pettit, G. R.; Singh, S. B.; Boyd, M. R.; Hamel, E.; Pettit, R. K.; Schmidt, J. M.; Hogan, F. *J. Med. Chem.* 1995, *38,* 1666–1672; Erratum in: *J. Med. Chem.* 1995, *38,* 2294.

[55] Pettit, G. R.; Singh, S. B.; Schmidt, J. M.; Niven, M. L.; Hamel, E.; Lin, C. *M. J. Nat. Prod.b*1988, *51,* 517-527.

[56] Horsman, M. R.; Murata, R.; Breidahl, T.; Nielsen, F. U.; Maxwell, R. J.; Stodkiled-Jorgensen, H.; Overgaard, *J. Adv. Exp. Med. Biol.* 2000, *476,* 311-323.

[57] Tozer, G. M.; Kanthou, C.; Parkins, C. S.; Hill, S. A. *Int. J. Exp. Pathol.* 2002, *83,* 21–38.

[58] (a) Tozer, G. M.; Kanthou, C.; Baguley, B. C. *Nat. Rev. Cancer* 2005, 5, 423-425; (b) Gaya, A. M.; Rustin, G. J. S. *Clin. Oncol.* 2005, *17,* 277-290.

[59] Pettit, G. R.; Temple, C., Jr.; Narayanan, V. L.; Varma, R.; Simpson, M. J.; Boyd, M. R.; Rener, G. A.; Bansal, N. *Anticancer Drug Res.* 1995, *10,* 299-309.

[60] McGown, A. T.; Fox, B. W. *Cancer Chemother. Pharmacol.* 1990, *26,* 79-81.

[61] (a) Dark, G. G.; Hill, S. A.; Prise, V. E.; Tozer, G. M.; Pettit, G. R.; Chaplin, D. *J. Cancer Res.* 1997, *57,* 1829-1834; (b) Liekens, S.; De Clerq, E.; Neyts, *J. Biochem. Pharmacol.* 2001, *61,* 253-270.

[62] (a) Young, S. L.; Chaplin, D. J. *Expert Opin. Investig. Drugs* 2004, *13,* 1171-1182; (b) Griggs, J.; Metcalfe, J. C.; Hesketh, R. *Lancet Oncol.* 2001, *2,* 82-87.

[63] Vincent, L.; Kermani, P.; Young, L. M.; Cheng, J.; Zhang, F.; Shido, K.; Lam, G.; Bompais-Vincent, H.; Zhu, Z.; Hicklin, D. J.; Bohlen, P.; Chaplin, D. J.; May, C.; Rafii, S. *J. Clin. Invest.* 2005, *115,* 2992-3006.

[64] Hori, K.; Saito, S. Br. *J. Cancer* 2003, *89,* 1334-1344.

[65] Guffroy, M.; Dally, C.; Vrignaud, P.; Beys, E.; Bissery, M. C. *Proc. Am. Assoc. Cancer Res.* 2004, *45,* Abstract 5438.

[66] Ohsumi, K.; Nakagawa, R.; Fukuda, Y.; Hatanaka, T.; Morinaga, Y.; Nihei, Y.; Ohishi, K.; Suga, Y.; Akiyama, Y.; Tsuji, T. *J. Med. Chem.* 1998, *41,* 3022–3032.

[67] Wang, L.; Woods, K. W.; Li, Q.; Barr, K. J.; McCroskey, R. W.; Hannick, S. M.; Gherke, L.; Credo, R. B.; Hui, Y.H.; Marsh, K.; Warner, R.; Lee, J. Y.; Zielinsky-Mozng, N.; Frost, D.; Rosenberg, S. H.; Sham, H. L. *J. Med. Chem.* 2002, *45,* 1697–1711.

[68] For reviews on the P-gp mediated MDR, see: (a) Fine, R. L.; Chambers, T. C.; Sachs, C. W. *Oncologist1996,1,* 261-268; (b) Sun, J.; He, Z. G.; Cheng, G.; Wang, S. J.; Hao, X. H.; Zou, M. *J. Med. Sci. Monit.*2004, *10,* RA5–RA14; (c) Robert, J. *Eur. J. Clin. Invest.*1999, *29,* 536-545; (d) Gottesman, M. M. *Annu. Rev. Med.* 2002, *53,* 615-627.

[69] Raderer, M.; Scheithauer, W. *Cancer* 1993, *72,* 3553-3563.

[70] Sarshar, S.; Zhang, C.; Moran, E. J.; Krane, E. J.; Rodarte, J. C.; Benbatoul, K. D.; Dixon, R.; Mjalli, A. M. M. *Bioorg. Med. Chem. Lett.* 2000, *10,* 2599-2601.

[71] For reviews on the MDR reversal agents, see: (a) Robert, J.; Jarry, C. *J. Med. Chem.* 2003, 46, 4805-4817; (b) Borowski, E.; Bontemps-Gracz, M. M.; Piwkowska, A. *ActaBiochim. Pol.* 2005, *52,* 609-627; (c) Ambukar, S. V.; Dey, S.; Hrycna, C. A.; Ramachandra, M.; Pastan, I.; Gottesman, M. M. Annu. Rev. Pharmacol. Toxicol. 1999, 39, 36-398; (d) Ford, J. M.; Hait, W. N. *Pharmacol. Rev.* 1990, *42,* 155-199.

[72] Zhang, C.; Sarshar, S.; Moran, E. J.; Krane, S.; Rodarte, J. C.; Benbatoul, K. D.; Dixon, R.; Mjalli, A. M. M. *Bioorg. Med. Chem. Lett.* 2000, *10,* 2603-2605.

[73] Newman, M. J.; Rodarte, J. C.; Benbatoul, K. D.; Romano, S. J.; Zhang, C.; Krane, S.; Moran, E. J.; Uyeda, R. T.; Dixon, R.; Guns, E. S.; Mayer, L. D. *Cancer Res.* 2000, *60,* 2964-2972.

[74] Chen, L.M.; Wu, X. P.; Ruan, J. W.; Liang, Y.J.; Ding, Y.; Shi, Z.; Wang, X.W.; Gu, L. Q.; Fu, L.W. *Oncol. Res.* 2004, 14, 355–362.

[75] Chen, L.M.; Liang, Y. J.; Ruan, J.W.; Ding, Y.; Wang, X.W.; Shi, Z.; Gu, L.Q.; Yang, X. P.; Fu, L.W. *J. Pharm. Pharmacol.* 2004, *56,* 1061–1066.

[76] Seung-Chul Lee, Dongkyu Shin, Joong-Myung Cho, Sunggu Ro, Young-GerSuh, *Bioorganic & Medicinal Chemistry Letters*, 2012, *22,* 1891-1894

[77] Lombardino, J. G.; Wiseman, E. H. *J. of Med.Chem.* 1974, *17(11),* 1182-1188.

[78] Rogers, S. A.; Lindsey, E. A.; Whitehead, D. C.; Mullikin, T.; Melander, C. *Bioorg. & Med. Chem. Lett.* 2011, *21(4),* 1257-1260.

[79] Soni, B.; Ranawat, M. S.; Sharma, R.; Bhandari, A.; Sharma, S. *E. J. of Med. Chem.*2010, *45(7)*, 2938-2942.

[80] Ansari, K. F.; Lal, C. *E. J. of Med. Chem.* 2009, *44,* 4028-4033.

[81] Khan, M. S.; Siddiqui, S. A.; Shaik, M. R.; Goswami, U.; Srinivasan, K. V.; Khan, M. I. *Chem. Biol. Drug. Des.* 2008, *72,* 197-204.

[82] PrabalBandyopadhya, ManishaSathe, S. Ponmariappan, Arti Sharma, Pratibha Sharma, A.K. Srivastava, M.P. Kaushik, *Bioorganic & Medicinal Chemistry Letters*, 2011, *21,* 7306–7309.

[83] Ozkay, Y.; Tunal, Y.; Karaca, H.; Isikdag, I. *E. J. of Med. Chem.*2010, *45(8),*3293-3298.

[84] Canan, K. *Turk J Chem.* 2003, 27, 35-39.

[85] Shen, H. C.; Ding, F. X.; Zhou, C.; Xiong, Y.; Verras, A.; Chabin, R. M.; Xu, S.; Tong, X.; Xie, D.; Lassman, M. E.; Bhatt, U. R.; Garcia-Calvo, M. M.; Geissler, W.; Shen, Z.; Chen, D.; Roy, R. S.; Hale, J. J.; Tata, J. R.; Pinto, S.; Shen D. M.; Colletti, S. L. *Bioorganic & Medicinal Chemistry Letters*, 2011, *21(5),* 1299-1305.

[86] Kwak, Y.; Coppola, G.; Forster, C. J.; Thomas A. Gilmore, Yongjin Gong, Aaron Kanter, Alan Neubert, Bryan Stroup, Paul Szklennik, Glowienke, S.; Stadelmann, P.; Bell, L.; Bickford, S.; Gangl, E.; Gunduz, M.; Jain, M.; Zhan J.; Serrano-Wu, M. H.; *Bioorg. & Med. Chem. Lett.* 2011, *21(5),* 1422-1424.

[87] Lee, S. H.; Kim, S.; Yun, M. H.; Lee, Y. S.; Cho, S. N.; Oh, T.; Kim, P. *Bioorg. & Med. Chem. Lett.* 2011, *21(5),*1515-1518.

[88] G. F.; Richter, G. M.; Benson, K. H.; Bleicher, D.; Blum, E.; Chaput, N.; Clemann, S.; Feng, C.; Gardes, U.; Grether, P.; Hartman, B.; Kuhn, R. E.; Martin, J. M.; Plancher, M. G.; Rudolph, F.; Taylor, S. *Bioorg. & Med. Chem. Lett.*2011, *21(4),* 1134-1140.

[89] Wiglenda, T.; Ott, I.; Kircher, B.; Schumacher, P.; Schuster, D.; Langer, T.; Gust, R. *J. Med. Chem.* 2005, *48,* 6516-6521.

[90] Dake S. A.; Khedkar, M. B.; Irmale, G. S.; Ukalgaonkar, S. J.; Thorat, V. V.; Shintre, S. A; Pawar, R. P. *Synthetic Communications,* 2012, *42,* 1509-1520.

[91] Padhy, A. K.; Chetia, B.; Mishra, S.; Pati, A.; Iyer P. K. *Tetrahedron Lett.* 2010, *51,* 2751-2753.

[92] Kong, L.; Lv, X.; Lin, Q.; Liu, X.; Zhou, Y.; Jia, Y. *Organic Process Research & Development,*2010, *14,* 902-904.

[93] Hojati, S. F.; Maleki, B.; Beykzadeh, Z.; *Monatsh Chem.,* 2010, *142(1),* 87-91.

[94] a) Safari, J.; Khalili S. D.; Banitaba, S. H. *J. Chem. Sci.,* 2010, *122(3),* 437-441; b) Safari, J.; Khalili, S. D.; Rezaei, M.; Banitaba, S. H. *MonatshChem,* 2010, *141,*1339-1345.

[95] Nalage, S. V.; Kalyankar, M. B.; Patil, V. S.; Bhosale, S. V.; Deshmukh, S. U.; Pawar, R. P. *The Open Catalysis Journal,* 2010, *3,* 58-61.

[96] Parveena, A.; Ahmed, A.; Ahmed, S. K. *RJPBCS,* 2010, *1(4),* 943-951.

[97] Tripathy, H.; ST, K.; Adhikary, L.; Chandrashekar, S. *RJPBCS,* 2010, *1,* 23-30.

[98] Joshi, R. S.; Mandhane P. G.; Dabhade S. K.; Gill,C. H. *Journal of the Chinese Chemical Society*, 2010, *57*, 1227-1231.

[99] Hasaninejad, A.; Zare, A.; Shekouhy, M.; Rad, J. A. *J. Comb. Chem.* 2010, *12,* 844-849.

[100] Samai, S.; Nandi, G. C.; Singh, P.; Singh, M. S.; *Tetrahedron,* 2009, 65, 10155-10161.

[101] Bratulescu,G. *Synthesis* 2009, *14,* 2319-2320.

[102] Pastor, I. M.; Yus, M. *Current Chemical Biology,* 2009, *3,*385-408.

[103] Rathod, S. B.; Gadekar, L. S.; Katkar, S. S.; Arbad, B. R.; Lande, M. K. *Bulletin of the Catalysis Society of India,* 2009, *8,* 16-24.

[104] Wang, L.; Cai, C.; *MonatshChem,* 2009, *140,* 541-546.

[105] Rostamizadeh, S.; Amani, A. M.; Aryan, R.; Ghaieni, H. R.; Norouzi, L. *Monatsh, Chem;* 2009, *140,* 547-552.

[106] Gadekara, L. S.; Mane, S. R.; Katkara, S. S.; Arbada, B. R.; Lande, M. K. *Cent. Eur. J. Chem.* 2009, *7,* 550-554.

[107] Shelke, K. F.; Sapkal, S. B.; Sonar, S. S.; Madje, B. R.; Shingate, B. B.; Shingare,M. S. *Bull. Korean Chem. Soc.*2009, *30(5),* 1057-1060.

[108] Shitole, N. V.; Shelke, K. F.; Sonar, S. S.; Sadaphal, S. A.; Shingate, B. B.; Shingare, M. S. *Bull. Korean Chem. Soc.*2009, *30(9),* 1963-1966.

[109] Ghorbani-Vaghei, R.; Hojat, V.*Mol Divers,* 2009, *14(2),*249-256.

[110] Rafiee, E.; Mahdavi, H.; Joshaghani, M.; *Mol Divers,* 2009, DOI 10.1007/s11030-009-9213-1.

[111] Karimi A. R.; Alimohammadi, Z.; Amini, M. M. *Mol Divers,* 2009, *14(4)*, 635-641.

[112] Sangshetti, J. N.; Kokare, N. D.; Kotharkar, S. A.; Shinde, D. B. *Chinese Chemical Letters,* 2008, *19,* 762-766.

[113] Khan, M. S.; Siddiqui, S. A.; Rafi M. S.; Siddiqui, A.; Goswami, U.; Venkatraman, K.; Muhammad, S.; Khan, I. *ChemBiol Drug Des*, 2008, *72,* 197 204.

[114] Khosropour, A. R.; *Can. J.Chem.,* 2008, *86,* 264-269.

[115] Shelke, K.; Kakade, G.; Shingate, B.; ShingareM. *Rasayan J. Chem.,* 2008, *1(3),* 489-494.
[116] Sangshetti, J. N.; Kokarea, N. D.; Sandeep, B.; Kotharkara A.; Shinde, D. B. *J. Chem. Sci.,* 2008, *120(5),* 463-467.
[117] Mohammed, A. F.; Kokare, N. D. Sangshetti, J. N.; Shinde, D. B. *Journal of the Korean Chemical Society,* 2007, *51(5),* 418-422.
[118] Parveen, A. M.; Ahmed, R. S.; Shaikh, K. A.; Deshmukh, S. P.; and Pawar, R. P. *Arkivoc,* 2007, *(xvi),* 12-18.
[119 Kokare, N. D.; Sangshetti, J. N.; Shinde,D. B. *Synthesis,*2007, *18,* 2829-2834.
[120] Khodaei, M. M.; Bahrami, K.; Kavianinia, I. *Journal of theChinese Chemical Society,*2007, *54,* 829-833.
[121] Xu, J. M.; Liu, B. K.; Wu, W. B.; Qian,C.; Wu, Q.; Lin,X. *J. Org. Chem.,*2006, *71,* 3991-3993.
[122] Pathan, M.Y.; Paike, V. V.; Pachmase, P. R.; More, S. P.; Ardhapure, S. S.; Pawar, R. P. *Arkivoc*, 2006, *(xv),* 205-210.
[123] Nagawade, R. R.; Shinde, D. B. *Chinese Chemical Letters,* 2006, *17(4),* 453-456.
[124] Bellina, F.; Cauteruccio, S.; Mannina, L.; Rossi, R.; Viel,S. *Eur. J. Org. Chem.,*2006, 693-703.
[125] Siamaki A. R.; Arndtsen,B. A. *J. Am. Chem. Soc.,* 2006, *128,* 6050-6051.
[126] Bellina, F.; Cauteruccio, S.; Mannina, L.; Rossi, R.; Viel, S. *J. Org. Chem.*2005, *70,* 3997-4005.
[127] Kidwai, M.; Saxena, S.; Rastogi, R.; Rastogi, S. *Bull. Korean Chem. Soc.,* 2005, *26(12),* 2051-2053.
[128] Sparks, R. B.; Combs,A. P. *Org. Lett.,* 2004, *6(14),*2473-2475.
[129] Golankiewicz, B.; Januszczyk, P.; Zeidler, J.; Popenda, M. *Nucleosides, Nucleotides & Nucleic Acids,* 2004, *23,* 127-136.
[130] Frantz, D. E.; Morency, L.; Soheili, A.; Murry, J. A.; Grabowski, J. J.; Tillyer, R. D. *Org. Lett.,*2004, *6(5),*843-846.
[131] Wolkenberg, S. E.; Wisnoski, D. D.; Leister, W. H.; Wang, Y.; Zhao, Z.; Lindsley, C. W. *Org. Lett.,*2004, *6(9),*1453-1456.
[132] Rao, A.; Chimirri, A.; Ferro, S.; Monforte, A. M.; Monforte, P.; Zappala, M. *Arkivoc,* 2004, *(v),* 147-155.
[133] Proskurnina, M. V.; Lozinskaya, N. A.; Tkachenko, S. E.; Zefirov, N. S. *Russian Journal of Organic Chemistry,* 2002, *38(8),* 1149-1153.
[134] Shilcrat, S. C.; Mokhallalati, M. K.; Fortunak, M. D.; Pridgen, L. N. *J. Org. Chem.*1997, *62,*8449-8454.

In: Bioactive Heterocycles
Editors: K. L. Ameta, R. P. Pawar and A. J. Domb
ISBN: 978-1-62257-636-4
© 2013 Nova Science Publishers, Inc.

Chapter 5

Synthesis and Biological Activity of Five-, Six- and Seven-membered *S,N*-Containing Saturated Heterocycles

V. R. Akhmetova[1,*], *E. B. Rakhimova*[1], *N. F. Galimzyanova*[2] *and A. G. Ibragimov*[1]
[1]Institute of Petrochemistry and Catalysis, Ufa, Russian Federation
[2]Institute of Biology, Ufa Scientific Centre, Ufa, Russian Federation

Abstract

One-pot synthesis of *N*-substituted 1,3,4-thiadiazolidines, 1,3,5-dithiazinanes, 1,3,5-thiadiazinanes, 1,5,3-dithiazepinanes *via* the cyclothiomethylation reaction of different amino substrates in the presence of two-component systems such as CH_2O-H_2S, RCHO-CS_2, CH_2O-thiols and their biological activity has been reviewed.

1. Introduction

Research prospects in the field of synthesis and transformations of *S,N*-heterocycles are determined by several factors. First, it is determined by an availability of sulfur-containing raw material and amines, second – by a possibility of obtaining unique heterocyclic bioregulators of living systems. Many research publications [1-4] are dedicated to the synthesis and pharmacological activity of S,N-containing five- and six-membered heterocycles. IN THIS CHAPTER approaches to a one-pot synthesis of S,N-containing five-, six- and seven-membered saturated heterocycles with potential biological activity are described, wherein H_2S, CS_2, S_2Cl_2 and thiols are utilized as sulfur-containing raw materials. These one-pot synthesis reactions are based on cyclothiomethylation of various amino substrates by the action of two-component systems, such as CH_2O-H_2S, RCHO-CS_2, and

[*] E-mail: ink@anrb.ru.

CH_2O-thiols. Studies in this area have allowed to trace back heterocyclization hemospecificity of various amines, hydrazines, amides and hydrazides of acids and to formulate methodological principles for the synthesis of 1,3,4-thiadiazolidines, perhydro-1,3,5-dithiazines, 1,3,5-thiadiazines and 1,5,3-dithiazepines, containing (*S,N*)-heteroatoms within cycles thereof in various combinations. The data on biological activity for some above-mentioned heterocycles are thus provided.

2. Discussion

2.1. One-pot Cyclothiomethylation Reaction

2.1.1. 1,3,4-Thiadiazolidines

Different methods for the synthesis of five-membered thiadiazoles exist, among those a one-pot synthesis of 1,3,4-thiadiazoles *1*, *2* by heterocyclization reaction of thiohydrazides of carboxylic acids [5, 6] and N-thiosemicarbazides [7, 8] with acid anhydrides is considered as one of the most efficient.

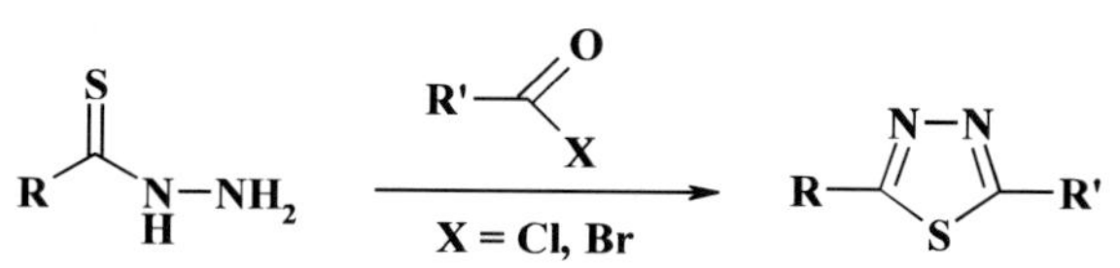

1

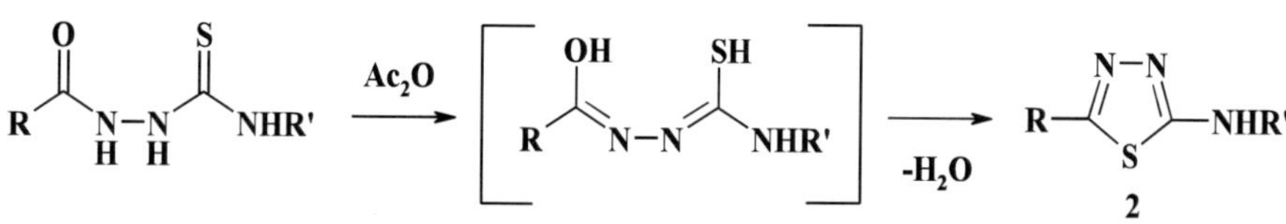

2

However 1,3,4-thiadiazolidines are saturated derivatives of aforesaid compounds *1*, *2*, are still little understood, although the synthesis of unsubstituted 1,3,4-thiadiazolidine *3* was implemented already in 1959 by one-pot reaction of hydrazine with formaldehyde and H_2S [9]. The reaction was carried out in N_2 atmosphere [9] since heterocycle *3* decays in air environment. Stable thiadiazolidines 4 and *5* were obtained by the reaction of hydrazine with aliphatic aldehydes or ketones and H_2S [10-12]. It is to be noted, that alkyl substituents preferably occupy *trans*-configuration (*5a*).

3

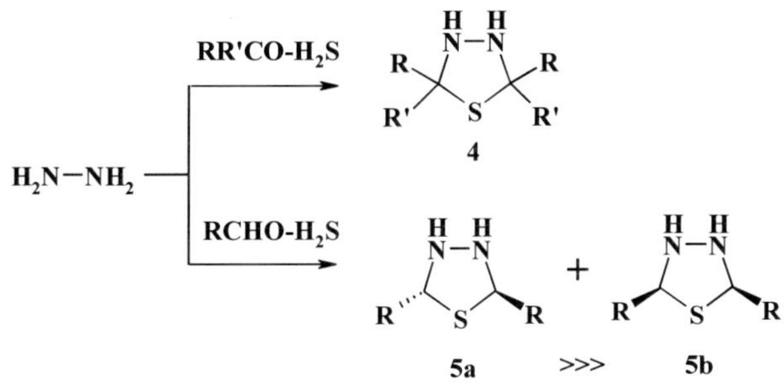

Given reaction proceeds chemoselectively with the formation of 1,3,4-thiadiazolidine 5 [9,10,12] and is not dependent on the order of mixing components (routes A or B).

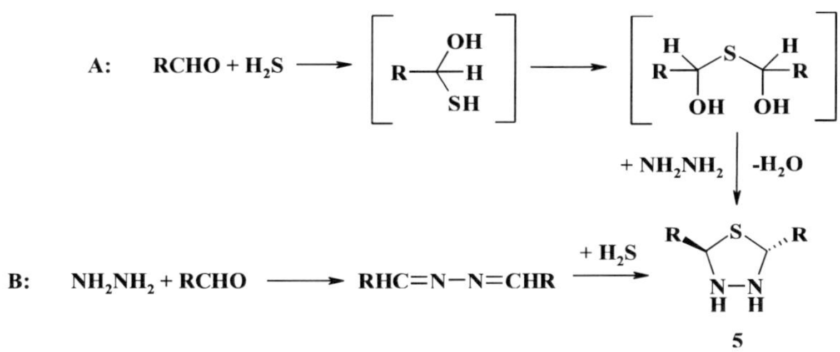

In the case of cyclothiomethylation of *N*,N'-diisopropylhydrazine unstable (*trans*)-3,4-diisopropyl-1,3,4-thiadiazolidine *6* may be obtained, which at 20 °C gradually transforms into a polymeric product [11].

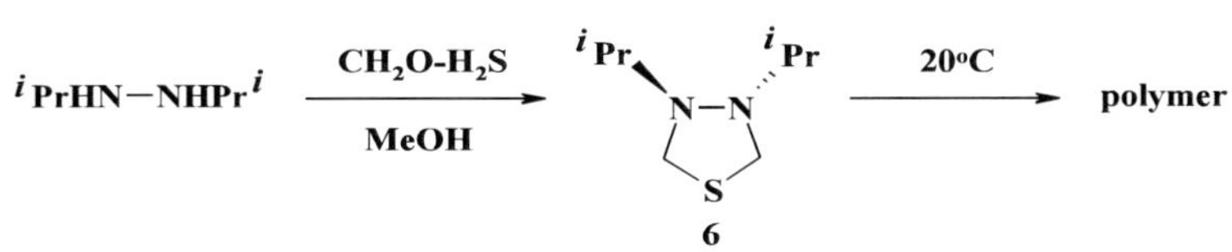

It has been shown [13] that N-aryl substituted hydrazines may form stable thiadiazolidines *7,* while a reaction of selective synthesis was carried out in presence of HCl (in the form of water-soluble hydrochlorides) at temperatures below 0 °C (kinetic control). Without HCI at higher temperatures thermodynamically stable six-membered 5-aminosubstituted 1,3,5-dithiazinanes *8* are formed along with the five-membered heterocycles *7*. Cyclothiomethylation reaction of aryl hydrazines with acetaldehyde and H_2S provides only 2,5-dimethyl-3-aryl-thiadiazolidine *7a,* probably due to the decrease in reactivity of acetaldehyde in comparison to CH_2O.

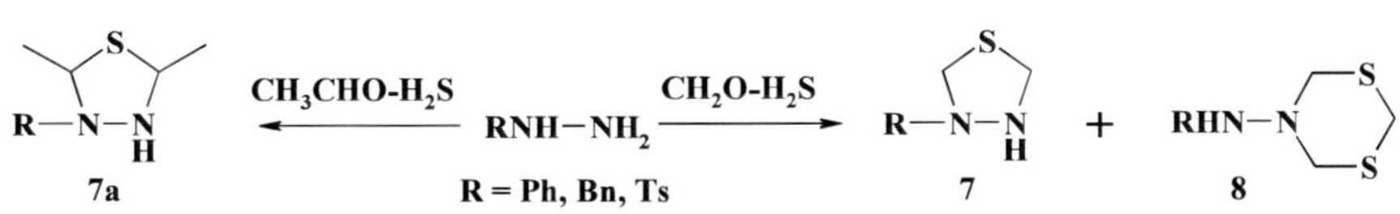

In afore-mentioned conditions cyclothiomethylation reaction of *N,N′*-diphenyl hydrazine with formaldehyde and H_2S leads to the formation of N,N'-diphenyl-1,3,4-thiadiazolidine *9* with less than 10% yield. Main reaction product is benzidine. However, in presence of BuONa the yield of heterocycle *9* increases up to 40% [13]. Obviously, *N,N′*-diphenyl substituent provides stability for the 1,3,4-thiadiazolidine cycle.

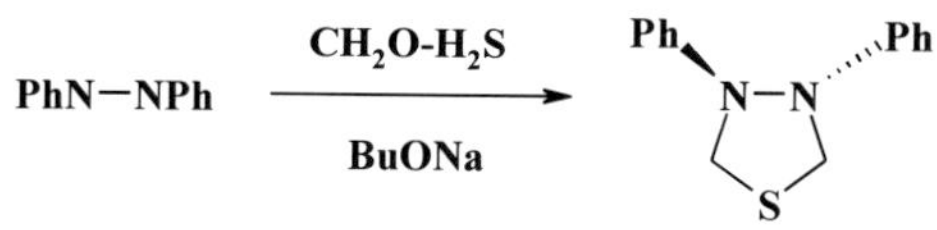

9

Present method was extended to hydrazides of carboxylic acids. Cyclothiomethylation of hydrazides of aryl-carboxylic acids is carried out similarly to that of aryl hydrazines with the formation of a mixture of five-and six- membered heterocycles, namely *N*-acyl-1,3,4-thiadiazolidines *10a-d* and N-(1,3,5-dithiazinan-5-yl)-amides *11a*-d [14]. By an examplary reaction of isonicotinic acid hydrazide it was established, that the yield of thiadiazolidine *10a* increases with decrease in reaction temperature (0, –10, –50 °C) to comprise 44%, 53% and 56%, respectively.

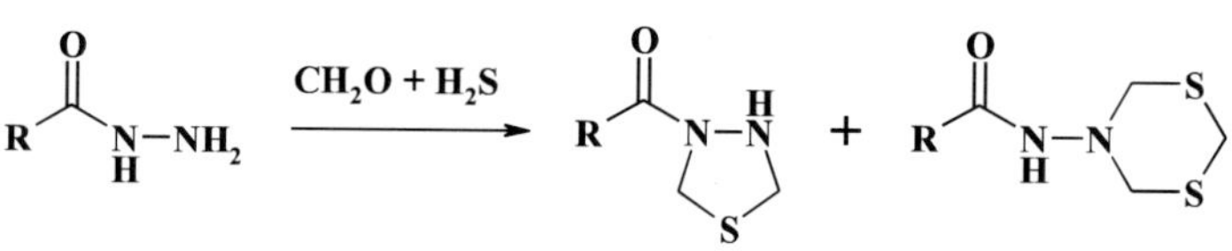

10 a-d **11 a-d**

R = C_6H_4N (a), C_6H_5 (b), *o*-CH_3O-C_6H_4 (c), HO-$C_6H_4CH_2$ (d)

Another example of hydrazine cyclothiomethylation is based on the reaction thereof with aromatic aldehydes and S_2Cl_2 in CH_2Cl_2 at 0–5 °C with the formation of 2,5-diaryl-1,3,4-thiadiazolidines 12 [15].

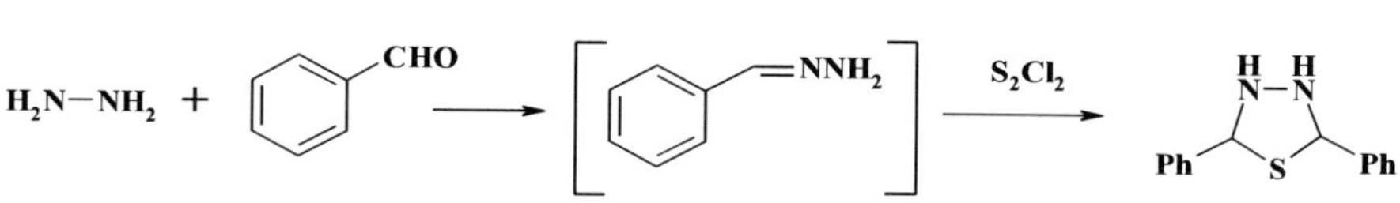

12

An alternative strategy of hydrazine derivatives cyclothiomethylation is based on the reaction thereof with aldehydes and CS_2 proceeding upon heating and leading to 5-aryl-1,3,4-thiadiazolidine-2-thiols 13 [16-18].

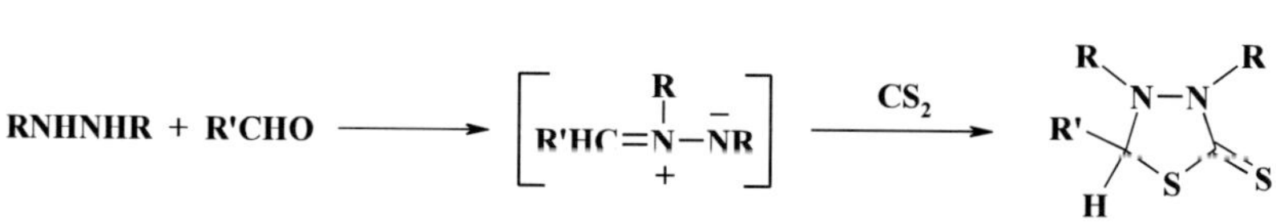

R = Me, Et, Pr; R' = Ph, *p*-CH_3O-C_6H_4, *p*-Cl-C_6H_4, *p*-NO_2-C_6H_4

13

Both reactions proceeds *via* the formation of hydrazones with subsequent cyclization thereof. Synthesis of fused 1,3,4-thiadiazolidine – perhydro-[1,3,4]-thiadiazol-[3,4-c] -1,3,4-thiadiazole *14* was for the first time carried out in 1959 by the reaction of hydrazine with CH_2O and H_2S at <0 C in 11% yield [10]. An excessive amounts of thiomethylating mixture CH_2O-H_2S has proved to be important for increase of bicyclic thiadiazole *14* yield up to 60% utilization of [19], while the presence of *n*-BuONa contributed selective formation of bicycle *14* in 80% yield [20].

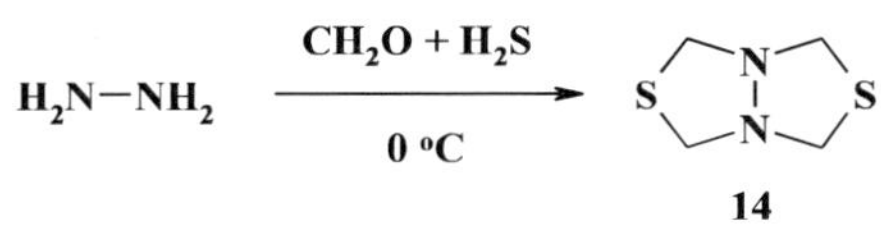

Taking into consideration the fact, that among 1,3,4-thiadiazolidine derivatives fungicidal compounds are present, in particular compounds with fungicidal activity in regards to fungi molds affecting timber and other building materials [21]. Water-soluble adducts *15-19* were obtained by interaction of perhydro-thiadiazoles *7a* and *14* with Lewis and Bronsted acids and CH_3I .

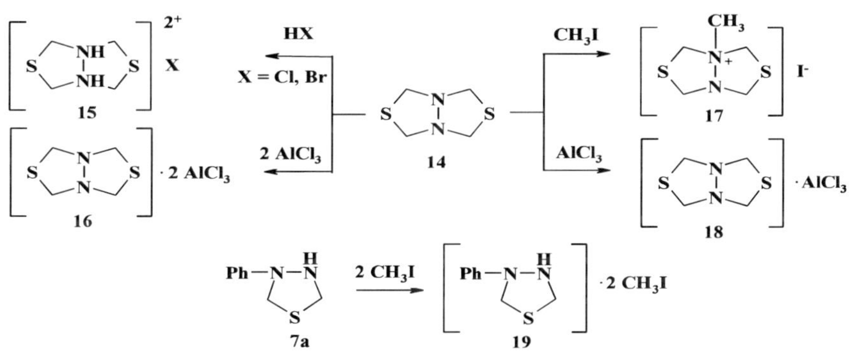

Present data are thus indicative of highly promising prospects for one-pot synthesis of 1,3,4-thiadiazolidines *via* three-component cyclothiomethylation reaction of hydrazines and hydrazides with carbonyl compounds and sulfur-containing agents H_2S, S_2Cl_2 and CS_2.

2.1.2. 1,3,5-Dithiazinanes and 1,3,5-thiadiazinanes

Most of methods for the synthesis of six-membered 1,3,5-dithiazinanes and 1,3,5-thiadiazinanes are based on multicomponent reactions of amino substrates with aldehydes and sulphur-containing nucleophiles, such as H_2S, NaHS, Na_2S, CS_2 and thiols.

One of the convenient approache for the synthesis of 1,3,5-dithiazinanes is cyclothiomethylation reaction of primary amines with formaldehyde and H_2S [22]. An example of aliphatic [23–25] and aromatic amines [26–29] had been shown that, the construction of 1,3,5-dithiazinanes *20* proceeds with high selectivity in condition that a molar ratio between amine, CH_2O and H_2S equals exactly 1:3:2 at 40–80 C (thermodynamic control). Whereas, with the ratio of 2:3:1 at temperatures below 0°C formation of 1,3,5-thiadiazinanes 21 occurs (kinetic control). Among various 1,3,5-thiadiazinane derivatives *21*,

N,N'-diaryl-substituted derivatives are most stable, so that the structure thereof was proved by X-ray analysis [29].

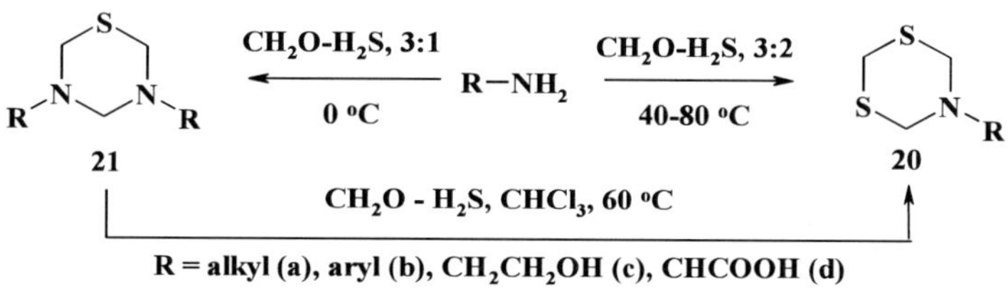

Transformation of 1,3,5-thiadiazinanes *21* into thermodynamically stable 1,3,5-dithiazinanes *20* easily occurs by reaction with thiomethylating mixture CH_2O-H_2S in boiling chloroform [29].

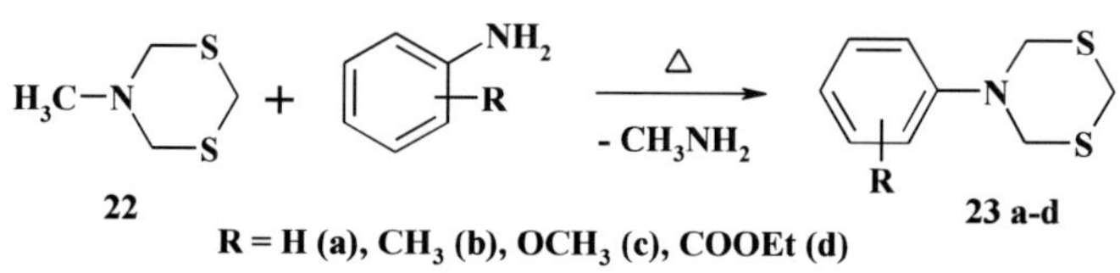

Since aromatic substituents in comparison with aliphatic substituents provides greater stability to six-membered heterocycles [29], thermal transamination of *N*-methyl-1,3,5-dithiazinane *22* with anilines was thus implemented with the formation of more stable *N*-aryl-1,3,5-dithiazinanes *23*.

Catalytic methods for the synthesis of *N*-aryl-1,3,5-dithiazinanes *23* based on transamination reaction of *N*-alkyl-1,3,5-dithiazinanes [30] and recyclization reaction of 1,3,5-trithiane [31] were developed later.

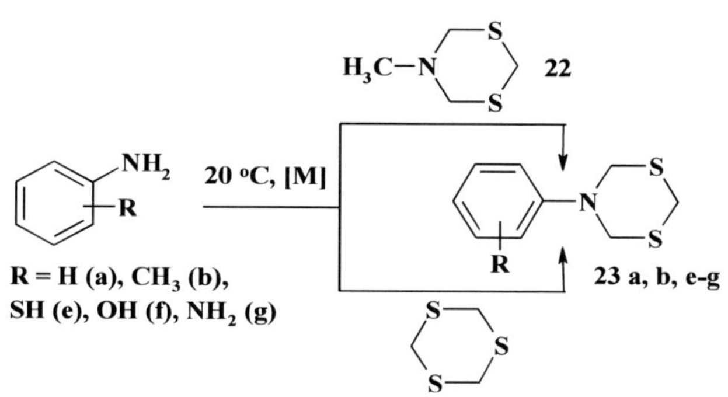

In certain systems as CH_2O-NaSH or CH_2O-Na_2S are applied as thiomethylating reagents for the synthesis of aminoalcohol-based 1,3,5-dithiazinanes [32, 33].

It should be noted, that amides and hydrazides of carboxylic acids are successfully subjected to cyclothiomethylation by primary amino group with the formation of 1,3,5-dithiazinanes *24* and *25* in presence of promoters, such as organic bases RONa [34, 35].

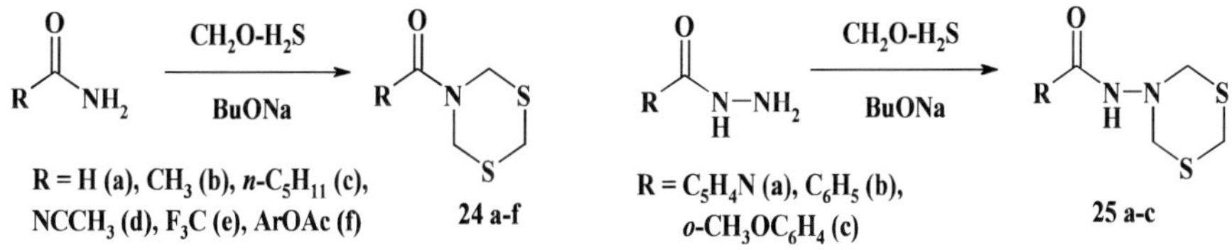

A necessary condition for intramolecular cyclocondensation of urea and thiourea by CH_2O-H_2S is presence of BuONa in 4-fold molar excess. As a result, said cyclothiomethylation reaction selectively leads to the formation of 1,3,5-thiadiazinan-4-(thi)ones 26 [36].

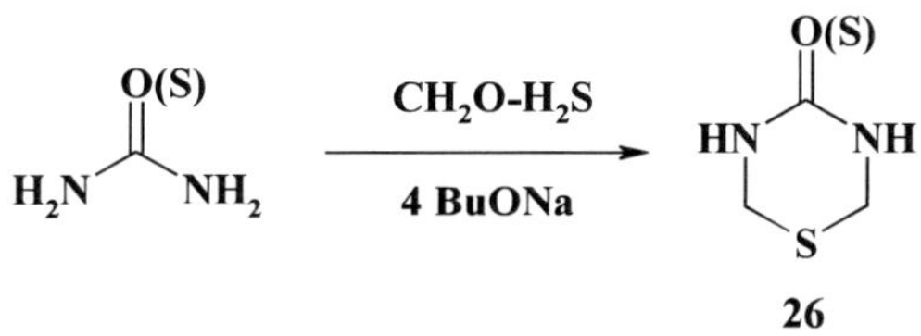

The crucial factors for the utilization of aliphatic di-and polyamines in the synthesis of bis-1,3,5-dithiazinanes *27* are considered to be reaction temperature of 40–80°C and an increase in reaction medium dilution level [37, 38]. It is also noted, that at 0°C an intramolecular cyclization with formation of bicyclo[3.n.1]thiadiazoalkanes 28 is characteristic for α,ω-diamines [39], while at 60 °C without dilution the formation of polymers *29* is observed.

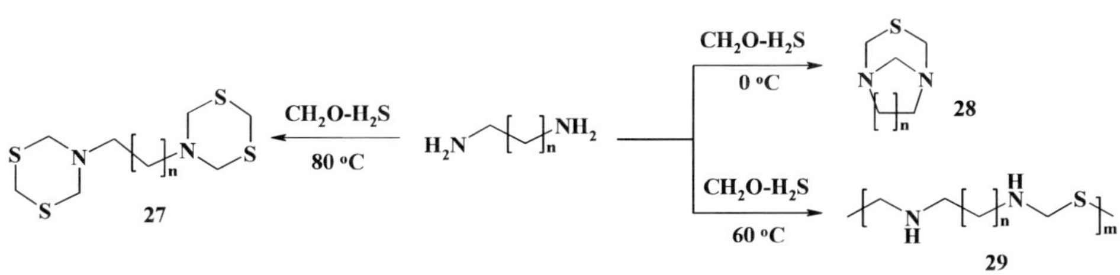

Alternative methods for cyclothiomethylation of amines with the formation of 1,3,5-thiadiazinan-2-thiones *31* are based on reactions with aldehydes and CS_2. Mechanism for such kind of the reaction is proposed through DFT-calculations and experimental results by several research groups [40-49]. It is assumed that amine along with CS_2 forms a dithiacarbaminic acid *30*, which in turn undergoes cyclization in presence of CH_2O and amine into a target heterocycle *31*.

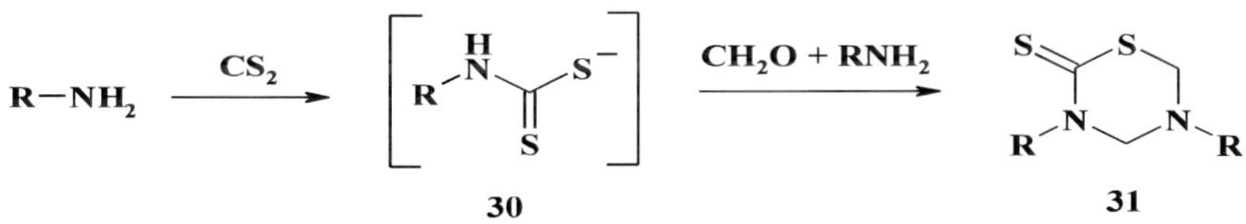

At the same time another heterocyclization pathway is possible, proceeding *via* the stage of imine formation followed by subsequent interaction with CS_2 and second imine molecule *32*.

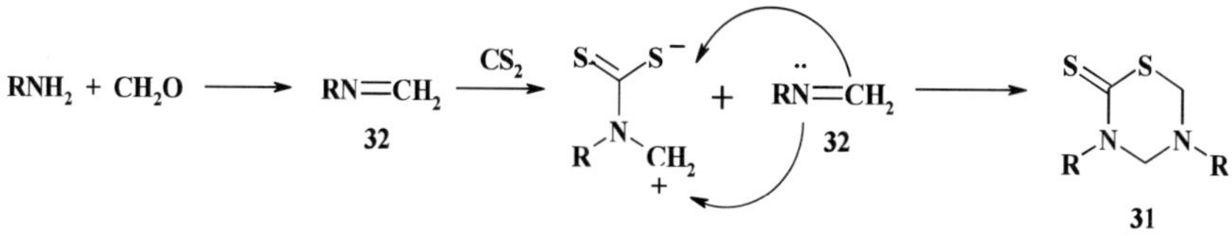

Examples of the heterocyclization reactions of amines with CH_2O and α-aminothiols *33* or thiourea *34* proceeding *via* an enol structure *34a* with the formation of 1,3,5-thiadiazinanes *35* and 36 are available [50-53].

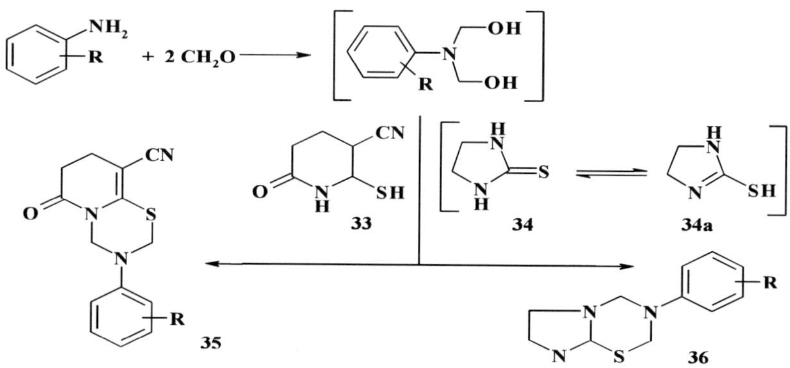

Similar result is obtained during the reaction of methylamine with formaldehyde and thiocarbaminic acid *37*, wherein, a thiol group participates in the formation of 1,3,5-thiadiazinane *38* as S-nucleophile. Probably stabilization of 1,3,5-thiadiazinanes *31*, *35*, *36* and *38* is ensured either by the presence of thionic groups in ring or N-aryl substituents.

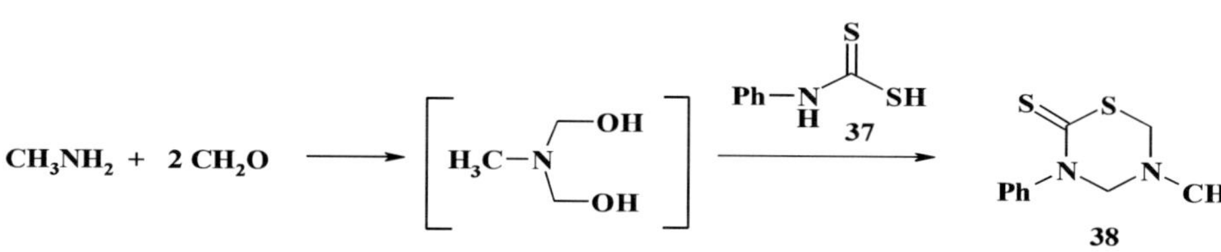

As shown above, several methods of one-pot synthesis of six-membered 1,3,5-dithiazinanes and 1,3,5-thiadiazinanes on the basis of amino substrates were thus developed. One-pot cyclothiomethylation of various amines with formaldehyde and H_2S is a rational approach for the production of 1,3,5-dithiazinanes. Among the effective synthetic approaches leading to the formation of 1,3,5-thiadiazinanes two routes may be drawn out. One of them is based on the reaction of amines with aldehydes and CS_2 and another is based on cyclocondensation of thiourea and its derivatives *via* Mannich reaction or cyclothiomethylation reaction with formaldehyde and H_2S in the presence of BuONa.

2.1.3. 1,5,3-Dithiazepinanes

One-pot multicomponent reactions of amino substrates with aldehydes and 1,2-dithiols opens a simple and effective pathway for the synthesis of seven-membered 1,5,3-dithiazepinanes *39* [54, 55].

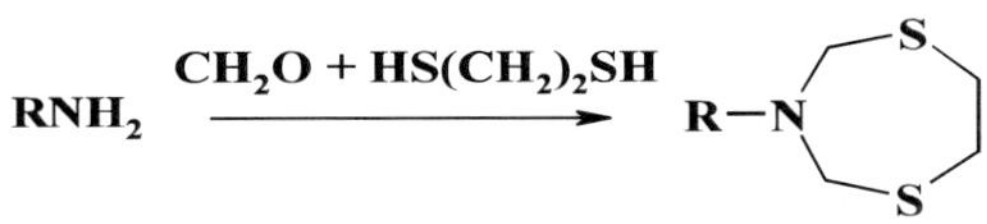

39

Present reaction is chemoselective for differing by nature thereof amino substrates, including bifunctional amines and hydrazines. Thus, hydrazine and α,ω-diamines do not undergo intermolecular thiomethylation with the formation of cyclic oligomers *40* and polymers *41*. *Bis*-1,5,3-dithiazepinanes *42a-b* are only formed during said reaction. This fact may lead to the conclusion of 1,5,3-dithiazepinane cycle being a thermodynamically more favorable structure, that is the key point of given methodology. In order to obtain water-soluble derivative the synthesis of *N*-coordinated adduct *43a* by the reaction of bis-1,5,3-dithiazepinane *42a* with CH_3I was implemented.

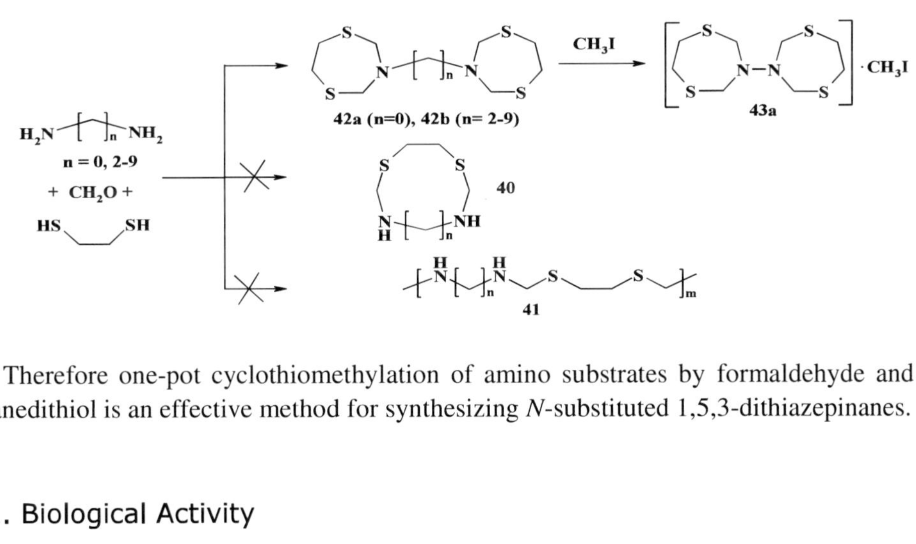

Therefore one-pot cyclothiomethylation of amino substrates by formaldehyde and 1,2-ethanedithiol is an effective method for synthesizing *N*-substituted 1,5,3-dithiazepinanes.

2.2. Biological Activity

2.2.1. 1,3,4-Thiadiazolidines

1,3,4-Thiadiazole-based compounds, such as acetozolamid, aethazolum and others, have been possess certain acitivity as antibacterial, antimicrobial, diuretic and antituberculous modes of action [56-60]. Saturated thiazolidine rings condensed with other ring constitute structural components of multiple drug substances including antibiotic penicillin series [1-4]. However, available data on saturated *N*-substituted 1,3,4-thiadiazolidines concerns only biological testing.

The fungicidal activity of thiadiazolidine-based compounds on microfungi has been established by bioassay [21]. Bioassay of microfungi species *Bipolaris sorokiniana* and *Fusarium oxysporum* were done, as the said species being known as crop pathogens.

Similarly, *Aspergillus fumigatus, Aspergillus niger* and *Paecilomyces variotii*, are able to cause fungal infections in humans and animals [61].

Authors of [62] consider ergosterol - a major cytoplasmic membrane component of fungi cells, to be the main target of antifungal agents. In particular for azole-based fungicides an ability to inhibit the activity of cytochrome P-450 dependent 14-α-demethylase – an ergosterol biosynthesis enzyme has been established [63, 64]. Due to the ability of micromycetes to develop resistance against fungicides applied, the search for new fungicides or to develop modifications thereof is necessary. Therefore it is convenient to study fungicidal activity of 1,3,4-thiadiazolidine compounds that are structurally close to azoles.

In *vitro* bioassay demonstrated, that thiadiazolidine bicycle *14* and its adduct *17* with CH_3I in concentrations of 0.04-0.13 % had completely suppressed the development of five test objects, namely *B. sorokiniana, F. oxysporum, A. fumigatus, A. niger* and *P.* variotii [21], whereas adducts *15* with HBr and *18* with $AlCl_3$ respectively possessed no activity.

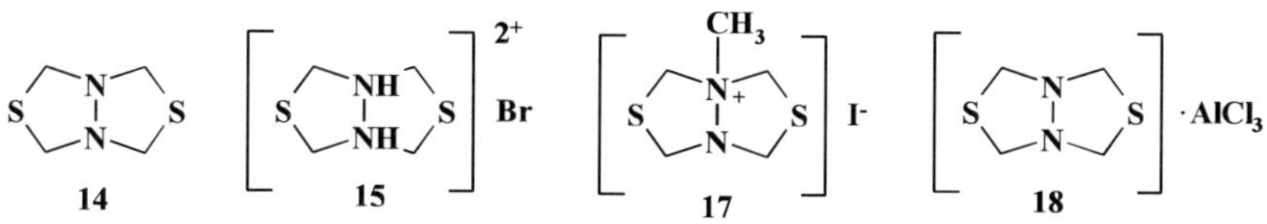

Inhibition activity of 2,4-dimethyl-3-phenyl-1,3,4-thiadiazolidine *7a* in regards to fungi development is selective, affecting only *Bipolaris sorokiniana* and Fusarium oxysporum [65].

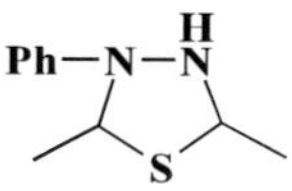

7a

2.2.2. 1,3,5-Dithiazinanes and 1,3,5-thiadiazinanes

1,3,5-dithiazinanes and 1,3,5-thiadiazinanes possess antibacterial, antimicrobial, herbicidal, insecticidal and miticidal properties [66-69]. In addition, some 1,3,5-dithiazinanes are also used as food additives that intensify smell and flavour of food [70, 71]. The data is available on biogenic origin of 2,4,6-trialkyl-1,3,5-dithiazinanes that have been found in beef broth [70], beef meat [71], sheepmeat [72], in Antarctic gull [73], in shrimp [74], as well as in cultivated soybean [75] and red pepper [76]. The biosynthetic precursors of 1,3,5-dithiazinanes are sulphur-containing amino acids, which undergoes subsequent biochemical processes to result in formation of NH_3, H_2S, acetaldehyde and CO_2 [77]. To prove this hypothesis research results were obtained during modeling of thermal or biochemical conversion of L-cysteine to dithiazinane *50a* provided in a mixture with thiadiazinane *50b* and trithiane 50c [78, 79], wherein dithiazinane *50a* is formed only at pH 7.1 [80].

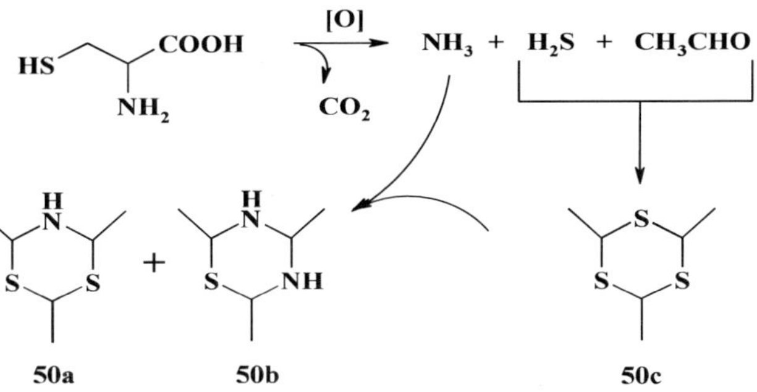

As reported in [81], study of canned pork (bacon), 2,4,6-triisobutyl-1,3,5-dithiazinane *50d* was isolated. The authors suggest, that the process proceeds with a contribution of leucine, which transforms into an isovaleraldehyde and NH_3, followed by formation of 1,3,5-dithiazinane *50d* in presence of biogenic H_2S. Further this heterocycle has been patented as a food additive to intensify the flavour of bacon, roasted nuts and poultry meat [82].

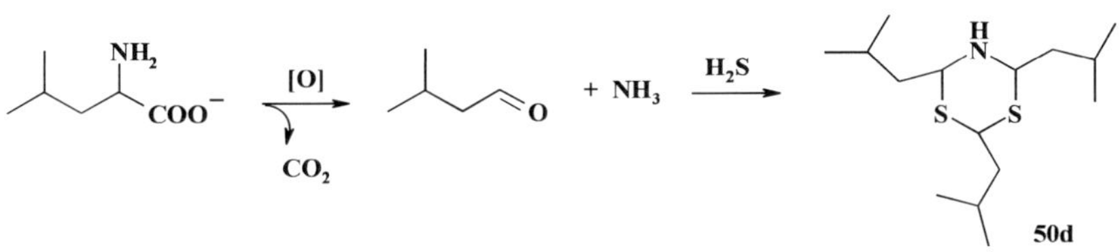

Substituted dimethyl-furyl-1,3,5-dithiazinanes *51a* and *51b* enhance the flavour of coffee, confectionery and bakery products [83].

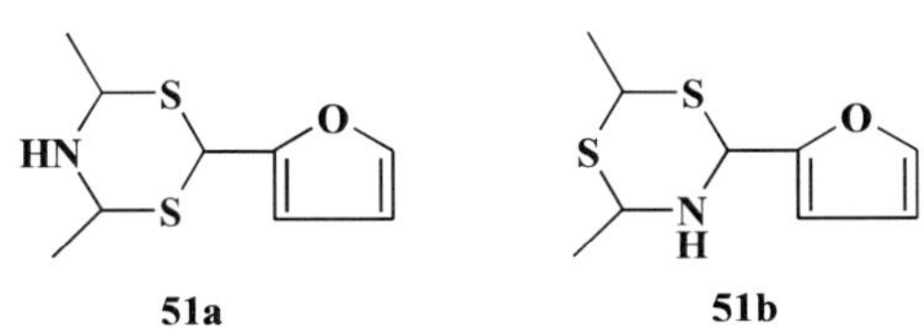

1,3,5-Dithiazinyl-5-ethanol *20c* and 1,3,5-dithiazinyl-5-acetic acid *20d* were established to demonstrate high biocidal activity in sulphate-reducing bacteria in various technological environments (from light industry to petroleum industry) [84]. These compounds are non-toxic and able to inhibit corrosion having a low solubility in petroleum.

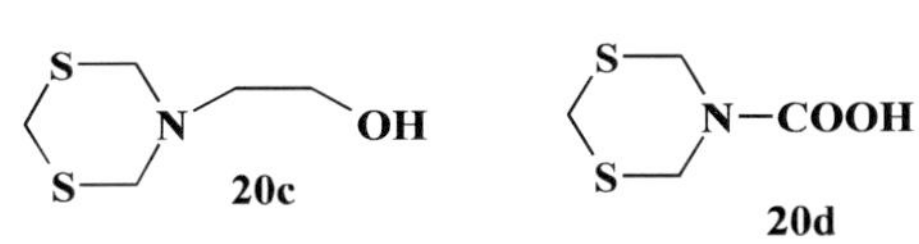

1,3,5-Dithiazinanes with pharmacophore substituents exhibit prolonged effects with preservation of pharmacological properties thereof. Thus, prone to cyclothiomethylation 4-aminoethylbenzoate is used in medical practice as an analgesic drug. Derivative ethyl-4-

(1,3,5-dithiazinyl)benzoate *23d,* is classified on the basis of mode of action on an organism as moderately toxic (LD 50 = 1100 mg/g) and reduces pain response about 50% [85] .

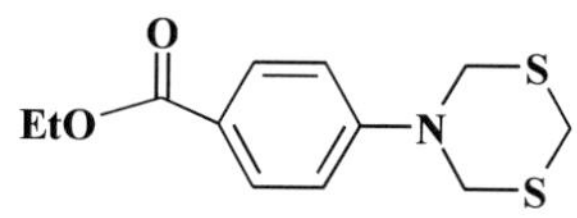

23d

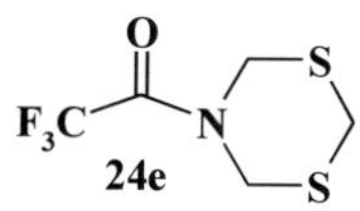

24e

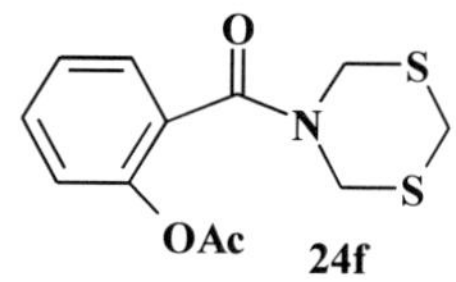

24f

A pronounced anti-inflammatory activity is demonstrated by 5-trifluoracyl-1,3,5-dithiazinane *24e* and 5-acetylsalicyloyl-1,3,5-dithiazinane 24f [86]. The greatest effect is observed for dithiazinane *24f.* An activity was found to be comparable with aspirin in 3 h and in 24 h an inflammatory effect of *24f* is still preserved, in contrast to that of aspirin.

5,5'-(3,6-Dioxaoctane-1,8-diyl)bis-1,3,5 dithiazinane *52* was found to possess high fungicidal activity in regards to microfungi *Bipolaris sorokiniana* and Aspergillus niger at concentration of 0.13% [87].

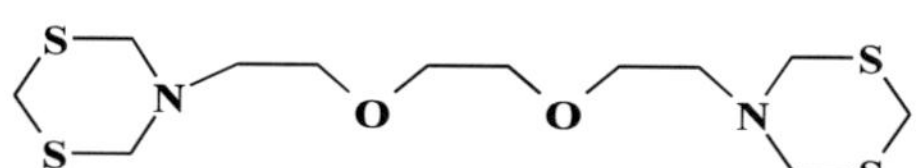

52

1,3,5-Thiadiazinanes *53, 54* were found to demonstrate antibacterial activity in regards to Salmonella, Staphylococicus ayreus, Bacteria coli [88] and *E. coli* [89]. 3,4-Ditolyl-1,3,5-dithiazinane *21b* were found to demonstrate fungicidal activity in regards to microfungi *B. sorokiniana*, *F. oxysporum*, *A.* fumigatus [90].

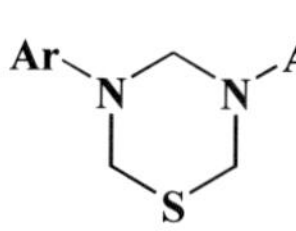

53

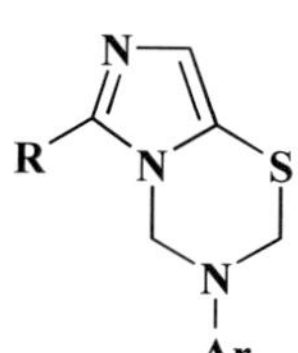

54

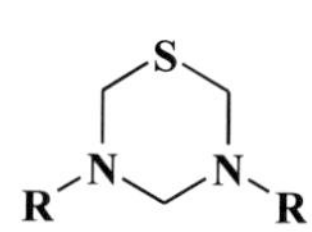

21b

Ar = 4-C_6H_4OEt, 4-C_6H_4Cl, 4-$C_6H_4CH_3$ **R = 4-$C_6H_4CH_3$**

N,N'-disubstituted 1,3,5-thiadiazinan-2-thiones 55 possess antibacterial and antifungal properties [91-94]. At the same time, N,N'-methyl-4,6-diaryl-1,3,5-thiadiazinan-2-thiones *56* are claimed as antisclerotic and antihyperlipidimic agents [95, 96]. On the basis of 1,3,5-thiadiazine-4-one *57* insecticide "Buprofezin" has been developed in order to control rice field pests planthopper (Nilaparvata lugens) [97].

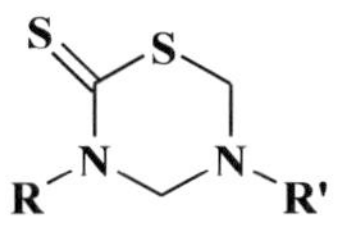

55

R = Bz, 4-ClC$_6$H$_4$CH$_2$, 4-MeOC$_6$H$_4$;
R' = Pr, CH$_2$CH$_2$OH, 4-MeOC$_6$H$_4$

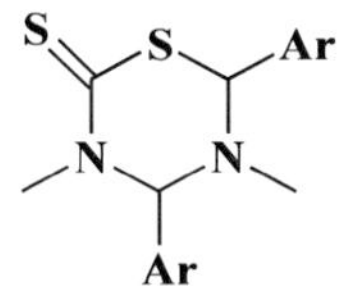

56

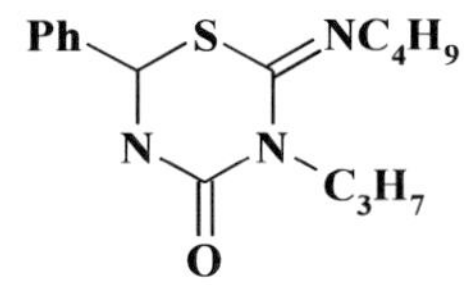

57

2.2.3. 1,5,3-Dithiazepinanes

Biological activity of seven-membered heterocycles containing three- and more S,N-heteroatoms in rings remains virtually unstudied . Authors of THIS CHAPTER, however, have conducted an *in vitro* bioassay for fungicidal activity of mono- and bis-1,5,3-dithiazepinanes *39*, *42* and *43* in regards to *B. sorokiniana*, *F. oxysporum*, *A. fumigatus*, *A. niger* and *P. variotii*.

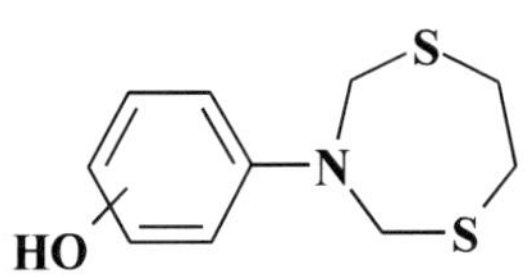

***o*-, *p*-isomers** **39**

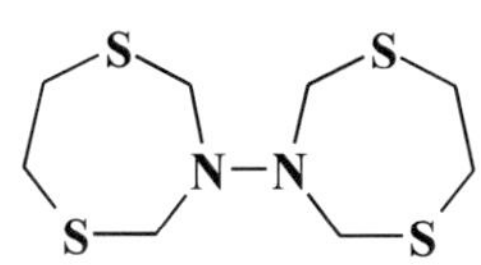

42

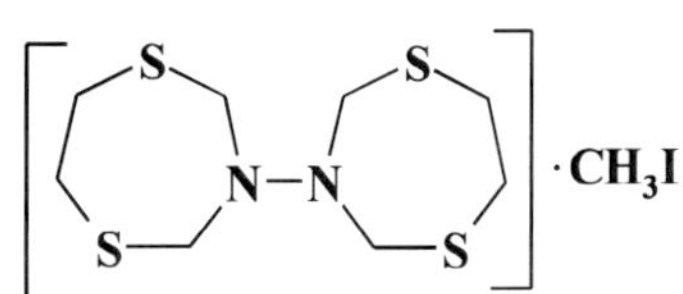

43

Among heterocycles tested, the compound *42* demonstrated the highest rate in ability to inhibit the development of microfungi. Water-soluble N-coordinated adduct *43* proved to be inactive. Mono-1,5,3-dithiazepinanes *39* were shown to inhibit the development solely of *Bipolaris sorokiniana*.

Conclusion

The data presented in this chapter on the synthesis and biological activity of saturated S,N-containing heterocycles illustrate methodological development of one-pot cyclothiomethylation reaction for the synthesis of bioactive 1,3,4-thiadiazolidines, 1,3,5-dithiazinanes, 1,3,5-thiadiazinanes and 1,5,3-dithiazepinanes *via* three-component reaction of amino substrates with carbonyl compounds and S-nucleophiles (H_2S, CS_2, thiols). These methods are simple, effective and of real interest for practical implementation. Six-membered 1,3,5-dithiazinanes and 1,3,5-thiadiazinanes derivatives are utilized as insecticides, fungicides, food additives and pharmaceuticals. Biological properties of five-membered 1,3,4-thiadiazolidines and seven-membered 1,5,3-dithiazepinanes are yet to become a subject of detailed study, which is reasonable when considering promising prospects of implementation thereof as highly effective agents for needs of agriculture, veterinary science and medicine.

References

[1] Pozharskii, A. F., Soldatenkov, A. T., Katritsky, A. A. *Heterocycles in Life and Society. 2011, John Wiley & Sons.*

[2] Katritsky, A. A., Pozharskii, A. F. Handbook of Heterocyclic Chemistry. 2000, *Pergamon: Amsterdam.*

[3] Pozharskii, A. F. Geterotsiklicheskie soedineniya v biologii i meditsine. *Sorovskiy obrazovatel'ny zhurnal, 1996, 6,* 25-32 (in Russian).

[4] Charushin, V. N. Chimiya v bor'be s infektsionnymi zabolevaniyami. *Sorovskiy obrazovatel'ny zhurnal, 2000, 3,* 64-72 (in Russian).

[5] Eilingsfeld, H. Competing ring closures of acylhydrazines. *Chem. Ser., 1965, 98,* 4, 1308-1321.

[6] Kurzer, F. Thiadiazoles. Part XI. Synthesis and cyclization of *N*-(thio-benzamido) guanidines and related compounds. *J. Chem. Soc., 1961*, 1617.

[7] Shakei, R. M., Mahmound, A. F., Abdel-Latif, F. F. Synthesis and biological activities of novel 1,4-bridged bis-1,2,4-triazoles, bis-1,3,4-thiadiazoles and bis-1,3,4-oxadiazoles. *Posph., Sulfur and Silicon and Relat. Elem., 2005, 180,* 2, 397-406.

[8] Abdel-Aal, M. T., El-Sayed, W. A., Abdel-Aleem, A. H., El-Ashry, E. S. H. Synthesis of some functionalized arylaminamethyl-1,2,4-triazoles, 1,3,4-oxa and thiadiazoles. *Pharmazie, 2003, 58,* 11, 788-792.

[9] Neureiter, N. P. Monomeric formaldazin – synthesis of 1,3,4-thiadiazolidine – a new heterocycle. *J. Am. Chem. Soc., 1959, 81,* 2910.

[10] Ruhlmann, K. Uber die Synthese von 1,3,4-Thiadiazolidinen und 1,3,4-Thiadiazolen. *J. Prakt. Chem., 1959, 5,* 6, 285-291.

[11] Buter, J., Wassenaar, S., Kellogg, R. M. Thiocarbonyl Ylides. Generation, Properties and Reaction. *J. Org. Chem., 1972, 37,* 25, 4045-4060.

[12] Kellogg, R. M., Noteboom, M., Kaiser, I. K. Thiocarbonyl Ylides. Stereochemical Properties of 4-*tert*-Butylcyclohexyl Derivatives. *J. Org. Chem., 1975, 40,* 17, 2573-2574.

[13] Akhmetova, V. R., Nadyrgulova, G. R., Tyumkina, T. V., Starikova, Z. A., Golovanov, D. G., Antipin, M. Yu., Kunakova, R. V., Dzhemilev, U. M. Cyclothiomethylation of aryl hydrazines with formaldehyde and hydrogen sulfide. *Russian Chemical Bulletin, 2006, 55,* 10, 1824-1834.

[14] Akhmetova, V. R., Khairullina, R. R., Tyumkina, T. V., Nelyubina, Yu. V., Smol'yakov, A. F., Bushmarinov, I. S., Starikova, Z. A., Abdullin, M. F., Kunakova, R. V. Cyclothiomethylation of carboxylic acid hydrazides with aldehydes and H_2S. *Russian Chemical Bulletin, 2010, 59,* 2, 425-433.

[15] Kentaro, O., Kazuko, N., Yasataka, N., Kento, K., Kasei, S. Synthesis of 1,3,4-thiadiazoles from aldehyde hydrazones. *Synthesis, 2004, 12,* 1929.

[16] Nalesnik, T. E. Thiadiazolidine additives for lubricants. *US Patent 2003, 6,*559, 107B2.

[17] Heugebaert, F. C., Willems, J. F. Certain 1,3,4-thiadiazolidine-2-thiones and their derivatives. *US Patent 1969, 3,*474, 108.

[18] Grashey, R., Huisgen, R., Kun Sun, K. 1,3-Dipolar Cycloadditions. XII. The Synthesis of 1,3,4-Thiadiazolidine-5-thiones. *J Org. Chem., 1965, 30,* 1, 74-79.

[19] Khafizova, S. R., Akhmetova, V. R., Tyumkina,T. V., Khalilov, L. M., Kunakova, R. V., Dzhemilev, U. M. Multicomponent heterocyclization of hydrazine, hydrogen sulfide, and formaldehyde. *Russian Chemical Bulletin, 2004, 53*, 8, 1717-1721.

[20] Akhmetova, V. R., Nadyrgulova, G. R., Khafizova, S. R., Khairullina, R. R., Paramonov, E. A., Kunakova, R.V., Dzhemilev, U.M. Cyclocondensation of hydrazine, formaldehyde, and hydrogen sulfide in the presence of acids and bases. *Russian Journal of Organic Chemistry, 2006, 42*, 1, 145-147.

[21] Akhmetova, V. R., Murzakova, N. N., Khabibullina, G. R., Galimzyanova, N. F. Synthesis of N- and N,N′-coordinated derivatives of 3,7-dithia-1,5-diazabicyclo[3.3.0]octanes and their fungicidal activity. *Russian Journal of Applied Chemistry, 2011, 84*, 2, 225-229.

[22] Wohl, A. (1886). Derivate des thioformaldehyds. *Berichte, 19,* 2344-2347.

[23] Le Fevre, C. S., Le Fevre, R. W. The interaction of ammonium and alkylammonium sulphides with aqueous formaldehyde. *J. Chem. Soc., 1932,* 1142-1148.

[24] Aleev, R. S., Dal'nova, Yu. S., Popov, Yu. N., Masagutov, R. M., Raphikov, S. R. Reaktsiya tiometilirovaniya. *Doklady Akademii Nauk, 1988, 4,* 873-875 (in Russian).

[25] Khafizova, S. R., Akhmetova, V. R., Korzhova, L. F., Tyumkina, T. V., Nadyrgulova, G. R., Kunakova, R. V., Kruglov, E. A., Dzhemilev, U. M. Multicomponent condensation of aliphatic amines with formaldehyde and hydrogen sulfide. *Russian Chemical Bulletin, 2005, 54*, 2, 432-436.

[26] Collins, D., Gramore, J. The action of hydrogen sulphide on certain aromatic amines in the presence of formaldehyde. *J. Chem. Soc., 1953*, 4089-4090.

[27] Cadenas-Pliego, G., Rosales-Hoz, J., Contreras, R., Flores-Parra, A. New chiral heterocycles: 5-[(R)-(+)-1′-methylbenzyl]-1,3,5-dithiazine and 3,7-[(R)-(+)-1′-methylbenzyl]-3,7-diaza-1,5-dithiacyclooctane. *Tetrahedron Asymm., 1994, 5,* 633-640.

[28] Khafizova, S. R., Akhmetova, V. R., Kunakova, R. V., Dzhemilev, U. M. Thiomethylation of aromatic amines: efficient method for the synthesis of heterocyclic compounds. *Russian Chemical Bulletin, 2003, 52,* 8, 1817-1821.

[29] Akhmetova, V. R., Niatshina, Z. T., Khabibullina, G. R., Bushmarinov, I. S., Borisova, A. O., Starikova, Z. A., Korzhova, L. F., Kunakova, R. V. Synthesis, crystal structure, and interconversions of new *N*-aryl-1,3,5-dithiazinanes, 1,3,5-thiadiazinanes, and 1,5-dithia-3,7-diazacyclooctanes. *Russian Chemical Bulletin, 2010, 59*, 5, 1002-1009.

[30] Niatshina, Z. T., Murzakova, N. N., Vasilieva, I. V., Rakhimova, E. B., Akhmetova, V. R., Ibragimov, A. G. Efficient method for a synthesis of *N*-substituted dithiazinanes via transamination of *N*-methyl-1,3,5-dithiazinane with arylamines and hydrazines. *Arkivoc, 2011, VIII,* 141-148.

[31] Murzakova, N. N., Rakhimova, E. B., Vasilieva, I. V., Prokof'yev, K. I., Ibragimov, A. G., Dzhemilev, U. M. A new method for the synthesis of N-substituted 1,3,5-dithiazinanes via the catalytic recyclizationof 1,3,5-trithiane with aryl(benzyl)hydrazines and aryl amines. *Tetrahedron Lett., 2011, 52,* 4090-4092.

[32] Galves-Ruiz, J. C., Noth, H., Flores-Parra, A. Organometallic aluminum compounds derived from 2-(1,3,5-dithiazinan-5-yl)ethanol ligands. *Inorg. Chem., 2003, 42,* 7569-7578.

[33] Galves-Ruiz, J. C., Jean-Gaspar, J. C., Castellanos-Azzora, I. G., Contreras, R., Flores-Parra, A. 2-(1,3,5-Dithiazinan-5-yl)ethanol Heterocycles, Structure and Reactivity. *Heterocycles, 2004, 63,* 2269-2285.

[34] Akhmetova, V. R., Khairullina, R. R., Nadyrgulova, G. R., Kunakova, R. V., Dzhemilev, U. M. Multicomponent heterocyclization of carboxamides with H_2S and CH_2O. *Russian Journal of Organic Chemistry, 2008, 44*, 2, 190-196.

[35] Akhmetova, V. R., Khairullina, R. R., Tyumkina, T. V., Nelyubina, Yu. V., Smol'yakov, A. F., Bushmarinov, I. S., Starikova, Z. A., Abdullin, M. F., Kunakova, R. V. Cyclothiomethylation of carboxylic acid hydrazides with aldehydes and H_2S. *Russian Chemical Bulletin, 2010, 59*, 2, 425-433.

[36] Akhmetova, V. R., Khairullina, R. R., Bushmarinov, I. S., Tyumkina, T. V., Yanybin, V. M. Multicomponent reactions of urea and its derivatives with CH_2O and H_2S in the synthesis of 1,3,5-thiadiazinane-4-(thi)ones and macroheterocycles. *Arkivoc, 2011, VIII,* 149-162.

[37] Akhmetova, V. R., Khabibullina, G. R., Rakhimova, E. B., Vagapov, R. A., Khairullina, R. R., Niatshina, Z. T., Murzakova, N. N. Multicomponent reactions of amines with aldehydes and H_2S as efficient route to heterocycles and thioaza macrocycles. *Mol. Divers., 2010, 14,* 463-471.

[38] Akhmetova, V. R., Rakhimova, E. B., Vagapov, R. A., Kunakova, R. V., Dzhemilev, U. M. Cyclothiomethylation of aliphatic polyamines with formaldehyde and hydrogen sulfide. *Russian Journal of Organic Chemistry, 2008, 44,* 4, 499-504.

[39] Akhmetova, V. R., Vagapov, R. A., Nadyrgulova, G. R., Tyumkina, T. V., Starikova, Z. A., Antipin, M. Yu., Kunakova, R. V., Dzhemilev, U. M. Synthesis of thiadiazabicyclane and bis-1,3,5-dithiazinane by cyclothiomethylation of aliphatic diamines with CH_2O and H_2S. *Tetrahedron, 2007, 63,* 47, 11702-11709.

[40] Perez, R., Suarera, M., Ochoa, E., Rodrigueza, H., Martin, N., Seoane, C., Novoa, H., Blatond, N., Pecters, O. M., Ranter, C. A joint theoretical and experimental structural study of 5-carboxyethyl-3-(2′-furfurylmethyl)tetrahydro-2H-1,3,5-thiadiazine-2-thione. *Tetrahedron, 2001, 57,* 7361-7367.

[41] Schorr, M., Durckheimer, W., Lammler, G., Rieche, A. Tetrahydro-1,3,5-thiadiazine-2-thione with effect against Opisthorchis felineus. *Naturwissenschaften, 1966, 53,* 156-157.

[42] Bialy Serry, A. A., Abdelal, A. M., El-Shorbagi, A., Kheira Samy, M. M. 2,3-Bis(5-alkyl-2-thiono-1,3,5-thiadiazin-3yl) propionic acid: one-pot domino synthesis and antimicrobial activity. *Archiv der Pharmazie, 2005, 338,* 1, 38-43.

[43] El-Shorbagi, A. Model for delivery of amines through incorporation into a tetrahydro-2H-1,3,5-thiadiazine-2-thione structure. *Eur. J. Med. Chem., 1994, 29,* 11-15.

[44] Abould-Fadl, T., Hussein, A. M., El-Shorbagi, A., Khallil, A. R. New 2H-tetrahydro-1,3,5-thiadiazine-2-thiones incorporating glycine and glycinamide as potential antifugal agents. *Archiv der Pharmazie, 2002, 335,* 438-442.

[45] Coro, J., Alvarez-Puebla, R., Montero, A. L., Suarez, M., Martin, N., Perez-Pineiro, R. A computational approach to the synthesis of 1,3,5-thiadiazinane-2-thiones in aqueous medium: theoretical evidence for water-promoted heterocyclization. *J. Mol. Model, 2008, 14,* 641-647.

[46] Molero, D., Coro, J., Perez, R., Suarez, M., Martinez-Alvarez, R., Herrera, A., Martin, N. ^{1}H and ^{13}C NMR spectral assignment of alkyl and polyamine linked bis(2 thioxo [1,3,5]-thiadiazinan-3-yl) carboxylic acids. *Magn. Reson Chem., 2007,45,* 93-98.

[47] Coro, J., Little, S., Yardley, V., Suares, M., Rodriges, H., Martin, N., Perez-Pineiro, R. Synthesis and antiprotozoal evaluation of new N^4-(benzyl)spermidyl-linked bis(1,3,5-thiadiazinane-2-thiones). *Archiv der Pharmazie, 2008, 341,* 708-713.

[48] Fujikawa, Y., Sakashita, M., Tsuruzoe, N. 3,5-Dialkyl-4,6-diaryl-tetrahydro-2H-1,3,5-thiadiazine-2-thione derivative, process for its production and antihyperlipidemic agent containing it. *US Patent 1987, 4 639 444.*

[49] Hisgen, R., Morikawa, M., Breslow, D. S., Grashey, R. 1,4-Dipolare Cycloadditionen, IV. Reaktionen der Azomethine mit Isothiocyanaten und Schwefelkohlenstoff. *Chem. Berichte, 1967, 100,* 1602-1615.

[50] Sexton, W. A., Harland, J. Insecticides and fungicides containing organic sulphur compounds. *Canadian Patent, 1953, 498 281.*

[51] Dotsenko, V. V., Frolov, K. A., Krivokolysko, S. G., Litvinov, V. P. The Mannich reaction in the synthesis of N,S-containing heterocycles 10. Recyclization of stable cyclic michael adducts, *N*-methylmorpholinium 6-R-6-hydroxy-1,4,5,6-tetrahydropyridine-2-thiolates, to pyrimido[4,3-*b*][1,3,5]thiadiazines under conditions of aminomethylation reaction. *Russian Chemical Bulletin, 2009, 58*, 7, 1479-1483.

[52] Frolov, K. A., Dotsenko, V. V., Krivokolysko, S. G., Litvinov, V. P. Mannich reaction in the synthesis of N,S-containing heterocycles. 1. Synthesis of novel derivatives of imidazo[2,1-*b*][1,3,5]thiadiazine. *Russian Chemical Bulletin, 2005, 54,* 9, 2226-2228.

[53] Dotsenko, V. V., Frolov, K. A., Krivokolysko, S. G., Chernega, A. N., Litvinov, V. P. Convenient Mannich-Type One-Pot Synthesis of Pyrimido[6,1-*b*][1,3,5]thiadiazines. *Monatshefte fur Chemie, 2006, 137,* 8, 1089-1098.

[54] Ito, K., Sekiya, M. Preparation of *N*-Alkylthiomethyl Derivatives of Hydroxylamines. *Chem. Pharm. Bull., 1979, 27,* 1691.

[55] Murzakova, N. N., Prokof'ev, K. I., Tyumkina, T. V., Ibragimov, A. G. Synthesis of N-aryl-1,5,3-dithiazepanes and N-aryl-1,5,3-dithiazocanes in the presence of samarium- and cobalt-containing catalysts. *Russian Journal of Organic Chemistry, 2012, 48,* 4, 588-593.

[56] Mashkovskii, M. D. *Lekarstvennye sredstva.* Moskva: Novaya volna (in Russian), *2008.*

[57] Kubo, H., Sato, R., Hamura, I., Ohi, T. Herbicidal activity of 1,3,4-thiadiazole derivatites. *J. Arg. Food Chem, 1970, 18,* 1, 60-65.

[58] Chapleo, C. B., Myers, M., Saville, J. F., Smith, A. C. B., Stilling, M. R., Tulloch, I. F., Walter, D. S., Welboum, A. P. Substituted 1,3,4-thiadiazoles with anticonvulsant activity. *J. Med. Chem., 1986, 29,* 2273-2280.

[59] Turner, S., Myers, M., Gadie, B., Hale, S. A., Horsley, A., Nelsen, A. J., Pape, R., Saville, J. F., Doxey, J. C., Berridge, T. L. Vasodilator activity of some 2-aryl-5-guanidino-1,3,4-thiadiazoles. *J. Med. Chem., 1988, 31,* 906-913.

[60] Oruc, E. E., Rollas, S., Kandemirli, F., Shvets, N., Dimoglo, A. S. Synthesis, structure elucidation, and structure – antiberculosis activity relationship investigation. *J. Med. Chem., 2004, 47,* 6760-6767.

[61] Saton, D., Fotergill, A., Rinaldi, M. *Opredelitel' patogennykh i uslovno-patogennykh gribov.* Moskva: Mir (in Russian), *2001.*

[62] Odds, F. C., Brown, A. J., Gow, N.A. Antifungal agents: mechanisms of action. A review. *Trends Microbiol., 2003, 11,* 6, 272-279.

[63] Becher, R., Hettwer, U., Karlovsky, P., Deising, H. B., Wirsel, S. G. Adaptation of Fusarium graminearum to tebuconazole yielded descendants diverging for levels of fitness, fungicide resistance, virulence, and mycotoxin production. *Phytopathology, 2010, 100,* 5, 444-453.

[64] Vicentini, C. B., Romagnoli, C., Manfredini, S., Rossi, D., Mares, D. Pyrazolo[3,4-c]isothiazole and isothiazolo[4,3-d]isoxazole derivatives as antifungal agents. *Pharm Biol., 2011,49,* 5, 545-552.

[65] Makhmudiyarova, N. N. Geterotsiklizatsiya gidrazinov s aldegidami i SH-kislotami. Dissertatziya na soiskanie uchenoi stepeni kandidata khimitcheskikh nauk. [Heterocyclization of hydrazines with aldehydes and SH-acids. Ph.D.Thesis]. Ufa, 26 p (in Russian), 2011.

[66] Kamal, A. Recent Advances in the Synthetic Uses of Chlorocarbonyl Isocyanate. *Heterocycles, 1990, 31,* 1377-1391.

[67] Yadav Lal Dhar, S., Vaish, A., Sharma, S. New Fungitoxic Fused-Ring Synthetics Incorporating Azoles and Azines in Different Combinations. *J. Agric. Food Chem., 1994, 42,* 811-813.

[68] Ozcelik, A. B., Ersan, S., Ural, A. U., Ozkan, S., Ertan, M. Synthesis of 3-substituted-5-(4-carboxycyclohexylmethyl)-tetrahydro-2H-1,3,5-thiadiazine-2-thione derivatives as antifibrinolytic and antimicrobial agents. *Arzneimittel Forschung, 2007, 57,* 554-559.

[69] Hussein, M. A., Hashem, M. Synthesis of New 3-Substituted-5-(2-hydroxyethyl)-3,4,5,6-tetrahydro-2H-1,3,5-thiadiazine-2-thione Derivatives with Potential Antimicrobial Activity. *Archiv der Pharmazie, 2008, 6,* 370-376.

[70] Brinkman, H. W., Copier, H., de Leuw, J. M., Tjan, S. B. Components Contributing to Beef Flavor: Analysis of the Headspace Volatiles of Beef Broth. *J. Agr. Food Chem., 1972, 20,* 2, 177-181.

[71] Mac Leod, G., Coppock, B. M. Comparison of the Chemical Composition of Boiled and Aromas of Heated Beef. *J. Agr. Food Chem., 1977, 25,* 1, 113-116.

[72] Tjan, S. B., Haakman, J. C., Teunis, C. J., Peer, H. G. Synthesis of 3,5-dialkyl-1,2,4-trithiolanes. Assignment of configuration and conformation analysis by PMR. *Tetrahedron, 1972, 28,* 13, 3489-3500.

[73] Kubota, K., Shijimaya, H., Kobayashi, A. Volatile Component of Roasted Shrimp. *Agric. Biol. Chem., 1986, 50,* 11, 2867-2873.

[74] Kubota, K., Nakamoto, A., Moriguchi, M., Kobayashi, A., Ishii, H. Formation of Pyrrolidino[1,2-e]-4H-2,4-dimethyl-1,3,5-dithiazine in the Volatiles of Boiled Short-Necked Clam, Clam and Corbicula.*J.Agric. Food Chem., 1991, 39,* 1127-1130.

[75] Kentaro, O. Recent Studies on the Reactions of Thioaldehydes and Thioketones. *Journal of Sulfur Chemistry, 2002, 23,* 2, 209-241.

[76] Buttery, P. J., Beekerton, A., Mitchell, R. M., Devis, K., Annison, E. F. Protein Turnover in Animals. *Biochem. J., 1975, 4,* 599-604.

[77] Mulders, E. J. Volatile components from the nonenzymic browning reaction of the cystein/cystine-ribose system. *Lebensm.Unters.-Forsch, 1973, 152,* 193-201.

[78] Ledl, F., Severin, T. Sulfur Compounds from Cystein and Xylose. *Chem.Mikrobiol. Tech.Lebensm., 1973, 2,* 155-159.

[79] Boelens, M., van der Linde, L. M., De Valois, P. J., van Dort, H. M., Takken, H. J. Organic Sulfur Compounds from Fatty Aldehydes, Hydrogen Sulfide, Thios, and Ammonia as Flavor Constituents. *J.Agr.Food Chem., 1974, 22,* 6, 1071-1076.

[80] Chi-Kuen, S., Chi-Tang, H. Effect of pH on the Volatile Formation from the Reaction between Cystein and 2,5-Dimethyl-4-hydroxy-3(2H)-furanone. *J.Agr.Food Chem. 1988, 36,* 801-803.
[81] Hinrichsen, L. L. and Andersen, H. J. Volatile Compounds and Chemical Changes in Cured Pork. *J. Agric. Food Chem., 1994, 42,* 7, 1537-1542.
[82] Shu, C., Mookherjee, B., Vock, M. Preparation of 2,4,6-tri-isobutyl-dihydro-1,3,5-dithiazine. *US Patent 4228278, 1981.*
[83] Bruning, J., Emberger, R., Guntert, M., Hopp, R., Kopsel, M., Werkhoff, P. A novel method for generation of thial S-sulfides from 2,4,6-trisubstituted 5,6-dihydro-1,3,5-dithiazines. *US Patent 5166342, 1992.*
[84] Khafizova, S. R., Akhmetova, V. R., Nadyrgulova, G. R., Rusakov, I. V., Kunakova, R. V., Dzhemilev, U. M. Tsiklotiometilirovanie aminokislot i ikh proizvodnykh s pomoshch'yu formal'degida i serovodoroda. *Neftekhimiya, 45,* 3, 1-6 (in Russian), *2005.*
[85] Niatshina, Z. T. Tiometilirovanie aril- i getarilaminov smesyu CH_2O-H_2S. Dissertatziya na soiskanie uchenoi stepeni kandidata khimitcheskikh nauk. [Thiomethylation of aryl- and hetarylamines with CH_2O-H_2S. Ph.D.Thesis]. Ufa, 26 (in Russian), 2010.
[86] Akhmetova, V. R., Khairullina, R. R., Parfenova, T. I., Sufiyarova, R. Sh., Bashkatov, S. A., Kunakova, R. V. Synthesis and anti-inflammatory activity of 5-acyl-1,3,5-dithiazinanes. *Pharmaceutical Chemistry Journal, 2011, 44,* 10, 534-535.
[87] Akhmetova, V. R., Rakhimova, E. B., Minnebaev, A. B., Kunakova, R. V., Galimzyanova, N. F. Cyclothiomethylation of heterochain α,ω-diamines with formaldehyde and H_2S. *Russian Journal of Organic Chemistry , 2012, 48,* 2, 202-208.
[88] Hughes, S. A., McCall, E. B. Thiazetidines, dihydrodithiazines, and tetrahydrothiadiazines as mycobacteriostats. *Brit. Patent 943273, 1963.*
[89] Dotsenko, V. V, Krivokolysko, S. G., Chernega, A. N., Litvinov, V. P. Synthesis and Structure of Pyrido[2,1-*b*][1,3,5]thiadiazine Derivatives. *Doklady Chemistry, 2003, 389,* 92-96.
[90] Akhmetova, V. R., Niatshina, Z. T., Burakaeva ,A. D., Galimzyanova, N. F., Kunakova, R. V. Synthesis and fungicidal activity of *N,N'*-dithol-4-yl-1,3,5-thiadiazinane and *N,N'*-[methylene-bis-(thiomethylene)]-bis-[4(5)-methyl-1,3-thiazole-2-amines]. *Russian Journal of Applied Chemistry, 2011, 84,* 3, 416-419.
[91] Traber, W. Tetrahydro-1,3,5-thiadiazine-3-thione. *US Patent 3475422, 1965.*
[92] Traber, W. Method and composition for controlling microorganisms with tetrahydro-1,3,5-thiadiazine-3-thiones. *US Patent 3513234, 1970.*
[93] Schorr, M., Durckheimer, W., Main, F., Lammler, G. Tetrahydro-1,3,5-thiadiazine-3-thione. *US Patent 3398146, 1965.*
[94] Eggensperger, H., Heinz, K. Tetrahydrothiadiazinethiones. *US Patent 3946006, 1975.*
[95] Sasaki, T., Fujikawa, Y., Sakoda, R., Sakashita, M., Hibi, M. Antiatherosclerotic agent compositions. *US Patent 4719209, 1986.*
[96] Sakashita, M., Toyoda, K., Kitahara, M., Tsuruzoe, N., Nakayama, S., Oguchi, K. Mechanism of the hypolipidemic effect of NIP-200 in rats. *Jpn. J. Pharmacol., 1993, 62,* 87-91.
[97] Kanno, H. An approach to a novel insect growth regulator buprofezin (Applaud). *Pure&Appl. Chem., 1987, 59,* 8, 1027-1032.

In: Bioactive Heterocycles
Editors: K. L. Ameta, R. P. Pawar and A. J. Domb
ISBN: 978-1-62257-636-4
© 2013 Nova Science Publishers, Inc.

Chapter 6

Synthesis and Bioactivity of Dihydropyrimidines

Thokchom Prasanta Singh and Okram Mukherjee Singh*
Department of Chemistry, Manipur University, Canchipur,
Imphal, India

Abstract

Dihydropyrimidines received considerable attention over the last few decades due to their interesting pharmacological properties like, antimicrobial, anti-tuberculosis, anti-inflammatory, calcium-channel blockers, mitotic inhibitions, anti-cancer activities, etc. In this chapter, we reviewed recent developments of DHPMs in the field of medicinal chemistry and pharmacology.

1. Introduction

Dihydropyrimidines comprise an important class of bioactive nitrogen containing heterocycles. They attracted considerable interest of medicinal chemists because of their wide spectrum of therapeutic and pharmacological properties. In 1893, Italian chemist Pietro Biginelli reported the first synthesis of 3, 4-dihydropyrimidin-2(1*H*)-ones *(4)* by simple one-pot condensation reaction of an aromatic aldehyde, urea and ethylacetoacetate[1]. The reaction was carried out by heating a mixture of three components in ethanol with a catalytic amount of HCl under reflux conditions. A representative example of the Biginelli reaction is shown in scheme-1.

* E-mail: ok_mukherjee@yahoo.co.in.

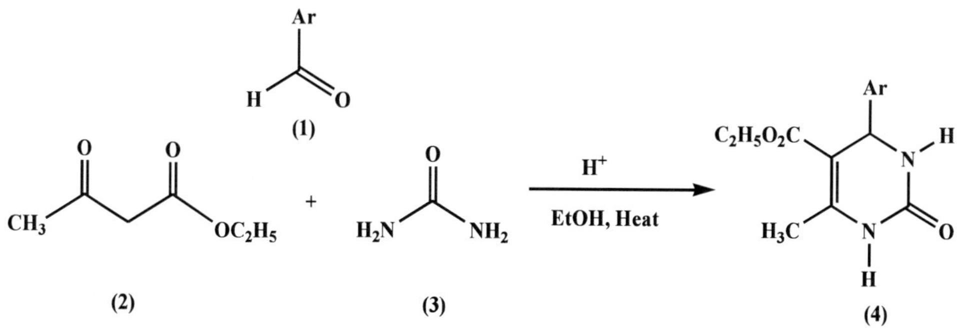

Scheme 1. Biginelli dihydropyrimidinone synthesis.

The scope of Biginelli dihydropyrimidine synthesis has now been broadened by changing all three building blocks, and thus accessing a large number of multifunctionalized pyrimidine derivatives [2-3].

2. Biological Activities of DHPMs

Dihydropyrimidines show a diverse range of biological activities. They are recognized for possessing antibacterial [4], antiviral (nitractin) [5], antitumor [6-9], analgesic and anti-inflammatory [10], antiplatelet aggregation [11], antihypertensive activity [12] and calcium-channel modulator [13] activities. Thus, the development of new methods for efficient bioactive dihydropyrimidine synthesis is always attracted the attention of pharmaceutical industry [14]. The various activities of recently prepared DHPMs are summarized as below.

2.1. Antifungal Activity

The potential utility of dihydropyrimidines as bioactive compounds is well known, however, their antifungal activities were rarely studied [15]. Singh *et al.* [16] prepared DHPMs under solvent-free condition using the catalytic amount of $CuCl_2.2H_2O$ and evaluated the antifungal activities of prepared compounds. The effect of the compounds on the radial growth of three fungal species *viz. Trichoderma hammatum, Trichoderma koningii* and *Aspergillus niger* was assessed by amending the compounds in potato dextrose agar (PDA) medium. DHPMs (*5-6*) showed significant *in vitro* antifungal activities with low MIC values in the range 0.025–0.35 μg/mL. Chitra *et al.* [17] reported the synthesis of several 4-aryl-5-isopropoxycarbonyl-6-methyl-3, 4-dihydropyrimidin-2(1H)-ones/thiones (*5-6*) by one-pot cyclocondensation of aldehydes, isopropyl acetoacetate and urea/thiourea in ethanol using strontium chloride hexahydrate catalyst. The prepared DHPMs were screened for antifungal activity against *Candida albicans, Aspergillus flavus, Rhizopus and Mucor.* The compound (*7*) showed the excellent antifungal activity. Shah *et al.* [18] reported an anti-fungal activity of heterocyclic *N*-Mannich bases of 3, 4-dihydropyrimidine-2(1*H*)-thione. Compound (*8*) showed excellent activity against *Aspergillus niger* and *Candida albicans.*

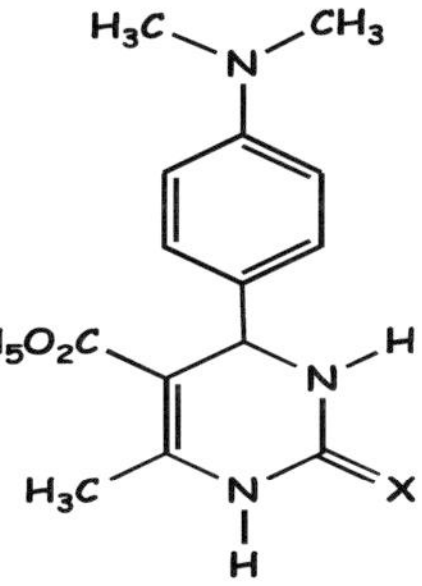

(5) X= O
(6) X= S

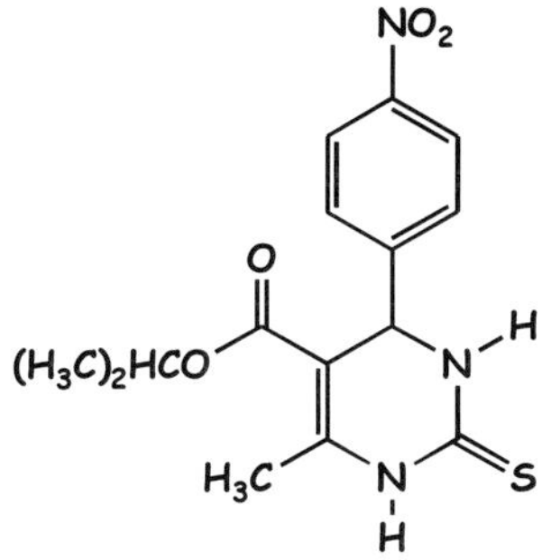

(7)

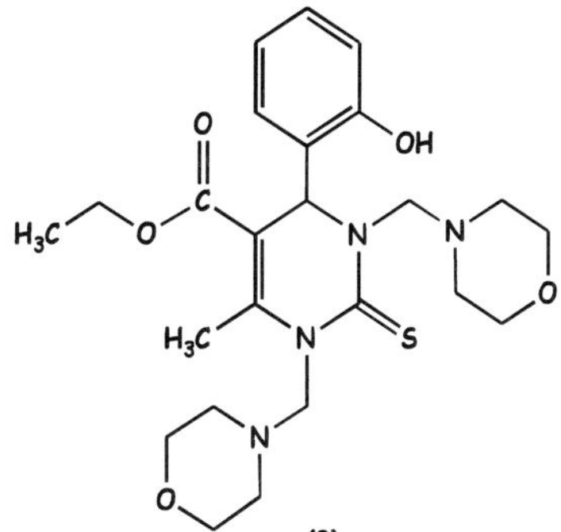

(8)

2.2. Antibacterial Activity

Adhikari *et al.* [19] reported the antibacterial activity of certain quinoline substituted dihydropyrimidines against *Escherichia coli*, *Staphylococcus aureus* and *Pseudomonas aeruginosa*. Compound (*9*) showed the highest anti-bacterial activity among the tested compounds at 10 mg/ml concentration. Ashok *et al.* [20] also reported the antibacterial activity of the newly DHPMs against *Escherichia coli, Staphylococcus aureus, Pseudomonas aeruginosa and Klebsiella pneumoniae* by disc diffusion method. Compound (*10*) showed excellent activity against *S. aureus, P. aeruginosa and E. coli*. It also showed moderate activity against *K. pneumoniae*. Singh *et al.* [21] prepared a series of 5-methylmercaptothio-carbonyl-4-aryl-3, 4-dihydropyrimidin-2(1*H*)-ones by Biginelli reaction and examined for antibacterial activities. Compound (*11*) showed excellent activities against bacteria *Bacillus cereus* and *Bacillus amyloliquefaceins* using agar disc diffusion method. Akhaja *et al.* [22] reported the antibacterial activity of newly prepared DHPMs consisting indole-2-one derivatives. Compound (*12*) showed better activity against bacteria *Escherichia coli, Pseudomonas aeruginosa, Klebsiella pneumoniae, Salmonella typhi, Staphylococcus aureus, Staphylococcus pyogenus and Bacillus subtilis.*

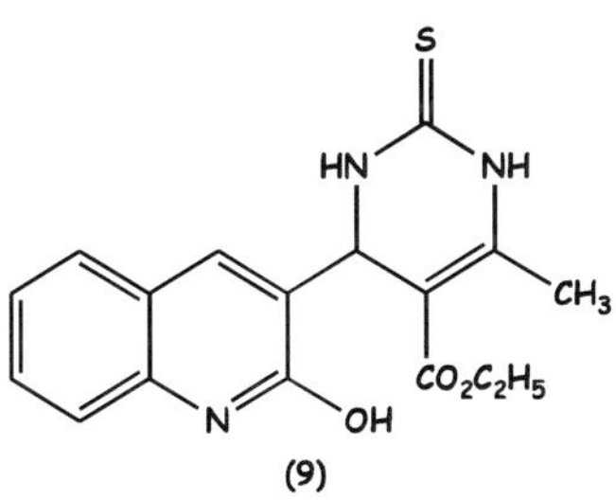

(9)

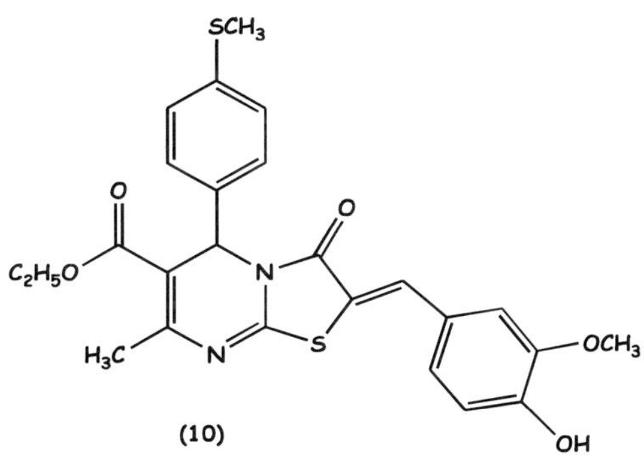

(10)

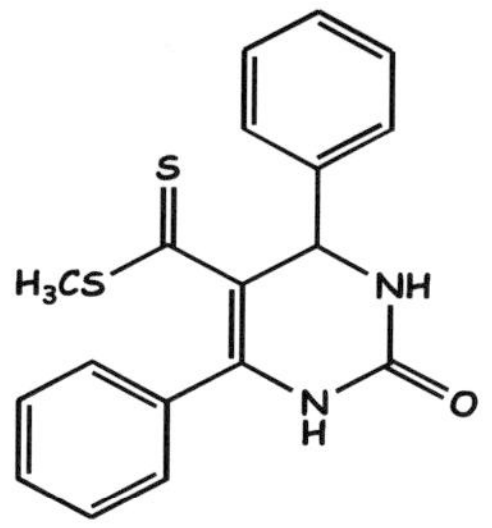

(11)

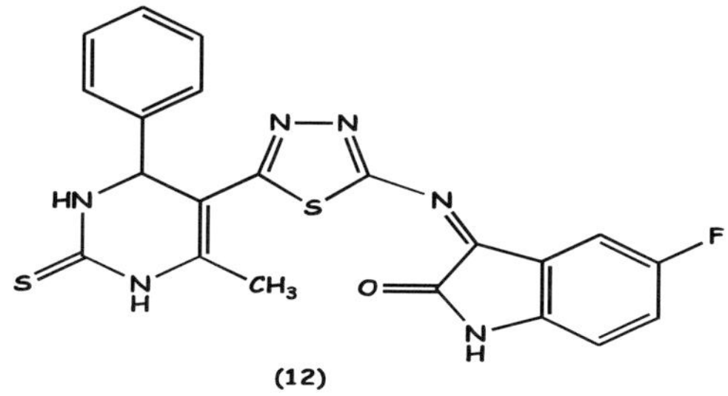

(12)

Mycobacterium Tuberculosis Activity

Dihydropyrimidines have been also reported as antitubercular agents [23]. Zalavadiya *et al.* [24] synthesized new DHPMs; among them compound (*13*) exhibited moderate antimycobacterial activity against *Mycobacterium tuberculosis* $H_{37}RV$. A series of dihydropyrimidines was synthesized and evaluated for *in vitro* antitubercular activity against *Mycobacterium tuberculosis* $H_{37}Rv$. Ethyl 4-[3-(4-fluorophenyl)-1-phenyl-1H-pyrazol-4-yl]-6-methyl-2-oxo-1,2,3,4-tetrahy-dropyrimidine-5 carboxylate (*14*) was found to be the most active compounds *in vitro* with MIC of 0.02 lg/mL against MTB. It was found to be more potent than drug isoniazid [25].

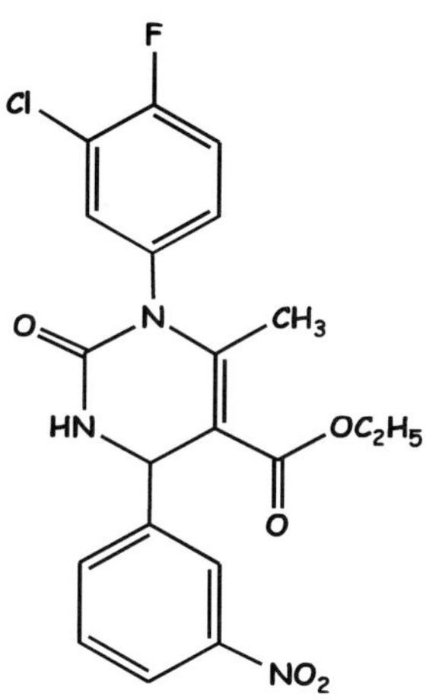

(13)

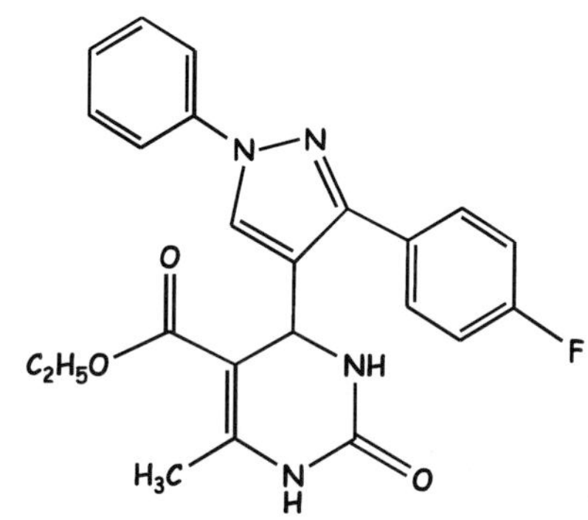

(14)

2.3. Anti-inflammatory Activity

DHPMs also exhibit anti-inflammatory activity. Chikhale *et al.* [26] reported a compound (*15*) exerting good anti-inflammatory activity at the 100 mg/kg dose level as compared to indomethacin. Aromatic substitution having methyl and hydroxyl analogues were less potent than corresponding methoxy group analogue (*15*). Rajesh *et al.* [27] also prepared a series of novel thiourea analogs of 3, 4-dihydropyrimidin-2(1-H)-one derivatives by Biginelli reaction followed by reduction reaction. The resulting amine formation takes place with different arylisothiocynates. All the synthesized compounds were screened against the pro-inflammatory cytokines (TNF-*a* and IL-6). Biological data revealed that among all the compounds screened, compound (*16*) found to be more promising anti-inflammatory (78% TNF-α and 96% IL-6 inhibitory activity at 10 lM). Compound (*16*) with -Cl and -CF_3 at 2- and 5-positions, respectively, exhibit highest TNF-α and IL-6 inhibitory activity. Whereas, presence of -Cl at 2-position and -CF_3 group at 6-position exhibits no TNF-α activity or low IL-6 inhibitory activity.

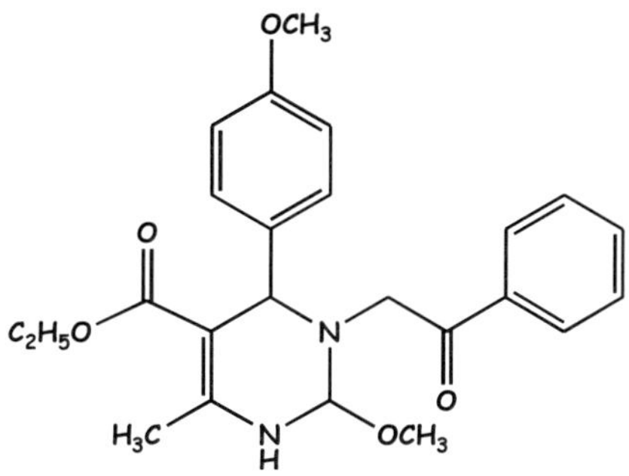

(15)

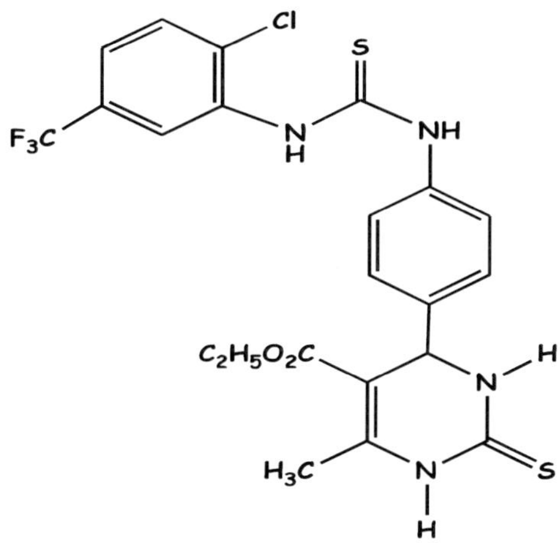

(16)

2.4. α $_{1a}$-Adrenergic Receptor Antagonists

Hypertension is one of the most serious health issues of the modern world with a continuous rise to a number of patients. Selective α_1-adrenoreceptor antagonists have many advantages and uses in the management of arterial hypertension. Their lacking at the level of α_1-adr subtypes lead to multiple side effects. Existence of multiple α_1-adr subtypes holds great promise for the discovery and development of more specific and selective drug molecules targeting only one α_1-adr subtype at a time and thus relatively free from side effects. Nagarathnam *et al.* [28] synthesized compound (*17*) which exhibited potent binding affinity and subtype selectivity for α_{1A}-adr. On the basis of cardiovascular effects and the superior pharmacodynamic profiles of these α_{1A}-selective compounds in the animal models, dihydropyrimidines offer a significant improvement over the current treatments of benign prostatic hyperplasia (BPH). Judicious alterations based on the sterric and electronic nature of the substituents on the phenyl ring of 4-phenylpiperazines led to the identification of 2-carboxamidophenylpiperazine moiety as a preferred side chain subunit having weaker binding affinity of the opioid receptors. Thus, Lagu *et al.* [29] synthesized new derivatives of dihydropyrimidinones containing substituted 4-phenylpiperazines. Compound (*18*) was identified as a leading compound with binding and functional profile comparable to the standard.

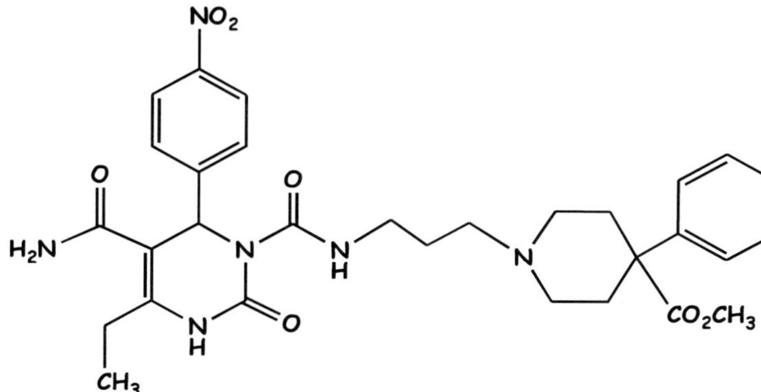

(17)

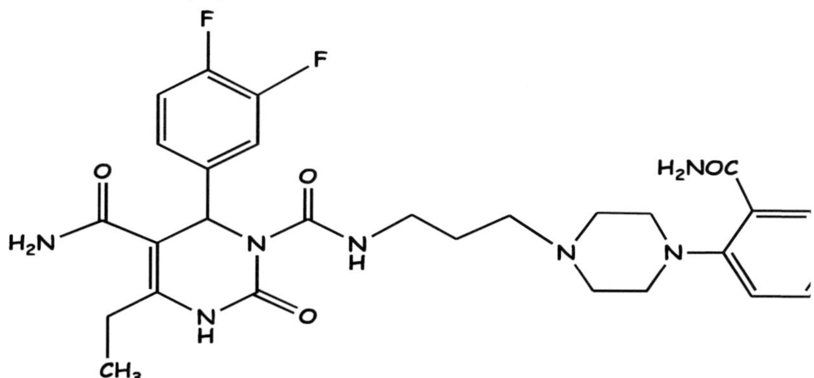

(18)

2.5. Anti-oxidant Activity

In pathologic conditions, an overproduction or scavenger diminution of the reactive oxygen species (ROS) can occur. In fact, ROS overproduction has been implicated in installation and/or progression of a variety of human diseases, including diabetes and various other neurodegenerative diseases [30]. Thus, the interest in natural and synthetic antioxidant compounds could potentially retarded the development of these diseases grown considerably in last decades. Stefani *et al.* [31] reported the synthesis of DHPMs which showed antioxidant activity. Analogues compounds (*19*) and (*20*) exhibited a strong activity against lipid peroxidation induced by Fe + EDTA, while compounds (*19*) and (*21*) were found to be more potent in reducing ROS levels. New hexahydropyrimido[5,4-c]quinoline-2,5-diones and 2-thioxohexahydropyrimido [5,4-c]quinoline-5-ones were prepared in two steps from ethyl 4-phenyl-6-methyl-2-oxo tetrahydropyrimidine-5-carboxylates or 4-phenyl-6-methyl-2-thioxotetrahydropyrimidine-5-carboxylates, using appropriate aldehydes, urea and ethyl acetoacetate. Their antioxidant properties were evaluated by two methods: scavenging effect on DPPH radicals and scavenging effect on hydroxyl radicals. The results showed that the compounds containing thiourea moiety (*22*) have better activity [32].

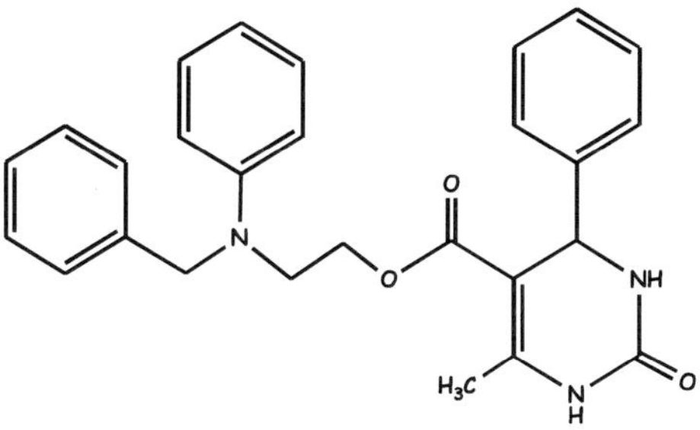

(19)

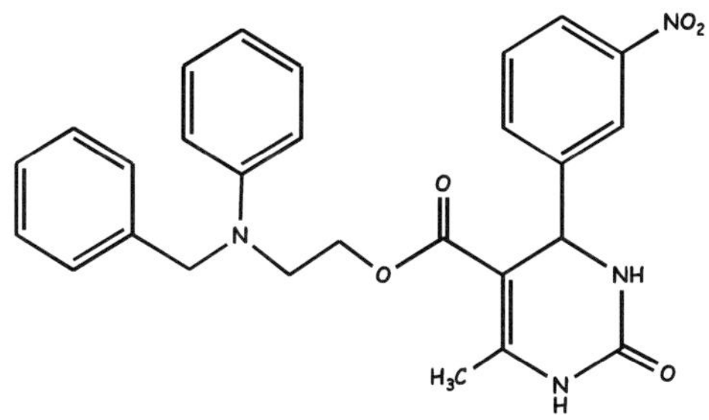

(20)

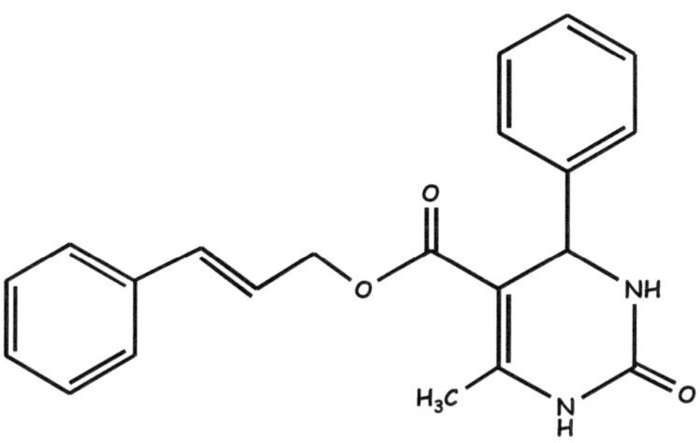

(21)

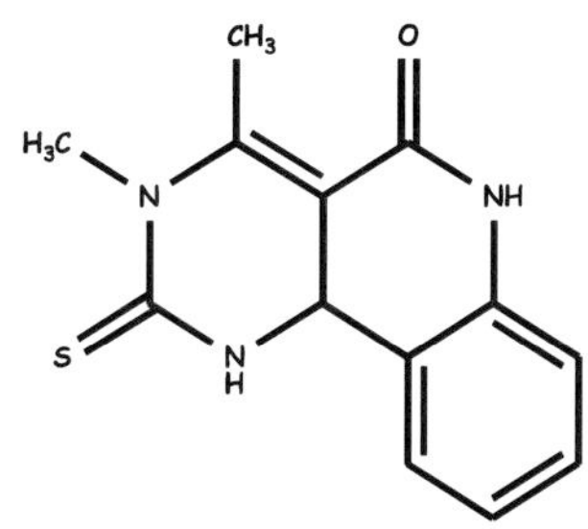

(22)

2.6. Anti-HIV Activity

A series of dihydropyrimidinone scaffolds was reported by Junwon *et al.* [33] and studied their inhibitory activities against HIV replication at low micromolar concentrations. Compound (*23*) exhibited significant inhibition of HIV-1 replication with a safety profile. Chiral separation of each enantiomer by fractional crystallization showed that only *S*-enantiomer retained anti-HIV activity. Compound *(24)*, a novel and potent DHPM analog served for further development and determination of the mechanism of action. Ji *et al.* [34] prepared a series of novel *S*-DABCO analogues and evaluated for inhibitors of type-1 (HIV-1). The activity of these compounds showed only micromolar potency against HIV-1 in MT-4 cells *in vitro*, though *25* and *26* were unusually potent (IC_{50}) 0.09 and 0.002 μM, respectively and selective (SI) 1500 and 4600, respectively.

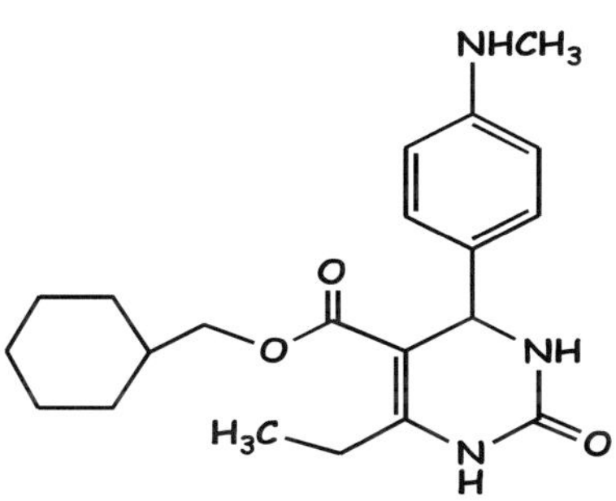

(23)

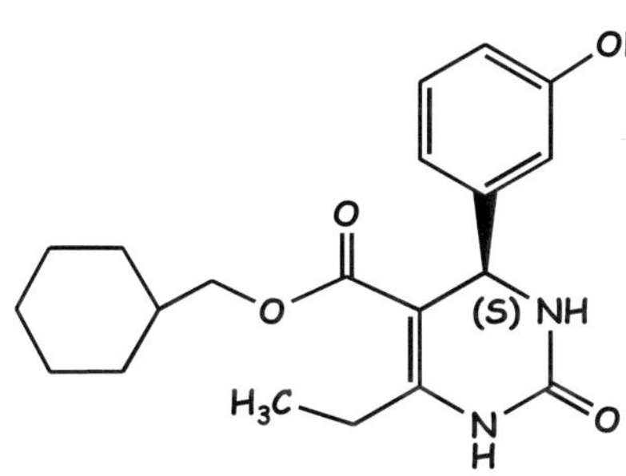

(24)

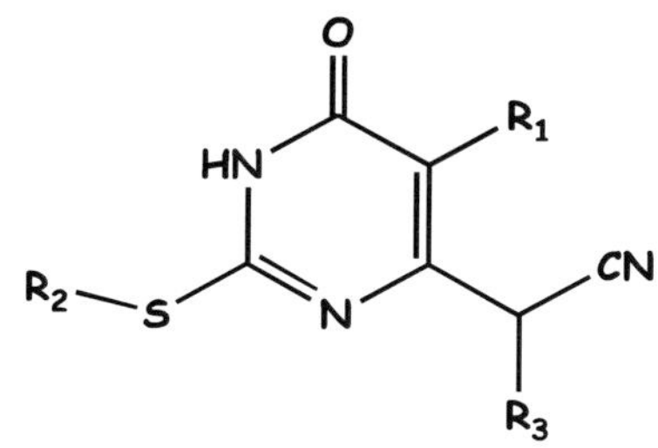

(25)R_1= Me, R_2=benzoylmethyl &
R_3= 1-naphthyl
(26)R_1= Me, R_2=i-Pr & R_3= Ph

2.7. Anticancer Activity

Kumar *et al.* [35] synthesized DHPMs and subjected for *in vitro* anticancer activity screening against MCF-7 human breast cancer cells. The title compounds were tested at the concentration of 10 mg. all the compounds exhibited activity ranging from weak to moderate and, from moderate to high in terms of percentage cytotoxicity. Among them, compound *(27)* exhibited significant anticancer activity. To elucidate the three-dimensional structure–activity relationships (3D QSAR) towards their anticancer activity, they subjected for comparative molecular similarity indices analysis (CoMSIA). Using comparative molecular field analysis (CoMFA), in rational drug designing to build activity model provides important information on possible improvements in ligand structure to increase the cytotoxicity. The developed CoMFA model is the most significant and reliable model toward the cytotoxicity having a substitution at fourth position of 1, 4-dihydropyrimidines. Further, Kumar *et al.* [36] synthesized compound *(28)* exhibiting potential cytotoxicity. Certain new DHPMs were synthesized and screened for antiproliferative effects in mdr1-gene transfected mouse lymphoma cell line (l5178 y). Among the newly synthesized compound *(29)* showed a potent antiproliferative activity [37]. Azizian *et al.* [38] conducted the cytotoxic activities of newly prepared DHPMs on human cancerous cell lines (Raji, HeLa, LS-180, SKOV-3 and MCF7). Their cytotoxic study indicated that they possessed a weak to moderate activity. Compound *(30)* showed the better activity among the other synthesized compounds against different cell lines. For instance, this compound was only having IC_{50} of lower than 100 lM in Raji and MCF7 cell lines. Furthermore, in all cell lines, its IC_{16} and IC_{50} were the lowest as compared to other ones. A structural analysis indicated that the key point of this is a presence of sulfur instead of oxygen or nitrogen at C2 position of pyrimidinone ring. This position plays an important role in cytotoxic effect of these compounds, which can be a helpful for future designing of drugs.

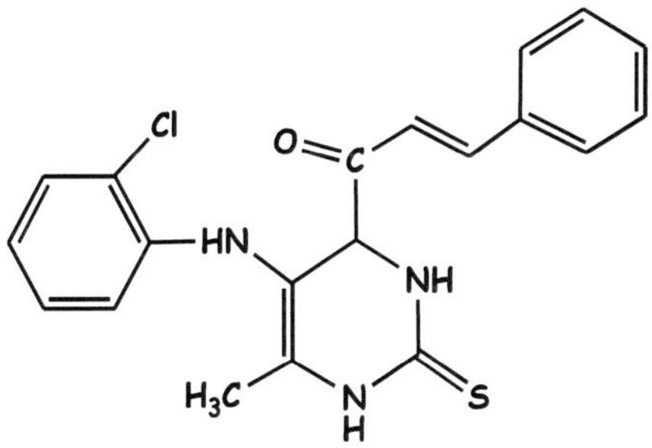

(27)

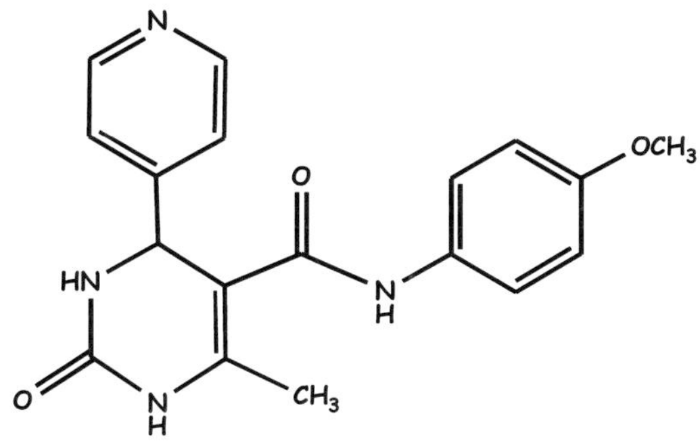

(28)

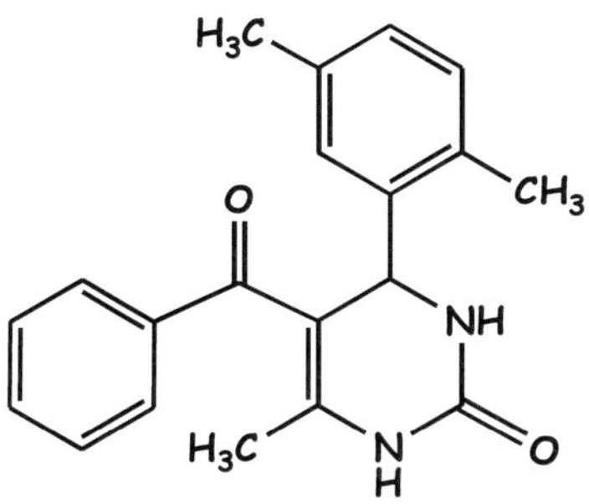

(29)

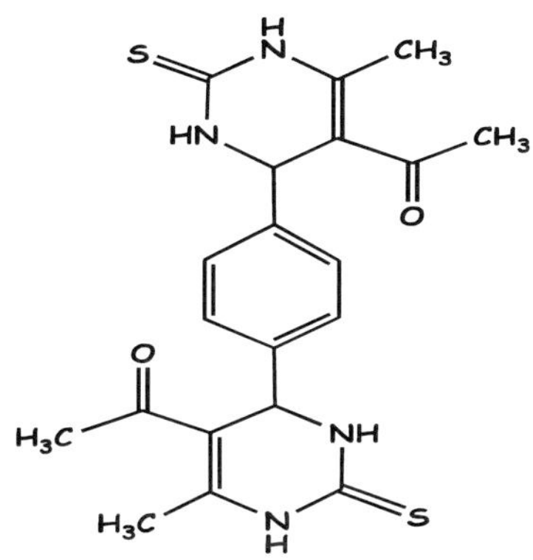

(30)

2.8. Cardiovascular Activity

Sehon *et al.* [39] described a series of potent and selective dihydropyrimidine which inhibits Rho-associated kinase isoform 1 (ROCK1) (*31*). A significant improvement has been made in overall profile for this series of ROCK1 inhibitors. Incorporation of substituent, next to the 4-pyridyl nitrogen provides the compounds with improved P_{450} profiles. In addition, indazole substitution provided very encouraging activity. Finally, compound (*32*), has been promising oral bioavailability (49% in rat and 19% in monkey), good half-life (1.8 h in rat and 2.2 h in monkey), good selectivity against a panel of 31 kinases (>100 fold), as well as RSK16 and p70S6K6 (RSK1 IC_{50}) 398 nM, p70S6K IC_{50}) 1000 nM), and a dramatically improved P450 profile (>2.2 μM at all isozymes tested). Oral administration of compound (*32)* gave a robust 25 mmHg reduction mean arterial pressure in spontaneously hypertensive rats at a single dose of 30 mg/kg. Sawant *et al.* [40] reported that 2-methylthio-1,4-dihydropyrimidines (*33*) and (*34*) showed significant cardioprotective activity against adrenaline-induced myocardial infarction in rats. Adrenaline-induced ECG alterations such as reduced R–R interval, increased heart rate, reduced P duration, and ST-segment elevation were brought to the normal values by pretreatment of compounds (*33*) and (*34*).

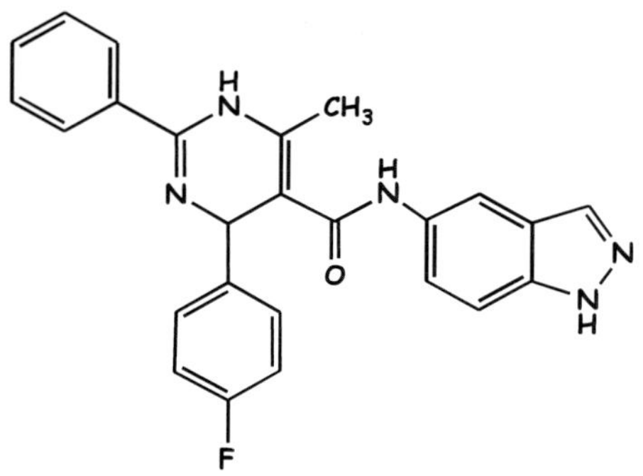

(31)

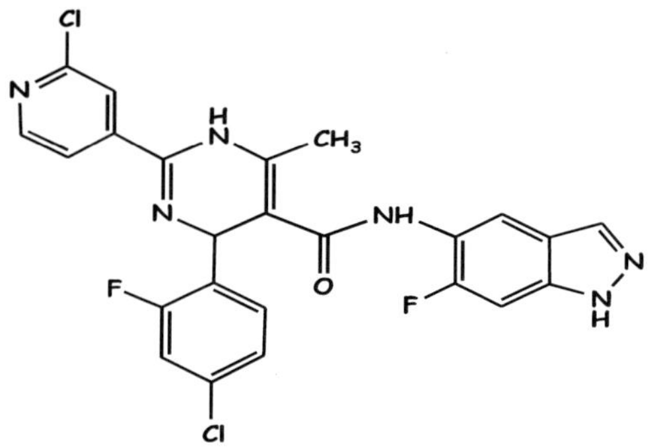

(32)

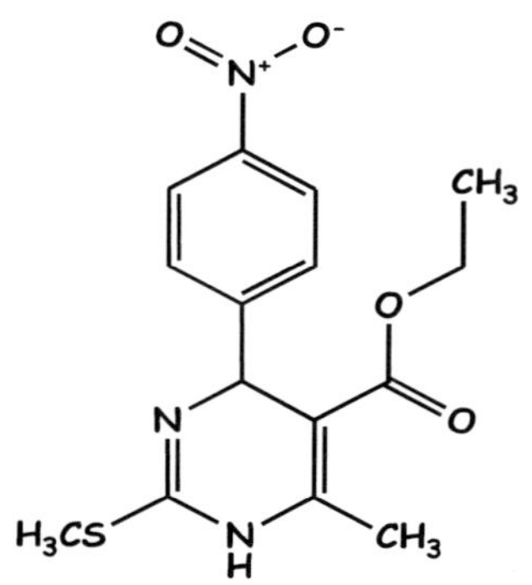

(33)

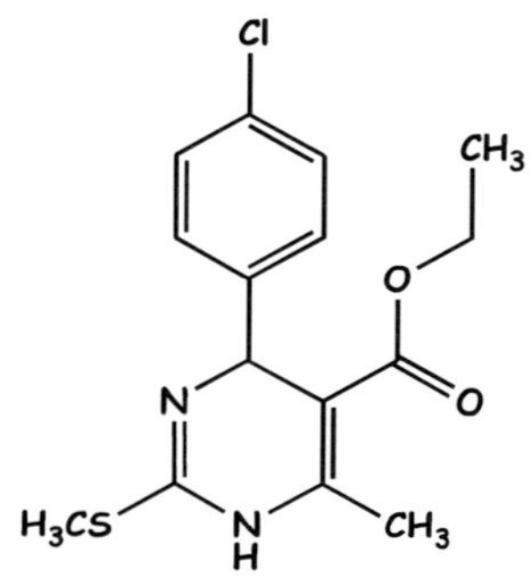

(34)

2.9. Inhibition of Eg5

The development of drugs that interrupt the cell cycle during the mitosis stage is a common strategy for cancer therapy. Compounds that perturb microtubule shortening (depolymerization) or lengthening (polymerization) causes arrest of the cell cycle in mitosis due to perturbation of the normal microtubule dynamics, necessary for chromosome movement. The structures of DHPM analogues: enastron (*35*), dimethylenastron (*36*), and fluorastrol (*37*) were taken for mitotic kinesin (Eg5) inhibitors test [41]. They observed that the 3-hydroxyphenyl group, the only common functional moiety, remains conformationally unchanged between the inhibitors and virtually identical interactions with the residues of Eg5. Addition of fluorines to lead compound has beneficial effects on the physicochemical properties of the drug and strengthened the protein-ligand interactions, as evident from other major fluorinated drugs: Prozac, an antidepressant, and Lipitor, a cholesterol-lowering drug [42]. While analogues with only one fluorine at either the meta-(2) or para-position (1) are more active than mon-97 in cell-based assays, the incorporation of both fluorines lead to the highly improved activity of fluorastrol. Drugs such as monastrol, enastron, and dimethylenastron belong to class I DHPM inhibitors that bind in the *S* configuration, whereas mon-97 and fluorastrol are class II DHPM inhibitors bind preferentially in the *R* configuration. With these new structural evidence for the stereoselectivity of DHPM inhibitors, the possibilities for future structure-based drug designing is considerably expanded.

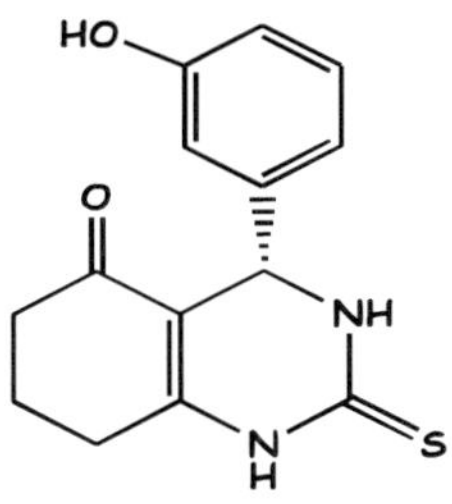

(35)

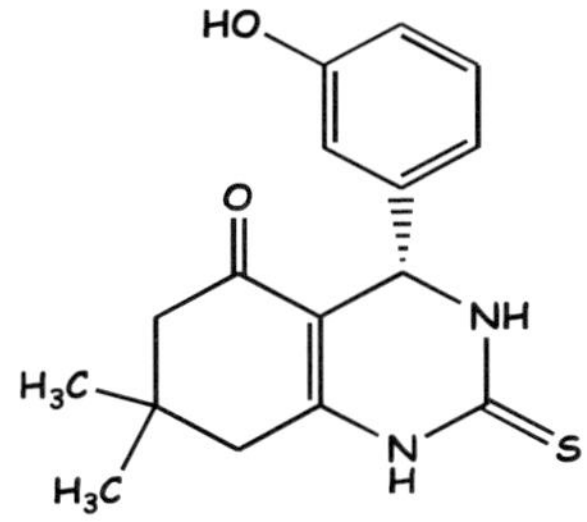

(36)

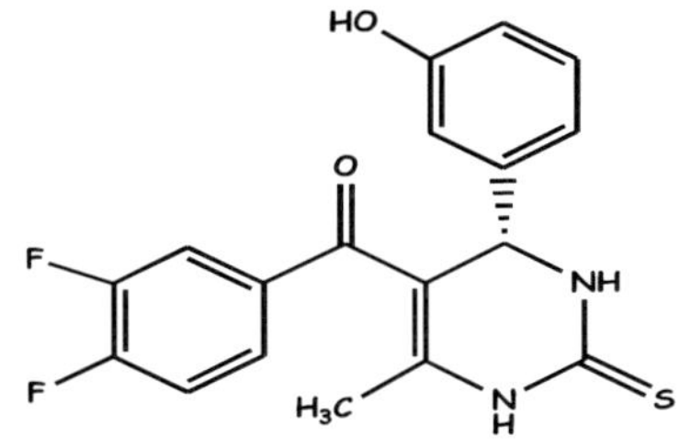

(37)

2.10. Antimalarial Activity

Plasmodium falciparum, the Apicomplexan parasite responsible for the most lethal forms of human malaria, was exposed to radically different environments and stress factors during its complex lifecycle. In any organism, Hsp70 chaperones (proteins) are typically associated with stress tolerance. Chiang *et al.* [43] reasoned the inhibition of *P. falciparum* Hsp70 chaperones which would adversely affect parasite homeostasis. To test this hypothesis, they measured whether pyrimidinone-amides, a new class of Hsp70 modulators, could inhibit the replication of pathogenic *P. falciparum* stages in human red blood cells. Compounds with IC_{50} values from 30 nM to 1.6 lM were identified. Each compound also altered the ATPase activity of purified *P. falciparum* Hsp70 in single-turnover assays, although higher concentrations of agents were required than was necessary to inhibit *P. falciparum* replication. Varying effects of these compounds on Hsp70s from other organisms were also observed. It indicated that pyrimidinone-amides compounds *38* and *39* constitute a novel class of anti-malarial agents.

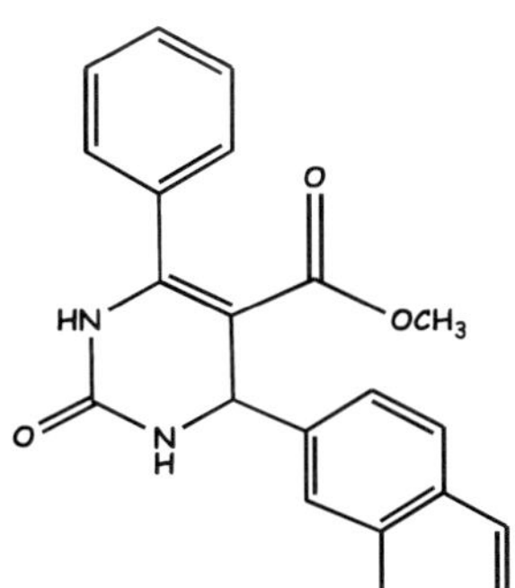

(38)

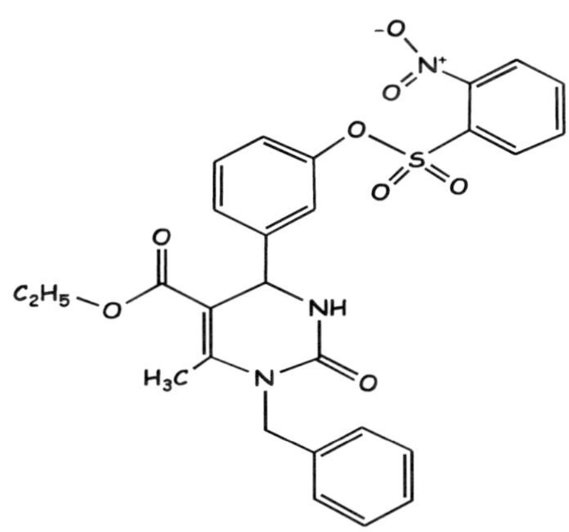

(39)

2.11. Calcium Channel Modulators

Dihydropyrimidines are potent mimics of calcium channel blockers (CCB) [44]. Structural variations of DHPMs elaborated for diversity oriented sites (C-6, C-4, N-3, N-1, N-1/N-3 and N1-C6 linked) have been studied using rational designs of synthetic methodologies

[45]. Singh *et al.* [46] prepared new compounds and screened for calcium channel blockers activity based on their ability to relax a membrane depolarization induced contraction of vascular smooth muscle and results were compared against standard nifedipine. Nifedipine completely relaxed the KCl induced contraction with an IC_{50} in the 10 μM range, among the tested DHPMs. Compound (*40*) relaxed the KCl induced contractions by only 40% with IC_{50s} ranging from 100-300 μM. Marvaniya *et al.* [47] found 3,4-dihydropyrimidin-2(1H)-one derivatives can exhibited calcium channel blocking activity. Compound (*41*) having highest activity with an IC_{50} value of 19.76. Zorkun *et al.* [48] also reported 4-aryl-3,4-dihydropyrimidin-2(1H)-thione derivatives. Compound (*42*) also possessed calcium channel blockers activity.

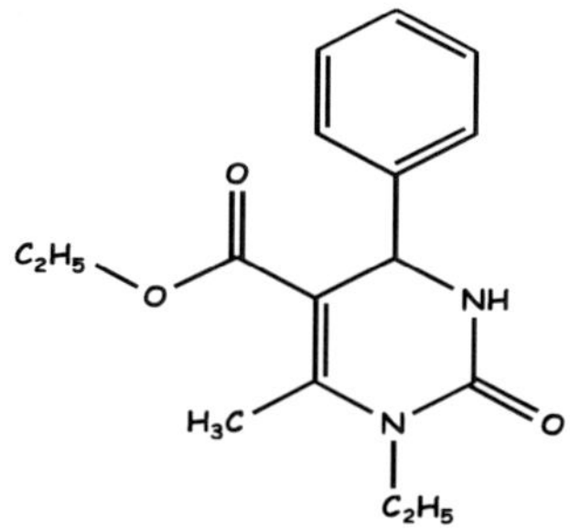

(40)

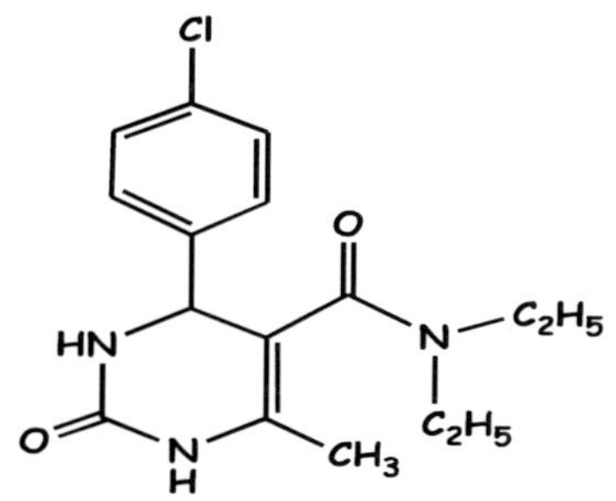

(41)

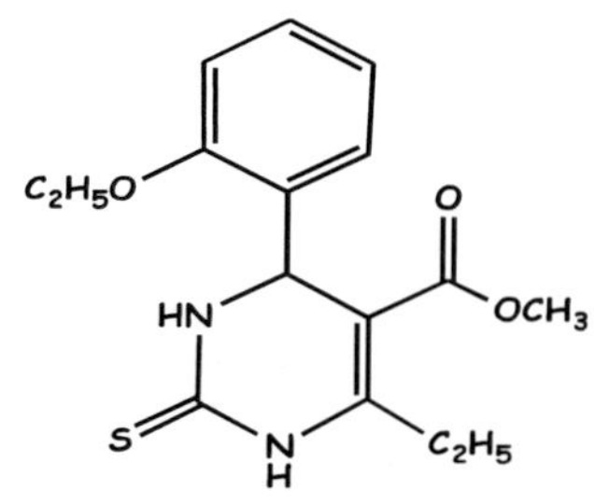

(42)

2.12. Larvicidal and Antifilarial Activity

A series of tetrahydro-1H-pyrimido[5,4-c]quinolin-5-ones have been synthesized [49]. Mosquito larvicidal activity of newly synthesized compounds was studied against fourth instar larvae *Culex quinquefasciatus*. Compounds (*43*) and (*44*) were proved to be lethal for mosquito larvae, which are useful and more toxic substances to kill mosquito larvae. The antifilarial activity of 2-sulfanyl-6-methyl-1,4-dihydropyrimidines series were evaluated against adult parasites of human lymphatic filarial parasite *Brugia malayi* (sub-periodic strain) *in vitro* and *in vivo* at various concentrations [50]. Compound (*45*) showed promising antifilarial activity *in vitro* and *in vivo* and exerted adulticidal activity along with female worm's sterilization. There were no apparent signs of any toxicity in gross health of treated and normal animals, which may serve as a prototype lead for further optimization and development of new antifilarial agents.

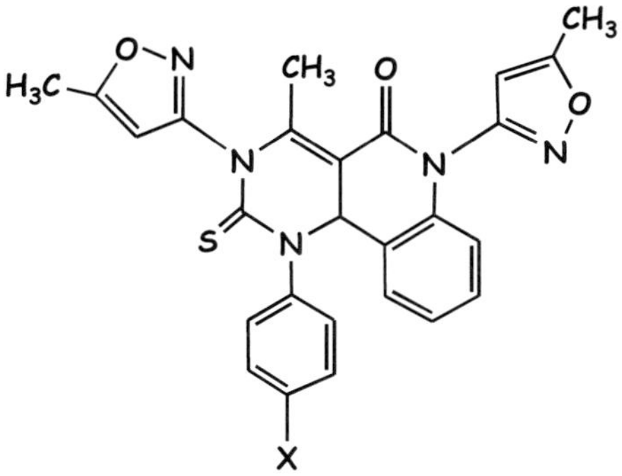

(43), X=CH_3
(44), X=Br

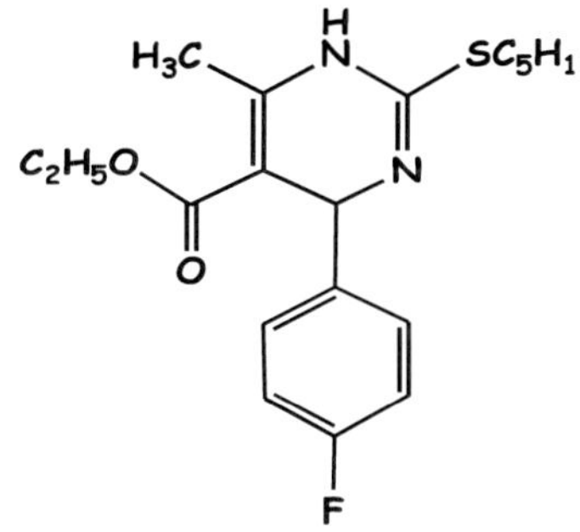

(45)

2.13. Others Activities

Blackburn *et al.* [51] reported the cell permeable 4-aryl dihydropyrimidinones (*46-47*), were inhibitors of fatty acid transporter [(*S*)-enantiomer approximately 100 times more potent than (*R*)-enantiomer]. Substituent like lipophilic ester at 5-position and presence of preferably *p*-NO_2 and CF_3 at 4-aryl group led to activation of compounds.

DHPM derivative *(50)* was identified as calcitonin mimetic (EC_{50} = 6 μM), active *in vivo* in the Weanling rat model when administered subcutaneously [52].

Compound *(51)* play an important role in the activation of molecular chaperones; responsible for a variety of protective, anti- apoptotic functions [53-54]. The activation of molecular chaperones blocks Aβ aggregation and has a favorable effect on neurodegenerative diseases like the Alzheimer disease.

DHPM also possessed antidiabetic activity. Compund (*52*) had high affinity towards ligand for peroxisomes proliferations activated receptor γ [55] and compounds (*53*) exhibited susceptibility allele for generalized epilepsy with febrile seizures plus [56].

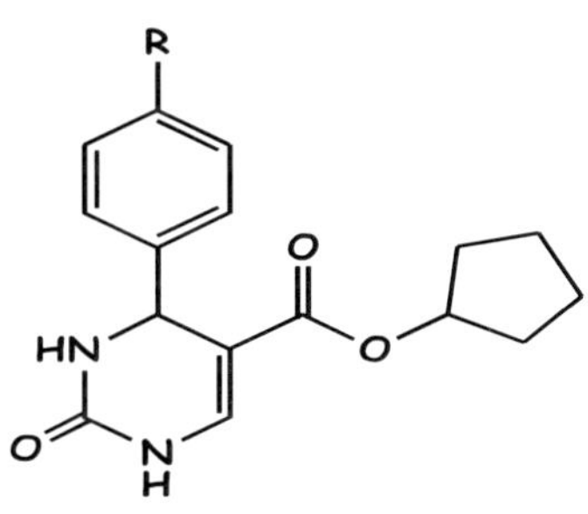

(46)R=NO_2
(47)R=CF_3

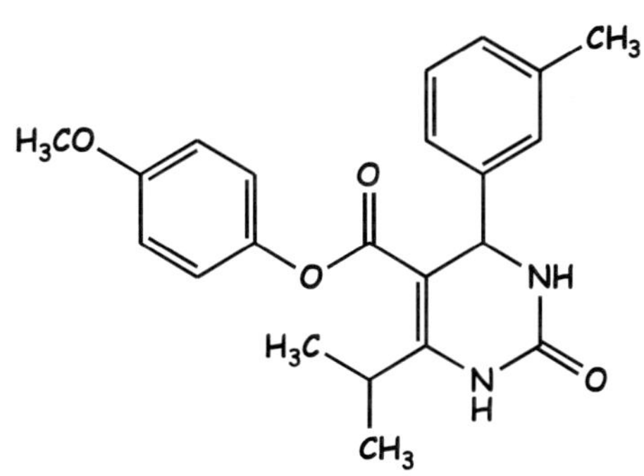

(48)

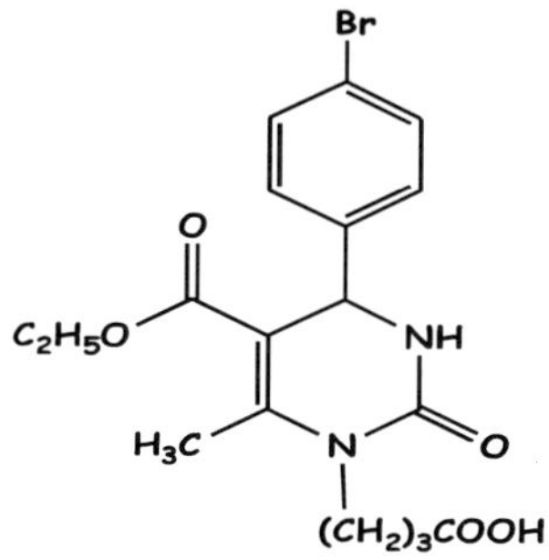

(49)

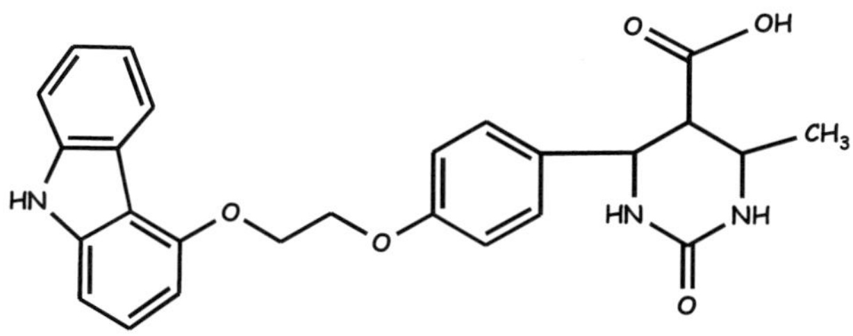

(50)

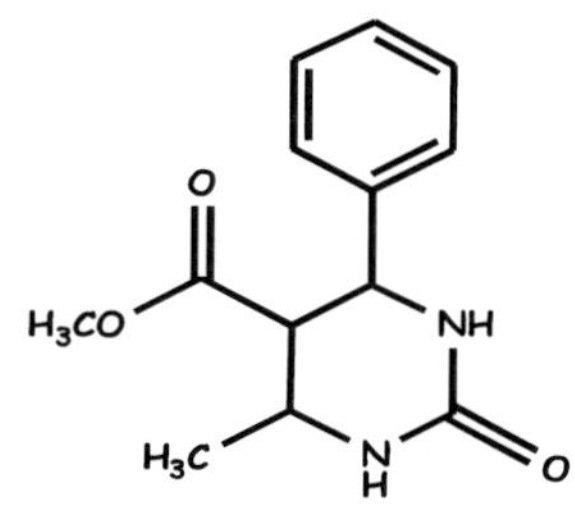

(51)

Conclusion

The present chapter highlights the latest developments on diverse biological activities of dihydropyrimidines. Over the past ten years there have been increasing reports concerning more on the newer synthetic methods of DHPMs. However, for the first time we have reviewed only the biological activities of recently reported DHPMs in this chapter. With the emergence of advanced techniques and methods, newer synthesis and biological activities of this class of compounds may be explored in near future.

Acknowledgments

The authors sincerely acknowledge the financial supports from CSIR, India (project No. 01(2387)/10/EMR-II) and DST, India (project No. SR/S1/OC-31/2009).

References

[1] Biginelli, P. *Gazz. Chim. Ital. 1893*, *23*, 360-413.
[2] Kappe, C.O. *Acc. Chem. Res. 2000*, *33*, 879-888.
[3] Singh, O.M.; Devi, N.S. *J. Org. Chem. 2009*, *74*, 3141-3144.
[4] Matsuda, T.; Hirao, I. *Nippon Kagaku Zasshi. 1965, 86*, 1195-1197.
[5] Hurst, E.W.; Hull, R.I. *J. Med. Pharm. Chem. 1961*, *3*, 215-229.

[6] Bokaeva, S.S.; Kaz, Tr.; Nauch-Issled. *Inst. Onkol. Radiol. 1967, 3* 305-309.

[7] Zidermane, A.; Duburs, G.; Zilbere, A.; Verpele, R.; Uldrlkis, J.; Kumsars, K. *Latv. PSR Zinat. Akad. Vestis. 1971*, 77-83.

[8] Kumsars, K.; Velena, A; Duburs, G; Uldrikis, J; Zidermane, A. *Biokhimiya. 1971, 36,* 1201-1209.

[9] Kato, T. Japan. Kokai Tokkyo Koho JP Patent 59,190,974, *1984*.

[10] Sadanandam, Y.S.; Shetty, M.M.; Diwan, P.V. *Eur. J. Med. Chem. 1992*, *27*, 87-92.

[11] Cooper, K. Pfizer Ltd., PCT Int. Appl., WO 11, *1990*. pp. 281-320.

[12] Kurono, M.; Hayashi, M.; Miura, K.; Isogawa, Y.; Sawai, K. Sanwa Kagaku Kenkyusho Co., Japan, Kokai Tokkyo Koho JP 62,267,272, *1987*.

[13] Atwal, K.S.; George, C.R.; David, K.; David, M.F.; Suzzane, M.; Swanson, B.N.; Joseph, S.; Smillie, K.M. *J. Med. Chem. 1990*, *30*, 2625-2635.

[14] Gallop, M.A.; Barret, R.W.; Dower, W.J.; Fodor, S.P.; Gordon, E.M. *J. Med. Chem. 1994*, *37,* 1233-1251.

[15] Chhillar, A.K.; Arya, P.; Mukherjee, C.; Kumar, P.; Yadav, Y.; Sharma, A.K.; Yadav, V.; Gupta, Y.; Dabur, R.; Jha, H.N.; Watterson, A.C.; Parmar, V.S.; Prasad, A.K.; Sharma, G.L. *Bioorg. Med. Chem. 2006*, 14, 973-981.

[16] Singh, O.M.; Singh, S.Y.; Devi, M.B.; Devi, L.N.; Singh, N.I.; Lee, S-G. *Bioorg. Med. Chem. Lett. 2008*, *18*, 6462-6467.

[17] Chitra, S.; Devanathan, D.; Pandiarajan, K. *Eur. J. Med. Chem. 2010*, *45*, 367-371.

[18] Shah, T.B.; Gupta, A.; Patel, M.R.; Chaudhari, V.S.; Patel, H.; Patel, V.C. *Indian J. Chem. B. 2010*, *49*,578-586.

[19] Adhikari, A.; Kalluraya, B.; Sujith, K.V.; Gouthamchandra,; Mahmood, R. *Saudi Pharmaceutical J. 2012*, *20*, 75-79.

[20] Ashok, M.; Holla, B.S.; Kumari, N.S. *Eur. J. Med. Chem. 2007*, *42*, 380-385.

[21] Singh, O.M.; Devi, N.S.; Devi, L.R.; Khumanthem, N. *Inter. J. Drug Design Dev. 2010*,*1*, 258-264.

[22] Akhaja, T.N.; Raval, J.P. *Eur. J. Med. Chem. 2011*, *46*, 5573-5579.

[23] Modha, J.; Datta, N.; Parekh, H. *Il Farmaco. 2001*, *56*, 641-646.

[24] Zalavadiya, P.; Tala, S.; Akbari, J.; Joshi, H. *Arch. Pharm. Chem. Life Sci. 2009*, *342*, 469-475.

[25] Amit, R.; Trivedi.; Vimal, R.; Bhuva.; Bipin, H.; Dholariya, Dipti, K.; Dodiya.; Vipul, B.; Kataria.; Viresh, H.S. *Bioorg. Med. Chem. Lett. 2010*, *20*, 6100-6102.

[26] Chikhale, R.V.; Bhole, R.P.; Khedekar, P.B.; Bhusari, K.P. *Eur. J. Med. Chem. 2009*, *44*, 3645-3653.

[27] Tale, R.H.; Rodge, A.H.; Hatnapure, G.D.; Keche, A.P.; Patil, K.M.; Pawar, R.P. *Med. Chem. Res.* DOI 10.1007/s00044-011-9943-3.

[28] Nagarathnam, D.; Miao, S.W.; Lagu, B.; Chiu, G.; Fang, J.; Murali Dhar, T.G.; Zhang, J.; Tyagarajan, S.; Marzabadi, M.R.; Zhang, F.; Wong, W.C.; Sun, T.D.; Wetzel, J.M.; Forray, C.; Chang, R.S.L.; Broten, T.P.; Ransom, R.W.; Schorn, T.W.; Chen, T.B.; O'Malley, S.; Kling, P.; Schneck, K.; Bendesky, R.; Harrell, C.M.; Vyas, K.P.; Gluchowski, C. *J. Med. Chem. 1999*, *42*, 4764- 4777.

[29] Lagu, B.; Tian, D.; Nagarathnam, D.; Marzabadi, M.R.; Wong, W.C.; Miao, S.W.; Zhang, F.; Sun, W.; Chiu, G.; Fang, J.; Forray, C.; Chang, R.S.L.; Ransom, R.W.; Chen, T.B.; O'Malley, S.; Zhang, K.; Vyas, K. P.; Gluchowski, C. *J. Med. Chem. 1999*, *42*, 4794- 4803.

[30] Haddad, J.J. *Free Radic. Biol. Med. 2002*, *33*, 907-926.

[31] Stefani, H.A.; Oliveira, C.B.; Almeida, R.B.; Pereira, C.M.P.; Braga, R.C.; Cella, R.; Borges, V.C.; Savegnago, L.; Nogueira, C.W. *Eur. J. Med. Chem. 2006*, *41*, 513-518.

[32] Ismaili, L.; Nadaradjane, A.; Nicod, L.; Guyon, C.; Xicluna, A.; Franc, J.; Robert, O.; Refouvelet, B. *Eur. J. Med. Chem. 2008*, *43*,1270-1275.

[33] Kim, J.; Park, C.; Ok, T.; So, W.; Jo, M.; Seo, M.; Kim, Y.; Sohn, J.H.; Park, Y.; Ju, M.K.; Kim, J.; Han, S.J.; Kim, T.H.; Cechetto, J.; Nam, J.; Sommer, P.; No, Z. *Bioorg. Med. Chem. Lett. 2012*, *22*, 2522-2526.

[34] Ji, L.; Chen, F.E.; Clercq, E.D.; Balzarini, J.; Pannecouque, C. *J. Med. Chem. 2007*, *50*,1778-1786.

[35] Kumar, B.R.P.; Sankar, G.; Baig, R.B.N.; Chandrashekaran, S. *Eur. J. Med. Chem. 2009*, 44, 4192-4198.

[36] Kumar, B.R.P.; Masih, P.; Karthikeyan, E.; Bansal, A.; Suja,; Vijayan, P. *Med. Chem. Res. 2010*, *19*, 344-363.

[37] Bariwal, J.J.; Malhotra, M.; Molnar, J.; Jain, K.S.; Shah, A.K.; Bariwal, J.B. *Med. Chem. Res.* DOI 10.1007/s00044-011-9925-5.

[38] Azizian, J.; Mohammadi, M.K.; Firuzi, O.; Mirza, B.; Miri, R. *Chem. Biol. Drug Des. 2010*, *75*, 375-380.

[39] Sehon, C.A.; Wang, G.Z.; Viet, A.Q.; Goodman, K.B.; Dowdell, S.E.; Elkins, P.A.; Semus, S.F.; Evans, C.; Jolivette, L.J.; Kirkpatrick, R.B.; Dul, E.; Khandekar, S.S.; Yi, T.; Wrigh, L.L.; Smith, G.K.; Behm, D.J.; Bentley, R.; Doe, C.P.; Hu, E.; Lee, D. *J. Med. Chem. 2008*, *51,* 6631-6634.

[40] Sawant, R.L.; Sarode, V.I.; Jadhav, G.D.; Wadekar, J.B. *Med. Chem. Res.* DOI 10.1007/s00044-011-9700-7.

[41] Kaan, H.Y.K.; Ulaganathan, V.; Rath, O.; Prokopcov, H.; Dallinger, D.; Kappe, C.O.; Kozielski, F. *J. Med. Chem. 2010*, *53*, 5676-5683.

[42] Muller, K.; Faeh, C.; Diederich, F. *Science. 2007*, *317*, 1881-1886.

[43] Chiang, A.N.; Valderramos, J.C.; Balachandran, R.; Chovatiya, R.J.; Mead, B.P.; Schneider, C.; Bell, S.L.; Klein, M.G.; Huryn, D.M.; Chen, X.S.; Day, B.W.; Fidock, D.A.; Wipf, P.; Brodsky, J.L. *Bioorg. Med. Chem. 2009*, *17*, 1527-1533.

[44] Wetzel, J.M.; Miao, S.W.; Borden, L.A.; Branhek, T.A.; Gluchowski, C. *J. Med. Chem. 1995, 38*, 1579-1581.

[45] Singh, K.; Arora, D.; Singh, K.; Singh, S. *Mini-Rev. Med. Chem. 2009, 9,* 95-106.

[46] Singh, K.; Arora, D.; Poremsky, E.; Lowery, J.; Moreland, R.S. *Eur. J. Med. Chem. 2009*, *44*, 1997-2001.

[47] Marvaniya, H.M.; Parikh, P.K.; Sen, D.J. *J. Appl. Pharm. Sci. 2011*, *1*, 109-113.

[48] Zorkun, I.S.; Sarac, S.; Celebi, S.; Erol, K. *Bioorg. Med. Chem. 2006*, *14*, 8582-8589.

[49] Rajanarendar, E.; Reddy, M.N.; Murthy, K.R.; Reddy, K.G.; Raju, S.; Srinivas, M.; Praveen, B.; Rao, M.S. *Bioorg. Med. Chem. Lett. 2010*, *20*, 6052-6055.

[50] Singh, B.K.; Mishra, M.; Nisha Saxena, N.; Yadav, G.P.; Maulik, P.R.; Sahoo, M.K.; Gaur, R.L.; Murthy, P.K.; Tripathi, R.P. *Eur. J. Med. Chem. 2008*, *43*, 2717-2723.

[51] Blackburn, C.; Guan, B.; Brown, J.; Cullis, C.; Condon, S.M.; Jenkins, T. J.; Peluso, S.; Ye, Y.; Gimeno, R.E.; Punreddy, S.; Sun, Y.; Wu, H.; Hubbard, B.; Kaushik, V.; Tummino, P.; Sanchetti, P.; Sun, D.Y.; Daniels, T.; Tozzo, E.; Balani, S.K.; Raman, P. *Bioorg. Med. Chem. Lett., 2006*, *16*, 3504-3509.

[52] Matthews, J.M.; Liotta, F.; Hageman, W.; Rivero, R.A.; Westover, L.; Yang, M.; Xu, J.; Demarest, K. *Bioorg. Med. Chem. Lett*. *2004*, *14*, 1155-1159.

[53] Wisen, S.; Androsavich, J.; Evans, C.G.; Chang, L.; Gestwicki, J.E. *Bioorg. Med. Chem. Lett*. *2008*, *18*, 60-65.

[54] Wright, C.M.; Chovatiya, R.J., Jameson, N.E.; Turner, D.M.; Zhu, G.; Werner, S.; Huryn, D.M.; Pipas, J.M.; Day, B.W.; Wipf, P.; Brodsky, J. L. *Bioorg. Med. Chem.* *2008*, *16*, 3291-3301

[55] Kumar, R.; Mittal, A.; Ramachandran, U. *Bioorg. Med. Chem. Lett*. *2007*, *17*, 4613-4618.

[56] Lewis, R.W.; Mabry, J.; Polisar, J.G.; Eagen, K.P.; Ganem, B.; Hess, G.P. *Biochem.* *2010*, *49*, 4841-4851.

In: Bioactive Heterocycles
Editors: K. L. Ameta, R. P. Pawar and A. J. Domb
ISBN: 978-1-62257-636-4
© 2013 Nova Science Publishers, Inc.

Chapter 7

Asymmetric 1,3-dipolar Cycloadditions of Acrylamides in the Synthesis of Bioactive Heterocycles

***Marie Kissane*[1] *and Anita R. Maguire*[2,*]**
[1]Department of Chemistry, Analytical and Biological Chemistry Research Facility, University College Cork, Cork, Ireland (current address Chemical Product R&D, Lilly Research Laboratories, Eli Lilly and Company, Indianapolis, Indiana, US)
[2]Department of Chemistry & School of Pharmacy, Analytical and Biological Chemistry Research Facility, University College Cork, Cork, Ireland

Abstract

The chapter focuses on biologically active heterocyclic compounds in which the key step is an asymmetric 1,3-dipolar cycloaddition with an acrylamide. The use of chiral acrylamides as dipolarophiles leads to high levels of stereocontrol, due to conformational constraint in the acrylamides. Employment of chiral tertiary acrylamides containing nitrogen heterocycles is particularly effective in controlling the stereoselectivity.

1. 1,3-Dipolar Cycloadditions

1,3-Dipolar cycloadditions offer a very useful method for the preparation of five-membered ring heterocycles. [1-3] This [4πs + 2πs] cycloaddition involves the reaction of a dipolarophile (*e.g.* alkenes, alkynes, carbonyls and nitriles) with a 1,3-dipolar compound (*e.g.* nitrones, nitrile oxides, azomethine ylides and diazoalkanes) (Scheme 1). The dipolarophile can be almost any double or triple bond, containing functionality such as C≡C, [4] C=C, [5] C≡N, [6] C=N, [7] C=O, [8] and C=S. [9] The π-bond may be isolated, conjugated [10] or part of a cumulene [11] system. The structural variety of dipolarophiles makes 1,3-dipolar

* Corresponding author. Tel.: +353 21 4901693; fax: +353 21 4274097. E-mail: a.maguire@ucc.ie.

cycloadditions valuable and versatile reactions in heterocyclic synthesis. The concept of 1,3-dipolar cycloadditions was initially suggested by Smith in 1938, [12] but it was only after the generalisation of the reaction by Huisgen in the 1960's that the reaction became widely applicable. [8] The research conducted in the area of 1,3-dipolar cycloadditions has been immense over the past 40 years, [13-52] and the reaction is now utilised in almost every area of chemistry, from materials chemistry [53] to drug discovery, [54] indicating its diversity.

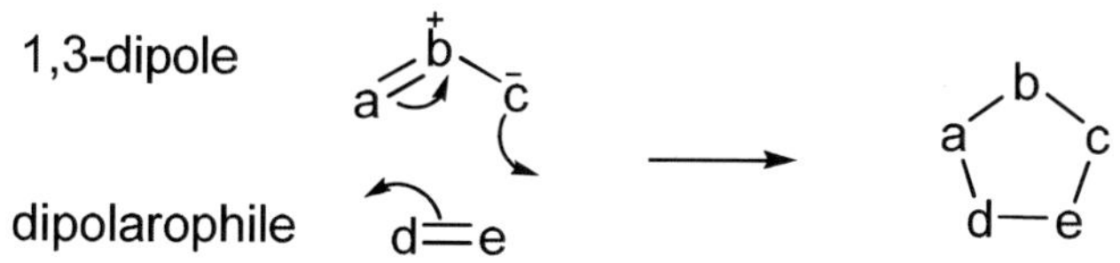

Scheme 1.

The regioselectivity of the 1,3-dipolar cycloaddition is predictable and dependant on electronic and steric effects, and the cycloaddition is highly stereoselective, with the stereochemistry of the original dipolarophile retained in the adduct. Provided that the cycloaddition reaction is significantly faster than the isomerisation of the dipole by rotation, then the addition is also stereoselective with respect to the dipole. Furthermore, if the dipole or dipolarophile bears a chiral auxiliary or if a chiral catalyst is used, then non-racemic cycloadducts can be produced. [13,55]

As up to four stereocentres can be introduced in a stereoselective manner in a single step; much attention has been devoted in recent years to the use of asymmetric 1,3-dipolar cycloadditions for the preparation of enantiomerically pure and biologically active five-membered ring heterocycles. [13,17,21-23,26] Control of the diastereo- and enantioselectivity in the addition step is the major challenge in asymmetric 1,3-dipolar cycloadditions. It is possible to control the diastereoselectivity by choosing the appropriate substrates or using a metal complex acting as a catalyst, and the enantioselectivity controlled by either choosing a chiral 1,3-dipole, a chiral dipolarophile or a chiral catalyst. Most of the reported studies on the control of stereoselectivity by the use of chiral dipolarophiles concern the use of α,β-unsaturated carbonyl compounds, and in particular acrylates. [56-63] However, in cycloadditions with chiral acrylates, in addition to controlling the direction of attack of 1,3-dipole, the rotameric preference of the acrylate must be controlled to achieve high levels of diastereoselectivity. The acrylate can exist in s-*cis* or s-*trans* conformation as depicted in Figure 1a. This is often controlled by the addition of Lewis acids, [63] but is not always feasible [for example in 1,3-dipolar cycloadditions with nitrile oxides]. [1]

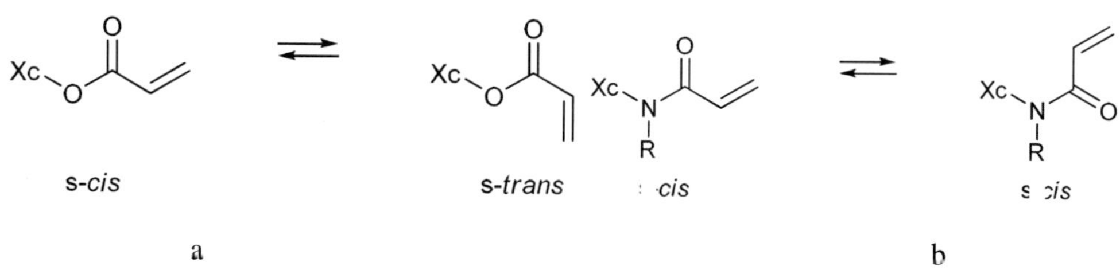

Figure 1.

The use of chiral tertiary acrylamides provides a solution to the rotamer problem encountered with esters, as the planar s-*trans* conformation is disfavored. [64] However, rotation about the C-N bond is possible leading to two low-energy s-*cis* conformers (Figure 1b). The development of chiral tertiary acrylamides containing nitrogen heterocycles such as *1* and *2* overcomes this problem. [64,65]

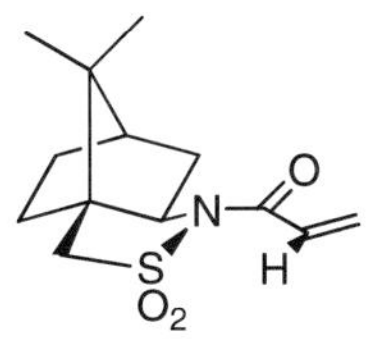

1

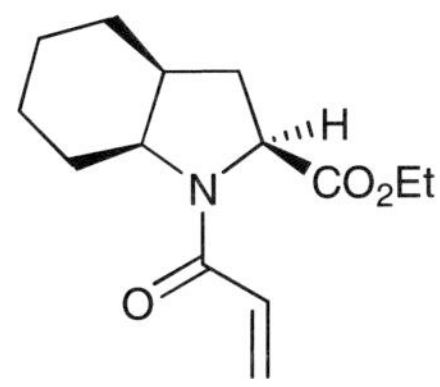

2

This chapter focuses on biologically active heterocyclic compounds in which the heterocycle moiety has been prepared via an asymmetric 1,3-dipolar cycloaddition with an acrylamide. The use of chiral acrylamides as dipolarophiles can lead to high levels of stereocontrol, due to conformational constraint in the acrylamides. Employment of chiral tertiary acrylamides containing nitrogen heterocycles is particularly effective on controlling the stereoselectivity of the process, as rotation around the C-N bond is disfavoured.

1.1. Cycloadditions with Azomethine Ylides

The 1,3-dipolar cycloaddition of azomethine ylides to alkenes and alkynes leads to the formation of pyrrolidines and pyrrolines respectively. Cycloadditions with azomethine ylides have been extensively investigated in recent years and these have been applied in asymmetric and natural product synthesis, as well as the syntheses of biologically interesting compounds. [2,13] As chiral pyrrolidines are common building blocks for many natural and unnatural compounds which possess important biological activity, the asymmetric 1,3-dipolar cycloaddition of azomethine ylides has attracted much attention. [41] Chiral azomethine ylides (both cyclic and acyclic), [66-70] chiral dipolarophiles [71-73] and chiral catalysts [71,74] have all been successfully implemented for asymmetric induction in azomethine ylide 1,3-dipolar cycloadditions.

Most of the reported studies on the control of the stereoselectivity by use of chiral dipolarophiles concern the use of acrylates. [72,75,76] However, excellent results have also been achieved with chiral acrylamides, and the synthesis of a number of biologically active compounds has been accomplished through their use. Garner investigated the cycloaddition of Oppolzer's chiral sultam derivative *1* to photochemically generated azomethine ylides as a route to the asymmetric synthesis of quinocarcin *3*, a potential antitumor antibiotic isolated from *Streptomyces* broths. [77-81]

Irradiation of the aziridine *4* yielded the azomethine ylide *5*, which subsequently undergo selective *exo* attack at the *si* face of Oppolzer's chiral sultam *1* to give the *exo*-substituted pyrrolidine *6* (diastereomeric ratio >25:1). As *1* is photochemically unstable, portionwise addition to the azomethine ylide *5* was necessary. The cycloaddition of chiral acrylates derived from menthol and 10-[dicyclohexyl(sulfonylamido)]isoborneol to a number of photochemically generated azomethine ylides was also conducted, however no facial selectivity was observed. [78] With four of the six stereogenic centres present in quinocarcin *3* now in place, the pyrrolidine *6* was transformed to *3* *via* a number of synthetic steps (Scheme 2). [80] . Following protection of the hydroxyl group as its MOM ether, benzylic bromination gave the monobromide which was subsequently converted to the phosphonium salt. The phosphonium salt was treated with potassium *t*-butoxide to give the phosphonium ylide and on heating in dimethylformamide, cyclisation occurred to yield the dihydroisoquinoline *7*. Cleavage of the chiral auxiliary was facilitated by reaction with lithium hydroxide to give the free acid, which was transformed to the ester *8* by treatment with diazomethane. Hydrogenation over Raney nickel gave the reduced product *9*. Exposure to sodium cyanide in water yielded the stable amino nitrile *10*, and on deprotection of the MOM ether with sodium iodide and trimethylsilyl chloride in acetonitrile, the free hydroxyl group was regenerated to give *11*. Quinocarin *3* was then obtained by the reaction of *11* with silver nitrate.

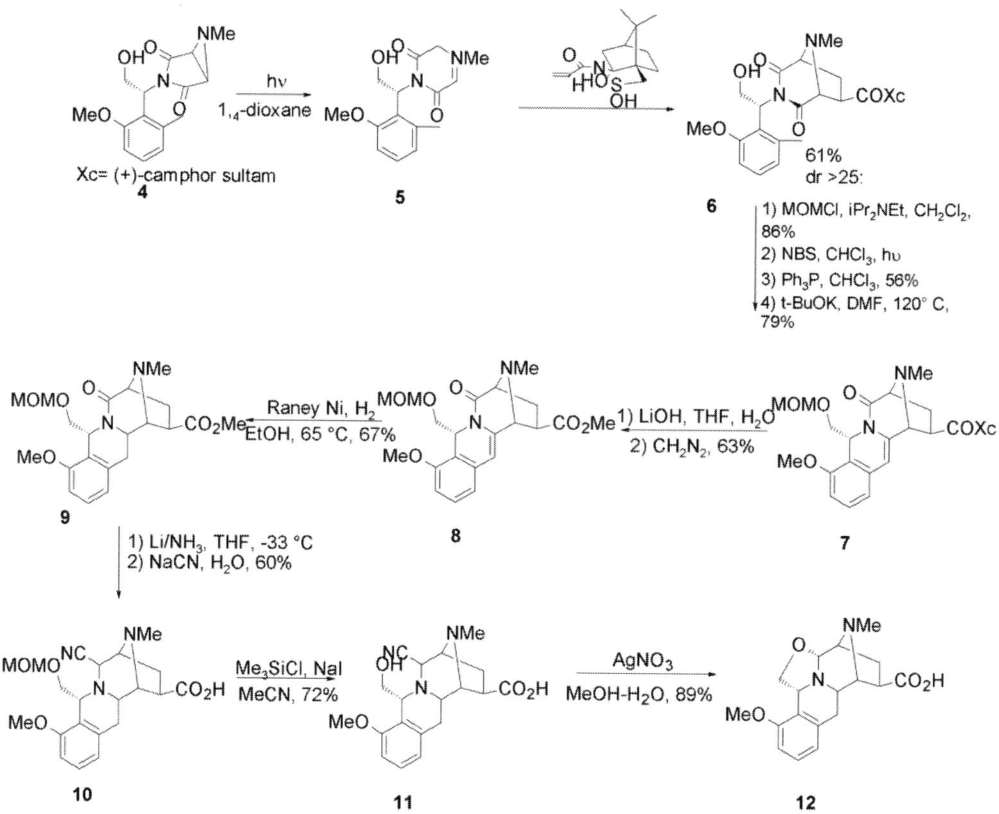

Scheme 2.

In 1997, Ma and co-workers utilised the asymmetric 1,3-dipolar cycloaddition of the azomethine ylide derived from *N*-benzyl-*N*-(methoxymethyl)-trimethylsilylmethylamine *12* with a range of chiral oxazolidinones as a route to a series of optically active *trans*-3-amino-4-alkylpyrrolidines, which is present at the *C*-8 position of 2-pyridones, novel DNA gyrase inhibitors that exhibit antibacterial activity (Scheme 3). [82] Reaction of a range of chiral oxazolidinones with the azomethine ylide derived from *12* afforded the diastereomeric pyrrolidines in ratios of 56:44 - 80:20. Hydrolysis of the major pyrrolidine to the β-amino acid was achieved using lithium hydroxide and hydrogen peroxide. Curtius rearrangement followed by debenzylation provided the desired chiral pyrrolidine. A series of optically active *trans*-3-amino-4-alkylpyrrolidines were prepared in yields of 46-75% for SAR study of the *C*-8 position of 2-pyridones using this route.

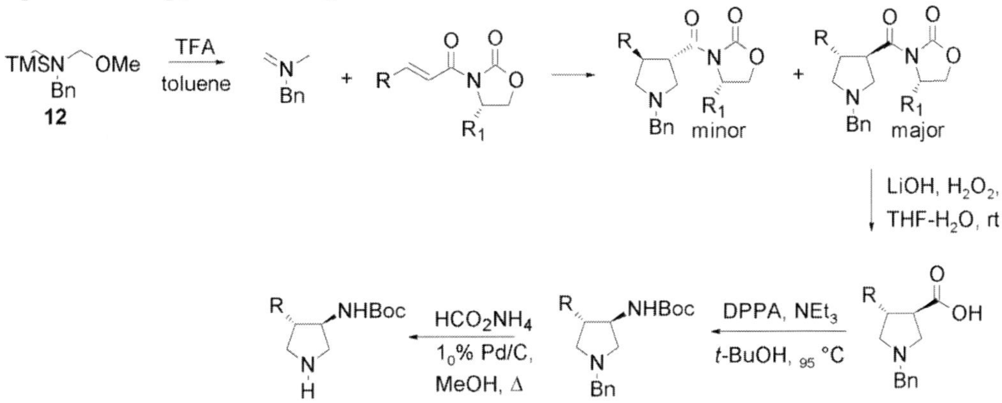

Scheme 3.

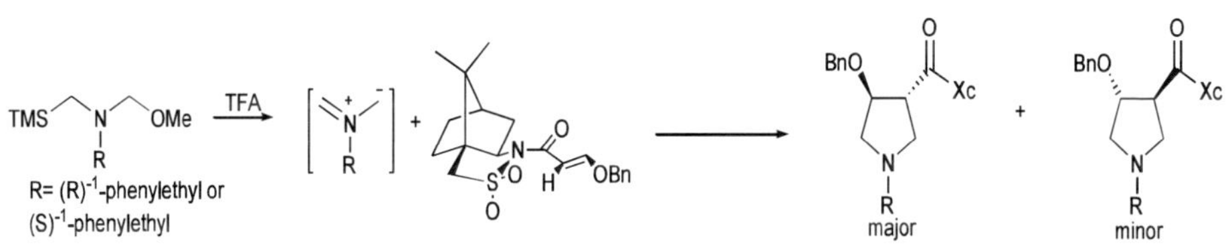

(–)-13

Solvent	*R*	*major:minor*
Toluene	(*R*)-1-phenylethyl	70:30
Toluene	(*S*)-1-phenylethyl	52:48
Acetonitrile	(*R*)-1-phenylethyl	88:12
Acetonitrile	(*S*)-1-phenylethyl	75:25

Scheme 4.

In 2001, Karlsson reported the doubly diastereoselective cycloaddition of chiral azomethine ylides to a range of chiral acrylamides. [83,84] On employment of oxazolidinone and camphorsultam derivatives, the camphorsultam derivatives furnished the greater diastereoselectivities, with diastereomeric ratios of up to 88:12 achieved. The diastereoselectivity was found to be solvent-dependant, with the employment of more polar solvents furnishing higher diastereoselectivities. The optimum diastereoselectivity was achieved by reacting the camphorsultam derivative (–)-*13* with the (*R*)-1-phenylethyl-derived azomethine ylide (Scheme 4).

The resulting pyrrolidines can act as chiral building blocks for the synthesis of enantiopure bioactive pyrrolidines, and Karlsson demonstrated this in his short synthesis of

(3*R*,4*R*)-4-(hydroxylmethyl)pyrrolidin-3-ol *14*, a known glycosidase inhibitor (Scheme 5). [83]

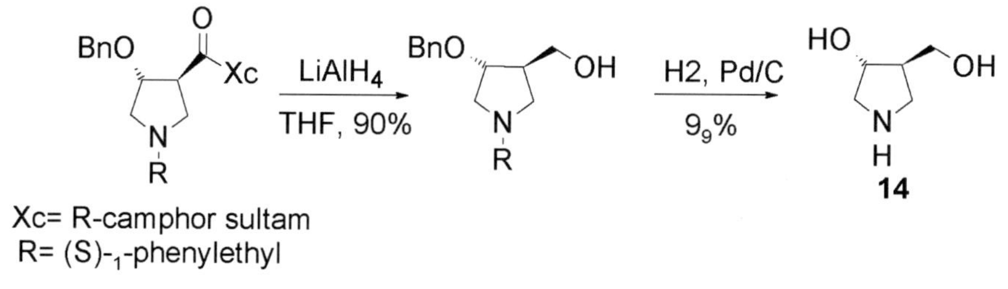

Scheme 5.

Karlsson extended this work to include monocyclic five-membered α,β-unsaturated acrylamides, and the cycloadduct was subsequently converted into a known precursor *15* of an antibacterial compound. [85,86] Reaction of the acrylamide *16* with (*S*)-1-phenylethyl-derived azomethine ylide provided the bicyclic adducts in a ratio of 68:32. The chiral auxiliary of the major isomer was cleaved on exposure to $Ti(OEt)_4$ in ethanol. The resulting ester was then transformed to the corresponding amide by the reaction with lithiated ammonia, and the amide was subsequently converted to the desired salt *15* (Scheme 6).

More recently, the large-scale synthesis of *14* has been reported by Chand *et al.*, in which they substituted chiral phenylethylamine for achiral benzylamine. Kilogram quantities of *14* have been prepared using this synthesis. [87]

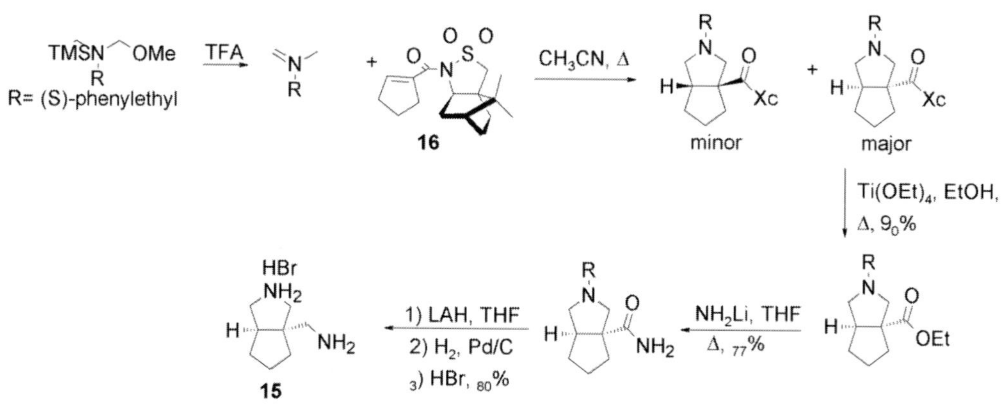

Scheme 6.

In 2001, Carey reported an efficient synthesis of (3*S*,4*R*)-ethyl 1-azabicyclo[2.2.1]heptane-3-carboxylate *17* in which the key step was the cycloaddition of an azomethine ylide with a camphorsultam derivative (Scheme 7). [88] 3-Carboxy-1-azabicyclo[2.2.1]heptane derivatives possess important physiological properties; for example, triazole and tetrazole derivatives have been identified as potent muscarinic agonists. [89] Cycloaddition of the camphorsultam derivative *18* with azomethine ylide derived from *12* yielded a diastereomeric mixture of pyrrolidines in 4:1 ratio, which were readily separated by column chromatography. Deprotection of the silyl ether group in the major isomer was readily achieved using aq. HCl, and reaction of the resulting alcohol with NEt_3 and MsCl afforded the quaternary ammonium salt. Hydrogenolysis followed by cleavage of the auxiliary group afforded *17*.

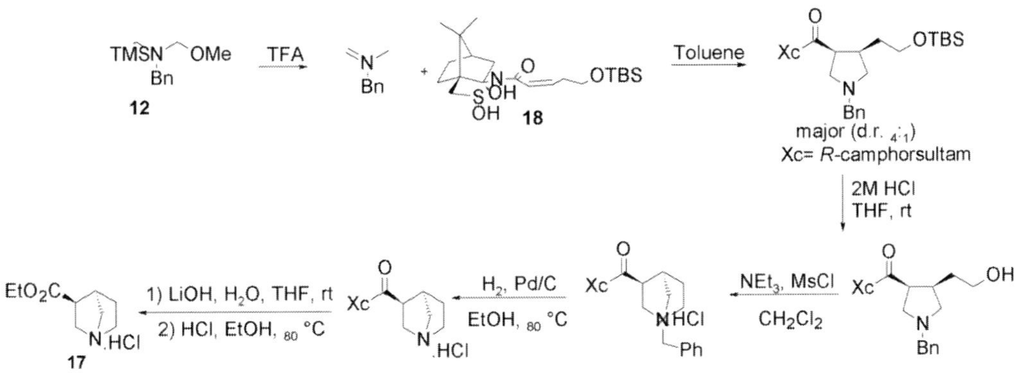

Scheme 7.

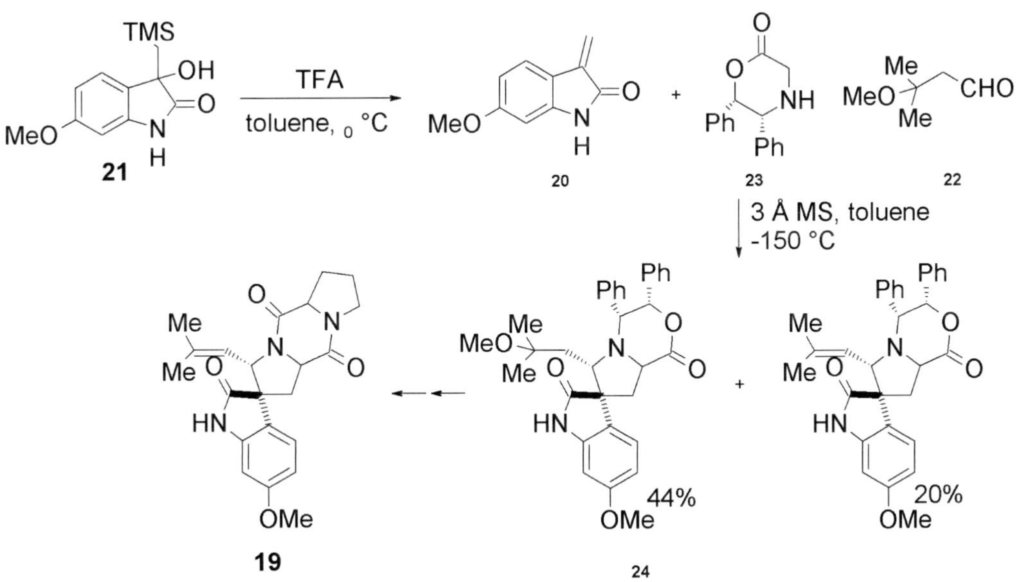

Scheme 8.

The cycloaddition of an azomethine ylide to an acrylamide derivative was also the key step in the Williams report of the asymmetric total synthesis of spirotryprostatin A *19*, which is a member of a promising class of antimitotic agents. [90,91] The unstable acrylamide dipolarophile *20* was prepared *in situ* by the treatment of *21* with trifluoroacetic acid in toluene. This was then added to the azomethine ylide generated from morpholinone *22* and aldehyde *23* to yield an approximately 2:1 mixture of cycloadducts. The major cycloadduct *24* was then subjected to further synthetic transformations to yield spirotryprostatin A *19* (Scheme 8).

The use of Lewis acid catalysts has been also reported to promote asymmetric 1,3-dipolar cycloadditions of azomethine ylides, with silver catalysts the most commonly employed. [13]

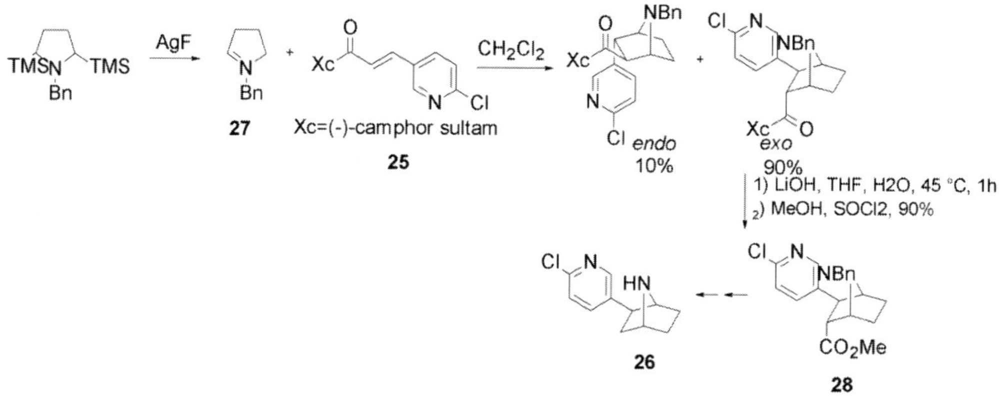

Scheme 9.

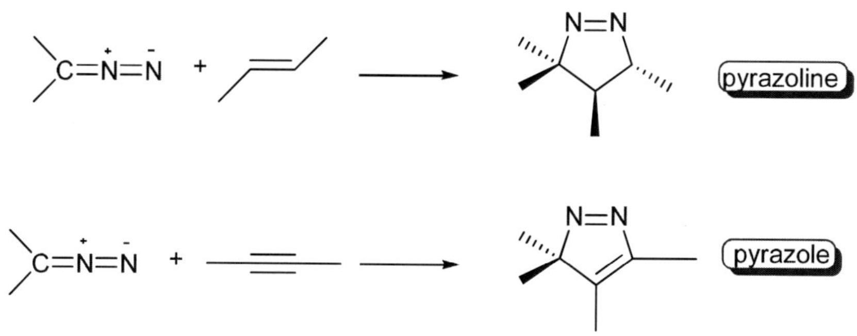

Scheme 10.

Diastereomeric ratios of up to 98:2 (*exo:endo*) were achieved by Pandey and co-workers on employment of cyclic azomethine ylides in silver fluoride catalysed cycloadditions with the Oppolzer chiral sultam derivative *25*. [92] This route was then utilised towards a formal synthesis of optically active *ent*-epibatidine *26*. Epibatidine is an important analgesic which is 200-500 times more potent than morphine. [93] The chiral dipolarophile *25*, which was synthesised by a Heck reaction, reacted with the cyclic azomethine ylide derived from *27* to give the pyrrolidine diastereomers in a 9:1 mixture. These were separated by chromatography, and the chiral auxiliary was cleaved from the major diastereomer by exposure to lithium hydroxide in tetrahydrofuran and water, followed by the reaction with thionyl chloride in methanol to yield *28* (Scheme 9). The conversion of *28* to *ent*-epibatidine *26* was reported earlier by the same group. [94]

1.2. Cycloadditions with Diazoalkanes

Diazoalkanes undergo 1,3-dipolar cycloadditions with alkenes and alkynes to yield pyrazolines and pyrazoles respectively (Scheme 10). [95]

The pyrazoline cycloadducts are too unstable to isolate and readily tautomerise to yield Δ^2-pyrazolines. Pyrazolines may also be transformed to pyrazoles by a 1,2-elimination reaction, and they can eliminate nitrogen on thermolysis or photolysis to yield cyclopropanes (Scheme 11). This method has been utilised in the synthesis of a number of cyclopropane natural products. [96]

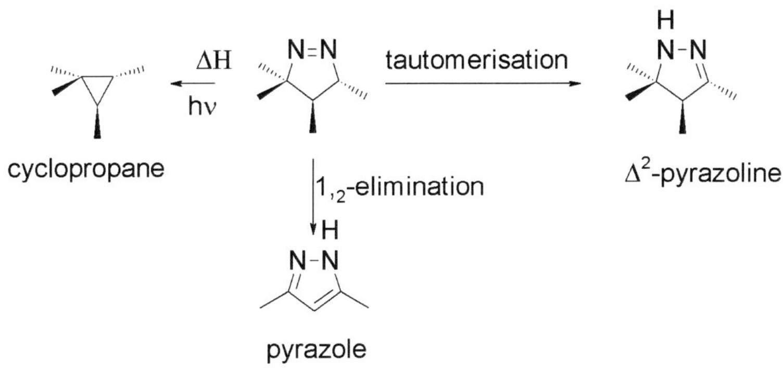

Scheme 11.

Synthetic applications of the pyrazoline cycloadducts obtained from the 1,3-dipolar cycloaddition of diazoalkanes to alkenes have not been extensively studied, and have usually been restricted to the preparation of the resulting cyclopropanes or pyrazoles. The difficulties associated with employing diazoalkanes in asymmetric dipolar cycloaddition arise because the diazoalkanes are not readily available, they cannot be stored for long periods and they may also be thermally unstable. Studies on the synthetic utility of diazoalkanes with acrylamides have been mainly concentrated on the commercially available trimethylsilyl-diazomethane.

Carreira has investigated the cycloaddition of trimethylsilyldiazomethane to a range of Oppolzer's chiral sultam derivatives. [97] 3-Trimethylsilyl-substituted pyrazolines were isolated in quantitative yield following the treatment with 2 M solution of trimethylsilyl-diazomethane in hexane. Tautomerisation with loss of the trimethylsilyl group on exposure to trifluoroacetic acid in dichloromethane yielded optically active pyrazolines as a single regioisomers with 90-94% diastereoselectivity. These were subsequently transformed to synthetically useful pyrazolidines by C=N reduction with sodium cyanoborohydride in acetic acid, chemoselective protection as the *N*-Cbz or *N*-Boc carbamates and auxiliary removal by treatment with dimethoxymagnesium in methanol (Scheme 12).

Azaproline analogues of the pyrazolines and the pyrazolidines were synthesised by chemoselective coupling reactions (Scheme 13). [97] It is anticipated that the azaproline analogues will find important applications as proline surrogates due to the presence of prolines in biologically important peptide sequences. [98, 99]

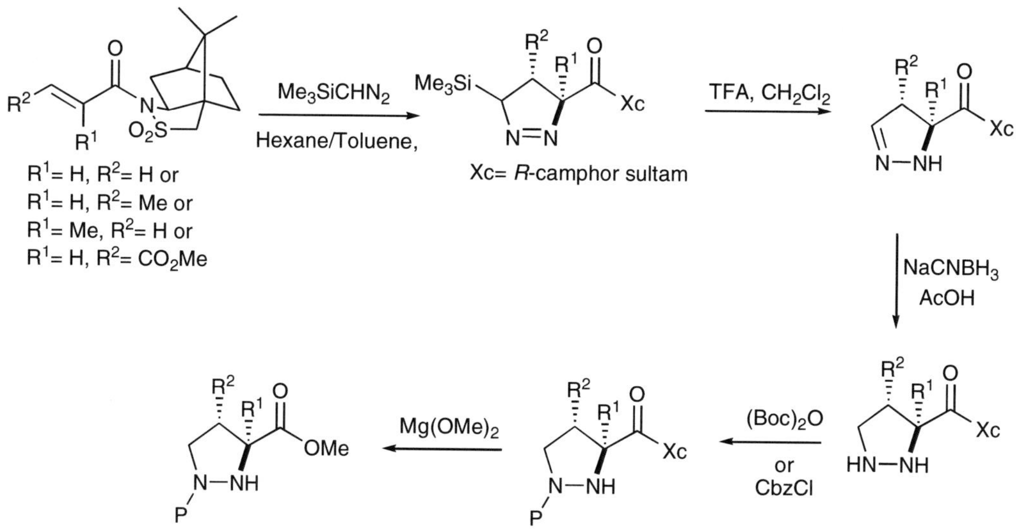

Scheme 12.

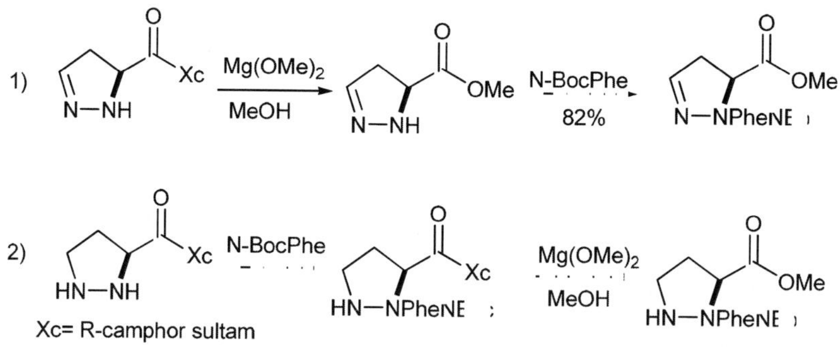

Scheme 13.

Carreira *et al.* demonstrated the versatility of this cycloaddition in the efficient synthesis of the marine metabolite stellettamide A *29*, which possesses anti-fungal activity and displays cytotoxicity against K562 epithelium cell lines. [100,101] Cycloaddition of the Oppolzer sultam derivative *30* with a solution of trimethylsilyldiazomethane led to the isolation of pyrazoline cycloadducts in quantitative yield as a 93:7 mixture of diastereomers. Desilylation was achieved on exposure to ethyl chloroformate and silver triflate to yield the diastereomeric pyrazolines *31* and *32* in a 92:8 ratio. These were separated by column chromatography and the auxiliary of the major diastereomer was cleaved by reaction with lithium aluminum hydride. Swern oxidation of the resulting alcohol facilitated oxidation to the aldehyde *33*, which reacted with the acetylide to yield the secondary propargyl alcohol as a 2:1 mixture of diasteromers. Mesylation of the alcohol yielded *34* and subsequent hydrogenolysis furnished the alkyne *35*. On hydrogenation of *35*, the (*Z*)-alkene *36* was isolated. The primary silyl ether and ethyl carbamate protecting groups were simultaneously cleaved by the treatment with aqueous barium hydroxide in dioxane to furnish *37*, which was selectively *N*-Boc protected and treated with mesyl chloride to yield *38*.

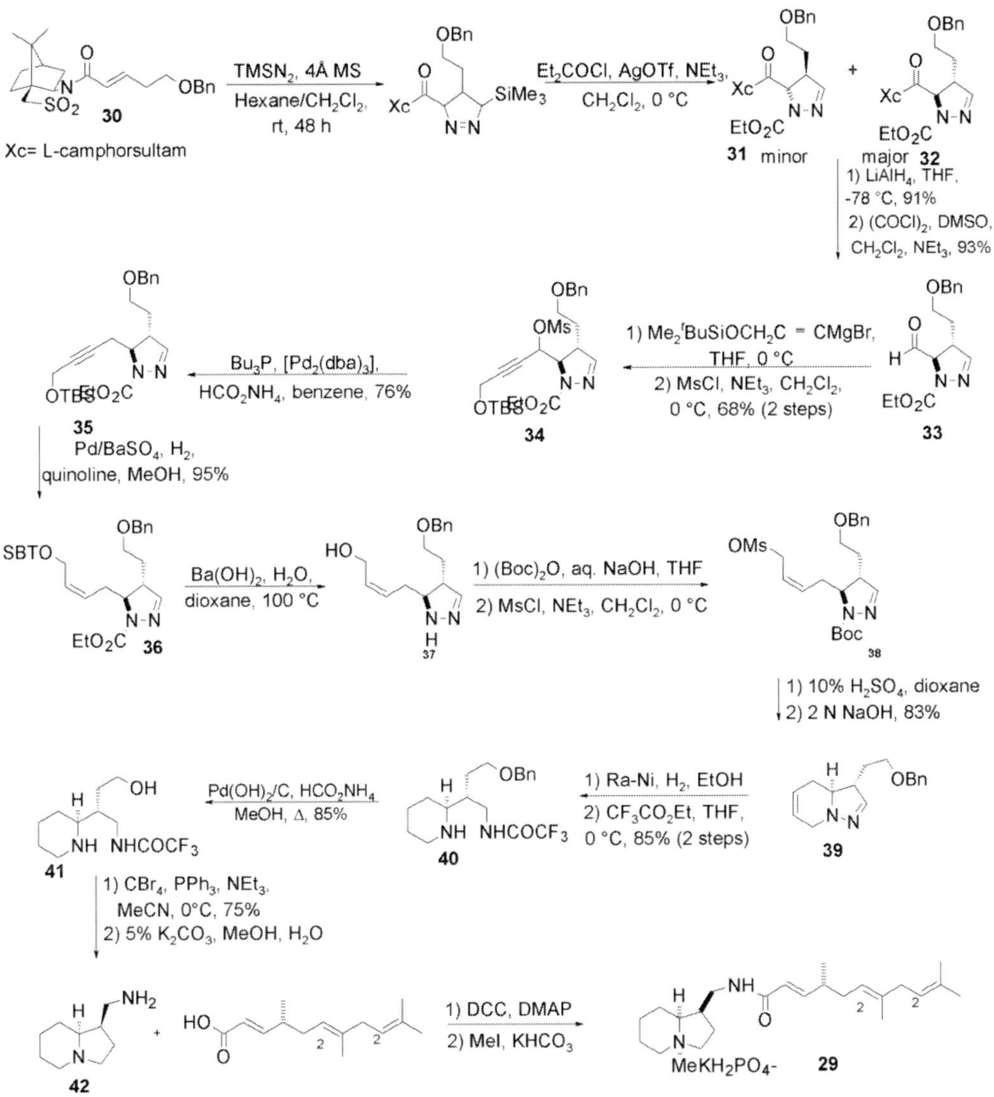

Scheme 14.

Deprotection of the *N*-Boc amide and cyclisation to the pyrazolidine was achieved simultaneously by reaction with concentrated sulfuric acid in dioxane followed by the addition of the reaction mixture to a 2 M solution of sodium hydroxide to give the pyrazolidine *39*. Treatment of *39* with Raney nickel under a hydrogen atmosphere resulted in cleavage of the N-N and reduction of the C=C and C=N bonds to yield a diamine, which was selectively protected as the *N*-trifluoroacetamide *40*. Hydrogenolysis resulted in cleavage of the benzyl group to furnish *41*, and ring closure was achieved by the treatment of primary alcohol with carbon tetrabromide and triphenyl phosphine. Alkaline hydrolysis of trifluoroacteamide yielded the indolizidine core of Stellettamide A *42*, which was subsequently coupled with the carboxylic acid *43* to furnish *ent*-Stellettamide A *29* (Scheme 14). [100,101]

1.3. Cycloadditions with Nitrile Oxides

Nitrile oxides are highly reactive 1,3-dipoles which undergo cycloadditions with a variety of dipolarophiles to form isoxazole or isoxazoline cycloadducts. 1,3-dipolar cycloaddition of nitrile oxides is a versatile reaction for the construction of stereoselective compounds. The isoxazoline ring can be readily cleaved, allowing stereocontrolled access to a variety of acyclic compounds, including γ-hydroxyketones and γ-amino alcohols (Scheme 15).

The employment of chiral nitrile oxides in the asymmetric 1,3-dipolar cycloaddition has not been widely reported; poor diastereoselectivity is achieved in most instances. [102,103] Asymmetric catalysis in the 1,3-dipolar cycloadditions of nitrile oxides by Lewis acids is also not commonly employed. The difficulties in controlling the stereoselectivity with metal complexes arise because the presence of base (such as triethylamine) used for the *in situ* generation of nitrile oxides may interfere with the metal catalyst. Also, nitrile oxides are strong Lewis bases (due to the high donor ability of the oxygen atom of the dipole) and their ready complexation with Lewis acids leads to deactivation of the dipole. [104] As a result, in the asymmetric 1,3-dipolar cycloadditions of nitrile oxides, cycloaddition of achiral nitrile oxides to optically active dipolarophiles such as acrylates [105] and acrylamides [64,65] has attracted most interest. The employment of acrylamides, and in particular tertiary acrylamides with chiral auxiliaries incorporated, is a particularly attractive route, and the synthesis of isoxazolines with high optical purities has been achieved using this approach.

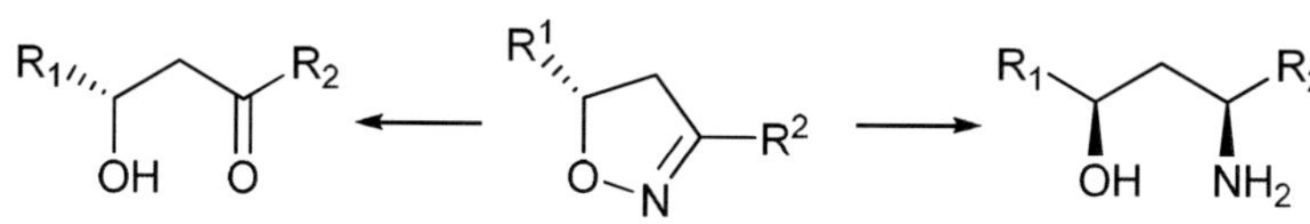

Scheme 15.

The use of chiral acrylamide derivatives in cycloadditions with nitrile oxides was firstly explored by Curran in 1988. [64] Curran had earlier studied the cycloaddition of a range of chiral acrylates with nitrile oxides, with modest degrees of asymmetric induction (up to 56% de) achieved. [56] Using Oppolzer's chiral sultam derivative *1*, diastereoselectivities of up to 95:5 were obtained (Scheme 16).

The usefulness of this asymmetric nitrile oxide cycloaddition was illustrated by Curran in total syntheses of (+)-hepialone *44* and (–)-(1*R*,3*R*,5*S*)-1,3-dimethyl-2,9-dioxabicyclo[3.3.1] nonane *45*. [106] These syntheses demonstrate that enantiomerically pure isoxazolines can be transformed to a variety of functional groups including β,γ-dihydroxy ketones, alcohols, 1,2 and 1,3-diols, 1,3,4-triols, 1,3-amino alcohols and 1,3,4-amino diols.

(+)-Hepialone is a sex pheromone isolated in 1984 from the hair pencils of the male moth *Hepialis californicus Bud.* [107] Curran's synthesis is outlined in Scheme 17. Cycloaddition of the nitro ketal *46* with Oppolzer chiral sultam derivative *1* was achieved by the Mukaiyama method to yield an 88:12 mixture of diastereomers. These were separated by flash chromatography, and the major diastereomer was reduced by the reaction with L-Selectride. The tosylate of the resulting alcohol was coupled with lithium dimethyl cuprate to give *47*. Raney nickel reduction of *47* yielded the β-hydroxy ketone, which was subsequently transformed to (+)-hepialone *44* by exposure to 3N hydrochloric acid.

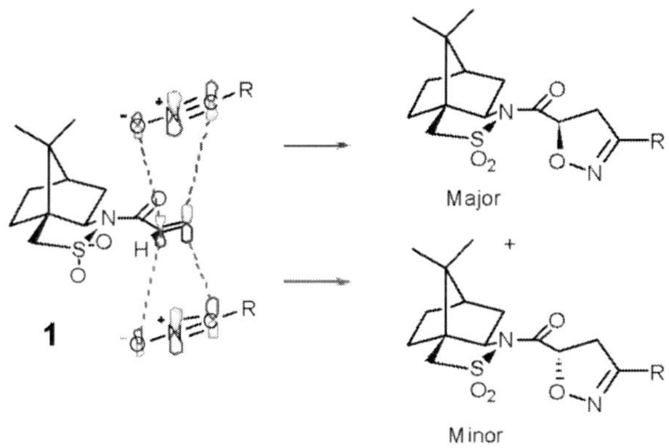

	Major:Minor
t-Bu	95:5
Ph	95:5
Me	90:10
Et	90:10
	88:12*

*This reaction was conducted in benzene; all others were conducted in hexane.

Scheme 16.

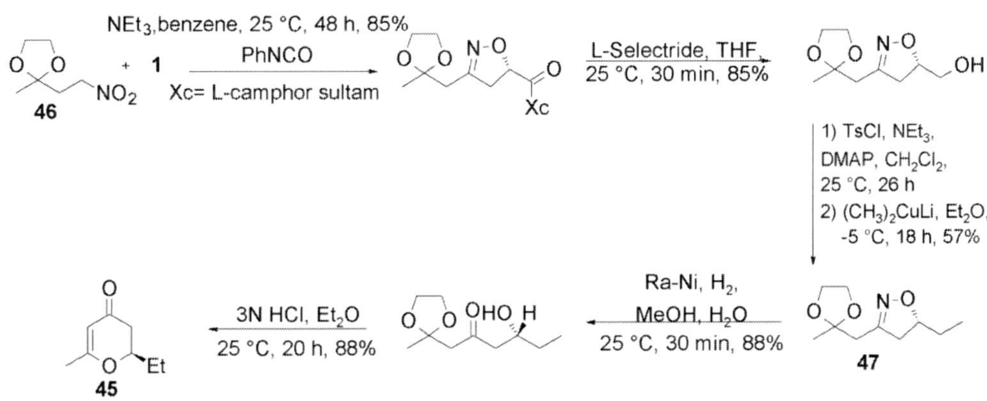

Scheme 17.

(–)-(1*R*,3*R*,5*S*)-1,3-Dimethyl-2,9-dioxabicyclo[3.3.1]nonane *45* is a host specific substance for an ambrosia beetle that infests the bark of the Norway Spruce. [108] Curran's synthesis is illustrated in Scheme 18. Cycloaddition of the oxime chloride *48* with *1* gave the optically active isoxazolines in a ratio of 92:8. Following separation of the diastereomers by column chromatography, the major isomer was reduced with L-Selectride and tosylated to yield *49*. Raney nickel reduction of the corresponding iodide yielded the β-hydroxy ketone, and this was further reduced to the *syn* 1,3-diol *50* by reaction with dimethoxyborane and sodium borohydride. *50* was subsequently converted to *45* by reaction with 3N hydrochloric acid.

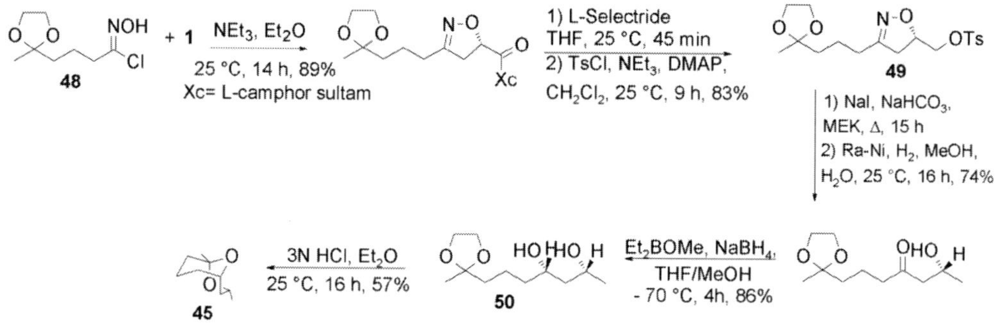

Scheme 18.

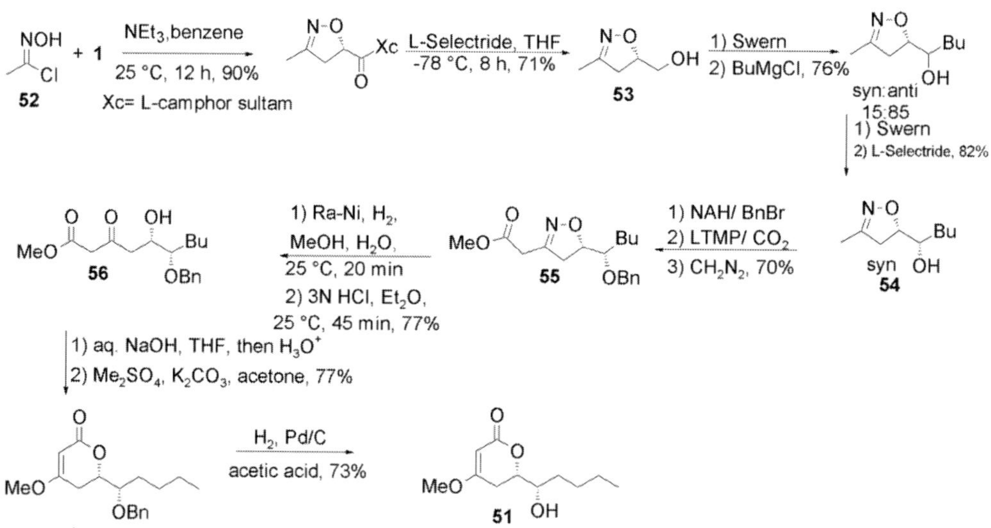

Scheme 19.

The synthesis of (–)-pestalotin *51*, a pyran-2-one metabolite that can act as a gibberellin synergist from nitrile oxide cycloaddition with *1* was communicated in a later paper by Curran. [109,110] Cycloaddition of acetohydroximoyl chloride *52* with *1* provided a diastereomeric ratio of 94:6 of the isoxazoline cycloadducts. The chiral auxiliary of the major isomer was cleaved as before with L-Selectride and Swern oxidation of the resulting alcohol *53*, followed by direct addition of butylmagnesium bromide yielded an 85:15 mixture of diastereomers. Unfortunately, the minor isomer was the isomer required for the (–)-pestalotin synthesis and the mixture was thus epimerized by Swern oxidation. L-Selectride reduction gave the required *syn* isomer *54*. The free alcohol was benzylated, and this was followed by *exo* deprotonation, carboxylation and treatment of the resulting acid with diazomethane to furnish *55*. Raney nickel reduction followed by brief exposure to 3N hydrochloric acid provided the β-hydroxy ketone *56*. Basic lactonisation, followed by *o*-methylation and debenzylation yielded ( ) pestalotin *51* (Scheme 19).

1.4. Cycloadditions with Nitrones

Nitrones are an important class of 1,3-dipole which undergo cycloadditions with alkenes and alkynes to yield isoxazolidines and isoxazolines respectively. The isoxazolidines formed from the reaction of alkenes with nitrones have proven to be very useful building blocks. [14,16,111] Cleavage of the isoxazoline system allows access to a variety of attractive compounds such as β-amino alcohols, with the configuration at the chiral centres retained upon reduction (Scheme 20). Employment of cyclic nitrones leads to much higher stereoselectivities as *E*/*Z* isomerisation is not possible.

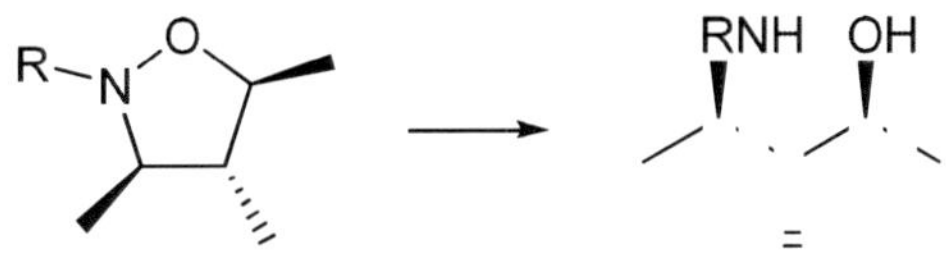

Scheme 20.

The preparation of non-racemic isoxazolidines has attracted much attention in the past 30 years. [13, 17, 27] Chiral acyclic nitrones in which the chiral substituent is located at the nitrogen atom or the carbon atom are commonly employed. [112] Chiral cyclic nitrones have been used for asymmetric induction in 1,3-dipolar cycloadditions. [113, 114] The use of optically active alkenes in asymmetric nitrone cycloadditions has also been extensively studied. Alkenes in which the chiral centre is vicinal to the double bond such as chiral allylic ethers, [115-118] chiral allylic amines [119, 120] and chiral vinyl sulfoxides [121, 122] are most frequently used, and chiral α,β-unsaturated carbonyl compounds such as acrylates [58, 123] and acrylamides [124,125] have also been successfully employed.

Tamura and co-workers have examined the cycloaddition of *C*-(3-pyridyl)nitrones and Oppolzer's chiral sultam derivative (2*S*)-*1* as a route to (+)-(3′*R*,5′*R*)-3′-hydroxycotinine *57*, which is one of the main metabolites of nicotine (Scheme 21). [126] Treatment of the L-gulose derived nitrone *58* with (2*S*)-*1* in the presence of $MgBr_2$ yielded the *trans*-isoxazolidine *59* as the major product (Scheme 21). When this reaction was attempted in the absence of Lewis acid catalyst, a complex mixture of isomers resulted. Hydrolytic cleavage of *59* with hydrochloric acid gave *N*-free isoxazolidine *60*. On reaction with formaldehyde in ethanol the *N*-(ethoxymethyl) isoxazolidine *61* was formed, which was subsequently reduced to the *N*-methylisoxazolidine *62* by reacting with triethylsilane-trifluoroacetic acid. Cleavage of the N-O bond followed by lactamisation was achieved by hydrogenolysis to give *63*, which is diastereomerically related to the desired *57*. On Mitsunobu reaction with benzoic acid followed by hydrolysis of the benzoate, *57* was isolated.

In 1993, Murahashi reported the cycloaddition of cyclic nitrones to acrylamide derivatives in the presence of zinc iodide as a route to the optically active β-amino alcohols (+)-sedridine *64* and (+)-hygroline *65*. [127] When 2,3,4,5-tetrahydropyridine *N*-oxide *66* was reacted with Oppolzer's sultam derivative *1* in the presence of zinc iodide, a diastereomeric ratio of *68a*:*69a* of 78:4 was achieved, along with 18% of other diastereomers. In the absence of metal catalyst, the ratio was 54:27:19. The major diastereomer *68a* was then transformed to (+)-Seridine by hydrolysis with lithium hydroxide to give the β-amino acid *70*, decarboxylation by Barton's method to furnish *71* and finally hydrogenation to yield (+)-Seridine *64* (Scheme 22).

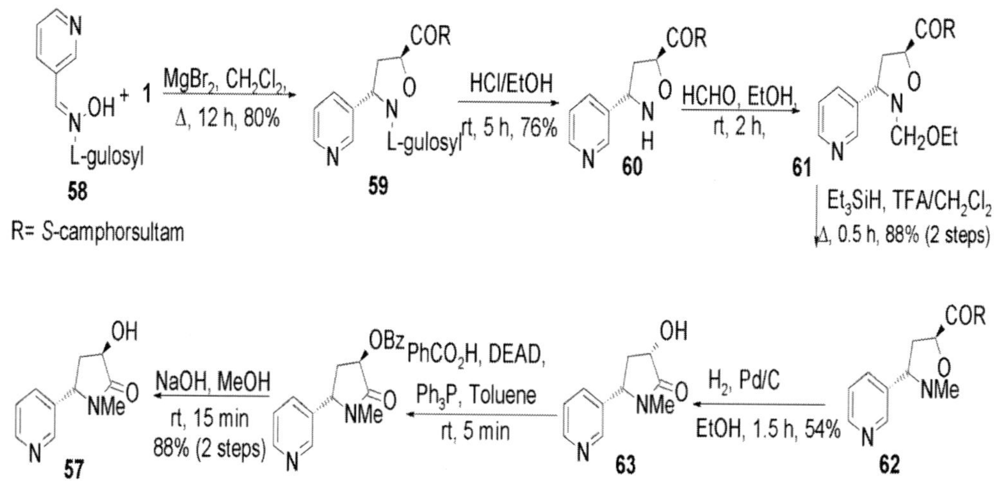

Scheme 21.

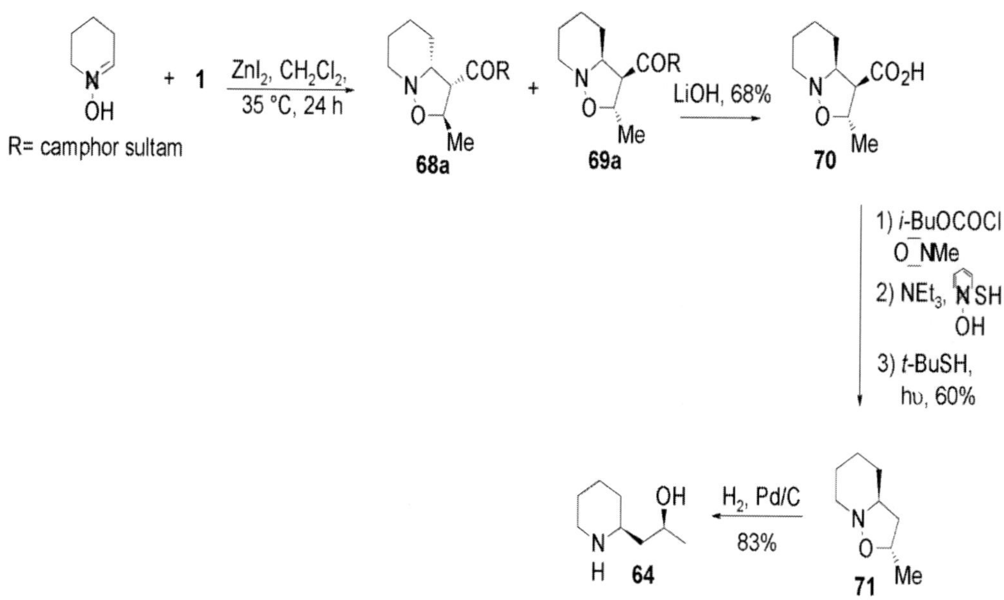

Scheme 22.

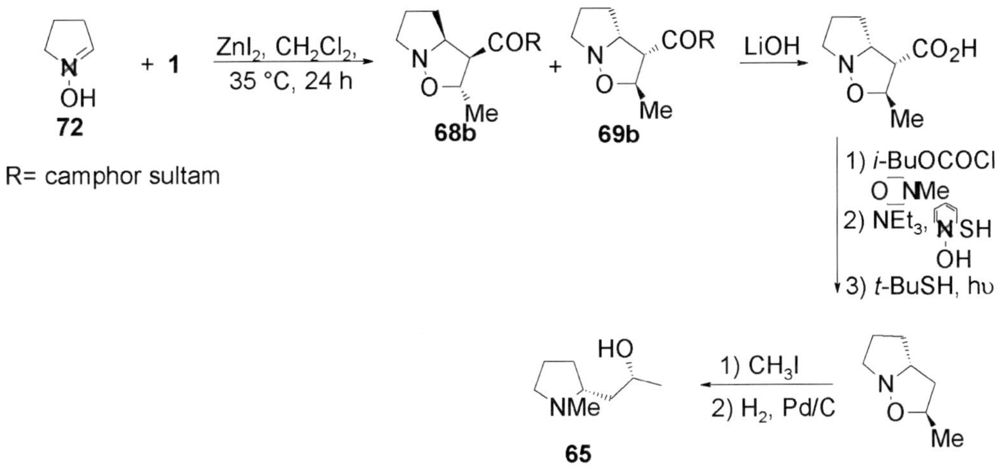

Scheme 23.

Oppolzer's sultam derivative *1* also underwent cycloaddition with 1-pyrroline *N*-oxide *72* in the presence of zinc iodide to yield the isoxazolidines *68b*:*69b* in ratio 77:14, along with 9% of other diastereomers. When the Lewis acid was not present, a ratio of 35:52:13 resulted. *69b* was then converted to (+)-Hygroline *65* using a similar synthetic sequence as outlined above (Scheme 23).

Conclusion

As up to four stereocentres can be introduced in a stereoselective manner in a single step, the asymmetric 1,3-dipolar cycloaddition is one of the most useful methods for the preparation of chiral five-membered ring heterocycles. This chapter focused on the employment of acrylamides as dipolarophiles in asymmetric 1,3-dipolar cycloadditions as the key step in the preparation of biologically active heterocycles. The synthetic utility of the cycloadducts was demonstrated in number of natural product syntheses. The employment of chiral acrylamides as dipolarophiles, particularly in 1,3-dipolar cycloadditions with nitrile oxides in which the use of chiral catalysts and chiral 1,3-dipoles is not an attractive option, will continue to be exploited in asymmetric 1,3-dipolar cycloadditions.

References

[1] M. Kissane, A. R. Maguire, *Chem. Soc. Rev.*, 2010, *39*, 845.

[2] A. Padwa, W. H. Pearson, *Synthetic Applications of 1,3-Dipolar Cycloaddition Chemistry Toward Heterocycles and Natural Products*, Wiley, New York, 2002.

[3] R. Huisgen, *Chem. Pharm. Bull.*, 2000, *48*, 757.

[4] U. Sirion, Y. J. Bae, B. S. Lee, D. Y. Chi, *Synlett*, 2008, 2326.

[5] X. C. Hang, Q. Y. Chen, J. C. Xiao, *Synlett*, 2008, 1989.

[6] R. Huisgen, H. Stangl, H. J. Sturm, H. Wagenhofer, *Angew. Chem.*, 1962, *74*, 31.

[7] M. S. Novikov, A. F. Khlebnikov, M. A. Egarmin, M. V. Shevchenko, V. A. Khlebnikov, R. R. Kostikov, D. Vidovic, *Russ. J. Org. Chem.*, 2006, *42*, 1800.
[8] R. Huisgen, *Angew. Chem., Int. Ed. Engl.*, 1963, *2*, 565.
[9] R. Huisgen, E. Langhals, *Heteroat. Chem.*, 2006, *17*, 433.
[10] C. L. Yoo, M. M. Olmstead, D. J. Tantillo, M. J. Kurth, *Tetrahedron Lett.*, 2006, *47*, 477.
[11] K. Kavitha, P. Venuvanalingam, *J. Chem. Soc., Perkin Trans.2*, 2002, 2130.
[12] L. I. Smith, *Chem. Rev.*, 1938, *23*, 193.
[13] H. Pellissier, *Tetrahedron*, 2007, *63*, 3235.
[14] A. Padwa, *1,3-Dipolar Cycloaddition Chemistry*, Wiley-Interscience, New York, 1984.
[15] C. Grundmann, P. Grunanger, *The Nitrile Oxides*, Springer, New York, 1971.
[16] K. B. G. Torssell, *Nitrile Oxides, Nitrones, and Nitronates in Organic Synthesis*, VCH Publishers, New York, 1988.
[17] K. V. Gothelf, K. A. Jorgensen, *Chem. Rev.*, 1998, *98*, 863.
[18] W. Carruthers, *Cycloaddition Reactions in Organic Synthesis*, Permagon Press, 1990
[19] H. W. Fruehauf, *Chem. Rev.*, 1997, *97*, 523.
[20] M. Bonin, A. Chauveau, L. Micouin, *Synlett*, 2006, 2349.
[21] S. Karlsson, H. E. Hogberg, *Org. Prep. Proced. Int.*, 2001, *33*, 103.
[22] S. Kanemasa, *Synlett*, 2002, 1371.
[23] K. V. Gothelf, *Cycloaddit. React. Org. Synth.*, 2002, 211.
[24] G. Broggini, G. Zecchi, *Synthesis*, 1999, 905.
[25] G. Broggini, G. Molteni, A. Terraneo, G. Zecchi, *Heterocycles*, 2003, *59*, 823.
[26] I. N. N. Namboothiri, A. Hassner, *Top. Curr. Chem.*, 2001, *216*, 1.
[27] M. Frederickson, *Tetrahedron*, 1997, *53*, 403.
[28] A. E. Koumbis, J. K. Gallos, *Curr. Org. Chem.*, 2003, *7*, 771.
[29] I. Coldham, R. Hufton, *Chem. Rev.*, 2005, *105*, 2765.
[30] S. Husinec, V. Savic, *Tetrahedron: Asymmetry*, 2005, *16*, 2047.
[31] D. M. Hodgson, F. Y. T. M. Pierard, P. A. Stupple, *Chem. Soc. Rev.*, 2001, *30*, 50.
[32] G. Molteni, *Heterocycles*, 2005, *65*, 2513.
[33] H. Zollinger, *Diazo Chemistry II*, VCH Publishers, New York, 1995.
[34] H. Pellissier, *Tetrahedron*, 2006, *62*, 5559.
[35] P. N. Confalone, E. M. Huie, *Org. React.*, 1988, *36*, 1.
[36] D. P. Curran, *Advances in Cycloaddition*, JAI Press, London, 1988.
[37] J. R. Liddell, *Nat. Prod. Rep.*, 1997, *14*, 653.
[38] J. J. Tufariello, *Acc. Chem. Res.*, 1979, *12*, 396.
[39] S. Patai, *The Chemistry of the Azido Group*, Interscience, London, 1971
[40] A. Nadin, *J. Chem. Soc., Perkin Trans.1*, 1998, 3493.
[41] C. Najera, J. M. Sansano, *Curr. Org. Chem.*, 2003, *7*, 1105.
[42] G. Pandey, P. Banerjee, S. R. Gadre, *Chem. Rev.*, 2006, *106*, 4484.
[43] M. B. Cid, J. Garcia Ruano, *Topics in Current Chemistry. Organosulfur Chemistry.*, Springer, Berlin, 1999.
[44] T. L. Gilchrist, R. C. Storr, *Organic Reactions and Orbital Symmetry*, 2nd, Cambridge University Press, 1979.
[45] B. M. Trost, I. Fleming, *Comprehensive Organic Synthesis*, Permagon Press, 1991.

[46] D. S. Wolfman, G. Linstrumelle, C. F. Cooper, *The Chemistry of Diazonium and Diazo Groups*, John Wiley and Sons, New York, 1978.
[47] C. J. Easton, M. M. Hughes, G. P. Savage, G. W. Sipmson, *Advances in Heterocyclic Chemistry*, Academic Press, 1994.
[48] D. S. Black, R. F. Crozier, V. C. Davis, *Synthesis*, 1975, 205.
[49] C. Najera, J. M. Sansano, *Angew. Chem., Int. Ed.*, 2005, *44*, 6272.
[50] M. Alvarez-Corral, M. Munoz-Dorado, I. Rodriguez-Garcia, *Chem. Rev.*, 2008, *108*, 3174.
[51] R. Huisgen, P. Poechlauer, G. Mloston, K. Polborn, *Helv. Chim. Acta*, 2007, *90*, 983.
[52] A. Ohta, K. Dahl, R. Raab, J. Geittner, R. Huisgen, *Helv. Chim. Acta*, 2008, *91*, 783.
[53] J. Zhu, B. M. Lines, M. D. Ganton, M. A. Kerr, M. S. Workentin, *J. Org. Chem.*, 2008, *73*, 1099.
[54] A. Krasinski, Z. Radic, R. Manetsch, J. Raushel, P. Taylor, K. B. Sharpless, H. C. Kolb, *J. Am. Chem. Soc.*, 2005, *127*, 6686.
[55] M. P. Sibi, L. M. Stanley, T. Soeta, *Org. Lett.*, 2007, *9*, 1553.
[56] D. P. Curran, B. H. Kim, H. P. Piyasena, R. J. Loncharich, K. N. Houk, *J. Org. Chem.*, 1987, *52*, 2137.
[57] V. C. Pham, J. L. Charlton, *J. Org. Chem.*, 1995, *60*, 8051.
[58] A. Carriere, A. Virgili, M. Figueredo, *Tetrahedron: Asymmetry*, 1996, *7*, 2793.
[59] P. Merino, S. Anoro, S. Franco, F. L. Merchan, T. Tejero, V. Tunon, *J. Org. Chem.*, 2000, *65*, 1590.
[60] A. Zhang, Y. Kan, B. Jiang, *Chin. J. Chem.*, 2000, *18*, 220.
[61] F. Pisaneschi, M. Gensini, M. Salvati, F. M. Cordero, A. Brandi, *Heterocycles*, 2006, *67*, 413.
[62] C. Najera, M. Gracia Retamosa, J. M. Sansano, *Tetrahedron: Asymmetry*, 2006, *17*, 1985.
[63] C. Najera, M. d. G. Retamosa, J. M. Sansano, A. de Cozar, F. P. Cossio, *Eur. J. Org. Chem.*, 2007, 5038.
[64] D. P. Curran, B. H. Kim, J. Daugherty, T. A. Heffner, *Tetrahedron Lett.*, 1988, *29*, 3555.
[65] Y. H. Kim, S. H. Kim, D. H. Park, *Tetrahedron Lett.*, 1993, *34*, 6063.
[66] A. S. Anslow, L. M. Harwood, H. Phillips, D. Watkin, *Tetrahedron: Asymmetry*, 1991, *2*, 997.
[67] A. S. Anslow, L. M. Harwood, H. Phillips, D. Watkin, *Tetrahedron: Asymmetry*, 1991, *2*, 169.
[68] J. F. Peyronel, S. Grisoni, B. Carboni, T. Courgeon, R. Carrie, *Tetrahedron*, 1994, *50*, 189.
[69] A. Padwa, Y. Y. Chen, U. Chiacchio, W. Dent, *Tetrahedron*, 1985, *41*, 3529.
[70] P. Garner, O. Dogan, *J. Org. Chem.*, 1994, *59*, 4.
[71] R. Grigg, *Tetrahedron: Asymmetry*, 1995, *6*, 2475.
[72] D. A. Barr, M. J. Dorrity, R. Grigg, S. Hargreaves, J. F. Malone, J. Montgomery, J. Redpath, P. Stevenson, M. Thornton-Pett, *Tetrahedron*, 1995, *51*, 273.
[73] R. Grigg, V. Sridharan, S. Suganthan, A. W. Bridge, *Tetrahedron*, 1995, *51*, 295.
[74] P. Allway, R. Grigg, *Tetrahedron Lett.*, 1991, *32*, 5817.
[75] D. M. Cooper, R. Grigg, S. Hargreaves, P. Kennewell, J. Redpath, *Tetrahedron*, 1995, *51*, 7791.

[76] R. Annunziata, M. Cinquini, F. Cozzi, L. Raimondi, T. Pilati, *Tetrahedron: Asymmetry*, 1991, *2*, 1329.
[77] P. Garner, W. B. Ho, *J. Org. Chem.*, 1990, *55*, 3973.
[78] P. Garner, W. B. Ho, S. K. Grandhee, W. J. Youngs, V. O. Kennedy, *J. Org. Chem.*, 1991, *56*, 5893.
[79] P. Garner, W. B. Ho, H. Shin, *J. Am. Chem. Soc.*, 1992, *114*, 2767.
[80] P. Garner, W. B. Ho, H. Shin, *J. Am.Chem. Soc.*, 1993, *115*, 10742.
[81] N. Hirayama, K. Shirahata, *J. Chem. Soc., Perkin Trans.2*, 1983, 1705.
[82] Z. Ma, S. Wang, C. S. Cooper, A. K. L. Fung, J. K. Lynch, F. Plagge, D. T. W. Chu, *Tetrahedron: Asymmetry*, 1997, *8*, 883.
[83] S. Karlsson, H. E. Hogberg, *Tetrahedron: Asymmetry*, 2001, *12*, 1977.
[84] S. Karlsson, H. E. Hogberg, *Tetrahedron: Asymmetry*, 2001, *12*, 1975.
[85] S. Karlsson, H. E. Hoegberg, *J. Chem. Soc., Perkin Trans.1*, 2002, 1076.
[86] M. Ogata, H. Matsumoto, S. Shimizu, S. Kida, H. Nakai, M. Motokawa, H. Miwa, S. Matsuura, T. Yoshida, *Eur. J. Med. Chem.*, 1991, *26*, 889.
[87] P. L. Kotian, T. H. Lin, Y. El Kattan, P. Chand, *Org. Process Res. Dev.*, 2005, *9*, 193.
[88] J. S. Carey, *J. Org. Chem.*, 2001, *66*, 2526.
[89] S. M. Jenkins, H. J. Wadsworth, S. Bromidge, B. S. Orlek, P. A. Wyman, G. J. Riley, J. Hawkins, *J. Med. Chem.*, 1992, *35*, 2392.
[90] T. Onishi, P. R. Sebahar, R. M. Williams, *Org. Lett.*, 2003, *5*, 3135.
[91] T. Onishi, P. R. Sebahar, R. M. Williams, *Tetrahedron*, 2004, *60*, 9503.
[92] G. Pandey, J. K. Laha, G. Lakshmaiah, *Tetrahedron*, 2002, *58*, 3525.
[93] T. F. Spande, H. M. Garraffo, M. W. Edwards, H. J. C. Yeh, L. Pannell, J. W. Daly, *J. Am. Chem. Soc.*, 1992, *114*, 3475.
[94] G. Pandey, T. D. Bagul, A. K. Sahoo, *J. Org. Chem.*, 1998, *63*, 760.
[95] M. Regitz, H. Heydt, *1,3-Dipolar Cycloaddition Chemistry*, Wiley-Interscience,New York, 1984
[96] G. Maas, *Synthetic Applications of 1,3-Dipolar Cycloaddition Chemistry Toward Heterocycles and Natural Products*, Wiley-Interscience,New York, 2003
[97] M. R. Mish, F. M. Guerra, E. M. Carreira, *J. Am. Chem. Soc.*, 1997, *119*, 8379.
[98] J. A. Simon, S. L. Schreiber, *Chemistry & Biology*, 1995, *2*, 53.
[99] M. R. Attwood, C. H. Hassall, A. Krohn, G. Lawton, S. Redshaw, *J. Chem. Soc., Perkin Trans. 1*, 1986, *6*, 1011.
[100] G. A. Whitlock, E. M. Carreira, *J. Org. Chem.*, 1997, *62*, 7916.
[101] G. A. Whitlock, E. M. Carreira, *Helv. Chim. Acta*, 2000, *83*, 2007.
[102] M. Zagozda, J. Plenkiewicz, *Tetrahedron: Asymmetry*, 2007, *18*, 1457.
[103] J. Romanski, J. Jozwik, C. Chapuis, J. Jurczak, *Helv. Chim. Acta*, 2007, *90*, 2116.
[104] S. Kanemasa, K. Onimura, *Tetrahedron*, 1992, *48*, 8631.
[105] T. Olsson, K. Stern, S. Sundell, *J. Org. Chem.*, 1988, *53*, 2468.
[106] D. P. Curran, T. A. Heffner, *J. Org. Chem.*, 1990, *55*, 4585.
[107] I. Kubo, T. Matsumoto, D. L. Wagner, J. N. Shoolery, *Tetrahedron Lett.*, 1985, *26*, 563.
[108] V. Heemans, W. Francke, *Naturwissenschaften*, 1976, *63*, 344.
[109] J. Zhang, D. P. Curran, *J. Chem. Soc., Perkin Trans.1*, 1991, 2627.
[110] Y. Kimura, A. Suzuki, S. Tamura, K. Mori, M. Oda, M. Matsui, *Plant Cell Physiology*, 1977, *18*, 1177.

[111] J. J. Tufariello, *1,3-Dipolar Cycloaddition Chemistry*, Wiley-Interscience,New York, 1984
[112] C. Berini, F. Minassian, N. Pelloux-Leon, J. N. Denis, Y. Vallee, C. Philouze, *Org. Biomol. Chem.*, 2008, *6*, 2574.
[113] K. Aouadi, S. Vidal, M. Msaddek, J. P. Praly, *Synlett*, 2006, 3299.
[114] K. Aouadi, E. Jeanneau, M. Msaddek, J. P. Praly, *Tetrahedron: Asymmetry*, 2008, *19*, 1145.
[115] M. Ito, C. Kibayashi, *Tetrahedron Lett.*, 1990, *31*, 5065.
[116] M. Ito, C. Kibayashi, *Tetrahedron*, 1991, *47*, 9329.
[117] M. Ito, M. Maeda, C. Kibayashi, *Tetrahedron Lett.*, 1992, *33*, 3765.
[118] H. Ina, M. Ito, C. Kibayashi, *J. Org. Chem.*, 1996, *61*, 1023.
[119] S. Mzengeza, R. A. Whitney, *J. Chem. Soc., Chem. Commun.*, 1984, 606.
[120] T. Naito, M. Ikai, M. Shirakawa, K. Fujimoto, I. Ninomiya, T. Kiguchi, *J. Chem. Soc., Perkin Trans.1*, 1994, 773.
[121] J. L. Garcia Ruano, J. Ignacio Andres Gil, A. Fraile, A. M. Martin Castro, M. R. Martin, *Tetrahedron Lett.*, 2004, *45*, 4653.
[122] J. L. Garcia Ruano, A. Fraile, A. M. Martin Castro, M. R. Martin, *J. Org. Chem.*, 2005, *70*, 8825.
[123] T. Olsson, K. Stern, G. Westman, S. Sundell, *Tetrahedron*, 1990, *46*, 2473.
[124] T. Gefflaunt, U. Bauer, K. Airola, M. P. Koskinen, *Tetrahedron: Asymmetry*, 1996, *7*, 3099.
[125] T. Tejero, A. Dondoni, I. Rojo, F. L. Merchan, P. Merino, *Tetrahedron*, 1997, *53*, 3301.
[126] O. Tamura, A. Kanoh, M. Yamashita, H. Ishibashi, *Tetrahedron*, 2004, *60*, 9997.
[127] S. Murahashi, Y. Imada, M. Kohno, T. Kawakami, *Synlett*, 1993, 395.

In: Bioactive Heterocycles
Editors: K. L. Ameta, R. P. Pawar and A. J. Domb
ISBN: 978-1-62257-636-4
© 2013 Nova Science Publishers, Inc.

Chapter 8

1,3-Dipolar Cycloadditions Approach to Bioactive Spiroheterocycles

Essam M. Hussein*
Department of Chemistry, Faculty of Science, Assiut University, Assiut, Egypt

Abstract

Spiro compounds represent an important class of naturally occurring substances characterized by highly pronounced biological properties. The most developed avenue for the synthesis of these compounds depends on the cycloaddition to an exocyclic bond. 1,3-dipolar cycloaddition reactions are considered to be one of the most useful processes for the construction of five membered heterocyclic ring systems in highly regio and stereoselective manner. This chapter describes 1) general methods for generation and preparation of most important 1,3-dipoles such as nitrones, nitrile oxides and azomethine ylides; and 2) the most recent examples of synthetic applications of 1,3-dipolar cycloaddition reactions to bioactive spiroheterocyclic compounds.

Keywords: 1,3-dipolar cycloaddition, spiro, nitrone, nitrile oxide, azomethine ylide, bioactive

1. Introduction

A "spirounion" is formed by a single atom, which is the only common member of two rings. A "free spirounion" constitutes the only union direct or indirect between two rings. The common atom is designated as the "spiro atom". According to the number of spiro atoms present, the compounds are distinguished as monospiro, dispiro, trispiro compounds, etc.

In 1900, Bayer created the first spiran described as a bicyclic hydrocarbon linked by a single carbon. Due to the tetrahedral nature of the spiro (linked) carbon, two ring planes are closely perpendicular to each other.

* E-mail: essam.hussein78@yahoo.com.

Spiro compounds having cyclic structures fused at a central carbon are of recent interest due to their interesting conformational features and their structural implications on biological systems. The asymmetric character of the molecule due to the chiral spiro carbon is an important criteria of the biological activities. The retention of neurotoxic properties of perhydrohistrionicotoxin, an analogue of a natural product (-)-histrionicotoxin, is evidence of the role of spiro carbon in navigating biological activity (Figure1) [1].

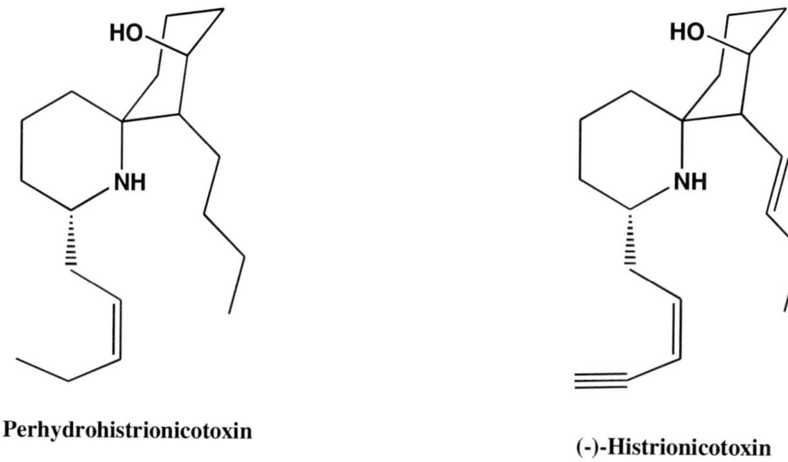

Figure 1. Structure of Perhydrohistrionicotoxin and (-)-Histrionicotoxin.

Spiroheterocyclic compounds are considered the most important and essential derivatives in the field of organic chemistry. Many outstanding chemists devoted a lot of efforts which paved the way for the synthesis of new spiroheterocyclic derivatives as well as for studying and understanding their valuable applications. The wide scope of the synthesis of spiroheterocyclic systems is due to a great number of new synthesized derivatives to be used as drugs and dyes. Recent literature surveys revealed that the syntheses and applications of spiroheterocyclic compounds have gained much more importance. Spirooxindolyl nucleus represents an important part of many naturally occurring alkaloids [2], such as strychnofoline [3], alstonisine [4], coerulescine [5], pteropodine [6], formosanine [7], spirotryprostatine A, spirotryprostatine B, (+)-elacomine, (-)-horsfiline and rychnophyilline [8], with highly pronounced pharmacological properties (Figure 2).

2. Bioactive Spiroheterocycles

The spiro functionality has been known for a long time to be presented in phytochemical either in alkaloids, lactones or terpenoids. The spirocyclic alkaloid (-)-histrionicotoxin , isolated from skin extracts of the poison dart frog, *Dendrobats histrionius*, found in Columbia, is a very potent nicotinic receptor antagonist [9]. Spiroketals are reported to be the subunits of many naturally occurring substances of biological interest such as insect pheromones, anti-feedants and polyether antibiotics [10]. A series of spiroketals (**1-4**) have been isolated from *Chrysanthemum coronanium*, a common vegetable of South China [11].

Some of these compounds are found to have antifeeding activity towards silkworm [12] and spasmolytic and antiphlogistic activity (Figure 3) [13, 14]. Unsaturated spiroacetals such as 1,6-dioxaspiro[4.4]nona-3,8-diene [15] and 1,6-dioxaspiro[4.5]decane [16] have also been isolated from *Artemisia sp.* The spiro[pyrrolidin-3,3'-indole] ring system is a recurring structural motif in a number of natural products such as vinblastine and vincristine, functioning as cytostatics and are of prime importance in cancer chemotherapy [17].

The derivatives of spiro-oxindole find very wide biological application as antimicrobial [18], antitumor, and antibiotic agents, and inhibitors of human NK-1 receptor [19, 20] and potent non-peptide inhibition of the p53–MDM2 interaction (Figure 4) [21].

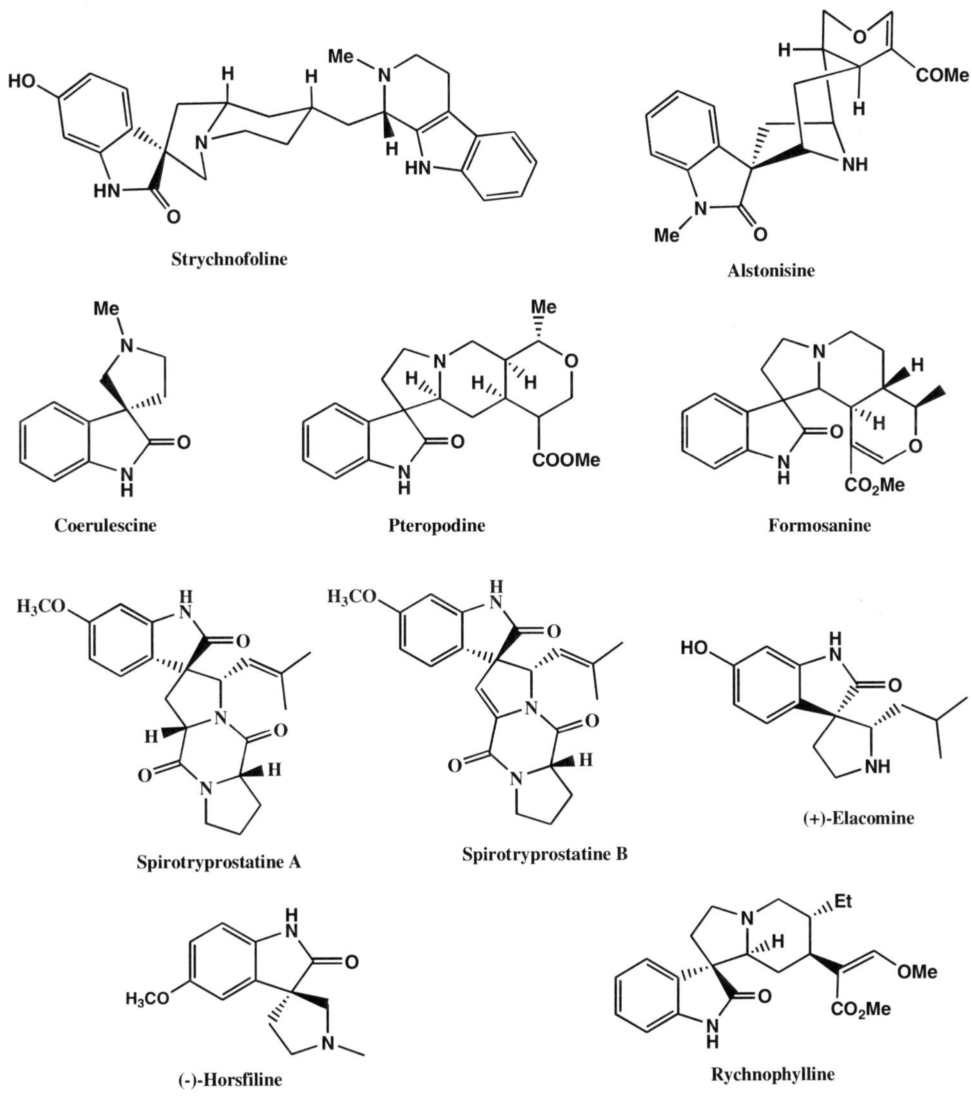

Figure 2. Representative naturally occurring spiropyrrolidinyl-oxoindole alkaloids.

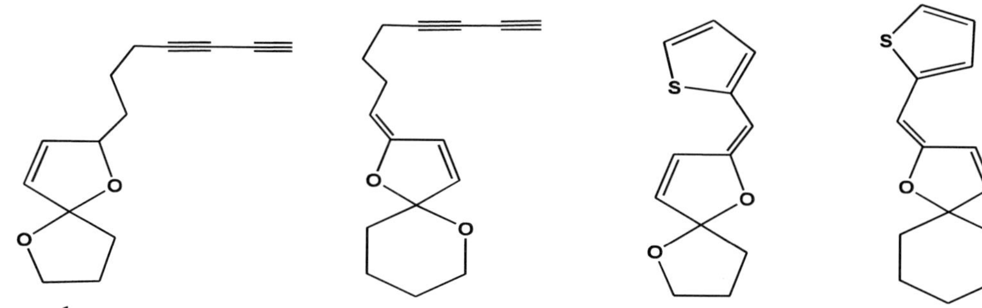

1 2 3 4

Figure 3. Structure of some naturally occurring spiroketals.

Horsfiline, an oxindole alkaloid containing a spiro-[indole-pyrrolidone] nucleus, has been isolated by Bodo and co-workers [22] from *Horsfieldia superba*, a tree from Malaysia. The extracts of *Horsfieldia superba* tree is commonly employed in local medicine. The saponaceolides (A–D), **5** are found to possess antitumor activity in 60 human cancer cell lines [23]. Each of these compounds contains a unit of tricyclic trioxaspiroketal (Figure 5).

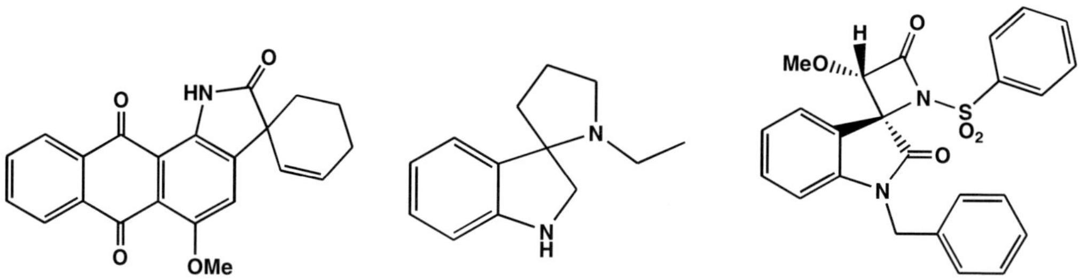

Figure 4. Representative bioactive spiro-oxindoles.

Wrobel *et al.* prepared spirosuccinimide incorporated with isoindolones (**6**) and benzisothiazole-1,1-dioxide moieties (**7**) as aldose reductase inhibitors and anti-hyperglycemic agents (Figure 6) [24].

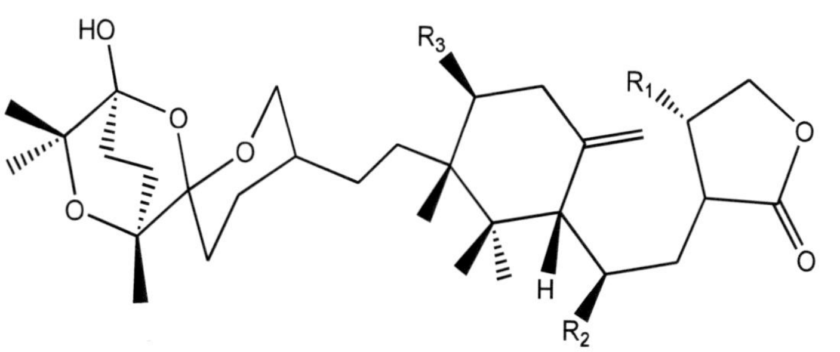

5: A: R_1= OH, R_2 = R_3 = H **C: R_1= R_2 = OH, R_3 = H**

B: R_1= R_2 = R_3 = H **D: R_1= R_3 = OH, R_2 = H**

Figure 5. Structure of Saponaceolides.

Zen *et al.* synthesized some spiro-isoxazolines (8) to be used as anticancer reagents (Figure 7) [25].

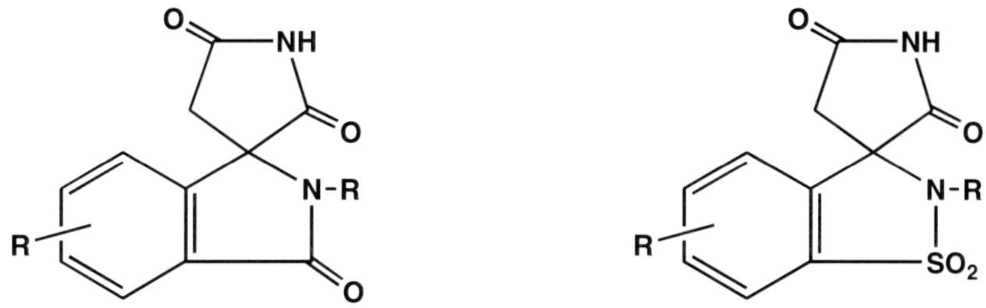

6 7

Figure 6. Structure of bioactive spirosuccinimide derivatives.

3. 1,3-Diploar Cycloaddition Reactions

The preparation of chiral compounds is an important and challenging area of contemporary synthetic organic chemistry. 1,3-dipolar cycloaddition, also known as the "Huisgen cycloaddition" is a classic reaction in organic chemistry consisting of the reaction of a dipolarophile with a 1,3-dipolar compound that allows the production of various five-membered heterocycles.

This reaction represents one of the most productive fields of modern synthetic organic chemistry. Most of dipolarophiles are alkenes, alkynes and molecules possessing related heteroatom functional groups (such as carbonyls and nitriles).

1,3-dipoles can be basically divided into two different types:

i. The allyl anion type nitrones, azomethine ylides, nitro compounds, bearing a nitrogen atom in the middle of the dipole, carbonyl ylides, or carbonyl imines, bearing an oxygen atom in the middle of the dipole.
ii. The linear propargyl/allenyl anion type such as nitrile oxides, nitrilimines, nitrile ylides, diazoalkanes, or azides.

Two π-electrons of the dipolarophile and four electrons of the dipolar compound contribute in a concerted pericyclic reaction pathway. The addition is stereo-conservative (*suprafacial*), and the reaction is therefore a $[\pi4_S+\pi2_S]$ cycloaddition (Figure 8).

However, the dipole might be stabilized by adjacent central heteroatom Y (nitrogen, oxygen, or sulfur) through resonance, and a non-concerted reaction pathway might also occur.

The transition state of concerted 1,3-dipolar cycloaddition reaction is controlled by frontier molecular orbitals of the substrates. Hence, the reaction of dipoles with dipolarophiles involves either a LUMO–dipole/HOMO–dipolarophile reaction or a HOMO–dipole/LUMO–dipolarophile interaction, depending on the nature of the dipole and the dipolarophile.

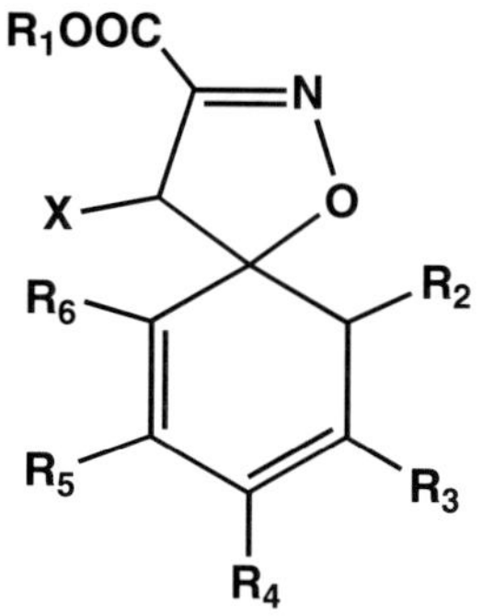

8

R_1 = H, lower cyclo alkyl, alkoxyalkyl, C_6H_5, arylalkyl; R_2 = halo, aryl X = halo

R_3-R_6 = H, lower alkyl; R_2,R_3,R_4,R_5 may from aromatic ring.

Figure 7. Structure of bioactive spiro-isoxazolines derivatives.

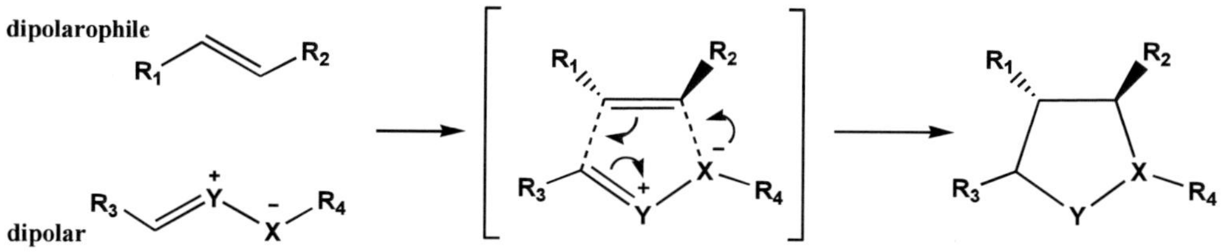

Figure 8. General concerted 1,3-dipolar cycloaddition.

1,3-Dipolar cycloaddition reactions constitute a highly versatile and powerful tool in the construction of pharmacologically important five-membered heterocycles at highly regio- and stereoselective manner through a 1,3-dipolar cycloaddition of nitrone, nitrile oxide, or azomethine ylides to exocyclic olefinic bond of heterocycles or carbocycles [26].

4. Nitrones

For the last 60 years many scientists have been drawn special attention to nitrones due to their successful application as building blocks in the synthesis of various natural and bioactive compounds. The reactions of nitrone dipoles play an important role in cycloaddition reactions. 1,3-dipolar cycloaddition of nitrones with olefinic dipolarophiles proceeds through a concerted mechanism yielding highly substituted isoxazolidines with generation of many three new adjacent chiral centres in a single step. 1,3-dipolar cycloaddition of exocyclic olefins with nitrones result in highly substituted spiro-isoxazolidines and they have also been transformed into complex heterocycles.

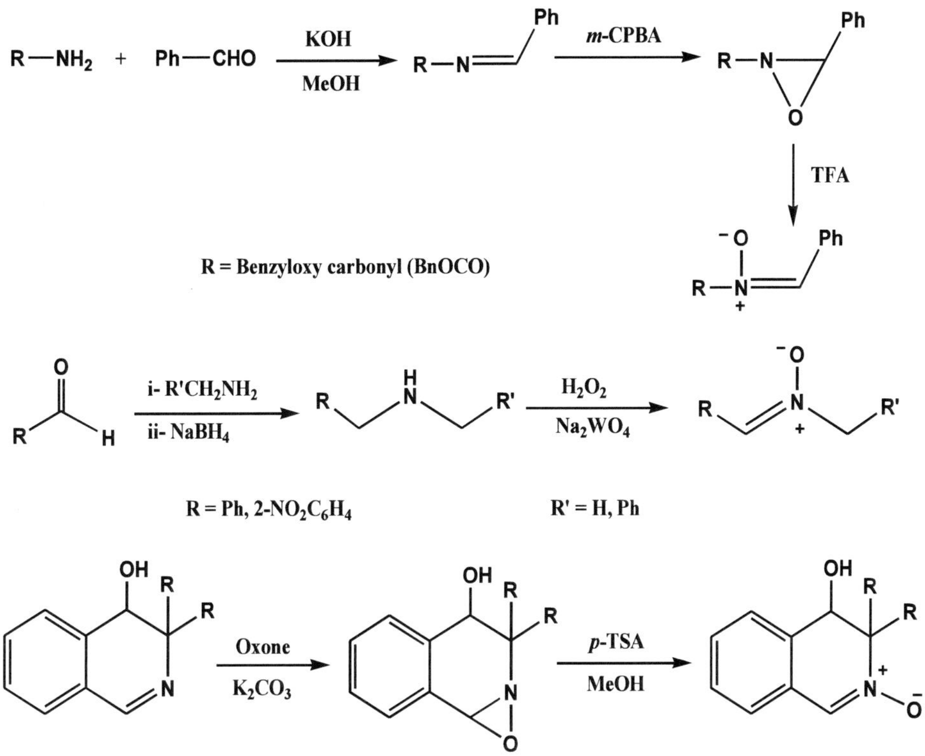

Scheme 1.

4.1. Synthesis of Nitrones

4.1.1. Oxidative Methods

4.1.1.1. Oxidation of Amines

Direct oxidation of primary amines usually leads to a mixture containing oximes, nitroso and nitro compounds. However, oxidation to nitrones can be accomplished after their conversion into secondary amines or imines. Occasionally, oxidation of secondary amines rather than direct imine oxidation seems to provide a more useful and convenient way of producing nitrones (Scheme 1) [27].

4.1.1.2. Oxidation of Hydroxylamines

Mild oxidation of hydroxylamines containing one or more protons at α-C seems to be one of the most convenient methods of nitrone fsynthesis (Scheme 2). Air, H_2O_2, m-chloroperbenzoic acid (*m*-CPBA), oxides of different metals (MnO_2, PbO_2, HgO, Ni_2O_3, etc.) can be used as oxidizing agents [28].

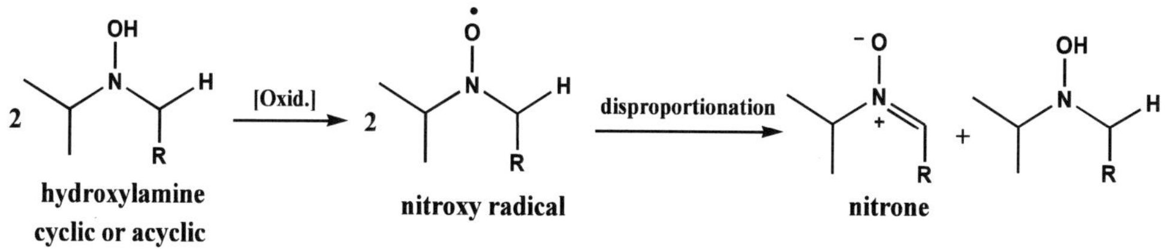

Scheme 2.

4.1.2. Non-oxidative Methods

4.1.2.1. Synthesis from Oximes

Alkylation of oximes at the nitrogen atom with various electron poor alkenes is one of the easiest and convenient methods for synthesizing nitrones.

The reaction known as Grigg's nitrone formation and the advantage of this method is that there is no need to use oxidants [29-31]. In most cases, the resulting nitrones quickly involved in a specific 1,3-cycloaddition reaction. Thus, the reaction of indol-oxime (9) with methyl acrylate, methyl vinyl ketone, acrylonitrile, and acrylamide gives indol-nitrones (10) (Scheme 3) [32, 33].

Bromo-intramolecular cyclization of γ,δ-unsaturated oximes (11) and (12) affords the corresponding bromomethylpyrroline-N-oxides (13) and (14) in 23% - 87% yield depending on the structure of the oxime (Scheme 4) [34].

4.2. Nitrone Cycloadditions Approach to Bioactive Spiroheterocycles

Isoxazolidines are potential precursors for biologically important compounds such as amino sugars [35], alkaloids [36, 37], β-lactams [37] and amino acids [38], and exhibit antibacterial and antifungal activities [39]. Among the dipoles, nitrones have been extensively used as they readily undergo both interand intra-molecular 1,3-dipolar cycloaddition with olefins [39].

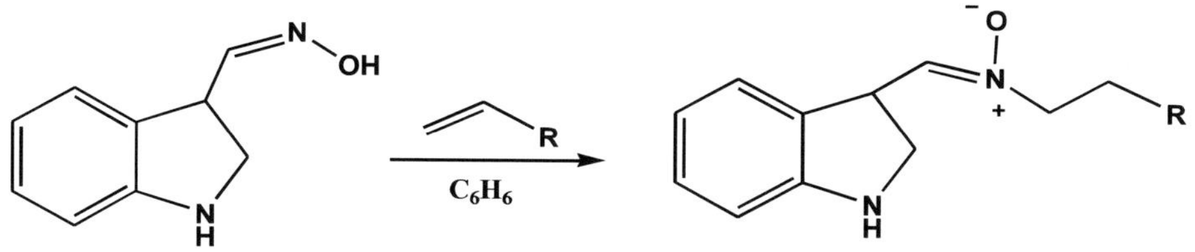

9

10

R = COMe, CO_2Me, $CONH_2$, CN

Scheme 3.

The 1,3-dipolar cycloaddition of exocyclic olefins with nitrones result in highly substituted spiro-isoxazolidines [40] and they have also been transformed into complex heterocycles [41].

4-(4'-Methoxyphenyl)-3-aryl-4H-spiro[isochromene-3',5-isoxazol]-4(1H)-one (21-25) was synthesized by reaction of p-R-benzadoxime (16-20) and 3-β-methoxy-benzylidene-isochroman-4-one (15) in chloroform and aqueous solution of sodium hypochlorite [42]. These compounds typically form the highly stable antitubercular pharmacophore site. Based on their pharmacological properties and thermal stability, these compounds may be useful as anti-tubercular agents (Scheme 5).

The 1,3-dipolar cycloaddition of 3,5-bis-(arylidene)-1-methylpiperidin-4-ones, 26a–o, with C-aryl-N-phenylnitrones, 27a and 27b (4 molar equiv), proceeds chemo-,regio- and stereoselectively affording mono- and bis-spiroisoxazolidines, the former predominating, with the electron-rich oxygen of nitrone attached to the more electron-deficient β-carbon of the benzylidene moiety, ascribable to the steric hindrance exerted by mono spiro-isoxazolidine ring for the second cycloaddition leading to bis- isoxazolidines (Scheme 6). The sterically bulky groups such as 1-naphthyl and o-anisyl in 26 lead to the formation of only mono-spiroisoxazolidines. The cycloaddition of mono-spiroisoxazolidines to furnish bis-spiroisoxazolidines displays facial diastereoselectivity ascribable to steric control [43].

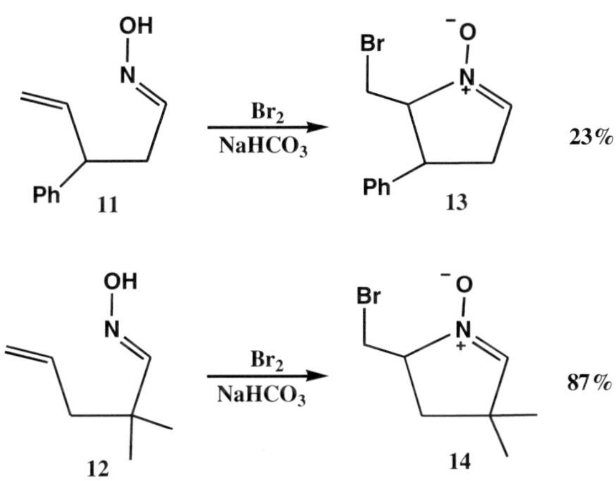

Scheme 4.

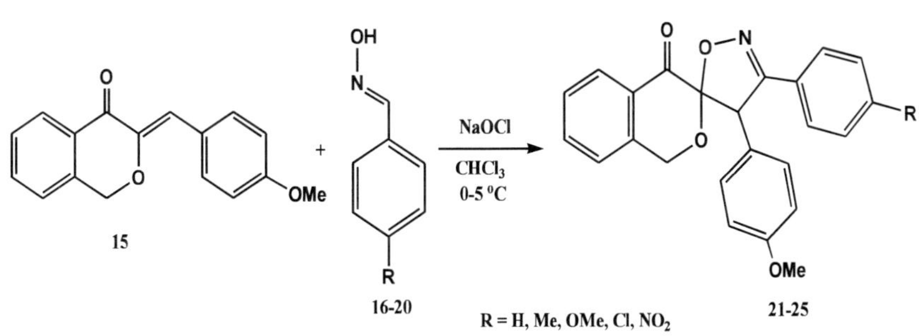

Scheme 5.

Reaction of *Z*-N-*tert*-butyl-α-phenylnitrone (**33**) with lithiated 3,3-diphenyl-2-oxazolinyloxirane (**34-Li**) affording diazadispiro[2.0.4.3]undecane derivative (**35**) in a good yield (Scheme 7) [44]. Lithiation of 3,3-diphenyl-2-oxazolinyloxirane (**34**) was performed using *s*-BuLi/ tetramethylethylenediamine (TMEDA) in THF [45].

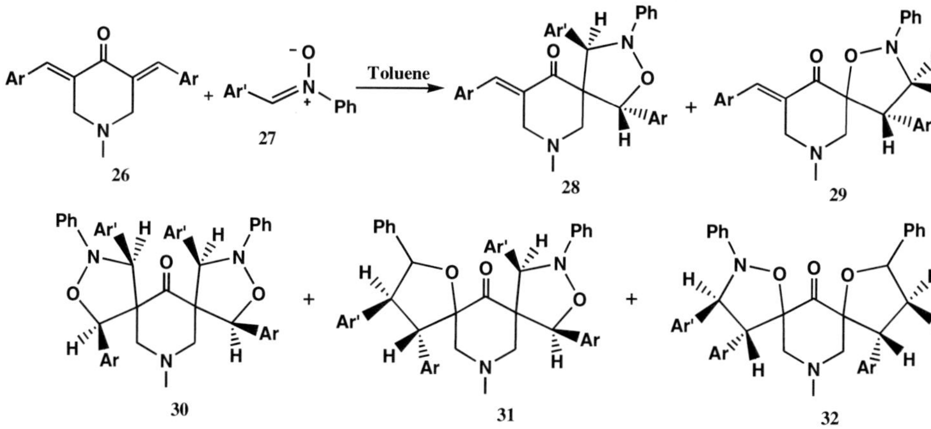

Ar = C_6H_5, *p*-ClC_6H_4, *p*-MeC_6H_4, *p*-$MeOC_6H_4$, *p*-FC_6H_4, *o*-ClC_6H_4, *o*-MeC_6H_4, *o*-$MeOC_6H_4$, 2-Thienyl, *m*-$O_2NC_6H_4$
Ar' = C_6H_5, *p*-MeC_6H_4

Scheme 6.

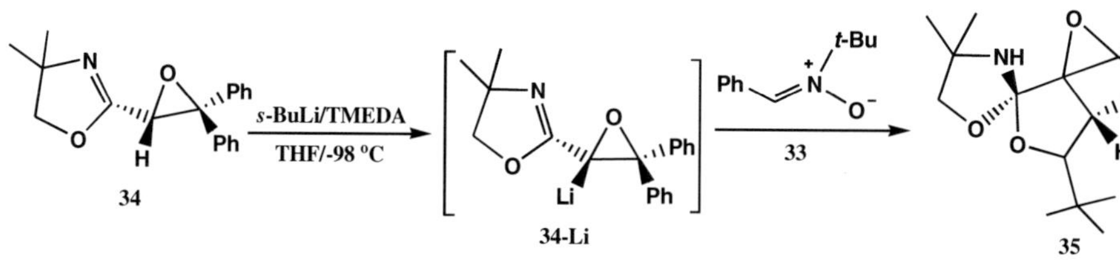

Scheme 7.

Cycloaddition of *C,N*-diphenylnitrone to α-methylene-γ-butyrolactone afforded two diastereomeric 5-spirosubstituted isoxazolidines with high selectivity [46]. The ratio of products formed in the cycloaddition depends on the reaction conditions, increases on lowering the temperature (Scheme 8). However, for practical reasons, the boiling temperature of benzene represents a good compromise between selectivity and reactivity: a good 9:1 ratio of products was obtained after 5 h.

A comparison of the possible transition states leading to the two diastereoisomers shows unfavorable steric interactions between the hydrogen at C-4 of the lactone pointing towards the nitrone and the N-phenyl group of nitrone in the *exo*-C=O transition state (Figure 9). The attack to the opposite face of lactone places the C=O group in the *endo* position, removing the strongest steric repulsions. Moreover, in the *endo*-C=O transition state the carbonyl group, facing the *N*-phenyl, is established for a stabilizing secondary orbital interaction, which might work together steric effects in determination of high stereoselectivity [47].

Heterocyclic aryl nitrone (41) was synthesized in 62-73% yield by bromination of 1-aryl-5-methoxy-2,3-dihydro-1H-2-pyrrolone (40) followed by the treatment with nitrosobenzene at room temperature. Reaction of arylnitrone (41) with ethylvinyl ether afforded spiroisoxazolidino-pyrrolidine derivative (42) (Scheme 9) [48].

1-Aryl-5-methoxy-pyrrol-2-one (40) was easily synthesized, in 74-82% yield, from the reaction of N-aryl-acetimidic ester (39) with chloroacetyl chloride (Scheme 10) [49].

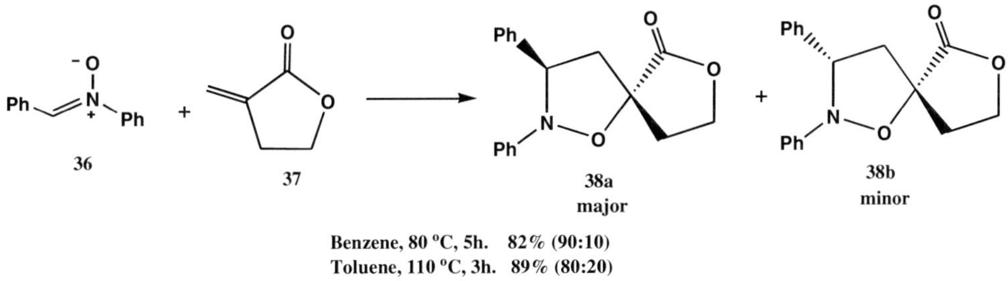

Scheme 8.

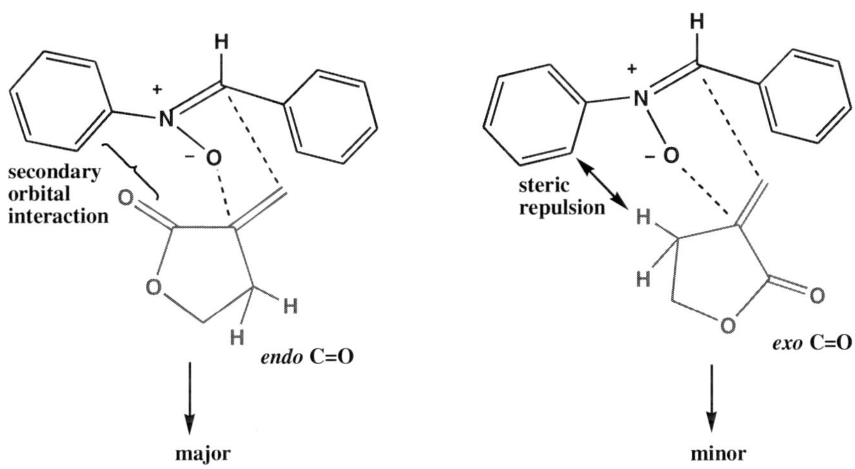

Figure 9. Transition state modes for formation of the two diastereoisomers **38a** and **38b.**

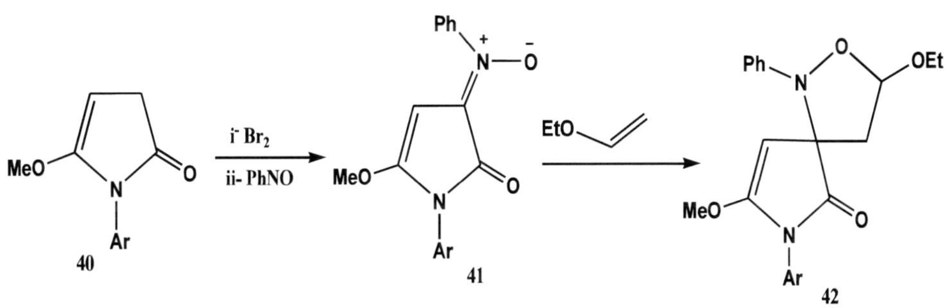

Scheme 9.

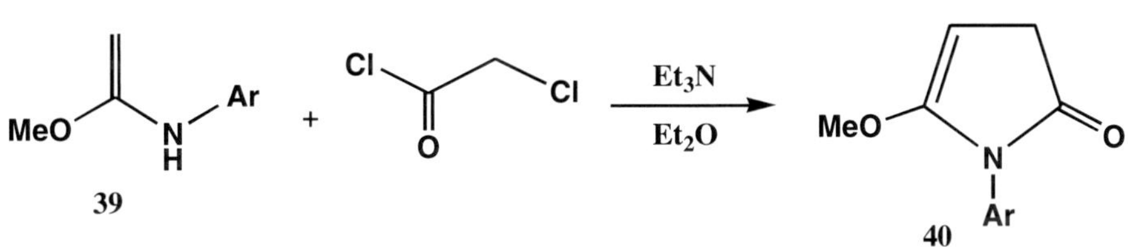

Scheme 10.

5. Nitrile Oxides

Nitrile oxides, RNCO, are the derivatives of *fulminic acid* (R=H). They can be named as *fulmido*-substituted parent molecules, but usually their names are derived from corresponding nitriles, for example, benzonitrile oxide, toluonitrile oxide, pyridine-4-carbonitrile oxide, mesitonitrile oxide. Specific properties of nitrile oxides depend on the structure of the functional group, which have highly polarized C-N and N-O bonds (Figure 10).

Figure 10. Electronic structure of nitrile oxides.

5.1. Synthesis of Nitrile Oxides

5.1.1. From Aldoximes

The conversion of aldoximes to nitrile oxides is a dehydrogenation process. It is only necessary to note here that the process is carried out mainly as halogenation-dehydrohalogenation. The reaction is convenient for both the generation of unstable nitrile oxides (in the presence of a dipolarophile) and the preparation of stable nitrile oxides. The intermediate hydroximoyl halide is frequently not isolated (Scheme 11). It is usually carried out in a two-phase water-organic solvent system with methylene chloride as the preferred solvent [50, 51].

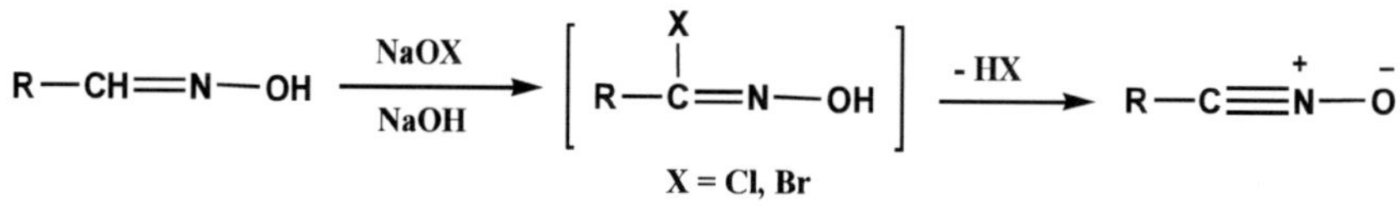

Scheme 11.

5.1.2. From Aliphatic Nitro Compounds (Mukaiyama Procedure)

Generation of nitrile oxides via dehydration of primary nitroalkanes with an aryl isocyanate, usually in the presence of Et_3N base, is one of the most important methods in nitrile oxide synthesis (Scheme 12). Many reagents, other than arylisocyanates, have been tested for the dehydration of nitroalkanes, such as $POCl_3$, $(CH_3CO)_2O$, CH_3COCl, CH_3SO_2Cl, p-toluenesulfonyl chloride. Dehydration of primary nitroalkanes results in unstable nitrile oxides, and, therefore, is limited by in situ transformation of nitrile oxide [52, 53].

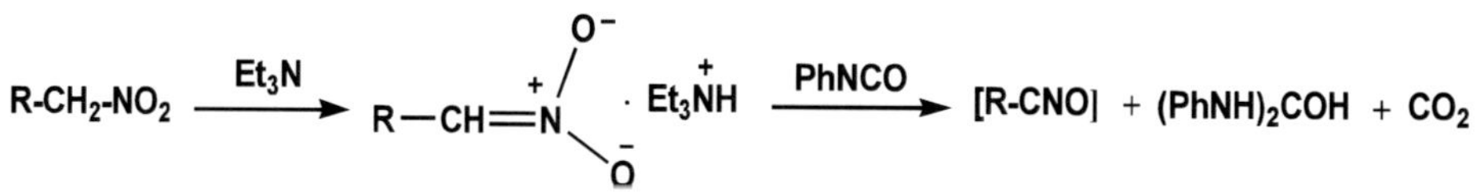

Scheme 12.

5.2. Nitrile Oxide Cycloadditions Approach to Bioactive Spiroheterocycles

1,3-Dipolar cycloadditions of mesitonitrile oxide to the exocyclic double bond of 7-methylenepyrrolo[1,2-c]pyrimidin-1(5H)-ones (41a-c) as dipolarophiles proceed with complete regioselectivity and lead to the spiroisoxazolinyl nucleosides (42a,b) and (43a,b) in good yields (Scheme 13) [54].

The approach of the dipole takes place predominantly from the less sterically hindered site of the dipolarophiles (R = H, CH_2Ph) providing a mixture of two 5,7-*cis* 42a,b and 5,7-*trans* 43a,b isomers. The [3+2] cycloaddition reactions of (*E*)-2-arylidene-(2*H*)-indanones (44) and (*E*)-2-arylidene-(2*H*)-3-methylindanones (47) with the arylnitrile oxides generated in situ from arylhydroxyaminoyl chlorides (45) and triethylamine led to single adducts in each case (Scheme 14 & 15). The reaction yielded regioselectively (100%) respectively a series of the spiro[3,4-diaryl-2-isoxazoline-5:2'-indanones] (46) and the spiro[3,4-diaryl-2-isoxazoline-5:2'-3'-methylindanones] (48) [55].

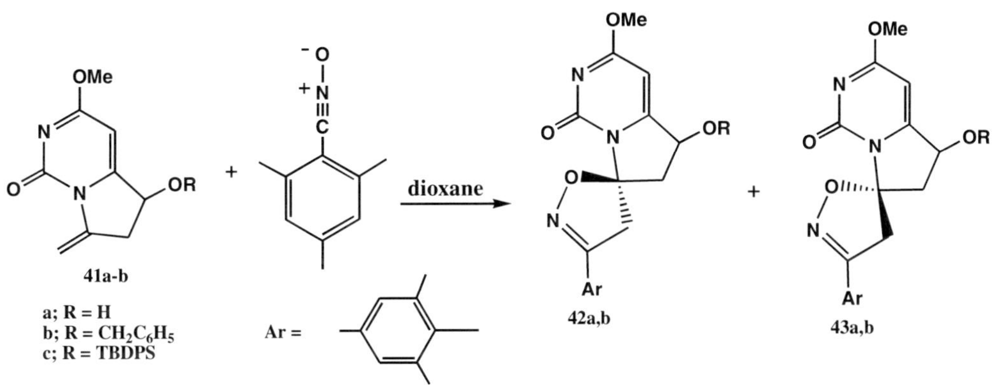

Hint: TBDPS = *tert*-Butyldiphenylsilyl

Scheme 13.

Face selectivity in the 1,3-dipolar cycloaddition reactions of benzonitrile oxide and its 4-substituted derivatives with 5-substituted adamantane-2-thiones (49) have been previously studied [56, 57]. In particular, X-ray single-crystal analysis confirmed the structure of the spiro-oxathiazoline derivative (50), resulting from the favored attack of nitrile oxide on the 5-fluoroadamantane-2-thione (Scheme 16) [58]. 1,3-Dipolar cycloaddition of 3-cyano-4*H*-1-benzopyran-4-thione (51) with benzonitrile oxide proceeded regioselectively to give cycloadduct (52) (involving the thione group). The unstable cycloadduct fragmented to yield 3-cyanochromone (53) and phenyl isothiocyanate (Scheme 17) [59].

The 1,3-cycloaddition reaction of 2,6-dichlorobenzonitrile oxide (55) with 2,10-bis (arylmethylene)-5-phenyl-2,3,6,7,8,9-hexahydro-5*H*-cyclohepta[1,2-*d*]thiazolo-[3,2*a*] pyrimidine-3-one (54) gave the corresponding isooxazoline derivative. The surprising cycloadducts of nitrile oxide to π-rich exocyclic double bond have been obtained as a couple of regioisomers (56 & 57) (Scheme 18) [60].

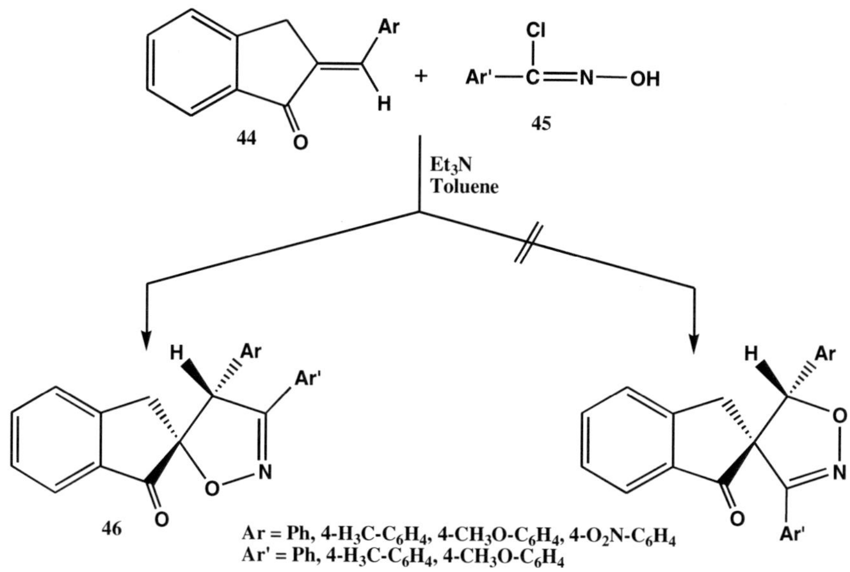

Scheme 14.

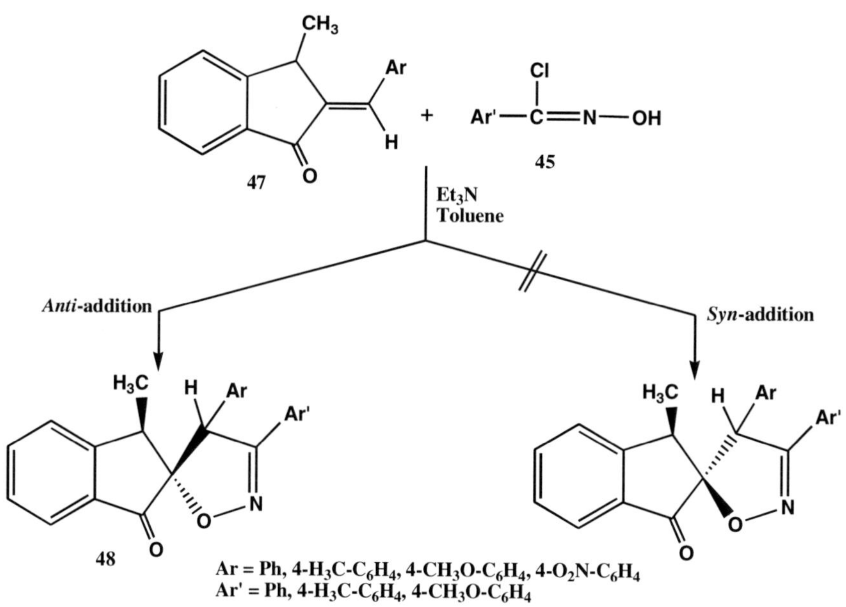

Scheme 15.

The (*R*)-1-(1-phenylethyl)-3-[(*E*)-arylmethylidene]tetrahydro-4(1*H*)-pyridinones (55) were synthesized in moderate yields (58–72%) through a solvent-free reaction of pyridinone (54) with aromatic aldehydes in the presence of pyrrolidine as basic catalyst under microwave irradiation (Scheme 19) [61].

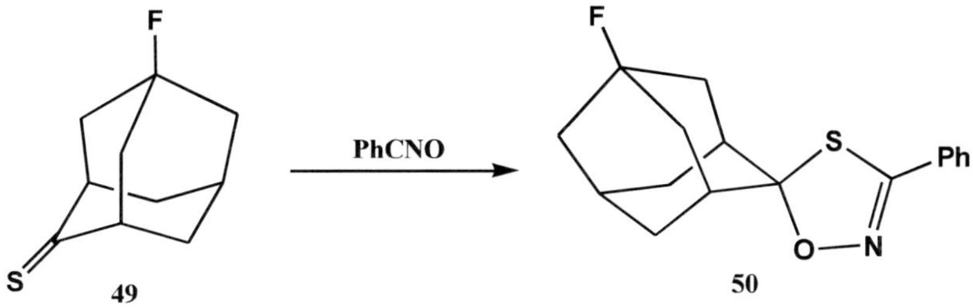

Scheme 16.

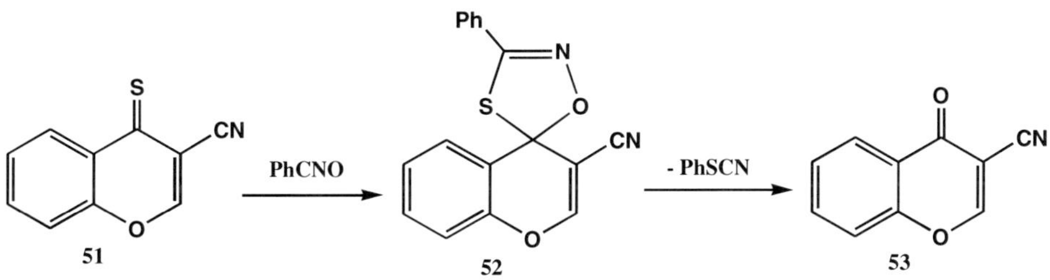

Scheme 17.

The 1,3-dipolar cycloaddition of nitrile oxide, generated *in situ* from 4-chlorobenzohydroximoyl chloride and triethylamine to (*R*)-1-(1-phenylethyl)-3-[(*E*)-arylmethylidene] tetrahydro-4(1*H*)-pyridinones (55) (Scheme 20), afforded two spiroisoxazolines, (56) and (57), in 52–56% and 8–10%, respectively. In the case of (55) with a 2-chlorophenyl ring, dispiro compound, (58), in 6-7% yields was also obtained [62].

The predominant formation of (56) in all the reactions shows that the cycloaddition proceeds as following:

- chemoselectively, with the nitrile oxide preferring to react with the C=C and not with C=O bond of (55),
- regioselectively, the oxygen of the nitrile oxide adding over the α-carbon of the C=C bond of (55), and
- stereoselectively, affording only one of the stereoisomers of the isoxazoline (56) as the major product.

5. Azomethine Ylides

Azomethine ylides are planar 1,3-dipoles composed of one nitrogen and two terminal sp^2 carbon atoms (Figure 11). In recent years, azomethine ylides have become the class of 1,3-dipoles thus, based on their cycloaddition chemistry, various methods for the synthesis of pyrrolidine derivatives have been developed [63]. Their cycloadditions to olefinic dipolarophiles provide a direct method for the synthesis of pyrrolidine derivatives.

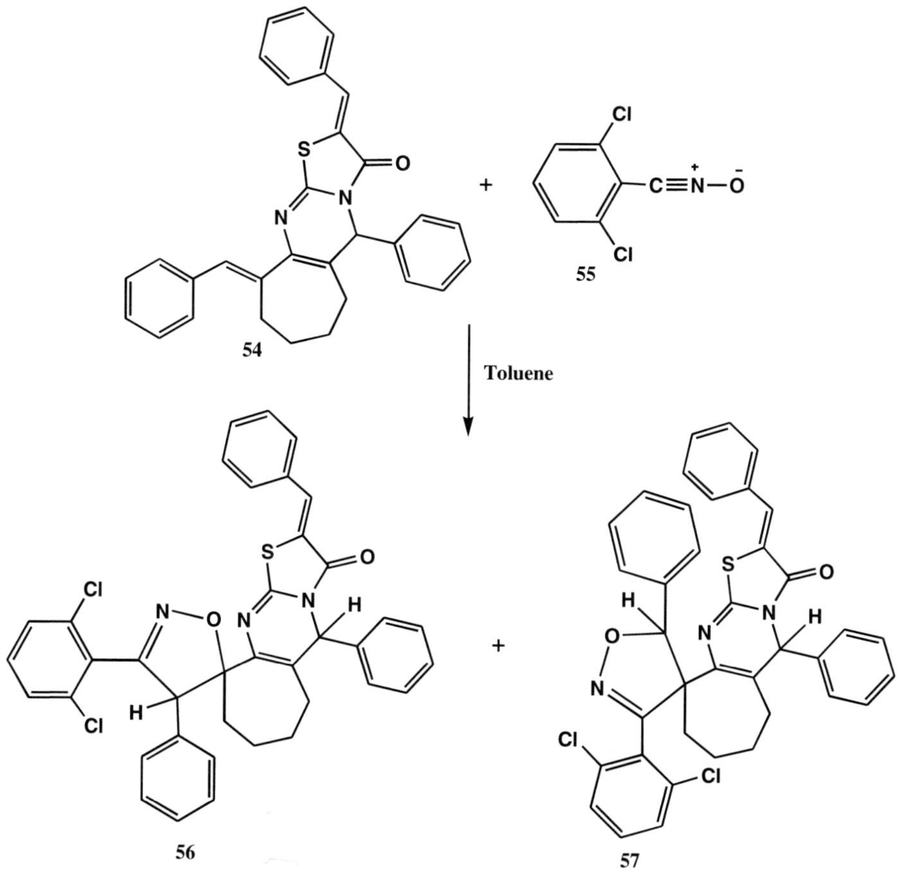

Scheme 18.

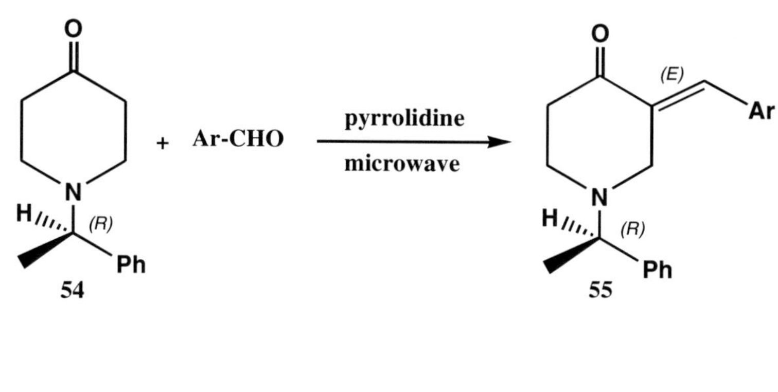

Scheme 19.

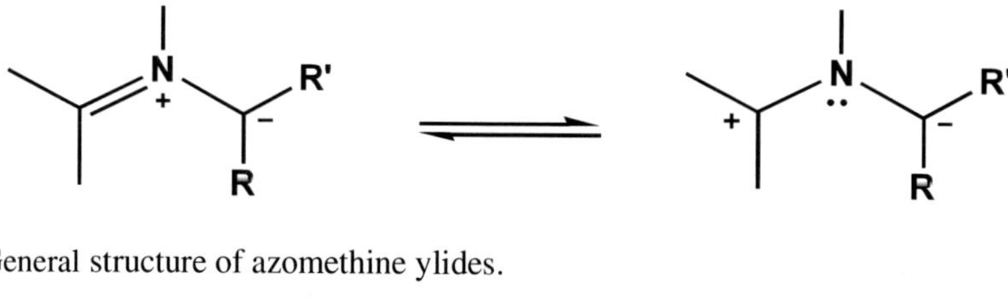

Figure 11. General structure of azomethine ylides.

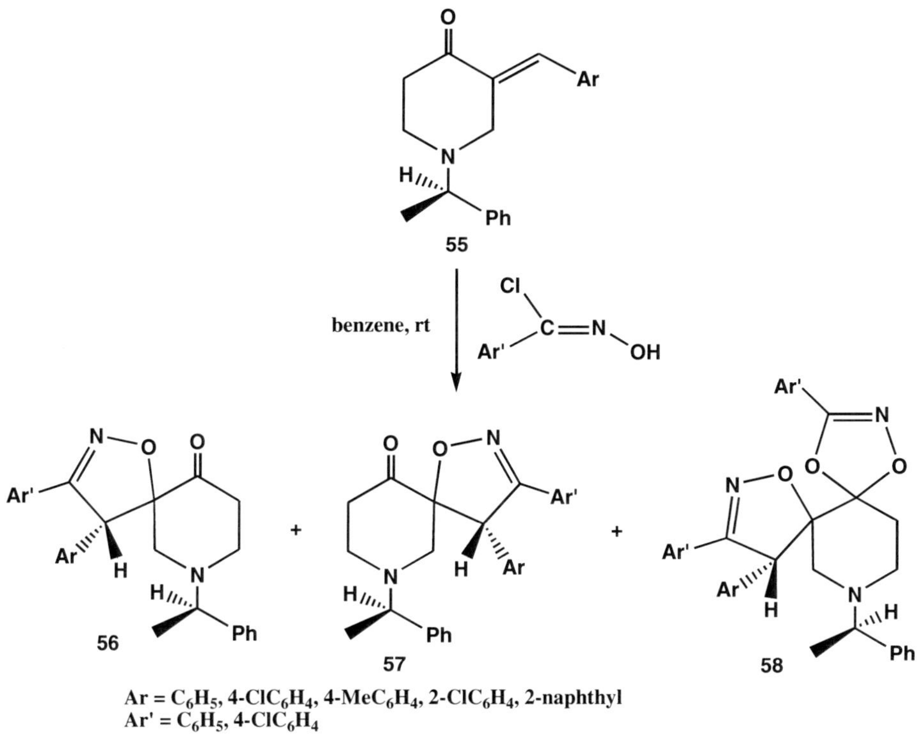

Scheme 20.

There are both stabilized and unstabilized azomethine ylides, depending on R and R` groups, although these are examples of stable, isolable azomethine ylides. They are normally generated in situ and trapped by almost any multiple C–C or C–X (X = N, O, or S) bond.

6. Generation Routes of Azomethine Ylides

A number of methods have been developed for their generation, including the ring opening of aziridines, the imine-tautomerization, and the decarboxylative condensation of amino acids.

6.1.1. Aziridine Route

The first example of the ring opening of an aziridine generating an azomethine ylide 1,3-dipole was reported earlier by Heine and peavy (Scheme 21) [64].

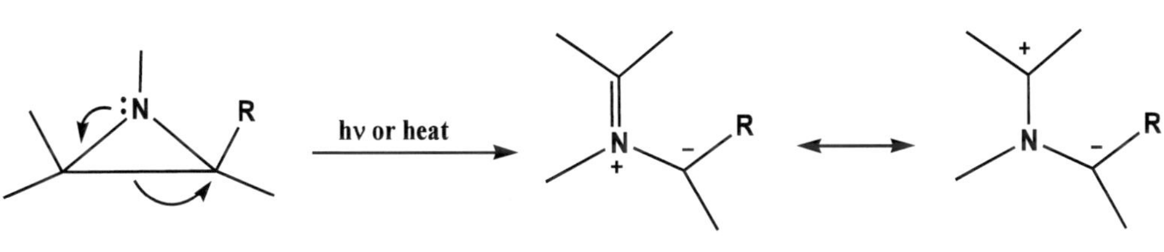

Scheme 21.

Aziridine rings open quite readily if the carbon atoms are substituted by electron-withdrawing groups. Ring-opening is stereospecific; however ylides can interconvert and must be trapped by reactive dipolarophiles (Scheme 22).

6.1.2. Carbene-Imine Route

Unstable azomethine ylides was obtained from imine and carbene (especially dihalocarbene) (Scheme 23) [65].

6.1.3. Tautomerization Route

The imine tautomerization method discovered by Grigg [66, 67], Joucla [68], and Tsuge [69] is one of the most commonly used methods for the generation of azomethine ylides stabilized by hydrogen bonding. One general approach to create azomethine ylides is the condensation of glycine derivative with an aldehyde (especially the aromatic ones), which gives stable imines. Tautomerization of imines results in the formation of intermediate azomethine ylides that can be trapped with dipolarophiles (Scheme 24) [70].

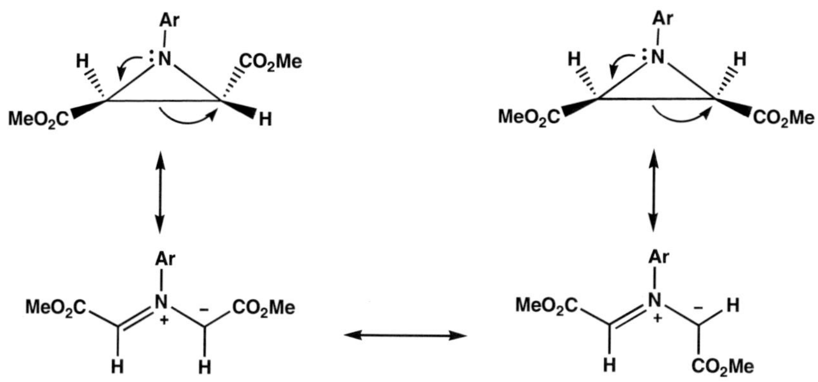

Scheme 22.

6.1.4. Decarboxylative Condensation Route

This method is one of the most important methods for preparing unstable azomethine ylides. Formation of this type of azomethine ylides almost generated in situ through condensation reaction of aldehydes or ketones with α-amino acids followed by losing the carbon dioxide (Scheme 25) [71, 72].

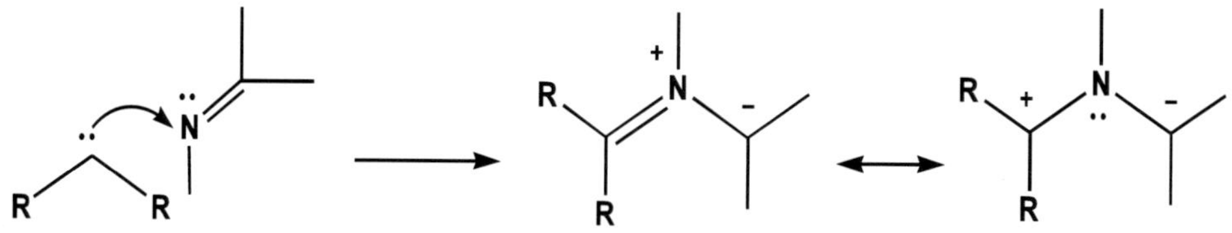

Scheme 23.

6.1.5. N-Metalation Route

The most recent advance in the chemistry of azomethine ylides is the use of chiral catalysts in the stereoselective synthesis of pyrrolidine derivatives via metallo-azomethine ylides [73], which was reviewed by Savic and Husinec, covering the literature until the beginning of 2005 [74]. Although a variety of Lewis acids have provided excellent results in the stereoselective metal-catalysed dipolar cycloaddition of azomethine ylides, arguably Ag^{+} salts (e.g. $AgOCOCH_3$, AgF) are the most effective Lewis acids in the cycloaddition reactions of metallo-azomethine ylides. The reaction times in these cases are generally short, requiring not more than few hours, and the products are normally isolated in excellent yields (Scheme 26).

6.1.6. Azomethine Ylides Cycloadditions Approach to Bioactive Spiroheterocycles

Advances in this area, over the last few decades, have made cycloaddition reactions of azomethine ylides a potent synthetic tool, extensively used in the synthesis of natural products as well as other biologically interesting compounds.

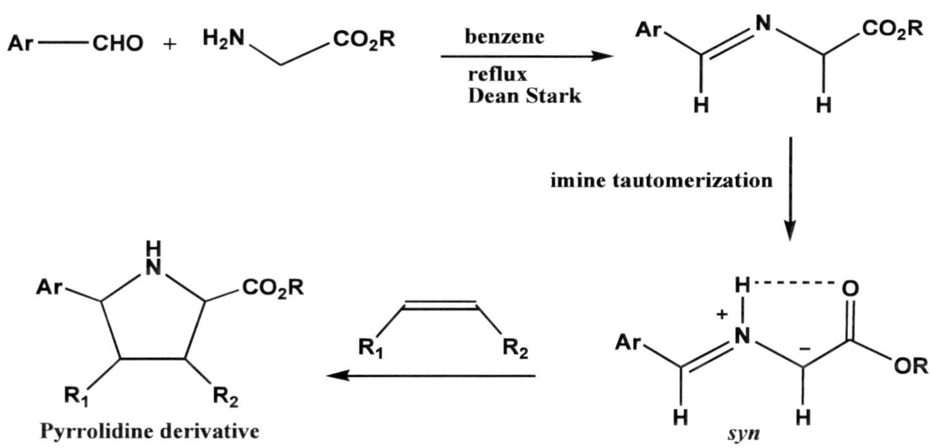

Scheme 24.

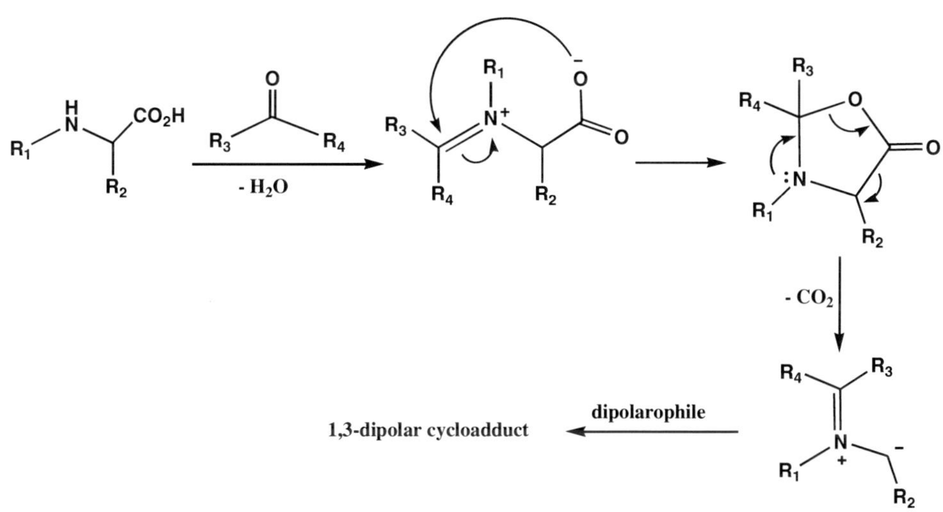

Scheme 25.

The azomethine ylide represents one of the most reactive and versatile class of 1,3-dipoles and is trapped readily by a range of dipolarophiles, either inter or intramolecularly, forming substituted pyrrolidines [75].

Pyrrolidine, pyrrolizidine and oxindole alkaloids constitute classes of compounds with significant biological activity and the spiro[pyrrolidine/oxindole] ring system is common to most oxindole alkaloids.

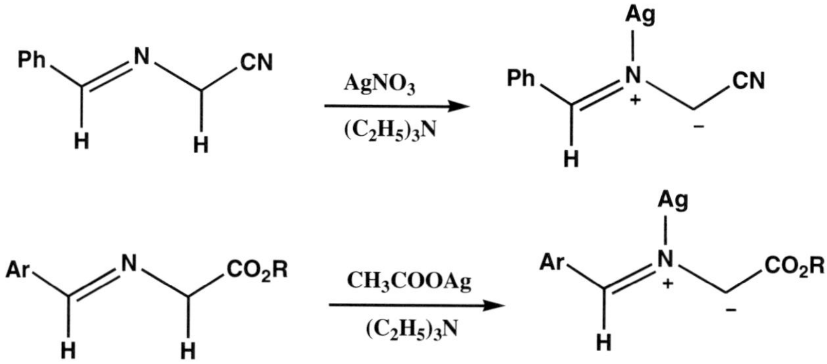

Scheme 26.

Dispiro[indane-2,3'-pyrrolidine-2',3"-indoline]-1,2",3-triones (62a-j) which reveal promising anti-inflammatory activities obtained by the reaction of 2-arylidene-indan-1,3-diones (59a-j) with non-stabilized azomethine ylides, generated in situ via decarboxylative condensation of sarcosine (61) and isatin (60) (Scheme 27) [76] in refluxing methanol afforded only one product in a highly regio- and stereoselective manner (Scheme 28) [77]. This cycloaddition is regioselective with the electron rich carbon of the dipole adding to the β-carbon of the α,β-unsaturated moiety of (59) and stereoselective affording only one diastereomer exclusively, despite the presence of three stereocentres in the product as found in many similar cycloaddition studies [78, 79].

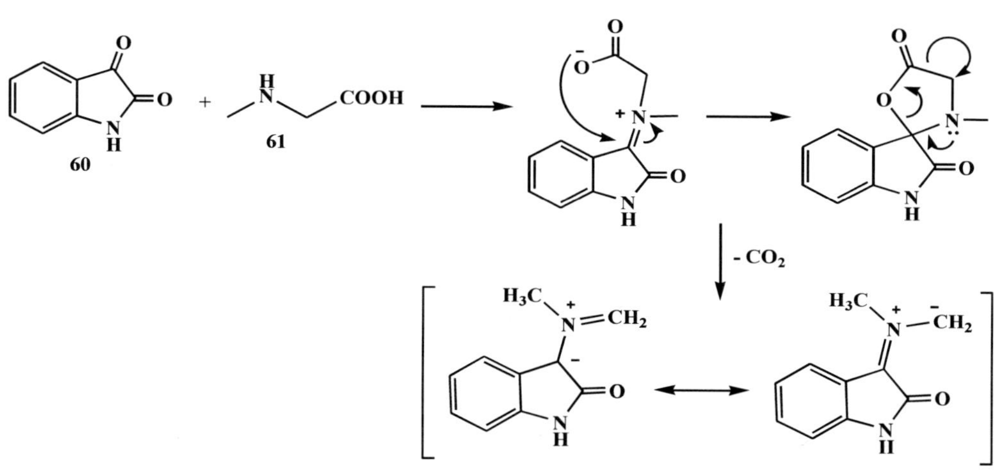

Scheme 27.

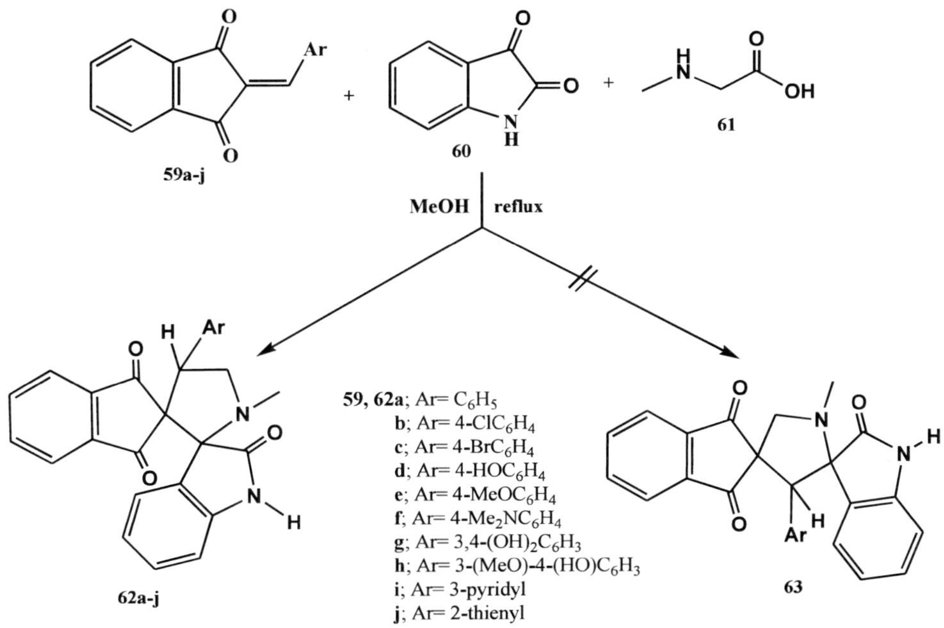

Scheme 28.

The regioselectivity in the product formation can be explained by considering secondary interaction of the orbitals of carbonyl group of dipolarophile (59) with those of azomethine ylide as shown in (Figure 12) [80]. Accordingly, the observed regioisomer (62) via path A is more favorable than (63) due to the secondary orbital interaction (SOI) which is not possible in path B. Hence, only one regioisomer (62) was formed.

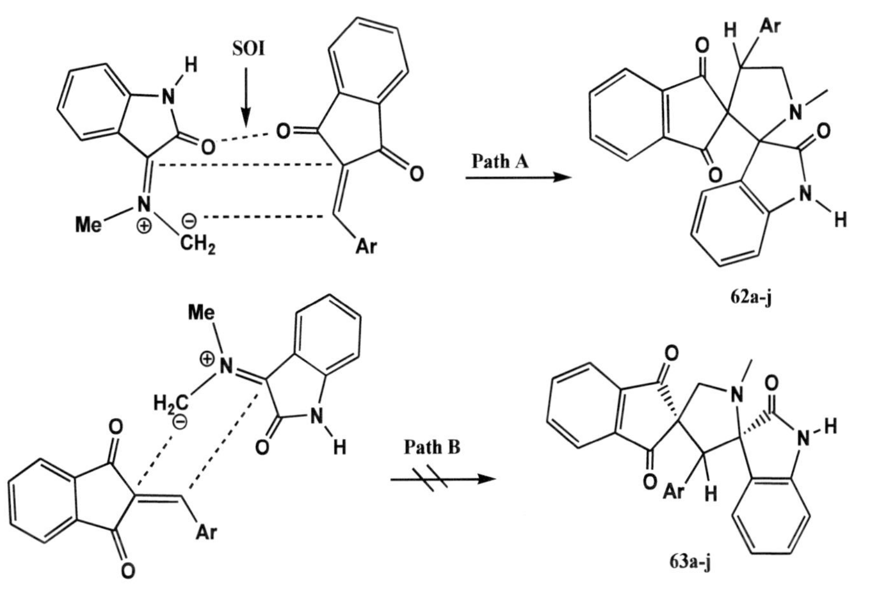

Figure 12.

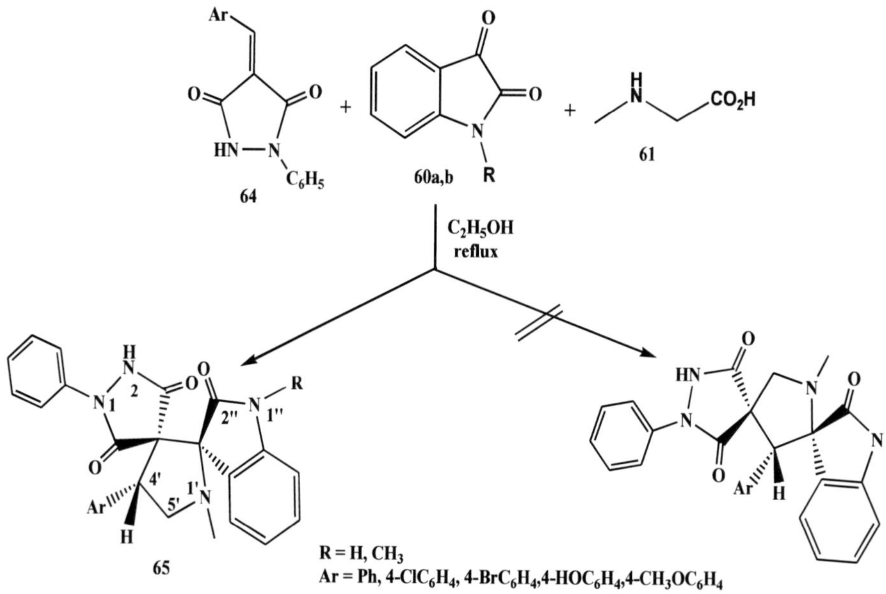

Scheme 29.

Novel dispiro[pyrazolidine-4,3'-pyrrolidine-2',3"-indoline]-2",3,5-triones (65) which reveal remarkable in vivo anti-inflammatory activities were obtained regioselectively by 1,3-dipolar cycloaddition reaction of 4-arylidene-1-phenylpyrazolidine-3,5-diones (64) as dipolarophiles with non-stabilized azomethine ylides, generated in situ via decarboxylative condensation of isatins 60a,b and sarcosine (61) in dry ethanol (Scheme 29) [81].

Tuberculosis (TB) caused by *Mycobacterium tuberculosis* bacteria (MTB) is one of the most prevalent diseases, responsible for the death of about one billion people during the last two centuries. A facile 1,3-dipolar cycloaddition of azomethine ylide generated *in situ* from the reaction of 1,3-thiazolane-4-carboxylic acid (67) and isatins (66) to 2-arylidene-1,3-indanediones (59) (Scheme 30) furnished novel dispiro- oxindolylpyrrolothiazoles (68) regio- and stereoselectively in moderate to good yields (60-92%). These dispiroheterocycles exhibited good *in vitro* antimycobacterial activity against *Mycobacterium tuberculosis* H37Rv (MTB) [82].

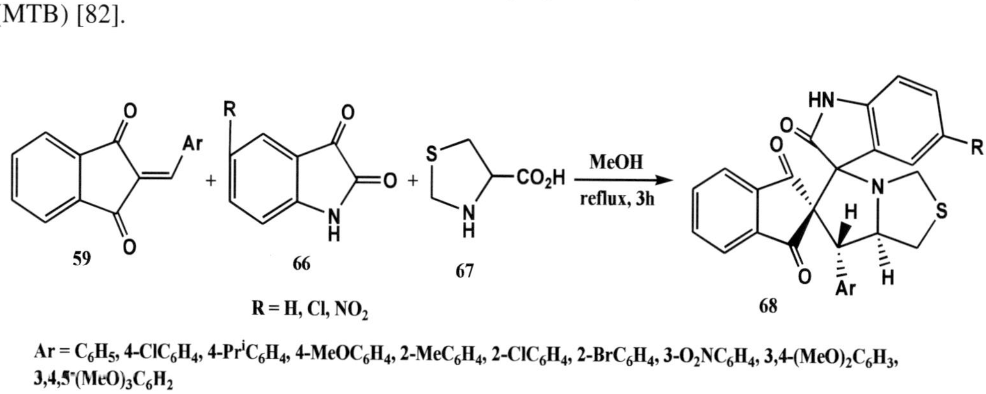

Ar = C_6H_5, 4-ClC_6H_4, 4-Pr^iC_6H_4, 4-MeOC_6H_4, 2-MeC_6H_4, 2-ClC_6H_4, 2-BrC_6H_4, 3-$O_2N$$C_6H_4$, 3,4-$(MeO)_2C_6H_3$, 3,4,5-$(MeO)_3C_6H_2$

Scheme 30.

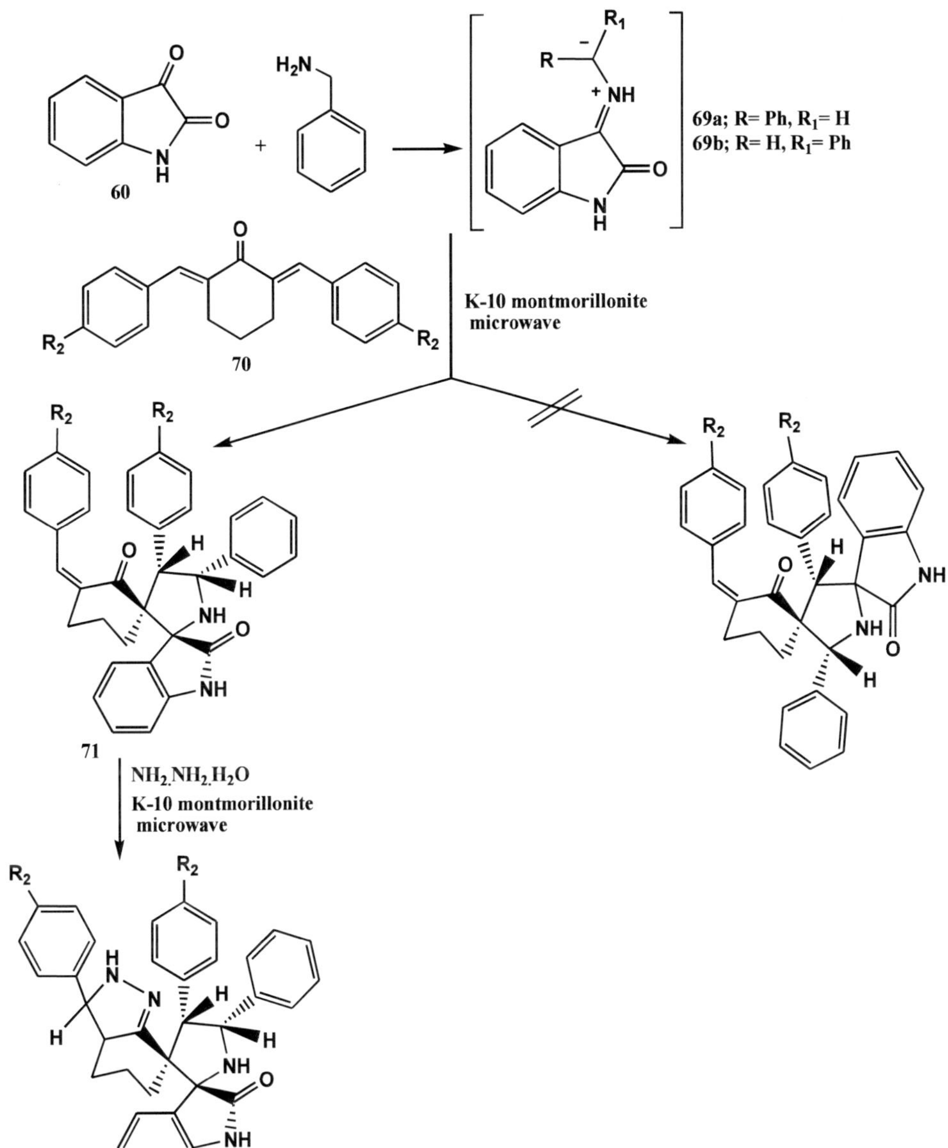

Scheme 31.

Condensation of benzylamine with isatin could give rise to two configurationally distinct azomethine ylides, 69a and 69b. The transition state leading to the azomethine ylide 69a is favored over 69b due to the steric interaction between the carbonyl moiety and the phenyl group [83]. Thus, 69a preferentially interacts with bis-arylmethylidenecyclohexanones (70) through 1,5-prototropic shift under solvent-free conditions, by grinding together the reactants with K-10 montmorillonite under microwave irradiation (600 W), to give a series of novel dispirooxindole derivatives (71) in a regioselective manner. The reaction had occurred at one of the exocyclic double bonds of (70), steric hindrance of the spiropyrrolidinyloxindole moiety prevents the attack of the 1,3-dipole on the other exocyclic double bond. The dispiro[oxindole/cyclohexanone]pyrrolidines were further annulated by grinding the mono-adduct and hydrazine hydrate with K-10 montmorillonite thoroughly and irradiating under microwave conditions to afford a series of novel 4-aryl-5-phenyl(spiro[2.3"]oxindole)-3'-aryl-3',3a',4',5',6',7'-hexahydro-2*H*-indazolospiro[7'.3]pyrrolidines (72) (Scheme 31) [84].

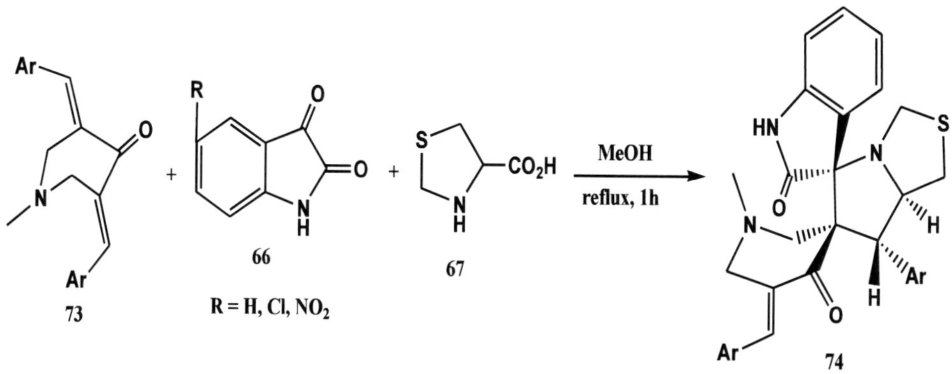

Ar = C_6H_5, 4-ClC_6H_4, 4-$Pr^iC_6H_4$, 4-$MeOC_6H_4$, 2-MeC_6H_4, 2-ClC_6H_4, 2-BrC_6H_4, 3-$O_2NC_6H_4$, 3,4-$(MeO)_2C_6H_3$, 3,4,5-$(MeO)_3C_6H_2$, 2-thienyl

Scheme 32.

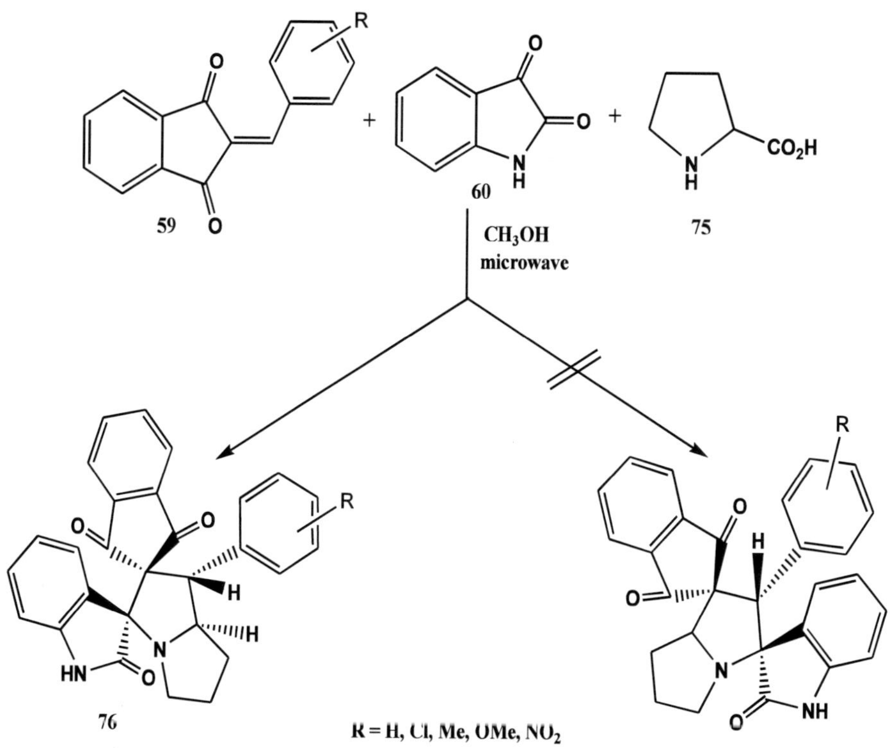

Scheme 33.

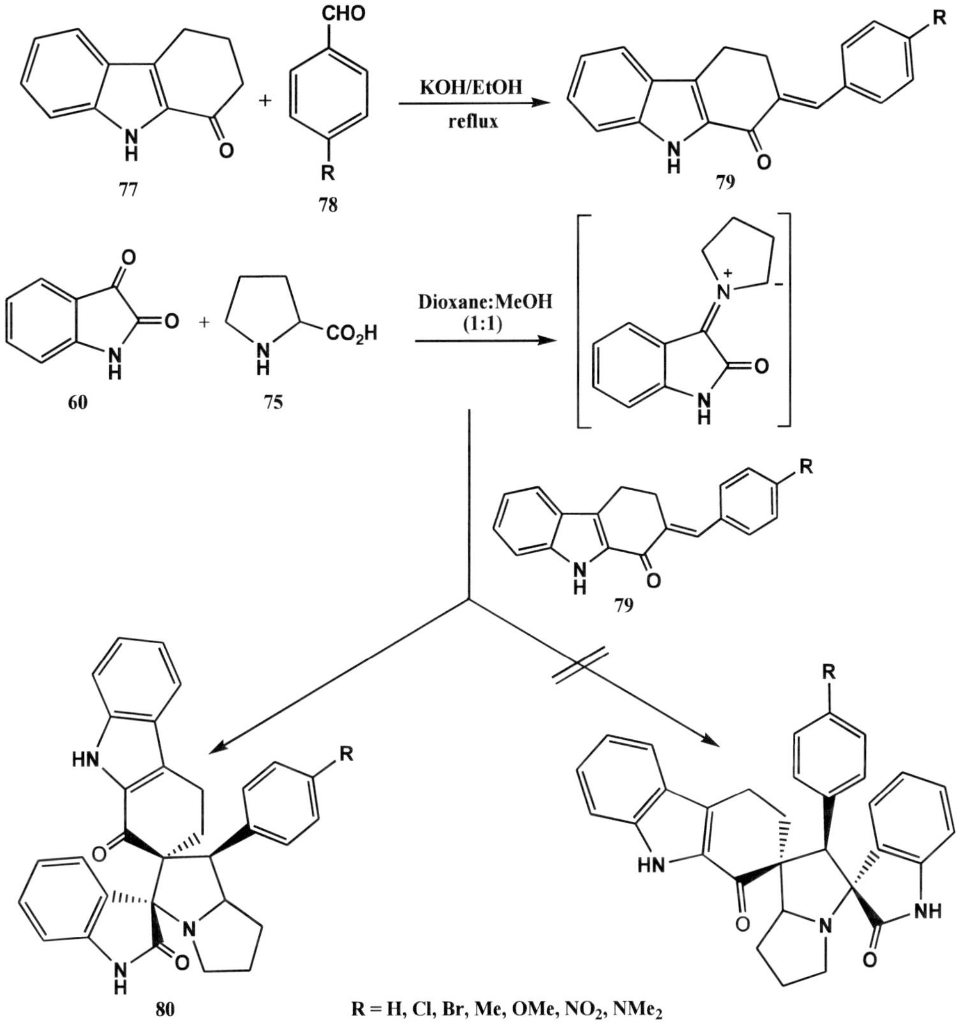

Scheme 34.

Spiro-pyrrolothiazoles (74) obtained by 1,3-dipolar cycloaddition of azomethine ylide generated in situ from substituted isatin (66) and 1,3-thiazolane-4-carboxylic acid (67) to 1-methyl-3,5-bis[(*E*)-arylmethyli-dene]- tetrahydro-4(1*H*)-pyridinones (73) in quantitative yields. These spiroheterocycles displayed good *in vitro* antimycobacterial activity against *Mycobacterium tuberculosis* H37Rv (MTB) and multi-drug resistant *M. tuberculosis* (MDR-TB). The antimycobacterial potency of these spiro heterocycles renders them valid leads for synthesizing new heterocycles endowed with enhanced activity (Scheme 32) [85]. Novel anti-microbial dispiro-oxindolopyrrolizidines (76) obtained in good yield under microwave irradiation *via* regioselective multicomponent 1,3-dipolar cycloaddition reaction of 2-arylidene-1,3-indanediones with the azomethine ylide derived from isatin and proline (75) (Scheme 33) [86]. A series of novel dispirooxindolopyrrolizidine (80) derivatives have been synthesized by Periyasami *et al.* through 1,3- dipolar cycloaddition reaction of azomethine ylide generated from proline (75) and isatin (60) with the dipolarophile (*E*)-2-arylidine-1-keto carbazoles (79). The synthesized cycloadducts were evaluated for antimicrobial activities showed relatively good antibacterial and antifungal activities (Scheme 34) [87]. The required dipolarophiles (*E*)-2-arylidine-1-keto carbazoles (77) prepared by the reaction of 1-keto carbazole [88] with substituted benzaldehydes (78) in the presence of alcoholic potassium hydroxide as basic catalyst.

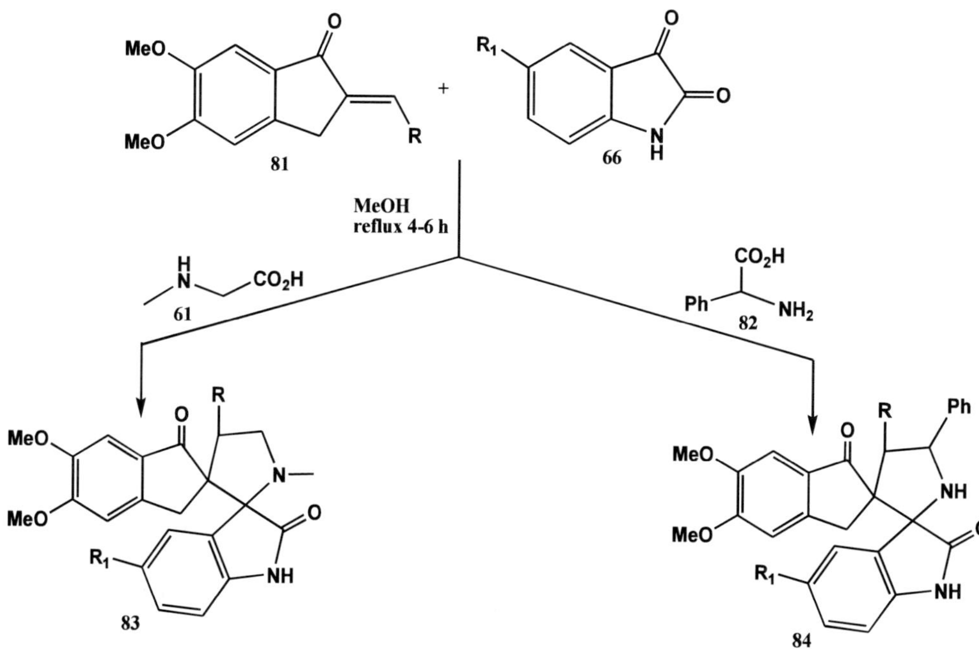

Scheme 35.

For a quarter of a century, the pathogenesis of Alzheimer's disease has been linked to a deficiency in the brain neurotransmitter acetylcholine. Novel *N*-methyl-spiro[2.3'] oxindolespiro[3.2"]-5,6-dimethoxy-1"-indanone-4-arylsubstituted)pyrrolidine (83) and spiro [2.3']oxindolespiro[3.2"]-5,6-dimethoxy-1'-indanone-4-(arylsubstituted)-5-phenyl pyrrolidine (84) were obtained by 1,3-dipolar cycloaddition of azomethine ylides, generated *in situ* via decarboxylative condensation of substituted isatins (66) and sarcosine (61) or phenylglycine (82) respectively to 2- (arylmethylene)-5,6-dimethoxy-2,3-dihydro-1H-inden-1-ones (81) (Scheme 35). The spiropyrolidine derivatives (83 & 84) exhibited potent activity as acetylcholinesterase (AChE) inhibitors [89].

Murugan et al. reported simple and efficient synthesis of regio- and stereocontrolled novel dispiropyrrolidine (87) derivatives which are found to exhibit attractive anti-diabetic properties, by 1,3-dipolar cycloaddition reaction with 5-arylidene-1,3-thiazolidine-2,4-dione and 5-arylidene-4-thioxo-1,3-thiazolidine-2-one derivatives (85 & 86) as dipolarophiles (Scheme 36) [90].

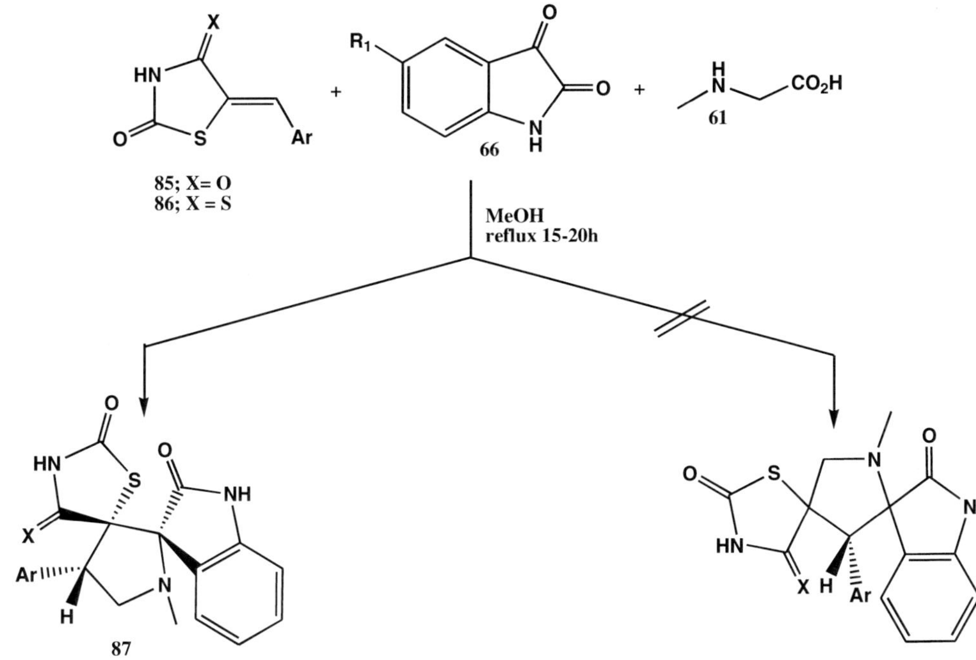

Scheme 36.

References

[1] Takahashi, K.; Witkop, B.; Brossi, A.; Maleque, M. A.; Albuquerque, E. X. *Helv. Chim. Acta, 1982*, *65*, 252-261.

[2] Trost, B. M.; Brennan, M. K. *Synthesis 2009*, 3003-3025.

[3] Bassleer, R.; Depauw-Gillet, M. C.; Massart, B.; Marnette, J-M.; Wiliquet, P.; Caprasse, M.; Angenot, L. *Planta Med. 1982*, *45*, 123-126.

[4] Garnick, R. L.; Le Quesne, P. W. *J. Am. Chem. Soc. 1978*, *100*, 4213-4219.

[5] Anderton, N.; Cockrum, P. A.; Colegate, S. M.; Edgar, J. A.; Flower, K.; Vit, I.; Willing, R. I. *Phytochemistry 1998*, *48*, 437-439.

[6] Kang, T. H.; Matsumoto, K.; Murakami, Y.; Takayama, H.; Kitajima, M.; Aimi, N.; Watanabe, H. Eur. J. Pharmacol. *2002*, *444*, 39-45.

[7] Marti, C.; Carreira, E. M. Eur. J. Org. Chem. *2003*, 2209-2219.

[8] Hilton, S. T.; Ho, T. C.; Pljevalijcic, G.; Jones, K. Org. Lett. *2000*, *17*, 2639-2641.

[9] Daly, J. W.; Kale, I. L.; Myers, C. W.; Tokueyama, T.; Waters, J. A.; Witkop, B. *Proc. Natl. Acad. Sci. U.S.A.*, *1971*, *68*, 1870-1875.

[10] Perron, F.; Albizati, K. F. *Chem. Rev.,* 1989, *89*, 1617-1661.

[11] Hegnauer, R. In Chemotaxonomie der Pflanzen, Vol. 3; Birkhauser: Basel, *1964*; 447-544.

[12] Tada, M.; Chiba, K. *Agric. Biol. Chem., 1984*, *48*, 1367-1369.

[13] Martinez, V.; Barbera, O.; Sanchez-Parareda, J.; Marco, J. A. *Phytochemistry, 1987*, *26*, 2619-2624.

[14] Breinclich, J.; Scharnagel, K. *Arznei,-Forsch., 1968*, *18*, 429-431.

[15] Weyerstahl, P.; Wahlburg, H.; Kaut, V. K.; Lochynski, S. *Liebigs Ann. Chem., 1992, 3*, 279-281.

[16] Tan, R. X.; Jia, Z. J.; Zhao, Y.; Feng, S. L. *Phytochemistry, 1992, 31*, 3405-3408.

[17] Cordell, G. A. *The Alkaloids: Chemistry and Biology; Academic: San Diego, 1998*; Vol. 5, 259-311.

[18] Okita, T.; Isobe, M. *Tetrahedron 1994, 50*, 11143-11152.

[19] Rosemmond, P.; Hossemi, M. M.; Bub, C. *Liebigs Ann. Chem., 1994, 2*, 151-158.

[20] Kornet, M. J.; Thio, A. P. *J. Med. Chem. 1976, 19*, 892-898.

[21] Skiles, J. W.; Mc Neil, D. *Tetrahedron Lett. 1990, 31*, 7277-7280.

[22] Jossang, A.; Jossang, P.; Hadi, H. A.; Sevenet, T.; Bodo, B. *J. Org. Chem., 1991, 56*, 6527-6530.

[23] Vidari, G.; Larfranchi, G.; Santori, P.; Serra, S. *Tetrahedron: Asymmetry, 1995, 6*, 2977-2990.

[24] Wrobel, J.; Dietrich, A.; Woolson, S. A.; Millen, J.; McCaleb, M.; Harrison, M. C.; Mohman, T. C.; Sredy, J.; Sullivan, D. *J. Med. Chem., 1992, 35*(24), 4613-4627.

[25] Zen, A.; Hiroya, K.; Iinuma, K. Jpn. Kokai Tokkyo Koho (Jp 05, 247, 006 [93, 247, 006] (Cl. CO7D261/20), 24 Sep. 1993, Appl. 92/44, 702, 2 Mar. 1992; p 7); *Chem. Abstr.,* 1994, *120*, 164156c.

[26] Huisgen, R. *Angew. Chem.*, Int. Ed. Engl. *1963, 10*, 565–598.

[27] Katritzky, A. R.; Cui, X.; Long, Q.; Yang, B.; Wilcox, A. L.; Zhang, Y. -K. *Org. Prep. Proced. Int . 2000*; *35*: 175–183.

[28] Bowman, D. F.; Gillan, T.; Ingold, K. *J. Am. Chem. Soc. 1971; 93:* 6555–6561.

[29] Coutouli-Argyropoulou, E.; Sabbas, I.; Konarski, S. *J. Heterocycl. Chem. 2000; 37*: 1055-1060.

[30] Grigg, R.; Dorrity, M. J.; Heaney, F.; Malone, J.F.; Rajviroongit, S.; Sridharan, V.; Surendrakumar, S. *Tetrahedron 1991*; *47*: 8297-8322.

[31] Grigg, R.; Hadjisoteriou, M.; Kennewell, P.; Markandu, J.; Thornton-Pett, M. *J. Chem. Soc., Chem. Commun. 1992*: 1388-1389.

[32] Ishikawa, T.; Tajima, Y.; Fukui, M.; Saito, S. *Angew. Chem., Int. Ed., Engl. 1996; 35:* 1863-1864.

[33] Malamidou-Xenikaki, E.; Stampelos, X. N.; Charambilis, T. A.; Karapostolou, C. C. *Tetrahedron 1997; 53:* 747-758.

[34] Grigg, R.; Hadjisoteriou, M.; Kennewell, P.; Markandu, J.; Thornton-Pett, M. *J. Chem. Soc., Chem. Commun. 1993:* 1340-1342.

[35] Frederickson, M. *Tetrahedron 1997*, 53, 403-425.

[36] Goti, A.; Fedi, V.; Nanneli, L.; De Sarlo, F.; Brandi, A. *Synlett. 1997*, 577-579.

[37] Hall, A.; Meldrum, K. P.; Therond, P. R.; Wightman, R. H. *Synlett 1997*, 123-125.

[38] Annuziata, R.; Chinquini, M.; Cozzi, F.; Raimondi, L. *Tetrahedron 1987*, 43, 4051-4056.

[39] Kumar, K. R. R.; Mallesha, H.; Rangappa, K. S. *Synth. Commun. 2003*, 33, 1545-1555.

[40] Hossain, N.; Papchikhin, A.; Plavec, J.; Chattopadhyaya, J. *Tetrahedron 1993*, 49, 10133-10156.

[41] Colombi, S.; Vecchio, G.; Gottarelli, G.; Samori, B.; Lanfredi, A. M. M.; Tiripicchio, A. *Tetrahedron 1978*, 34, 2967-2976.

[42] Al Houari, G.; Kerbal, A.; Bennani, B.; Baba, M. F.; Daoudi, M.; Ben Hadda, T. *Arkivoc 2008* (xii), 42-50.

[43] Kumar, R. R.; Perumal, S.; Kaganb, H. B.; Guillot, R. *Tetrahedron 2006, 62,* 12380–12391.
[44] Luisi, R.; Capriati, V.; Degennaro, L.; Florio, S. *Tetrahedron 2003, 59*, 9713–9718.
[45] Abbotto, A.; Capriati, V.; Degennaro, L.; Florio, S.; Luisi, R.; Pierrot, M.; Salomone, A. *J. Org. Chem. 2001, 66*, 3049–3058.
[46] Cacciarini, M.; Cordero, F. M.; Faggi, C.; Goti. A. *Molecules 2000, 5*, 637-647.
[47] Tufariello, J. J. *In 1,3-Dipolar Cycloaddition Chemistry*; Padwa, A., Ed.; John Wiley & Sons: New York, *1984*, pp. 83-168.
[48] Abd El-Nabi, H. A. Tetrahedron *2000, 56*, 3013-3020.
[49] DeWolfe, R. H. *J. Org. Chem. 1962*, 27, 490-493.
[50] DeSarlo, F.; Brandi, A.; Guarna, A.; Goti, A.; Corezzi, S. *Tetrahedron Lett. 1983*; *24*, 1815-1816.
[51] DeSarlo, F.; Guarna, A.; Brandi, A.; Goti, A. *Tetrahedron 1985*; *41*, 5181-5185.
[52] Adams, J. P.; Paterson, J. R. *J. Chem. Soc., Perkin Trans. 1, 2000*, 3695-3705.
[53] Adams, J. P. *J. Chem. Soc., Perkin Trans. 1, 2002*, 2586-2597.
[54] Fischer, R.; Hýrosová, E.; Drucková, A.; Fisera, L.; Hametner, C.; Cyrański, M. K. *Synlett. 2003, 15*, 2364.2368.
[55] Boudriga, S.; Wannassi, N.; Askri, M.; El Baker Rammah, M.; Strohman, C. *J. Soci. Chim. Tun., 2009, 11*, 29-36.
[56] Chung W-Sh.; Tsai T-L.; Ho Ch-Ch.; Chiang, M. Y. N.; le Noble, W. J. *J. Org. Chem. 1997, 62*, 4672-4676.
[57] Tsai T-L.; Chen, W-Ch.; Yu Ch-H.; le Noble, W. J.; Chung, W-Sh. *J. Org. Chem. 1999*, 64, 1099-1107.
[58] Bonini, B. F.; Maccagnani, G.; Mazzanti, G.; Abdel-Lateef, G, Atwa A, Zani P, Barbaro G. *Heterocycles 1990*, 31, 47-57.
[59] Sain, B.; Prajapati, D.; Mahajan, A. R.; Sandhu, J. S. *Bull. Soc. Chim. Fr. 1994*, 131, 313-316.
[60] Hu, X.; Feng, Y.; Su, Q. A Study of the 1,3-Dipolar Cycloaddition Reactions of Nitrile Oxide with Unusual Dipolarphiles, *2005, The Proceedings of the 3rd International Conference on Functional Molecules.*
[61] Kuehne, M. E.; Matson, P. A.; Bornmann, W. G. *J. Org. Chem. 1991, 56*, 513–528.
[62] Kumar, R. S.; Perumal, S.; Kaganb, H. B.; Guillot, R. *Tetrahedron: Asymmetry 2007, 18,* 170-180.
[63] Najera, C.; Sansano, J. M. *Curr. Org. Chem. 2003, 7*, 1105–1150.
[64] Heine, H. W.; Peavy, R. Tetrahedrone Lett. *1965*, 6, 3123-3126.
[65] Padwa, A.; Dean, D. C.; Hertzog, D. L.; Nadler, W. R.; Zhi, L. *Tetrahderon, 1992, 48,* 7565-7580.
[66] Grigg, R.; Kemp, J. *J. Chem. Soc., Chem. Commun. 1977*, 125-131.
[67] Grigg, R.; Kemp, J. *J. Chem. Soc., Chem. Commun. 1978*, 109-111.
[68] Joucla, M.; Hamelin, J. *Tetrahedron Lett. 1978, 19*, 2885-2888.
[69] Tsuge, O.; Ueno, K. O. *Chem. Lett. 1979*, 1407-1410.
[70] Dogan, O.; Koyuncu, H.; Kaniskan, U. *Turk. J. Chem. 2001, 25*, 365-371.
[71] Amal Raj, A.; Raghunathan, R. *Tetrahedron 2001*, 57, 10293-10298.
[72] Grigg, R.; Aly, M. F.; Sridharan, V.; Thianpatanagul, S. *J. Chem. Soc., Chem. Commun. 1984*, 182-183.
[73] Najera, C.; Sansano, J. M. *Angew. Chem., Int. Ed. 2005, 44*, 6272–6276.

[74] Husinec, S.; Savic, V. *Tetrahedron: Asymmetry 2005*, *16*, 2047–2061.

[75] Coldham, I.; Hufton, R. *Chem. Rev. 2005*, *105*, 2765-2810.

[76] Suresh Babu, A. R.; Raghunathan, R.; Gayatri, G.; Narahari Sastry, G. *J. Heterocycl. Chem. 2006*, *43*, 1467-1472.

[77] Hussein, E. M.; Abdel-Monem, M. I. *Int. Res. J. Pharm. Pharmacol. 2012*, *2*(3), 45-51.

[78] Karthikeyan, S. V.; Devi Bala, B.; Alex Raja, V. P.; Perumal, S.; Yogeeswari, P.; Sriram, D. *Bioorg. Med. Chem. Lett. 2010, 20,* 350-353.

[79] Ranjith Kumar, R.; Perumal, S.; Manju, S. C.; Bhatt, P.; Yogeeswari, P.; Sriram, D. Bioorg. *Med. Chem. Lett. 2009*, *19*, 3461-3465.

[80] Pardasani, R. T.; Pardasani, P.; Chaturvedi, V.; Yadav, S. K.; Saxena, A.; Sharma, I. *Heteroat. Chem. 2003*, *14*, 36-41.

[81] Hussein, E. M.; Abdel-Monem, M. I. *Arkivoc 2011*, *x*, 85-98.

[82] Maheswari, S. U.; Balamurugan, K.; Perumal, S.; Yogeeswari, P.; Sriram, D. *Bioorg Med Chem Lett. 2010*, *20*, 7278-7282.

[83] Ardill, H.; Dorrity, M. J. R.; Grigg, R.; Leon-Ling, M. S.; Malone, J. F.; Sridharan, V.; Thianpatanagul, S. *Tetrahedron 1990*, *46*, 6433-6448.

[84] Jayashankaran, J.; Manian, R. R. S.; Venkatesan, R.; Raghunathan, R. *Tetrahedron 2005*, *61,* 5595–5598.

[85] Karthikeyan, S. V.; Bala, B. D.; Raja, V. P. A.; Perumal, S.; Yogeeswari, P.; Sriram, D. *Bioorg Med Chem Lett. 2010*, *20,* 350–353.

[86] A. R. Suresh Babua, R. Raghunathan, K. Kumaresanb, N. Raaman. *Current Chemical Biology 2009, 3,* 432-443.

[87] Periyasami, G.; Raghunathan, R.; Surendiranb, G.; Mathivanan, N. *Bioorg Med Chem Lett. 2008*, *18,* 2342–2345.

[88] Andreeva, N. I.; Farmak, T. *Chem. Abstr. 1957*, 36, 713.

[89] Ali M. A.; Ismail, R.; Choon, T. S.; Yoon, Y. K.; Wei, A. C.; Pandian, S.; Kumar, R. S.; Osman, H.; Manogaran, E. *Bioorg Med Chem Lett. 2010, 20,* 7064-7066.

[90] Murugan, R.; Anbazhagan, S.; Narayanan, S. S. *Eur. J. Med. Chem. 2009*, *44*, 3272–3279.

In: Bioactive Heterocycles
Editors: K. L. Ameta, R. P. Pawar and A. J. Domb
ISBN: 978-1-62257-636-4
© 2013 Nova Science Publishers, Inc.

Chapter 9

Bioactive Purines: Fused Ring Heterocycles

Ambadas B. Rode[1], Sunil U. Tekale[2], Sushma S. Kauthale[3], Namrata D. Erande[3], Vinod V. Thorat[3], Vinod K. Ahirrao[3] and Rajendra P. Pawar[3*]

[1] Department of Chemistry, Kongju National University, Kongju, Chungnam, Republic of Korea

[2] Department of Chemistry, Shri Muktanand College, Gangapur, MS, India

[3] Department of Chemistry, Deogiri College, Aurangabad, MS, India

Abstract

Since long ago, purines have been investigated for their vital role in many metabolic processes. Recently they have attracted the attention of medicinal chemists because of their biological activities. The present chapter provides an overview of the synthesis and pharmacological as well as the biological significance of compounds containing pyrazoline, which are pharmaecuticals of immense importance.

1. Introduction

Purine **(1)** is a bicyclic heterocycle in which an imidazole ring is fused with the pyrimidine ring at 4th - 5th positions. The name 'purine' was first coined by the scientist Emil Fischer; a German chemist and recipient of the Nobel Prize (1902) in chemistry. He synthesized this colorless crystalline weak base from uric acid for the first time in 1899.

[*] Corresponding Email: rppawar@yahoo.com

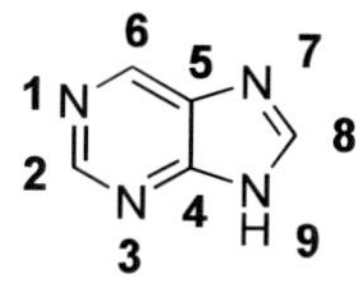

(1)

Purine derivatives have attracted significant attention owing to their ability to mediate several biological processes. In nature there are 4 different purines: Adenine - 6-amino purine (2), Guanine - 2-amino-6-oxy purine (3), Hypoxanthine - 6-oxy purine (4) and Xanthine - 2, 6-dioxy purine (5).

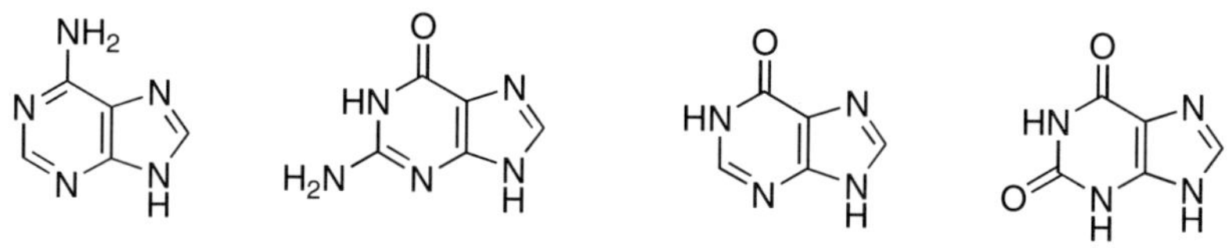

(2) (3) (4) (5)

The most important purines in nature are those that constitute the central core of the genetic material and include adenine (2) and guanine (3), two of the nitrogenous bases present in biopolymer DNA (Deoxyribonucleic acid) and RNA (Ribonucleic acid). Erwin Chargaff's found out that nucleic acids comprise equal amounts of pyrimidine and purine nucleotides [1]. This fact marks a milestone in purine chemistry and has become the basis for elucidation of the double helical structure of DNA [2]. Purine, along with pyrimidine nucleotides, plays a vital role in DNA and RNA synthesis. In addition to the free bases, a number of purines occur as the N-glycosides [nucleosides (6), (7), (8), (9), (10)]; some of which are of therapeutic use.

Purines constitute an important class of heterocyclic compounds in biology that are involved in many functional biomolecules such as energy storage, nucleic acids, coenzymes, and constituents in metabolic processes, etc. These are ubiquitous molecules existing at relatively high concentrations in almost all living organisms. Thus, purines may be regarded as the most widely distributed N-heterocyclic compounds in nature.

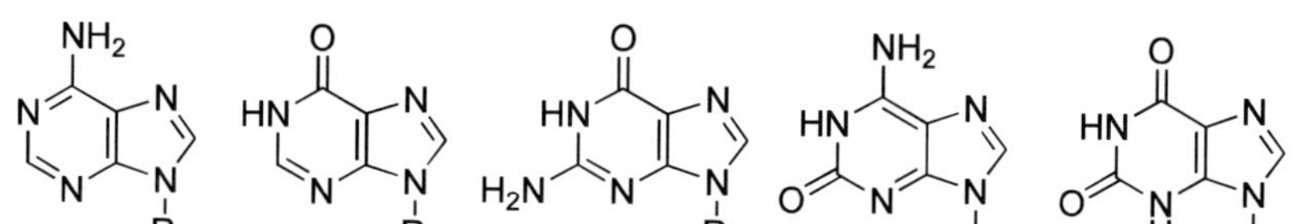

(6) (7) (8) (9) (10)

a R= H, b R= β-D-ribofuranosyl or 2'deoxy-β-ribofuranosyl canonical bases

Purines are also present in beans, lentils, pork, poultry beef, fish and other seafood, asparagus, etc. Furthermore the purine ring is an integral part of several molecules of pharmaceutical and biological interest.

Neplanocin F (11) is a natural carbocyclic nucleoside possessing moderate anti-HIV activity in human lymphocytes without any marked cytotoxicity and can be easily synthesized from D-ribonolactone [3]. Sinefungin analogs (12), having structural similarity with Sinefungin, contain purine moiety in their structures which have many biological activities including antifungal, antibacterial, antiviral, antiparasitical, etc. [4]. 9-[(1′*R*, 2′*R*, 3′*S*)-bis (hydroxymethyl) thietan-1′-yl] adenine, 3′-thio (13) is a purine containing analogues of oxetanocin A [5]. (-)-Aristeromycin (14) is a carbocyclic analogue of adenosine isolated from fermentation of Streptomyces citricolor [6].

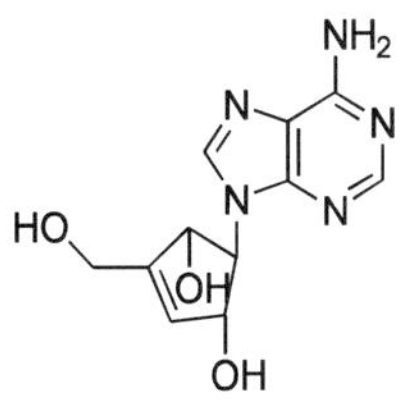

(11)

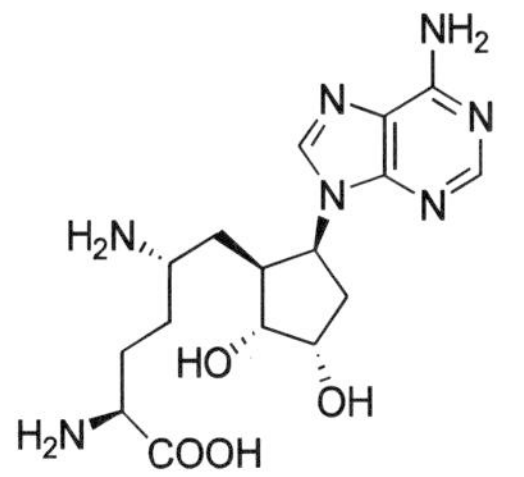

(12)

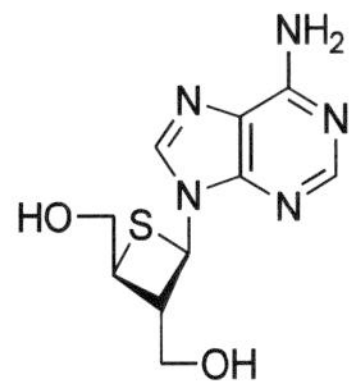

(13)

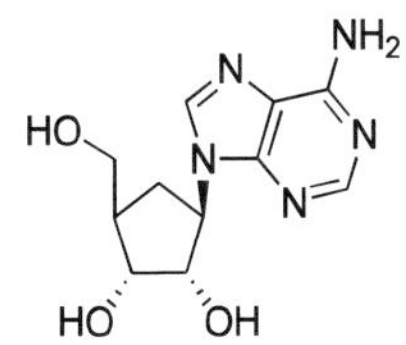

(14)

2. Purine: A Molecule of Nature

Various substituted purines and their derivatives are biologically important heterocyclic compounds, which are a widely distributed class of heterocycles occurring in nature. Caffeine (15) present in coffee, tea and other beverages, theobromine (16) present in cocoa plants, and uric acid (17) present in urine of all carnivores, are some examples of naturally occurring purine derivatives.

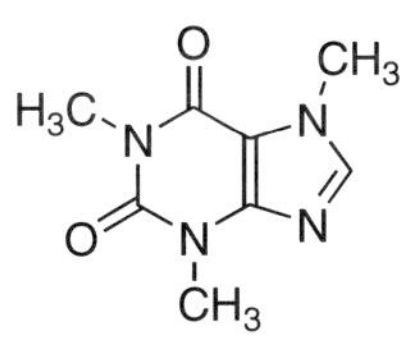

(15)

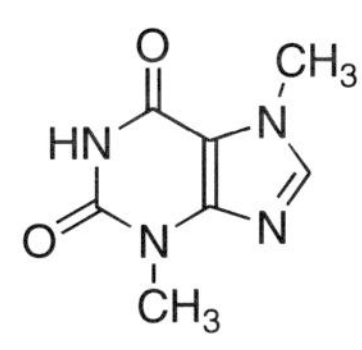

(16)

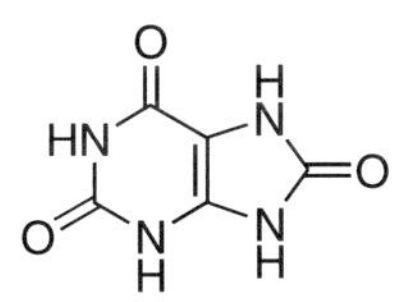

(17)

Two unusually purine-*N* 7-glycosylated cytokinines are raphnatin (18) and 6-(benzylamino)-7-(β-D-glucopyranosyl) purine (19) were isolated from de-rooted radish seedlings [7]. Mucronatine (20) is an *N*-methyl purine derivative. It was isolated from the marine sponge *Stryphnusmucronatus [8]*.

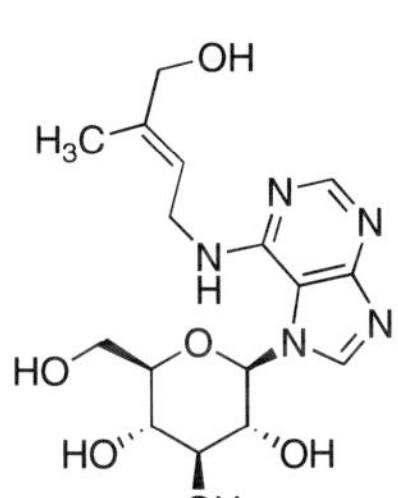

(18)

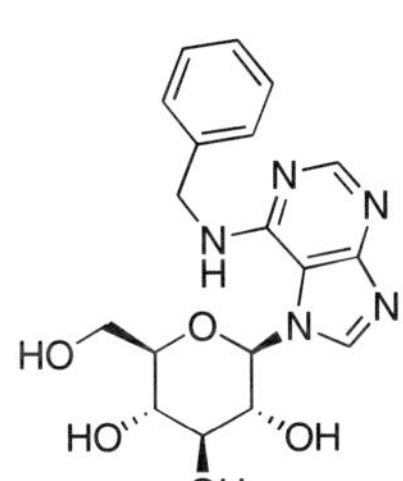

(19)

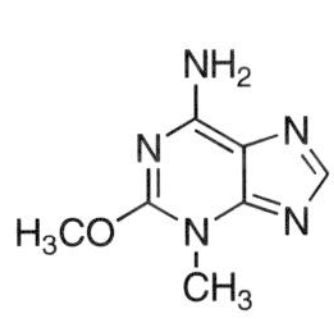

(20)

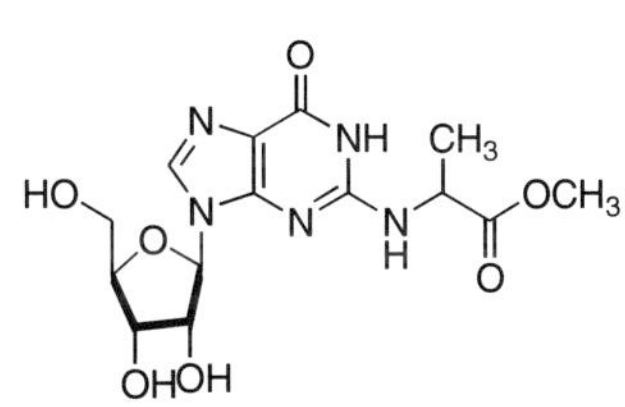

(21)

The first report on natural nucleosides where α-amino acid is bonded to a nucleobaseaglycone (21) by a C–N bond was given by Y. L. Chi [9]. The compound was isolated from the poisonous mushroom i.e. Amanita exitialis.

Most of the purines and their nucleosides exist in nature as mono- or polymethylated derivatives. 1-methylxanthine (22) is one of the most highly concentrated purine components observed in human urine besides uric acid.

In tea plants, caffeine (15) is found together with theophylline (1, 3-dimethylxanthine; (23). Both substances are effective as diuretics. Theophylline (23) is often prescribed as an anti-asthma drug, particularly for acute asthma attacks. In the cocoa bean, theophylline (23) is replaced by theobromine (3, 7-dimethylxanthine;) as the main xanthine component [10].

Some purines carry several methyl groups on the imidazole ring. The first derivative of this kind to be described, herbipoline (24)-a 7, 9-dimethylguanine with betaine structure, was first isolated from a giant silica sponge [11]. A nucleoside, kumusine, from an Indonesian sponge *Theonella* species is 2-chloro-9-(5′-deoxy-2′-methyl-β-D-ribofuranosyl)-adenine (**25**) [12].

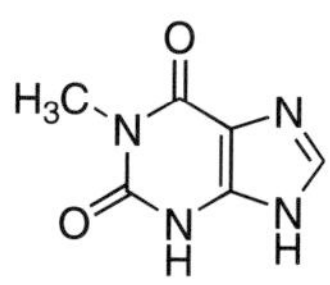

(22)

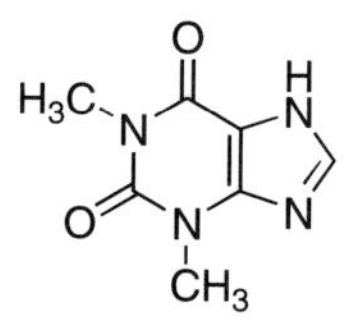

(23)

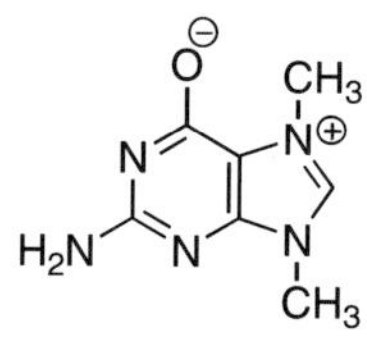

(24)

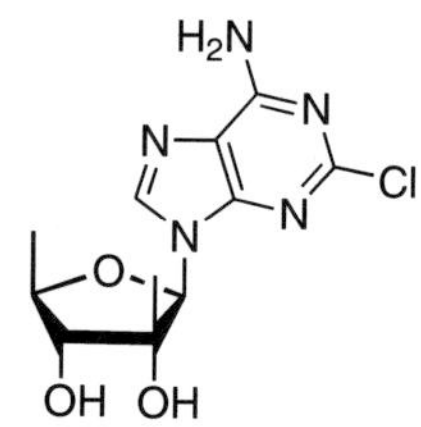

(25)

3. Synthesis of Purine and its Derivatives

On account of the bioactivities and pharmacological significance associated with purine derivatives; the development of efficient, convenient and environmentally friendly methods for the synthesis of 2, 6, 9-substituted purines is an important task. The coupling reaction has emerged as a valuable method for the preparation of 9-arylpurines, but longer reaction times are required. The synthetic methods of purines may be categorized under two headings:

Microwave assisted Green synthesis of purines
Synthesis of purines by conventional methods

3.1. Microwave Assisted Green Synthesis of Purines

Recently, "Green Chemistry" has emerged as an excellent tool in the field of organic chemistry. It aims to avoid side product formation, saving time and reducing costs. The application of microwave radiation provides an alternative route to heating the reaction mixtures to significantly reduce reaction times and increase the yields of chemical reactions. Rapid reactions, shorter reaction times, high yields and clean reaction profiles are advantages associated with microwave-assisted synthesis. Several microwave assisted synthetic methods of purine are well documented in literature.

Microwave assisted green protocol for the formation of C_6-alkylthio-substituted purine derivatives (27) in the absence of metal catalysts was developed by H-Y. Niu [13]. Chloropurines (26) required a shorter reaction time and the products obtained in excellent yield.

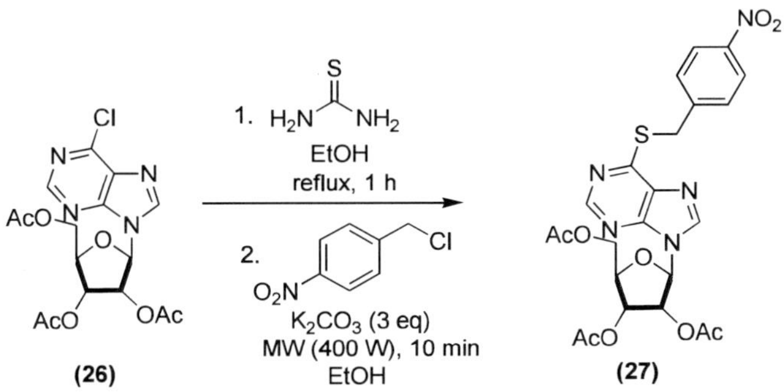

(26) (27)

6-[*N*, *N*-bis (2-hydroxyethyl) amino] purines (30) were synthesized [14] from the coupling of 6-chloropurines (28) with diethanolamine (29) under microwave assisted reaction conditions,using water as the reaction medium. The yield was excellent with a high product purity. The reactions required a shorter time for completion (10 min).

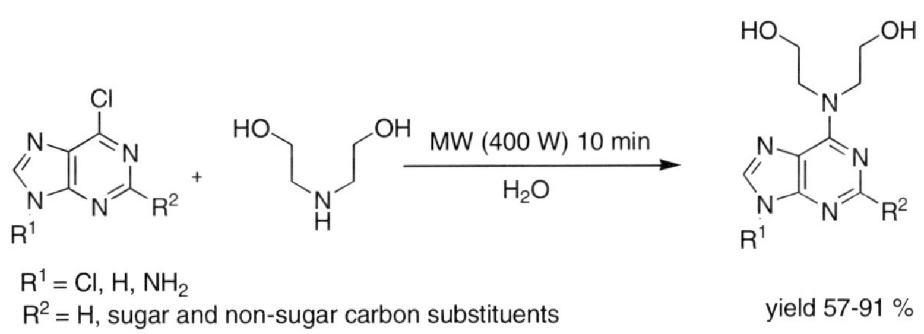

(28) (29) (30)

A green protocol for the synthesis of purines (33) involving the application of microwave irradiation from the chloropurines (31) and piperidines (32) in water as the reaction medium was developed by Gui-Rong Qu [15]. The method affords several advantages such as rapidness and high efficiency.

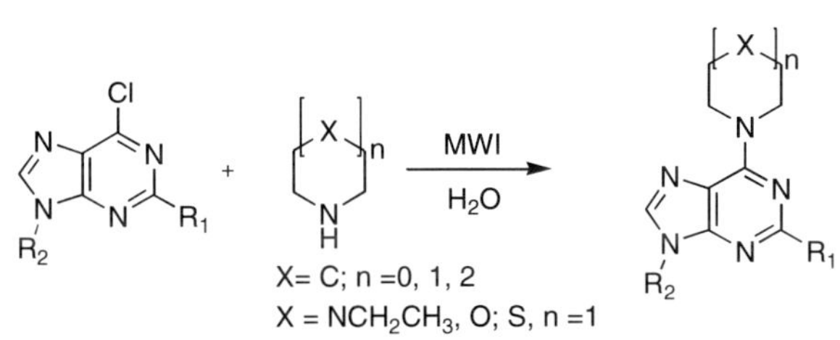

(31) (32) (33)

2, 6, 9-trisubstituted purines (41) were synthesized through a series of reactions by R. E. Austin [16]. Microwave irradiation was suitable for enhancing the nucleophilic displacement of halogens by amines at the C-2 position of purine nucleus.

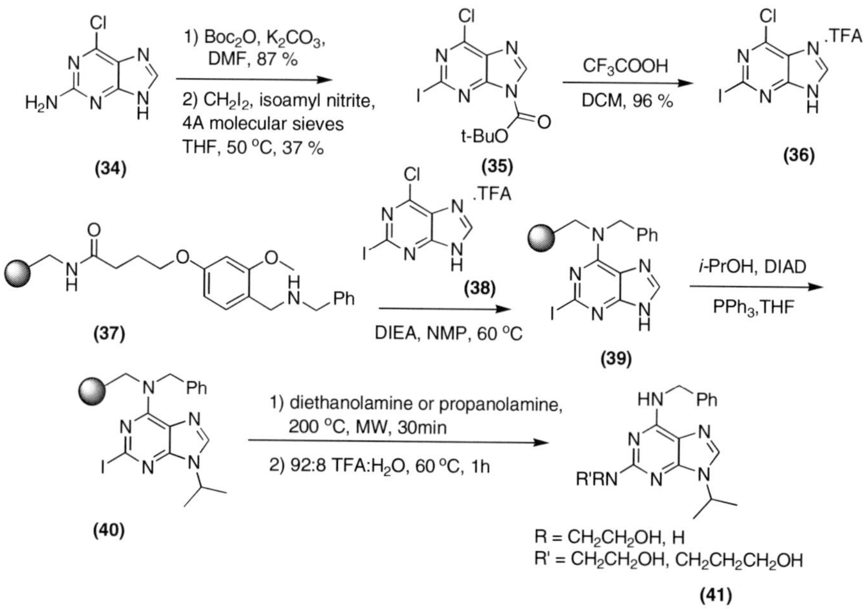

The phosphorylation of chloro purines (42) using trialkylphosphite (43) under microwave irradiation to afford the new purines (44) was carried out by G-R. Qu etal [17]. The reaction could be performed under solvent free conditions and required no catalyst or any other additive.

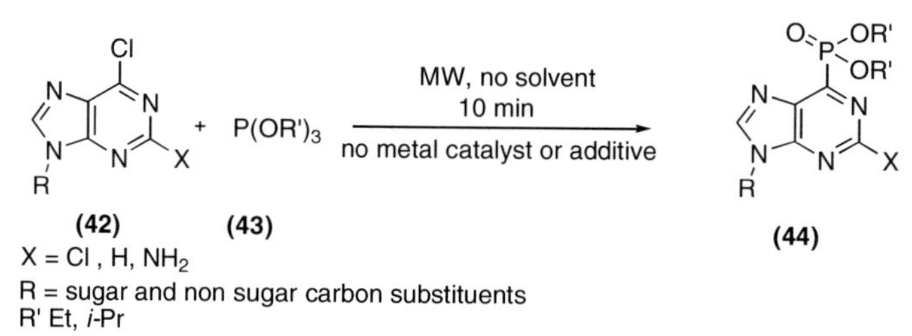

An efficient synthesis of polyfunctionalized purines (47) involving microwave assisted conditions was developed by R.Vabre and co-workers [18].

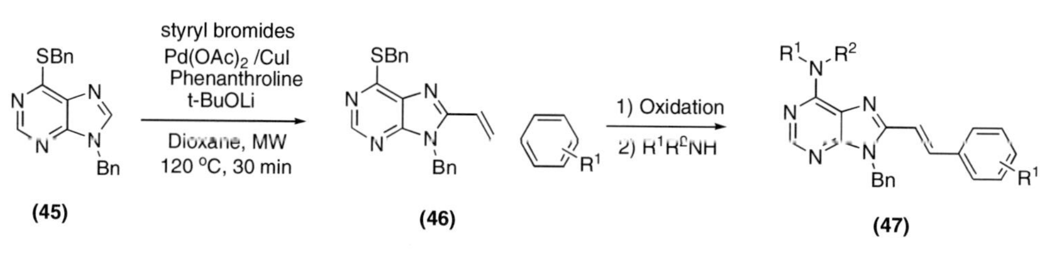

3.2. Synthesis of Purines by Conventional Methods:

Purine (49) is simply obtained in good yield when formamide (48) is heated in an open vessel at 170°C for 28 hours [19].

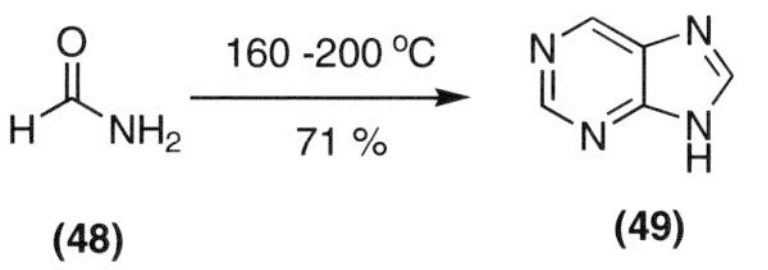

Synthesis of purines (51) by the chemoselective N[1]-arylation using copper acetate catalyzed Buchwald protocol in the presence of pyridine and pyr-N-oxide was developed by J. J. Strouse et al [20].

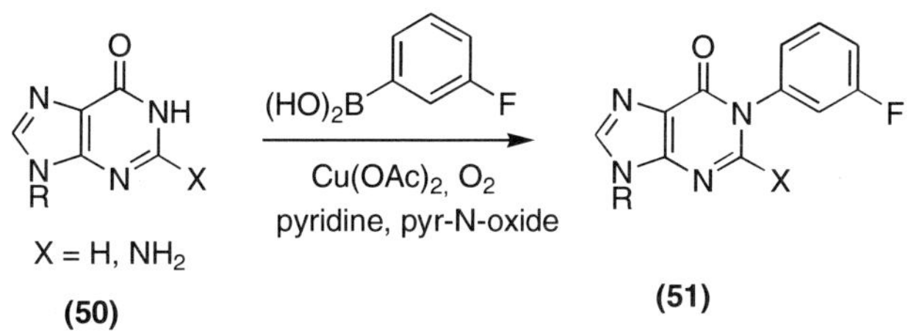

A ligand free protocol for the C–H arylation of adenines (52) using copper iodide and Pearlman's catalyst (Pd $(OH)_2/C$) resulting in the formation of adenines (53) was reported by S. Sahnoun [21]. It was observed that the $-NH_2$ group did not require protection strategy during the reaction. The reaction afforded high yield in the case of aryl iodides rather than bromides and chlorides.

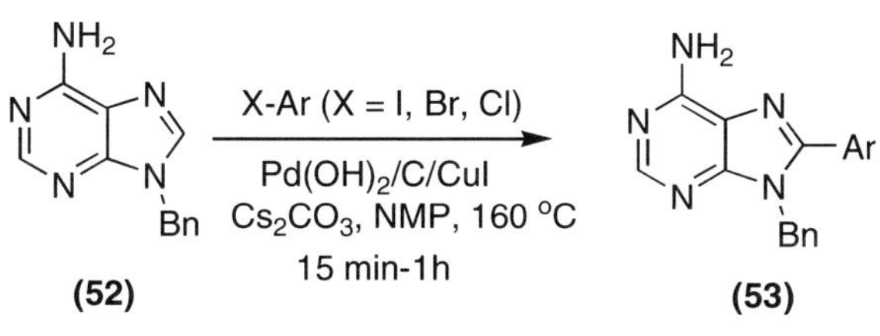

In 1999 Havelkova [22], developed the first palladium catalyzed Suzuki-Miyaura cross coupling reaction on halogenated purine (54). He standardized two protocols for this purpose:

1) Anhydrous conditions: 2.5 mol % Pd $(PPh_3)_4$, K_2CO_3, MePh, 100°C
2) Aqueous conditions: 2.5 mol % Pd $(PPh_3)_4$, aq. K_2CO_3, DME, 85°C.

It was shown that both of these methods were effective for C_6-arylation with phenylboronic acid to afford the corresponding phenylpurines (55) with excellent yield.

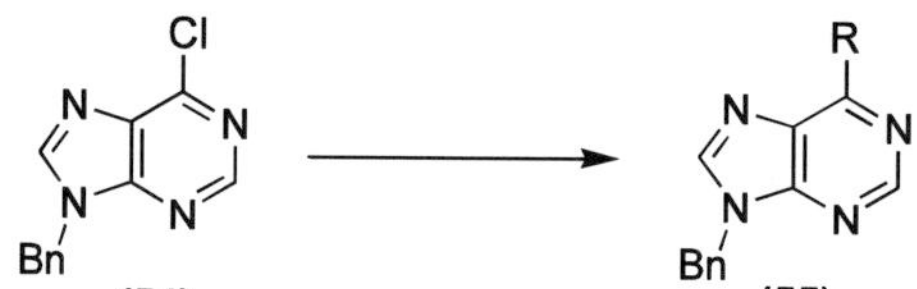

(54) (55)

Anhydrous conditions: 2.5 mol% $Pd(PPh_3)_4$, K_2CO_3, MePh, 100 °C

Aqueous conditions: 2.5 mol% $Pd(PPh_3)_4$, aq. K_2CO_3, DME, 85 °C

The cross coupling of 6-chloropurines (56) with the Reformatsky reagent (57) followed by the reduction with hydride donor and treatment with MnO_2 leading to the formation of (purin-6-yl) acetates and 6-(2-hydroxyethyl)purines (59) was developed by Z. Hasnik [23].

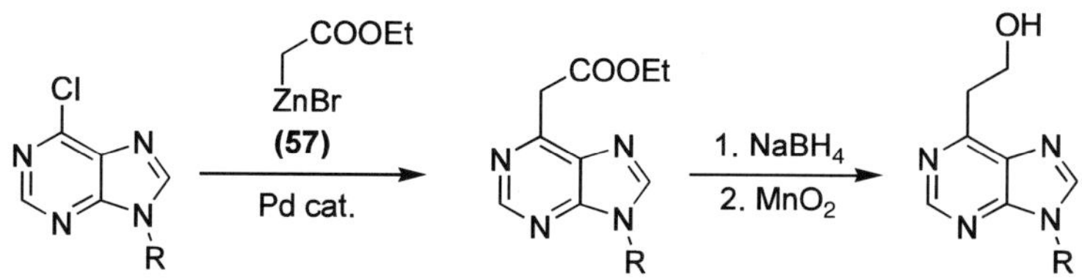

(56) (58) (59)

A palladium catalyzed cross coupling reaction between 6-halopurines (60) with acyloxymethylzinc iodides at room temperature conditions was reported by P. Silhar et al [24]. Such compounds (62) possess a significant cytostatic activity and serve as the inhibitors of adenosine deaminase.

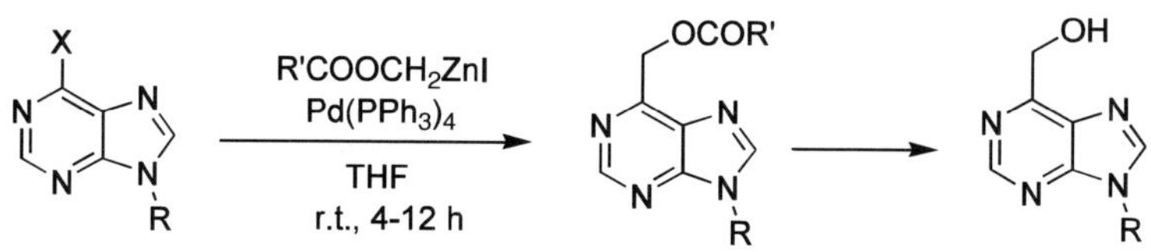

(60) (61) (62)

The regioselective synthesis of 2-chloro-6-methylpurines (64) by the coupling reaction between 9-substituted or protected 2, 6-dichloropurines (63) and Grignard reagent catalyzed by iron catalyst $Fe(acac)_3$ was suggested by M. Hocek [25] .

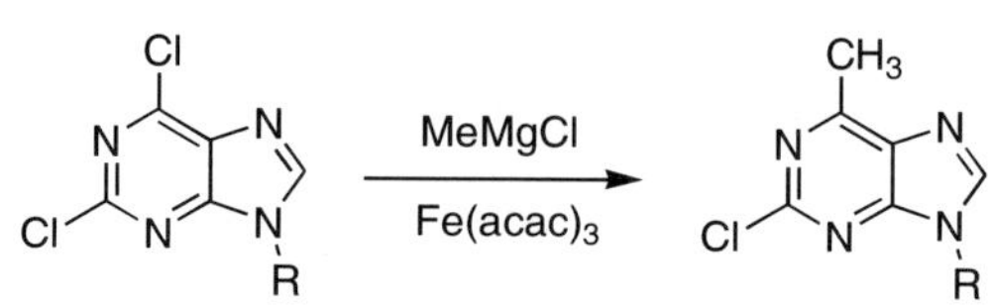

(63) (64)

The purines (68) function as amino acid–nucleobase conjugates, which can be synthesized by the palladium catalyzed cross coupling reaction between 6-iodopurines (65) and organozinc compounds (66) through a series of reactions [26].

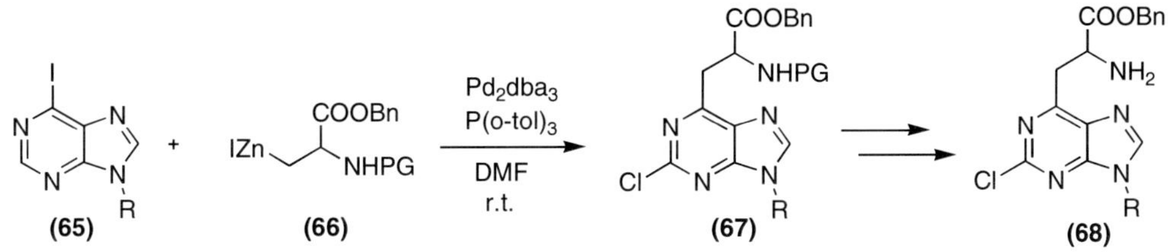

The transformation of 9-benzyl-6-iodopurines (69) into the corresponding purines (70) involving the use of Grignard reagent under cooling conditions (0°C) in THF solvent followed by treatment with aldehydes was successfully achieved by T. Tobrman and D. Dvorak [27]. Studies have revealed that other functionalities such as ketone, nitrile, ester etc. instead of aldehyde could not result in the formation of the desired product.

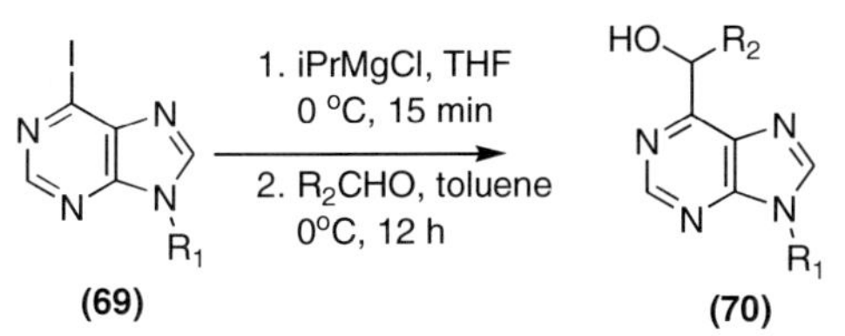

A room temperature synthesis of purines (77), (78) from the 2-amino-6-chloropurines (71) was reported by S. Fletcher [28]. The reaction sequence involved the Mitsunobu reaction as the key step.

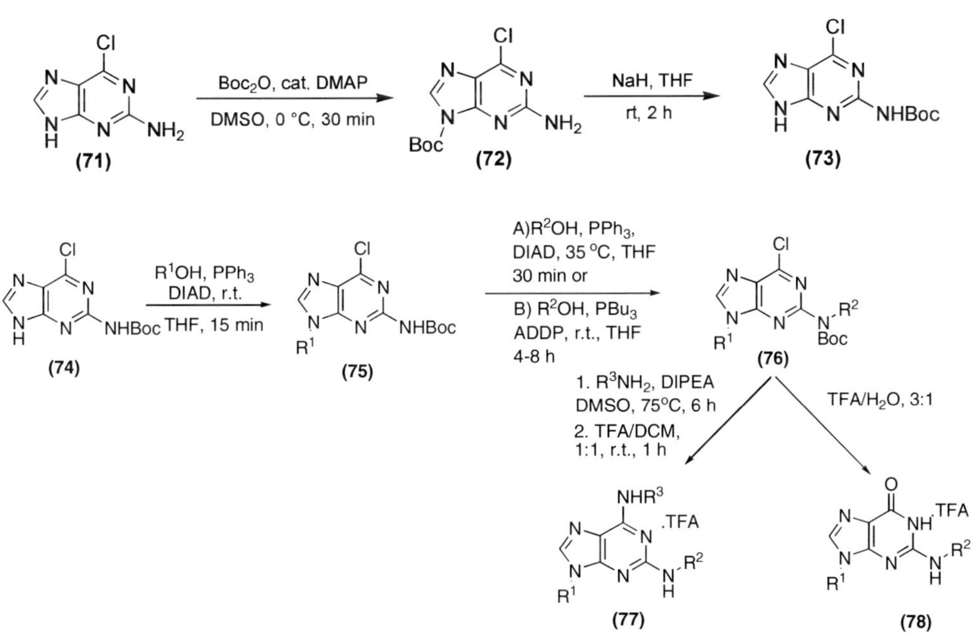

The copper-catalyzed amidation reaction for the synthesis of 6, 7, 8-trisubstituted purines (81) was suggested by N. Ibrahim [29]. The synthesis was carried out from the readily available iodo precursors (79). The synthesized purines were subjected for biological screening.

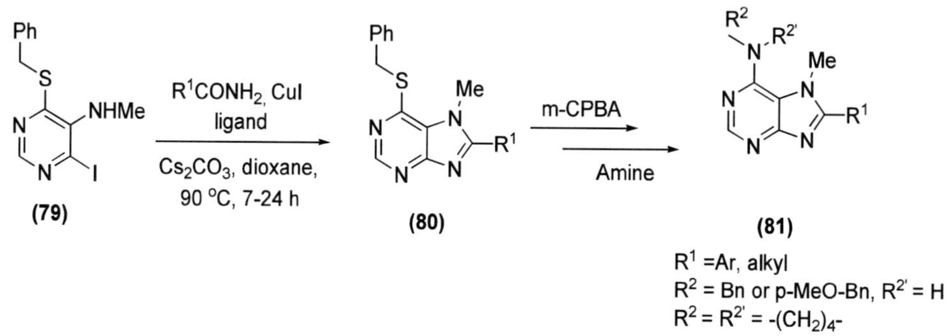

The regioselective arylation of 9-allyl-9*H*-purine (82) under palladium acetate catalyzed Heck-reaction conditions for the synthesis of purines (84) is also well documented in literature [30].

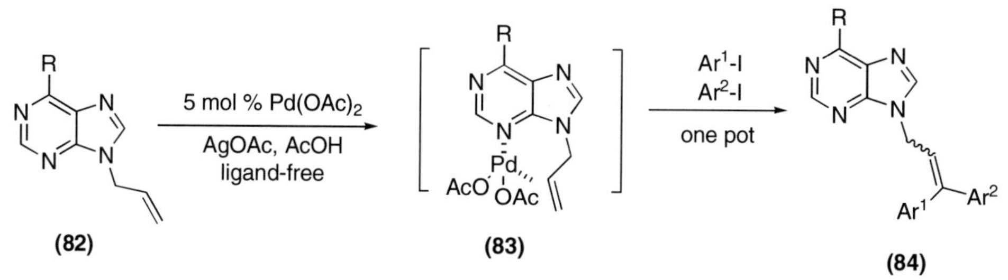

A simple protocol for the synthesis of 6-aminated purines (86) from the corresponding bromo purines involving the nucleophilic substitution on bromo compounds (85) with amines in methanol under reflux conditions was developed by E. A. Veliz and P. A. Beal [31].

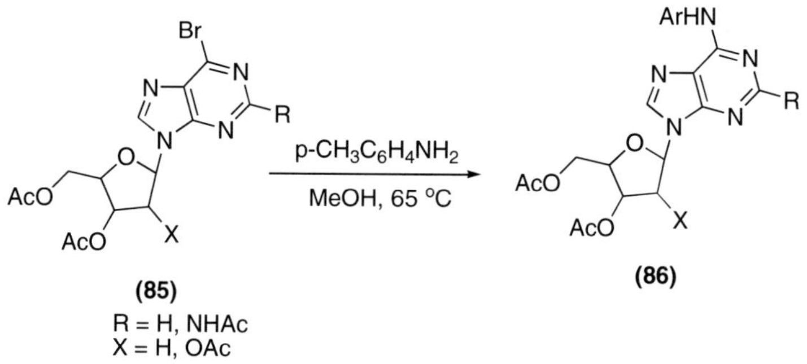

S. Ostrowski [32] has demonstrated the synthesis of imidazole ring containing purines (89) from oximes (87). The cyclization step was performed in a sealed tube.

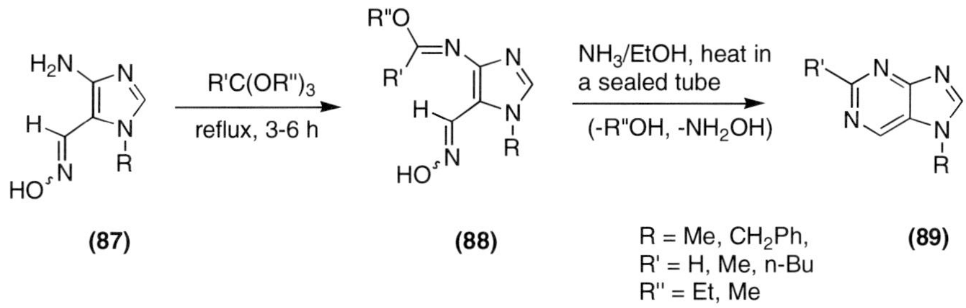

Wan et al reported the synthesis of 6-aminopurines (91) from unprotected inosine (90) involving amination reaction in the presence of BOP and DMF/ethanol solvent [33]. The reactions required excessive amounts of amine reagent (1.2-5 eq).

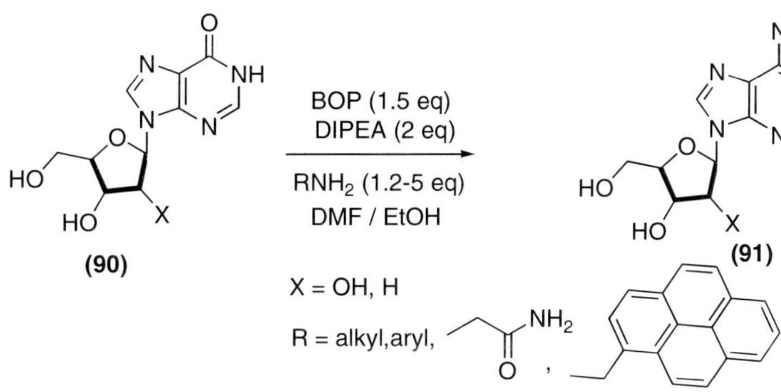

The first application of a 6-arylsulfonate (92) for the C-C bond formation reactions was reported in 2005 [34]. The cross-coupling of arylsulfonate with boronic acid to form the purine derivatives (93) was carried out at room temperature in the presence of the palladium catalyst and an appropriate ligand.

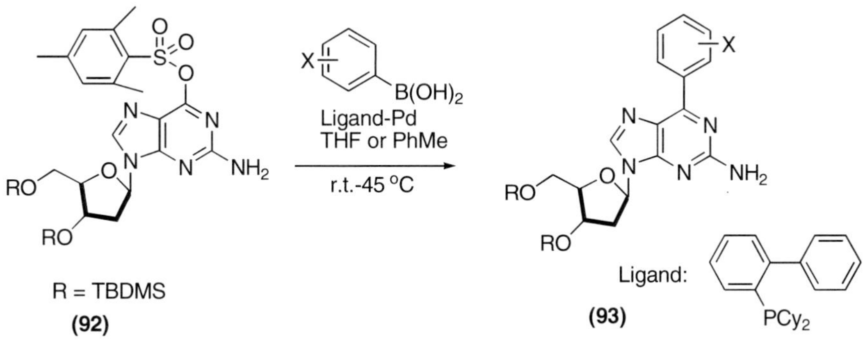

The purines (95) were synthesized by Stille coupling on halopurines (94) and organotin compounds in the presence of palladium catalysts [35]. These resultant compounds were reported for their antimycobacterial activity.

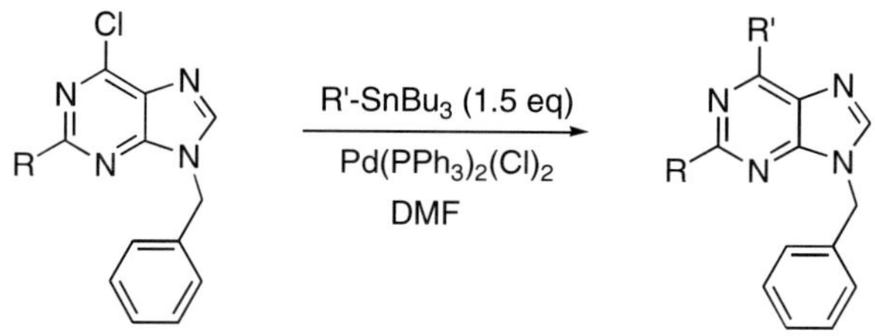

(94) R = H, NH_2

(95) R' = Ph-C≡—, Ph, 2-furyl

2, 3-diamino pyrimidines containing thioether (96) scaffold were cyclized [36] to purines (97) using either acid or aldehydes, which on treatment with m-CPBA followed by reaction with amines results in the synthesis of purines (98). The protocol resulted in high purity products with excellent yields.

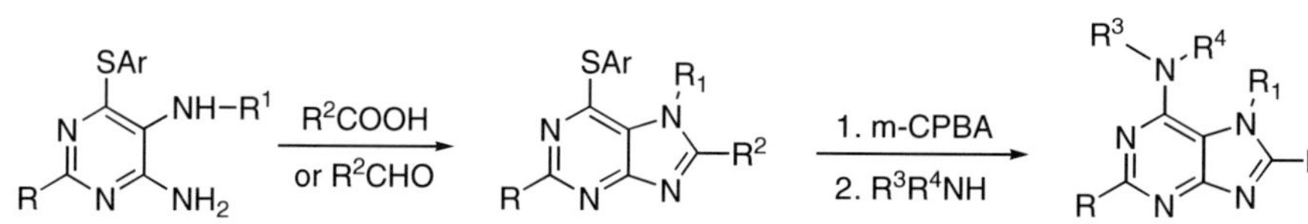

(96) **(97)** **(98)**

Purino [8, 9-*f*] phenanthridines (100) were synthesized from the intramolecular reaction of fused purines (99) [37]. In 2007, the successful amidation of 2, 6-dihalopurines (101), (102) leading to the formation of new purines (103), (104) was carried out by S. Piguel and M. Legraverend [38]. The reactions required temperatures of 0-100°C depending upon the nature of the substituent and the structure of reactants.

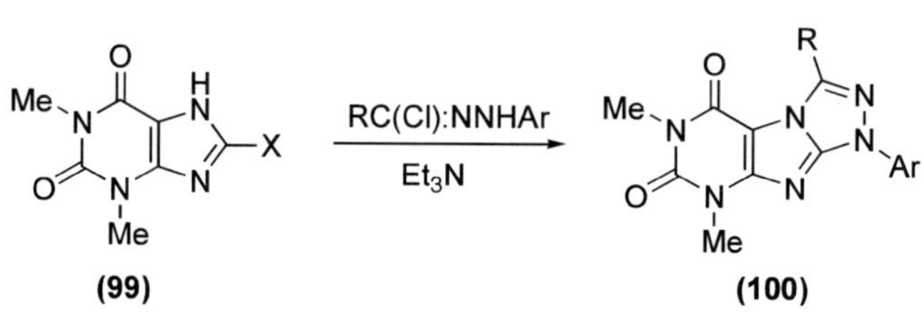

(99) **(100)**

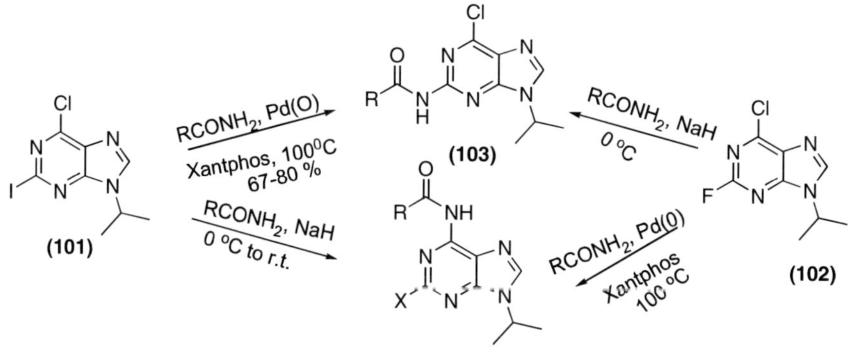

(101) **(103)** **(102)** **(104)** X = F, I

The regiospecific alkylation of 6-(Heteroaryl) purines (105) in the presence of sodium hydride or potassium carbonate base followed by the treatment with iodides was reported by M. Zhong [39].

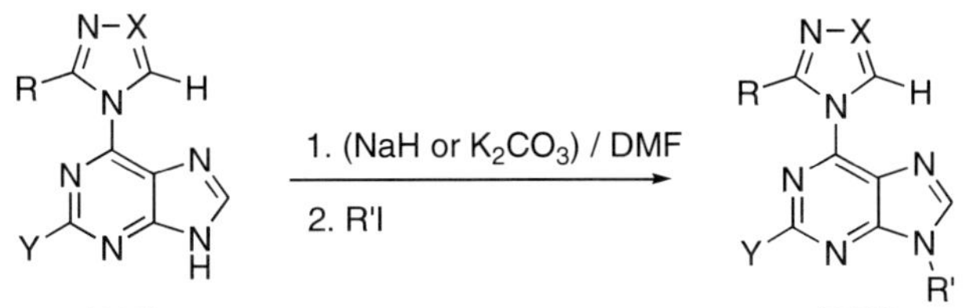

(105) (106)

2, 6, 8-trisubstituted purines (108) were obtained by arylation of the protected purines (107) using aryl halides in the presence of copper iodide and palladium acetate, piperidine base and DMF solvent [40]. The reaction required a prolonged time (24 h) and a high temperature. The deprotection was carried out using TFA in DCM at room temperature to afford the desired 2, 6, 8-trisubstituted purines.

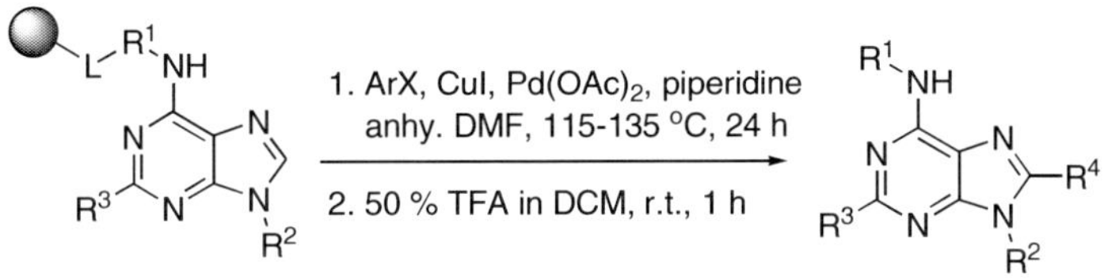

L =Wang linker

(107) (108)

The purines (110), (111), (112) bearing carbon substituent at C-6 position were obtained from 6-halopurines (109) under catalyst free basic (K_2CO_3) or Cs_2CO_3) conditions by the treatment with β-keto ester [41].

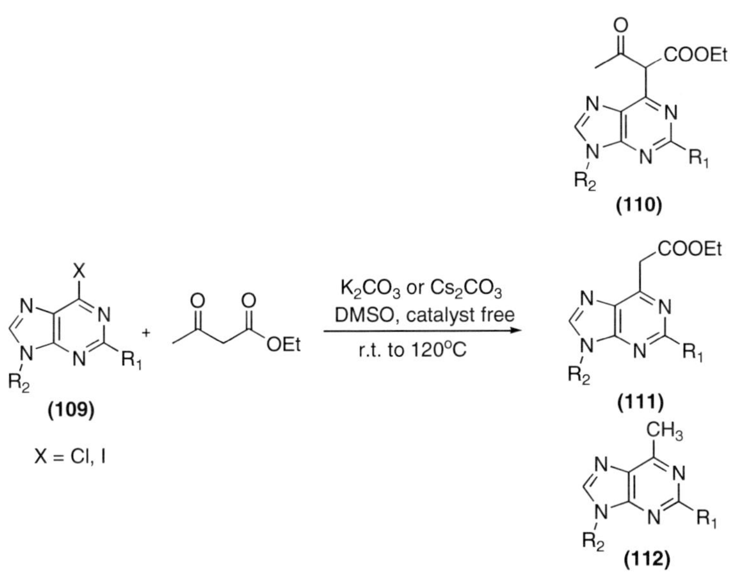

In 2001, S. Ding developed a copper acetate catalyzed route for the synthesis of N-aryl substituted purines (114) from the dichloropurines (113) in the presence of triethyl amine in excellent yield [42]. The chloro substituents were observed to remain intact even in the presence of boronic acids.

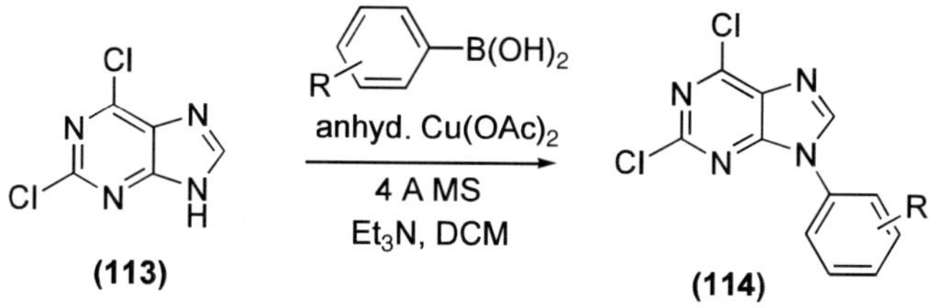

The synthesis of 2-amino-6-chloropurines (119) via the formation of 2,5- diamino-4,6-dichloropyrimidine is well reported in literature [43].

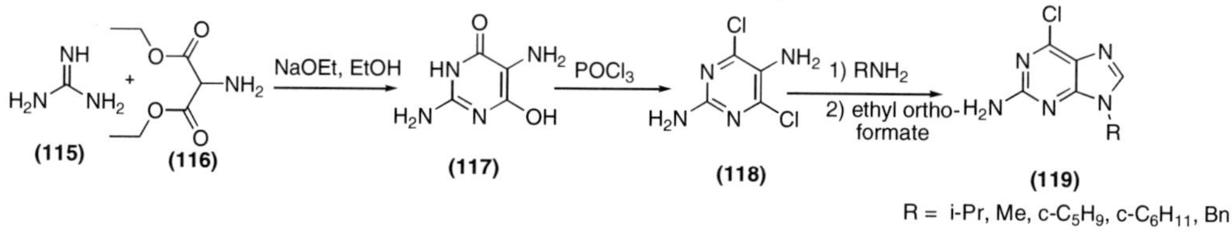

Sulfoximine containing purines (123) were synthesized by R. E. Dolle and David McNair [44]. These compounds were shown to be antiherpetic in nature.

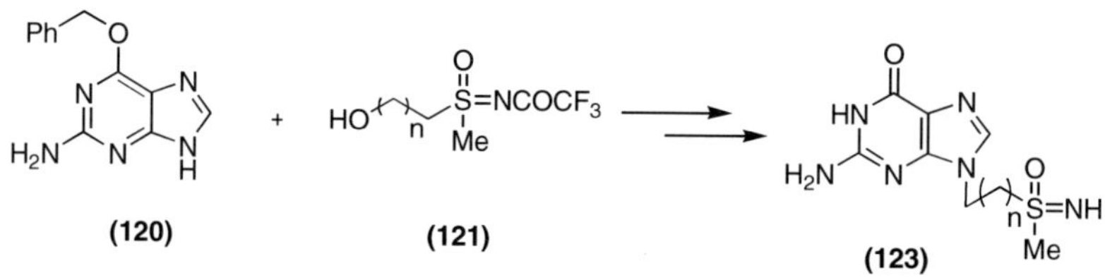

9-(2, 3-dideoxy-2-fluoro-β-D-*threo*-pentofuranosyl) adenines (126) were synthesized from 3′-*O*-benzoyl-5′-*O*-tritylriboside (124) in the presence of triethylaminetrihydrofluoride ($Et_3N{\cdot}3HF$) [45]. The protocol can be applied for large scale synthesis.

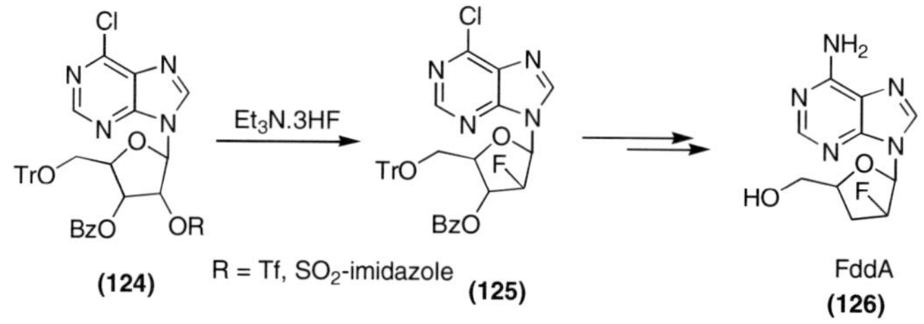

Thioether linkage bearing purines (127) were oxidized to sulfoxides (128), which on treatment with amines results in the formation of new purines (129) [46].

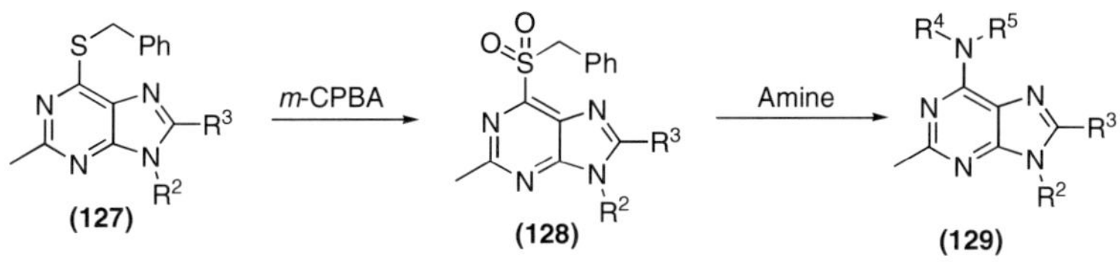

Purines (130) and diaryliodonium salts (131) were cross-coupled to afford the corresponding purines (132) by copper catalyzed reaction in the presence of base in refluxing DCM within a few hours [47]. This method offers several advantages such as excellent yield, short reaction duration etc. These simple conditions also provide an alternative to the synthesis of 9-arylpurines with arylboronic acids as the arylating agents.

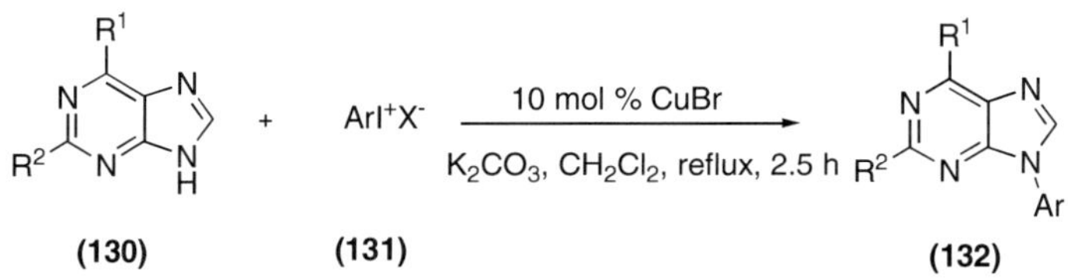

The purines (135) were efficiently obtained from the S_NAr displacement reaction between 6-cyclohexylmethoxy-2-fluoropurine (133) and variously substituted amines (134) in the presence of trifluoroacetic acid in 2, 2, 2-trifluoroethanol solvent [48].

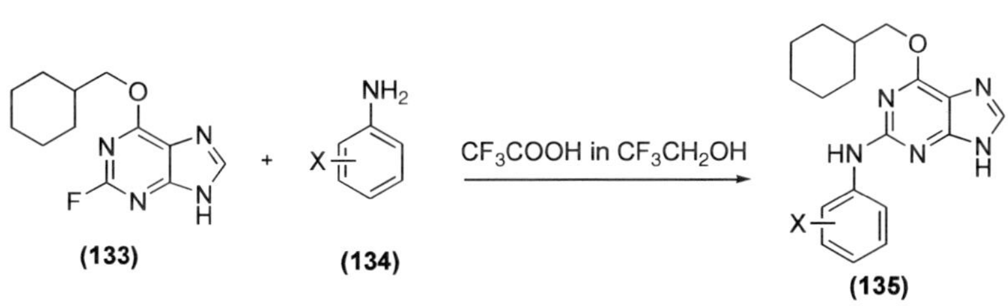

An efficient regioselective method for the activation of C-Ar–H bond leading to the synthesis of purines (137), (138) was recently developed by H. M. Guo and coworkers [49]. The transformation was carried out in the presence of palladium catalysts.

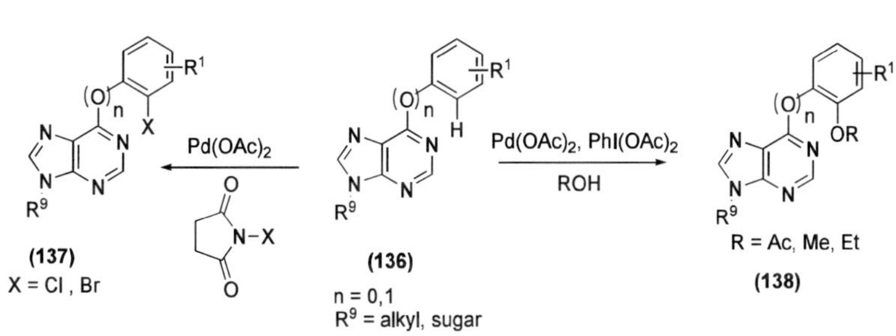

2-Arylamino-6-alkylaminopurines (141) were successfully obtained from the corresponding 2, 6-dichloro purines (139) using isopropyl ethyl amine followed by the reaction with TMS-Cl [50].

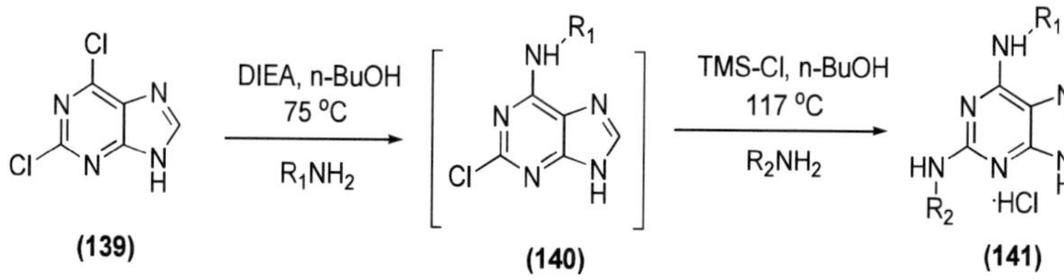

4. Bioactivity of Purines

Purine derivatives represent the scaffolds for many biosynthetic, regulatory and signal transduction proteins including G proteins, cellular kinases etc. These also constitute the potent inhibitors, sulfotransferases, phosphodiesterases etc. Thus, the literature survey reveals that the purine nucleus has become the privileged molecule for medicinal chemistry due to its pharmacological and biological activities. Presently, the chemistry of purine derivatives has become an area of great interest because of the presence of these purine ring systems as the core structures in many heterocyclic compounds covering a wide range of pharmacological applications.

a) Antimicrobial purines
b) Cyto-toxic and anticancer purines
c) Enzyme inhibitor purines
d) Anti HIV purines
e) Miscellaneous activities of purine derivatives

4.1. Antimicrobial Purines:

3, 7-dimethylisoguanine (142) exhibits antibacterial properties [51]. Many isolated N-glycosylated purines are lead structures for the development of new chemotherapeutic agents. The purine antimetabolite (143-145), are all strong antifungal compounds. Shortly after their discovery in 1977 [52], it was recognized that amipurimycin (143) and miharamycin (145) contain the 2-aminopurine, which is not found in nature. The sugar moiety remained unidentified until 1995, when it was recognized to be hepturonic acid [53].

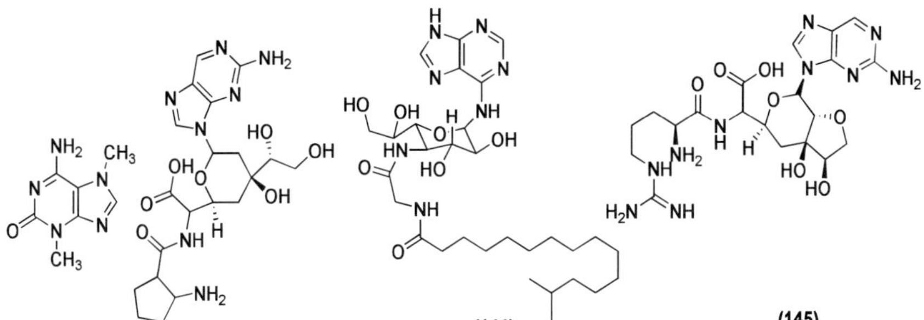

(142) (143) (144) (145)

Some highly functionalized purines (149) were obtained from the 6-chloropurines (146) by M. Jeni and the compounds were subjected for their antimicrobial activity on different microorganisms including gram-positive and gram-negative bacteria and fungi such as E. coli, P. aeruginosa C. albicans, S. aureus, etc. [54]. These compounds exhibited good to moderate activity against pathogenic microorganisms. The activity of these compounds was compared with standard drugs such as ciprofloxacin, chloramphenicol, ampicillin etc.

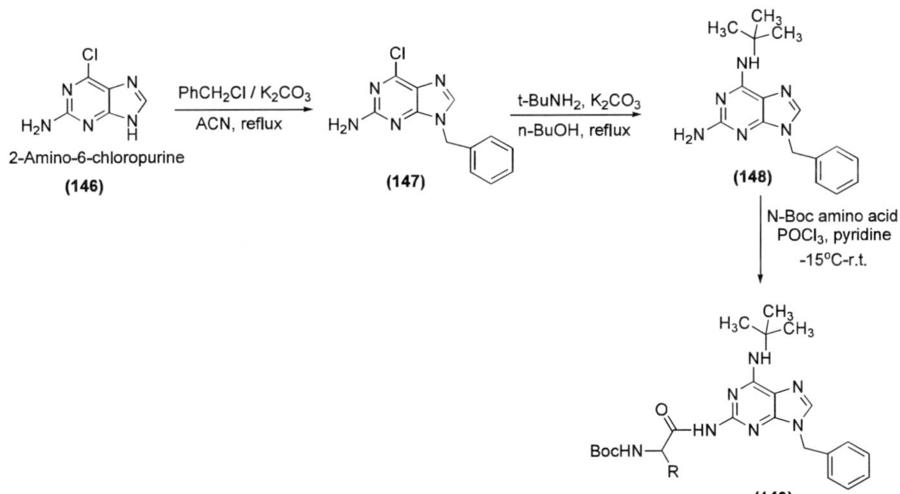

R = –CH₃, -CH₂CH₃, -CH(CH₃)₂, -CH₂CH(CH₃)₂, -CH₂Ph

The purine derivative (150) was found as a potential agent against DNA viruses and/or retroviruses [55].

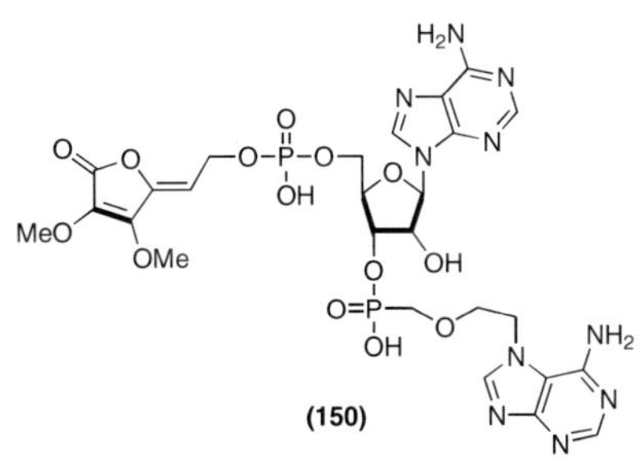

(150)

4.2. Cytotoxic and Anticancer Purines:

Cancer is the uncontrolled cell growth at a particular organ of body. It is becoming one of the most serious and hazardous diseases for mankind. As per the WHO report, about 7.6 million people died from cancer in the year 2005 and the number is expected to rise to 11.5 million by 2030. Hence, the development of novel, potent and selective anticancer agents has become the most intensely pursued goal in the area of drug development. Several purine derivatives actively functioned against the cell growth and hence are anticancer in nature.

Myosevrin (151) is well known for inhibiting microtubule and arresting the cell cycle in mitosis. Some derivatives of Myosevrin (152) were synthesized and evaluated for their anticancer activity [56]. An important compound containing purine skeleton is the Roscovitine (153). It is a highly potent selective inhibitor of cyclin-dependent kinases. Observation reveals that the treatment of roscovitine with the cultured cells leads to cause cell cycle arrest or apoptosis.

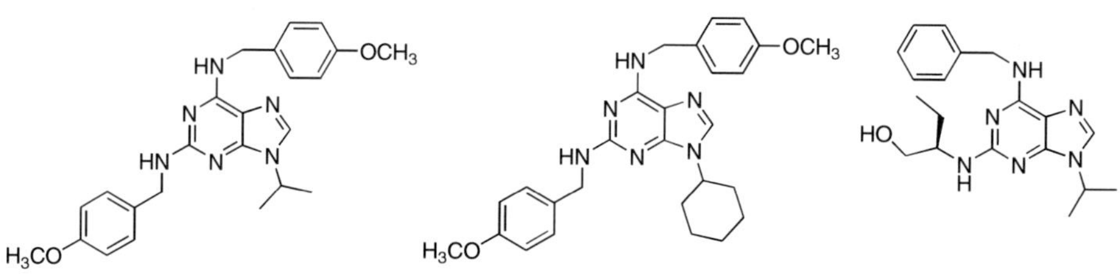

(151) **(152)** **(153)**

From the aerial parts of *Heterostemmabrownii* three new purine derivatives, heteroamines 2-amino-6-methoxy-7,9-dimethylpurinium(154),6-methoxy-7,9-dimethyl-2-methylaminopurinium chloride (155) and 6-methoxy-7,9-dimethyl-2-dimethylaminopurinium chloride (156) [57], have been isolated from the Chinese climber Heterostemma brownie, which are used in Taiwan as folk medicines for the treatment of certain tumors.

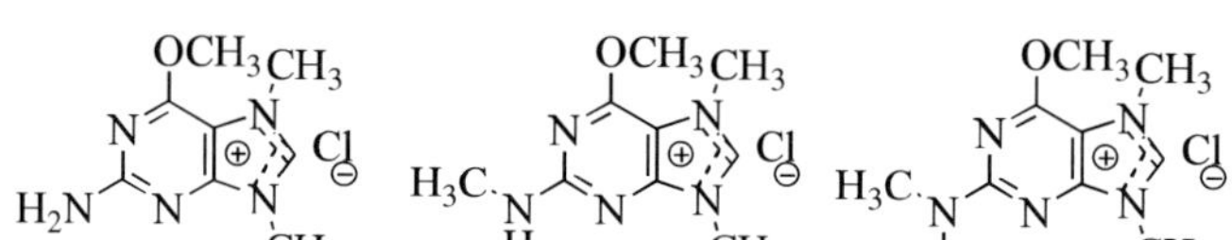

(154) **(155)** **(156)**

Between 1998 and 2002, a new group of interesting purine alkaloids, tricyclic purine derivatives (157), (158), (159) known as asmarines, were isolated from the red sea sponge Raspailia [58, 59]. Of the six members of this family thus far described, all show significant cytotoxicity against various tumor cell lines. All asmarines contain the [1,4] diazepino-[1,2,3-gh]-purine skeleton as well as an carbocyclic hydrocarbon from the small group of naturally occuring bicyclic sesquiterpenes.

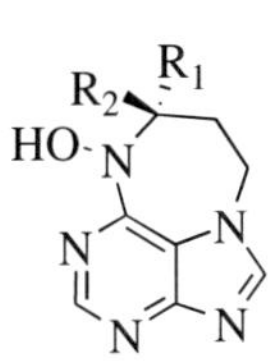 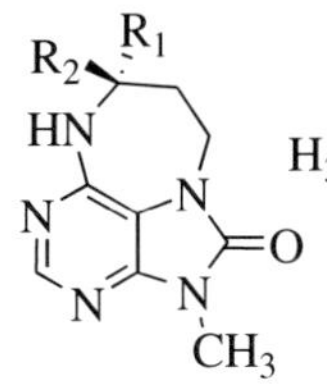 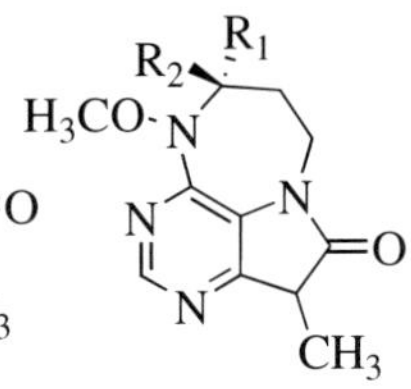

(157) **(158)** **(159)**

The cytostatic activity of some 6-(4-X-substituted phenyl)purine ribonucleosides (160) were evaluated against CCRF-CEM, HeLa, and L1210 cell lines. It was observed that the compounds with X = H, F, Cl and OR were active while the 6-phenylpurine and 2-amino-6-phenylpurine bases as well as 2-amino-6-phenylpurine ribosides were completely inactive against these cell lines [60]. A series of relatively simple methylated purine derivatives, isolated from marine organisms 1, 3-dimethylisoguanine is pharmacologically active (161). It shows activity against human ovarian cancer [61].

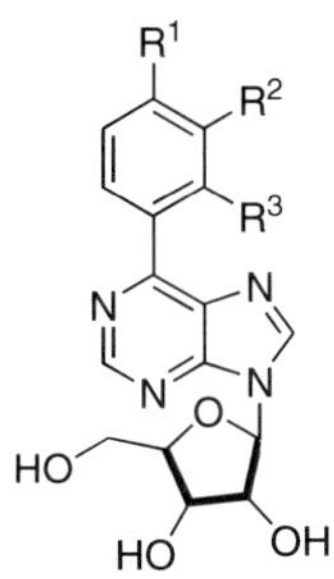 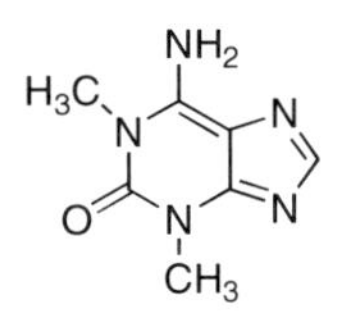

(160) **(161)**

4.3. Enzyme inhibitor Purines

An enzyme inhibitor is a molecule that binds to enzymes and decreases their activity. As a result of blocking the enzymatic activity pathogens are killed and many drugs function as enzyme inhibitors. Enzyme inhibition helps to control many diseases. Several purine derivatives also serve to be the potent enzyme inhibitors.

1-methylherbipoline (162) acts as an inhibitor of collagenase [62]. Forodesine (163) is a potential therapeutic agent. Research is going on to develop Forodesine and evaluate it for controlling the proliferation of T-cells [63].

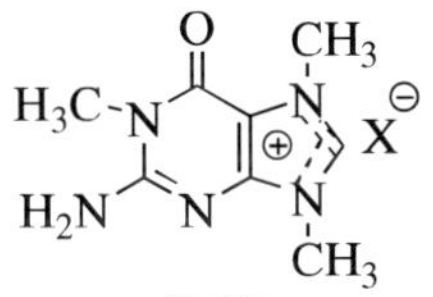 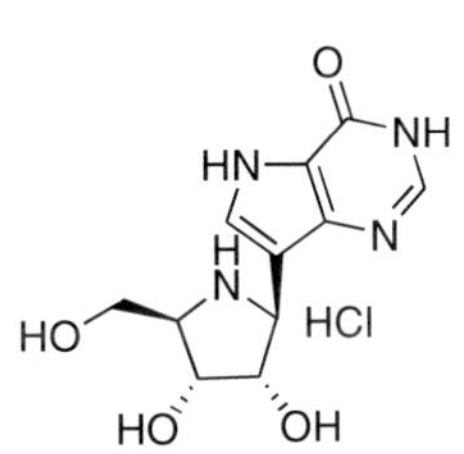

(162) **(163)**

The acycloneplanocin A (164), **a** potential *S*-adenosylhomocysteine hydrolase inhibitor was synthesized *via* reductive cleavage of 2′,3′-*O*-isopropylideneadenosine (165) by DIBALH [64].

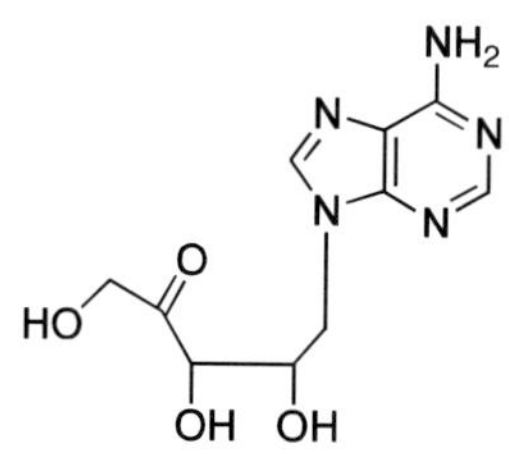

(164)

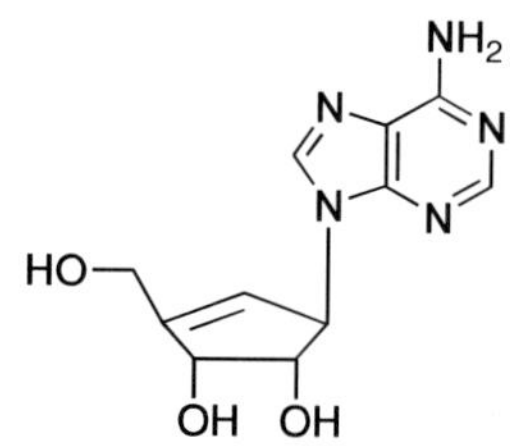

(165)

The purines bearing 9-(arenethenyl) substituents (166) and (167) were observed to possess highly potent activity as Scr/Abl kinase inhibitors [65].

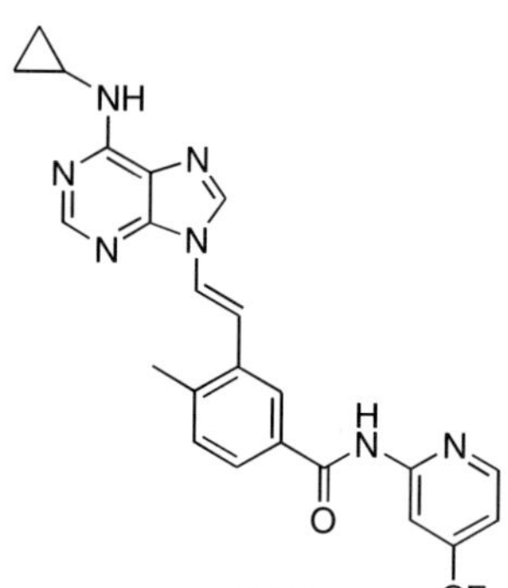

(166)

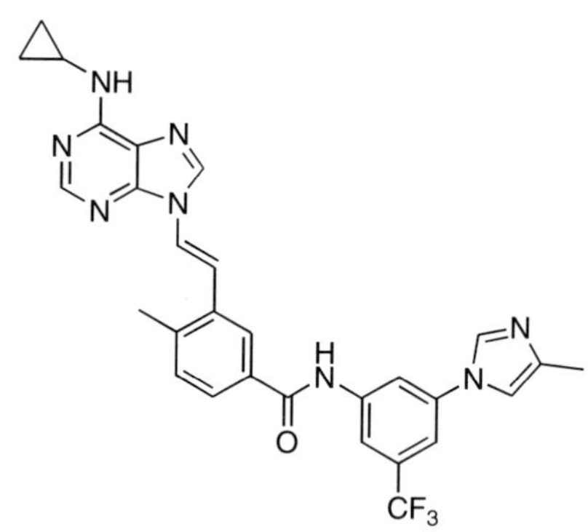

(167)

Agrocin (168) was first described in 1977, but its structure has not yet been completely elucidated [66]. It is unique, as it is the only known 6-N-phosphoramidate to which a glucofuranose is bound. It is an efficient inhibitor of the induction of crown gall tumors by Agrobacterium tumefaciens [67].

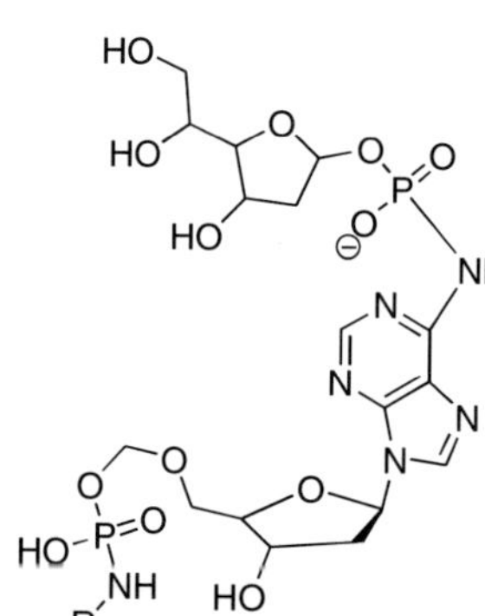

(168)

R= pentose

4.4. Anti-HIV Purines

The Acquired Immune Deficiency Syndrome; abbreviated as (AIDS) caused due to HIV-1 infection is among the most serious threats to mankind. It is estimated that more than 25 million people were killed since it was first recognized. In the fight against AIDS, first and second generation non-nucleoside reverse transcriptase inhibitors (NNRTIs) are now established as a part of highly active antiretroviral therapy (HAART) for treating HIV infection.

(–)-BCA (169) having significant anti-HIV activity was synthesized from (-)-Corey lactone and its absolute configuration was determined [68]. It is potent and has selective effects against human immune deficiency virus. Iso-ddA (**170**) represents a new class of potential anti-AIDS drugs [69].

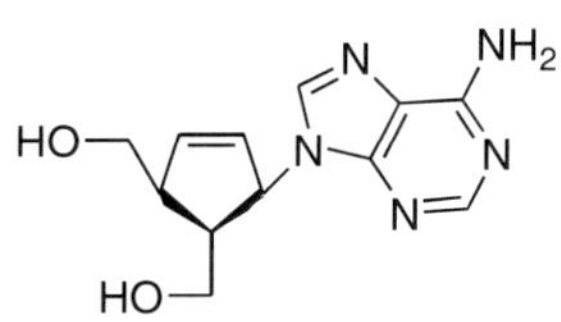

(169)

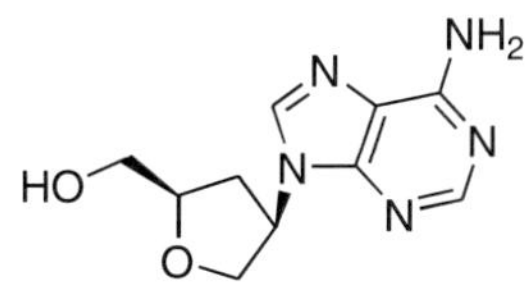

(170)

In 2000, H. Ohrui and coworkers [70] evaluated the anti-HIV activity of purine compounds (171). 2',3'-didehydro-2',3'-dideoxy-2'-fluoro-4'-thionucleosides bearing purine derivatives (172) were reported to be active against HIV-1 and HBV viruses [71].

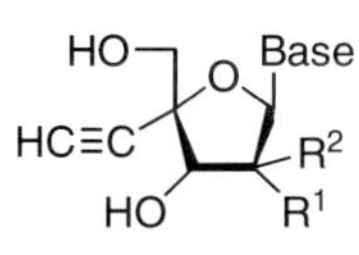

(171)

R^1 = H, R^2 = OH, or R^1 = R^2 = H
Base = Purine or pyrimidine

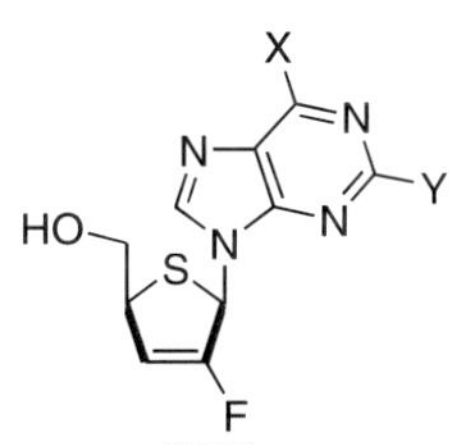

(172)

X = NH_2, OH
Y = H, F, NH_2

4.5. Miscellaneous Activities of Purines and their Derivatives

The cytokinines [72] are a very special group of phytohormones that promotes the cell growth as well as cell differentiation [73]. It is assumed that cytokinines bind to receptor proteins in the cell to carry out cytokinine-specific effects. Interestingly, cytokinines have also been found in animals such as in the salivary glands of the moth's stigmellaargyopeza and S. Argentipedella. Chemically cytokinines are the substituted adenines and appear either as free bases or as nucleosides or nucleotides, bound, e.g. to ribose or glucose. Such sugar-linked cytokinines show only weak or no activity and have to be cleaved at their glycosylic bond. Most probably these conjugates represent transport or depot forms.

A purine ring is present in Reversine (173); a potent selective human A_3 adenosine receptor antagonist.

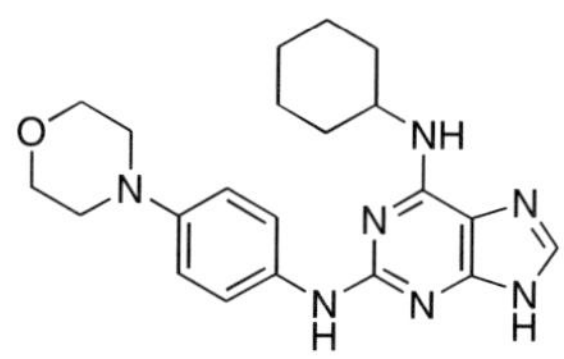

(173)

All the purines existing in nature are substituted derivatives [74]. Many purine derivatives, especially adenine derivatives, are involved in numerous metabolic processes. Adenosine 5'-triphosphate (174) is used for the storage of energy in all living cells [75]. Adenosine 3',5'-cyclophosphate (cyclic AMP; cAMP; (175) acts as a second messenger for controlling the activation of protein kinases and the K^+ levels in cell, as well as in transcription and other metabolic processes [76]. Nicotinamide adenine dinucleotide (176) (NAD; NAD^+) [77] and flavin adenine dinucleotide (177) (FAD) [78] are coenzymes involved in cellular reduction-oxidation processes.

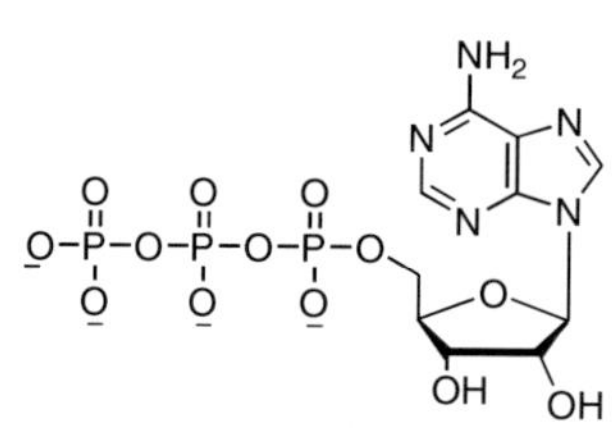

(174)

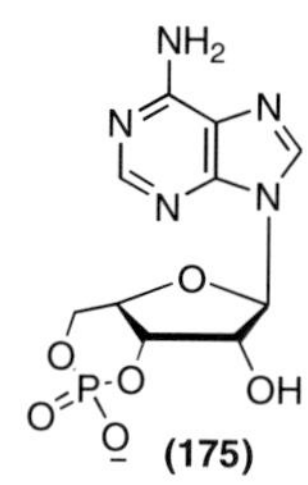

(175)

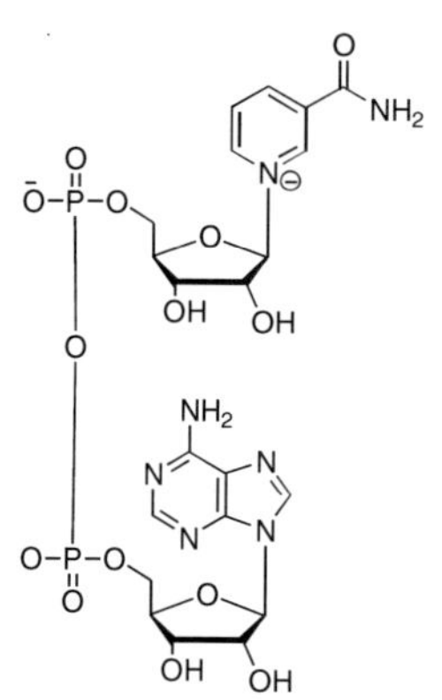

(176)

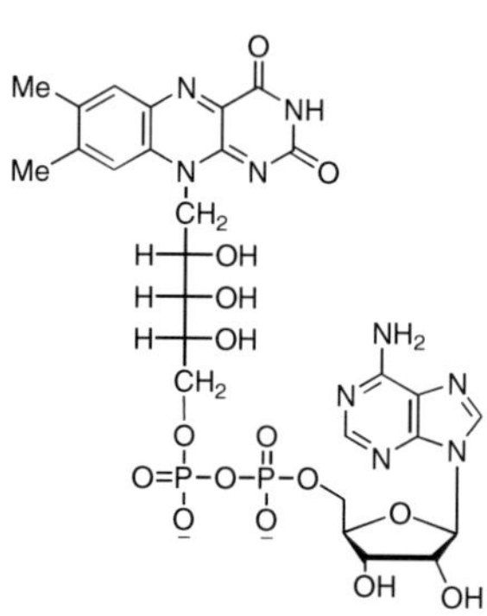

(177)

Very often in medicinal chemistry, substances isolated from natural sources function as the lead structures for new pharmacologically active compounds. Many isolated purine derivatives also exhibit some biological activity. Furthermore, purine plays a diverse role in the construction of many naturally occurring compounds. H. Rosemeyer recently reviewed the chemo diversity of this heterocyclic as a constituent of natural products [79]. Another group of purine derivatives isolated mainly from marine sponges *Tedaniadigitata* and *anisodorisnobilis* is doridosine (1-methylisoguanosine; (178) [80-81]. This purine nucleoside has attracted attention because it is an A_1-receptor agonist and also binds to the human benzodiazepine receptor. Its effect includes blood pressure reduction and slowing of the heart rate.

Zeatin (179) was the first cytokinin isolated from Zea mays [82]. Only the (E)-configured compound exhibits marked phytohormone activity.

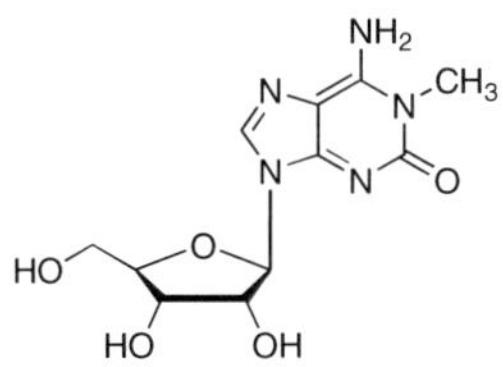

(178)

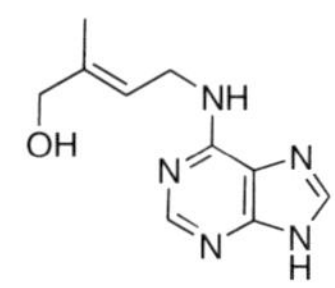

(179)

Also septacidin (180) is unusal, being glycosylated at the exocyclic amino group of adenine. The sugar moiety is an aminoaldoheptose to which a branched C12-C18 fatty acid is coupled (e.g. isopalmitic acid) [83]. Such branched fatty acids are relatively rare in nature and occur mainly in the skin sebum as a constituent of triglycerides.

Isopentenyladenine (181) was isolated as free base from Coryne bacterium fasciens [84-85] . Its riboside is found in t-RNA on the 3'-side of those anticodons that pair with codons beginning with uridine. Those codons that start with adenosine carry a hydrophilic nucleotide such as (182) on the anticodon on the 3'-side:

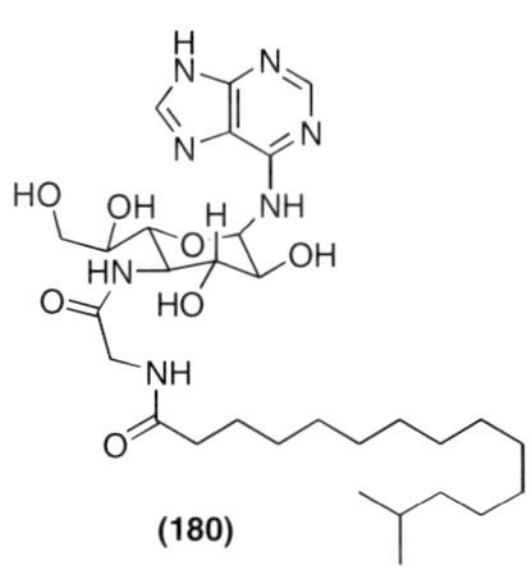

(180)

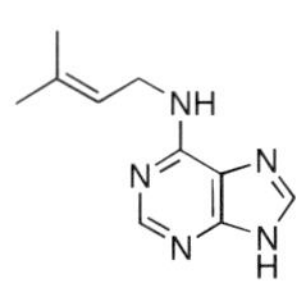

(181)

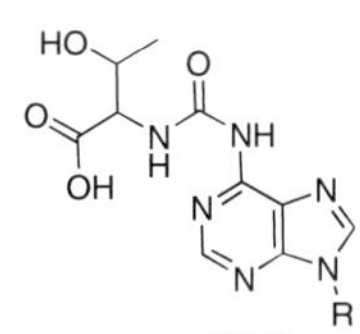

(182)

Eritadenine (183) in particular demonstrates a significant hypocholesteremic effect [86] Therefore, in Japan and elsewhere, the shiitake mushroom is looked upon as a life elixir. Eritadenine improves the cholesterol transport along the blood vessels [87]. Many eritadenie derivatives e.g. (184) and (185) have been synthesized in the past.

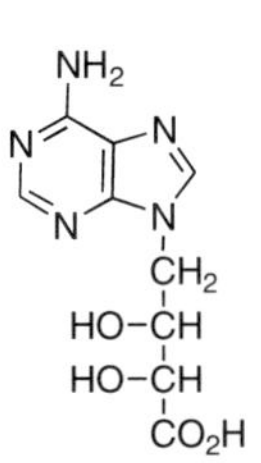

(183)

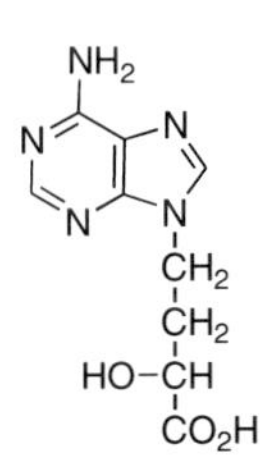

(184)

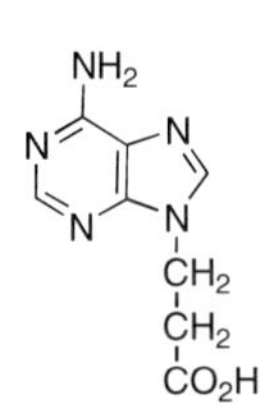

(185)

Compound (186) (9-[3-(4-Chlorophenoxy)-2-hydroxypropyl]adenine) demonstrates the highest hypocholesteremic effect [88]. The purine system is an integral part of higher-alkylated adenine alkaloids. The first alkaloid of this kind to have been detected is triacanthine (187), isolated in 1950 from the North American tree Gleditsia triacanthos and the structure was verified by total synthesis in 1962 by Leonard and Deyrup [89].

Inspired by the narrative history of native peoples of the midwesern United States regarding the hallucinogenic action of the leaves of this tree, several groups have been investigating the alkaloid activity of triacanthine (187) for more than 10 years without any success. Triacanthine (187) appears to have no biological activity.

Interestingly, the hydrochloride but not the parent compound, forms the tricyclic pyrotriacanthine (188) upon heating. Another, as yet unknown compound in the leaves must be responsible for the hallucinogenic effect.

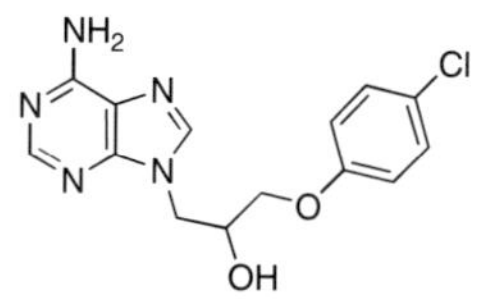

(186)

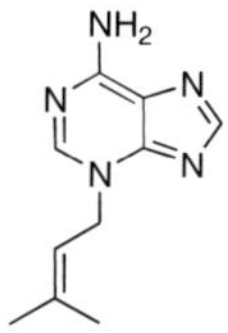

(187)

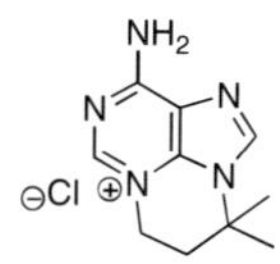

(188)

The synthesis of pyrrolo[2,1-*f*][1,2,4]triazine adenosines (189) was reported by S. A. Patiletal [90]. *In vitro* growth inhibitory activities of leukemic cell lines of these compounds were compared with the standard drug 9-deazaadenosine. The compound (190) was developed for its affinity for the adenosine receptors (AdoRs) [91].

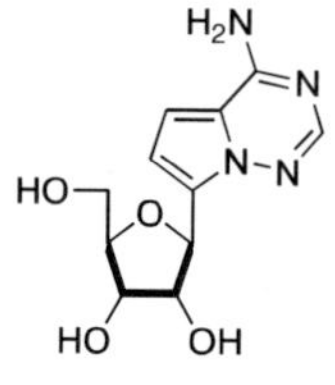

(189)

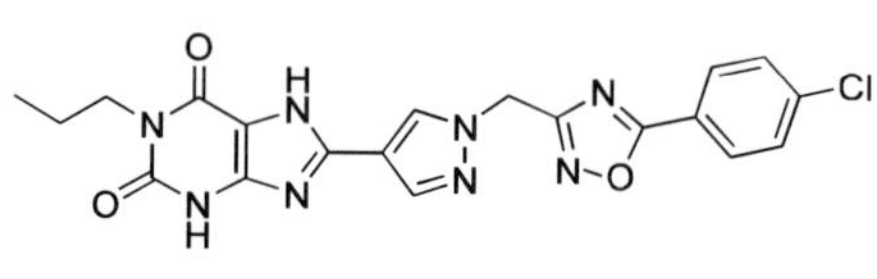

(190)

Conclusion

In summary; the present chapter describes briefly the natural occurrence, synthesis and bioactivity of purine scaffolds. Research is under progress to develop new purine derivatives which will function as the promising chemotherapeutic agents in pharmacology and medicinal chemistry. This provides a thrust area and a broad platform for researchers in the field of biotechnology, pharmacology and medicinal chemistry.

References

[1] Liu, J. K. (2005). N-Containing compounds of macromycetes. *Chemical Review, 105*, 2723-2744.

[2] Chargaff, E. (1963). *Essays on nucleic acids*. Amsterdam, Elsevier.

[3] Bodenteich, M. & Marquez, V. E. (1989). Total synthesis of (±)-neplanocin F, *Tetrahedron Letters, 30*, 4909-4912.

[4] Yin, X.; Zhao, G. & Schneller, S.W. (2007). Carbocyclic sinefungin. *Tetrahedron Letters, 48*, 4809–4811.

[5] Ichikawa; Yamamura, S. & Kato, K. (1999). Synthesis of enantiomerically pure 9[(1′*R*, 2′*R*, 3′*S*)-bis (hydroxymethyl) thietan-1′-yl] adenine, 3′-thio analog of oxetanocin A. *Tetrahedron*, 40, 7385–7388.

[6] Wolfe, M. S.; Borcherding, D. R. & Borchardt, R. T. (1989). A 9-step enantiospecific synthesis of (-)-aristeromycin from D-ribonic acid γ-lactone. *Tetrahedron Letters, 30,* 1453-1456.

[7] Duke, C. C.; Liepa, A. J.; MacLeod, J. K.; Letham, D. S. & Parker C. W. (1975). Synthesis of raphantin and its 6-benzylaminopurine analogue. *Journal of Chemical Society, 24,* 964-965.

[8] Bourguet-Kondracki, M-L.; Martin, M-T.; Vacelet, J. & Guyot, M. (2001). Mucronatine, a new N-methyl purine from the French mediterranean marine sponge Stryphnusmucronatus. *Tetrahedron Letters, 42*, 7257-7259.

[9] Chi, Y. L.; Zhang, H. Y.; Xue, J. H.; Hao, J.; Liu, M. F. & Wei, X. Y. *(2009).* N^2-(1-Methoxycarbonylethyl)guanosine, a new nucleoside coupled with an amino acid derivative from *Amanita exitialis.* Chinese chemical Letters, *20,* 830-832.

[10] Wieland, T. & Bauer, L. (1951). Purine and aminosauren. Angew Chem, 63, 511-513.

[11] Pfleiderer, W. (1961). Purine, IV.über die Methylierung des 9-methyl-guanins und die Struktur des Herbipolins. *Liebigs Annalen der Chemie, 647*, 167-173.

[12] Ichiba, T.; Nakao, Y. & Scheuer, P. J. (1995). Kumusine, a chloroadenineriboside from a sponge *Theonella* sp. *Tetrahedron Letters, 36*, 3977–3980.

[13] Niu, H-Y.; Xia, C.; Qu, G-R.; Wu, S.; Jiang, Y.; Jin, X. & Guo, H-M. (2012). Microwave-promoted one-pot synthesis of 4-nitrobenzylthioinosine analogues using thiourea as a sulfur precursor. *Chemistry – an Asian Journal, 7,* 45–49.

[14] Qu, G-R. ; Wu, J.; Wu, Y-Y. ; Zhang, F. & Guo, H-M. (2009). Synthesis of novel 6-[*N*, *N*-bis (2-hydroxyethyl) amino] purine nucleosides under microwave irradiation in neat water. *Green Chemistry, 11*, 760-762.

[15] Qu, G-R. ; Zhao, L.; Wang, D-C. ; Wu, J. & Guo, H-M. (2008). Microwave-promoted efficient synthesis of C6-cyclo secondary amine substituted purine analogues in neat water. *Green Chemistry, 10*, 287-289.

[16] Austin, R. E.; Okonya, J. F.; Bond, D. R. S. & Al-Obeidi, F. (2002). Microwave-assisted solid-phase synthesis (MASS) of 2, 6, 9-trisubstituted purines. *Tetrahedron Letters, 43,* 6169–6171.

[17] Qu, G-R. ; Xia, R.; Yang X-N. ; Li, J-G. ; Wang, D-C. & Guo, H-M. (2008). Synthesis of novel C_6-phosphonated purine nucleosides under microwave irradiation by SNAr−Arbuzov reaction. Journal of Organic Chemistry, 73, 2416–2419.

[18] Vabre, F. C.; Legraverend, M. & Piguel, S. (2011). Microwave-assisted Pd/Cu-catalyzed C-8 direct alkenylation of purines and related azoles: An alternative access to 6,8,9-trisubstituted purines. Journal of Organic Chemistry, 76, 9542–9547.

[19] Yamada, H. & Okamoto, T. (1972). A one-step synthesis. *Chemical & Pharmaceutical Bulletin, 20*, 623-624.

[20] Strouse, J. J.; Jeselnik, M.; Tapaha, Jonsson, C. B.; Parker, W. B. & Arterburn, J. B. (2005). Copper catalyzed arylation with boronic acids for the synthesis of N^1-aryl purine nucleosides. *Tetrahedron Letters, 46,* 5699-5702.

[21] Sahnoun, S.; Messaoudi, S.; Peyrat, J-F. ; Brion, J-D. & Alami, M. (2008). Microwave-assisted Pd $(OH)_2$-catalyzed direct C–H arylation of free-(NH_2) adenines with aryl halides. *Tetrahedron Letters*, *49*, 7279–7283.

[22] Havelkova, M.; Hocek, M.; Cesnek, M. & Dvorak, D. (1999). The Suzuki-Miyaura cross-coupling reactions of 6-halopurines with boronic acids leading to 6-aryl- and 6-alkenylpurines. *Synlett, 7,* 1145–1147.

[23] Hasnik, Z.; Silhar, P. & Hocek, M. (2007). Synthesis of (purin-6-yl) acetates and 6-(2-hydroxyethyl) purines via cross-couplings of 6-chloropurines with the Reformatsky reagent. *Tetrahedron Letters, 48*, 5589-5592.

[24] Silhar, P.; Pohl, R.; Votruba I. & Hocek, M. (2004). Facile and efficient synthesis of 6-(hydroxymethyl)purines. Organic Letters, *6*, 3225–3228.

[25] Hocek, M. & Dvorakova, H. (2003). An efficient synthesis of 2-substituted 6-methylpurine bases and nucleosides by Fe- or Pd-catalyzed cross-coupling reactions of 2, 6-dichloropurines. Journal of Organic Chemistry, *68*, 5773–5776.

[26] Capek P.; Pohl, R. & Hocek, M. (2004). A facile and efficient synthesis of (purin-6-yl)alanines. Journal of Organic Chemistry, *69*, 7985–7988.

[27] Tobrman, T. & Dvorak, D. (2003). 6-Magnesiated purines: Preparation and reaction with aldehydes. Organic Letters, *5*, 4289–4291.

[28] Fletcher, S.; Shahani, V. M. & Gunning, P. T. (2009). Facile and efficient access to 2, 6, 9-trisubstituted purines through N9, N2 Mitsunobu reactions. *Tetrahedron Letters, 50,* 4258-4261.

[29] Ibrahim, N. & Legraverend, M. (2009). Synthesis of 6, 7, 8-trisubstituted purines via a copper-catalyzed amidation reaction. Journal of Organic Chemistry, *74*, 463–465.

[30] Guo, H-M. ; Rao, W-H. ; Niu, H-Y. ; Jiang, L-L. ; Liang, L.; Zhang, Y. & Qu, G-R. (2011). Chelation-assisted palladium-catalyzed high regioselective Heck diarylation reaction of 9-allyl-9*H*-purine: Synthesis of 9-(3, 3-diaryl-allyl)-9*H*-purines, *RSC Advances*, *1*, 961-963.

[31] Veliz, E. A. & Beal, P. A. (2001). 6-Bromopurine nucleosides as reagents for nucleoside analogue synthesis. Journal of Organic Chemistry, 66, 8592–8598.

[32] Ostrowsk, S. (1999).Synthesis of *N*-7-Substituted purines from imidazole precursors. *Molecules, 4,* 288-309.

[33] Wan, Z-K.; Binnun, E.; Wilson, D. P. & Lee, J. (2005). A highly facile and efficient one-step synthesis of N^6-adenosine and N^6-2'-deoxyadenosine derivatives. *Organic Letters, 7,* 5877–5880.

[34] Lakshman, M. K.; Gunda, P. & Pradhan, P. (2005). Mild and room temperature C–C bond forming reactions of nucleoside C-6 arylsulfonates. *Journal of Organic Chemistry, 70*, 10329–10335.

[35] Bakkestuen, A. K.; Gundersen, L.-L.; Langli, G.; Liu, F. & Nolsoe, J. M. (2000). 9-Benzylpurines with inhibitory activity against *mycobacterium tuberculosis*. *Bioorganic and Medicinal Chemistry Letters, 10,* 1207–1210.

[36] Liu, J.; Dang, Q.; Wei, Z.; Shi, F. & Bai, X. (2006). An efficient and regiospecific strategy to N^7-substituted purines and its application to library of trisubstituted purines. Journal Combinatorial Chemistry, *8*, 410–416.

[37] Shawali, A. S.; Mosselhi, M. A. N. & Tawfik, N. M. (2001). A simple one-pot synthesis of novel [1, 2, 4]triazolo[3,4-*f*]purines. Journal Organic Chemistry, 66, 4055–4057.

[38] Piguel & Legraverend, S. M. (2007). Selective amidation of 2, 6-dihalogenopurines: application to the synthesis of new 2, 6, 9-trisubstituted purines. Journal Organic Chemistry, 72, 7026–7029.

[39] Zhong, M. & Robins, M. J. (2006). Regiospecific N9 alkylation of 6-(heteroaryl) purines: Shielding of N7 by a proximal heteroaryl C–H[1]. Journal Organic Chemistry, *71*, 8901–8906.

[40] Vankova, B.; Krchnak, V.; Soural, M. & Hlavac, J. (2011). Direct C–H arylation of purine on solid phase and its use for chemical libraries synthesis. Combinatorial Science, *13*, 496–500.

[41] Qu, G-R. ; Mao, Z-J. ; Niu, H-Y. ; Wang, D-C. ; Xia, C. & Guo, H-M. (2009). Straightforward and highly efficient catalyst-free one-step synthesis of 2-(purin-6-yl)acetoacetic acid ethyl esters, (purin-6-yl)acetates, and 6-methylpurines through S_NAr-based reactions of 6-halopurines with ethyl acetoacetate. Organic Letters, 11, 1745–1748.

[42] Ding, S.; Gray, N. S.; Ding, Q. & Schultz, P. G. (2001). Expanding the diversity of purine libraries. *Tetrahedron Letters, 42*, 8751–8755.

[43] Legraverend, M.; Boumchita, H. & Bisagni, E. (1990). A new route to 2, 5-diamino-4, 6-dichloropyrimidine: A key precursor of 9-substituted guanines. *Synthesis*, *7*, 587–589.

[44] Dolle, R. E. & McNair, D. (1993). 9-(Sulfoximinoalkyl) guanine nucleosides as potential antiherpetic agents. *Tetrahedron Letters, 34*, 133-136.

[45] Takamatsu, S.; Maruyama, T.; S. Katayama, N. Hirose & K. Izawa. (2001). Improved synthesis of 9-(2, 3-dideoxy-2-fluoro-β-D-threo-pentofuranosyl) adenine (FddA) using triethyl amine trihydrofluoride. *Tetrahedron Letters, 42*, 2321-2324.

[46] Liu, J.; Dang, Q.; Wei, Z.; Zhang, H. & Bai, X. (2005). Parallel solution-phase synthesis of a 2, 6, 8, 9- tetrasubstituted purine library via sulfur intermediate. Journal of Combinatorial Chemistry, 7, 627–636.

[47] Niu, H-Y. ; Xia, C.; Qu, G-R. ; Zhang, Q.; Jiang, Y.; Mao, R-Z. ; Li, D-Y. & Guo, H-M. (2011). CuBr catalyzed C–N cross coupling reaction of purines and diaryliodonium salts to 9-arylpurines. *Organic Biomolecular Chemistry*, *9*, 5039-5042.

[48] Whitfield, H. J. ; Griffin, R. J. ; Hardcastle, I. R. ; Henderson, A. ; Meneyrol, J. ; Mesguiche, V. ; Sayle, K. L. & Golding, B. T. (2003). Facilitation of addition–elimination reactions in pyrimidines and purines using trifluoroacetic acid in trifluoroethanol. *Chemical Commuications*, *22*, 2802-2803.

[49] Guo, H-M. ; Rao, W-H. ; Niu, H-Y. ; Jiang, L-L. ; Meng, G.; Jin, J-J. ; Yang, X-N. & Qu, G-R. (2011). Palladium-catalyzed C–H bond functionalization of C_6-arylpurines. *Chemical Communications*, *47*, 5608-5610.

[50] Ciszewski, L.; Waykole, L.; Prashad, M. & Repic, O. (2006). Practical synthesis of 2-arylamino-6-alkylaminopurines from 2, 6-dichloropurine. *Organic* Process Research and Development, 10, 799–802.

[51] Cafieri, F.; Fattorusso, E.; Mangoni, A. & Taglialatela-Scafati, O. (1995). Longamide and 3, 7-dimethylisoguanine, two novel alkaloids from the marine sponge agelaslongissima. *Tetrahedron Letters, 36*, 7893-7896.

[52] Harada, S.; Kishi, T. (1977). Isolation and characterization of a new nucleoside antibiotic, amipurimycin. *Journal of Antibiotics*, *30*, 11-6.

[53] Fairbanks, A. J. & Sinay, P. (1995). Synthesis of the bicyclic moiety of the miharanycins by samarium (II) iodide induced ring closure. *Synthetic Letters, 3*, 277-279.

[54] Jeni, M. M.; Nathani, B. R. & Pandya, K. S. (2011). Synthesis, characterization and anti-microbial activity of some novel trisubstituted purine bearing amino acid. *Der Pharma Chemica*, *3*, 20-28.

[55] Hakimelahi, G. H. ; Ly, T. W. ; Moosavi-Movahedi, A. A.; Jain, M. L. ; Zakerinia, M. ; Davari, H. ; Mei, H-C. ; Sambaiah, T.; Moshfegh, A. A.; & Hakimelahi, S. (2001). Design, synthesis and biological evaluation of novel nucleoside and nucleotide analogues as agents against DNA viruses and/or retroviruses. Journal of Medicinal Chemistry, *44*, 3710–3720.

[56] Chang, Y-T. ; Wignall, S. M. ; Rosania, G. R. ; Gray, N. S. ; Hanson, S. R. ; Su, A. I.; Merlie, J.; Moon, Jr., H-S. ; Sangankar, S. B.; Perez, O.; Heald, R. & Schultz, P. G. (2001). Synthesis and biological evaluation of myoseverin derivatives: Microtubule assembly inhibitors. Journal of Medicinal Chemistry, *44*, 4497–4500.

[57] Lin, Y. L.; Lee, H. P.; Ou, J. C. & Kuo, Y. H. (1996). Three New purinium Derivatives, Heteromines A, B and C from Heterostemma brownie. *Heterocycle*, *43*, 781-786.

[58] Yosief, T.; Rudi, A.; Stein, Z.; Goldberg, I.; Gravalos, G. M. D.; Schleyer, M. & Kashman,Y. (1998). Asmarines A-C. Three novel cytotoxic metabolites from the marine sponge raspailia sp. *Tetrahedron Letters,39*, 3323-3326.

[59] Ohba, M. & Tashiro, T. (2002). Preparatory Study for the Synthesis of the Marine Sponge Alkaloids Asmarines A-F: Synthesis of their heterocyclic portions. *Heterocycles, 57*, 1235-1238 and references cited therein.

[60] Hocek, M.; Holy, A.; Votruba, I. & Dvorakova, H. (2000). Synthesis and cytostatic activity of substituted 6-phenylpurine bases and nucleosides: Application of the Suzuki−Miyaura cross-coupling reactions of 6-chloropurine derivatives with phenylboronic acids. Journal of Medicinal Chemistry*, 43*, 1817–1825.

[61] a). Mitchell, S. S.; Whitehill, A. B.; Trapido-rosenthal, H. G. & Ireland, C. M. (1997). Isolation and characterization of 1, 3-dimethylisoguanine from the Bermudianspongeamphimedonviridis. *Journal of Natural Products, 60*, 727-728.

[62] Yagi, H.; Matsunaga, S. & Fusetanix, N. (1994). Isolation of 1-methylherbipoline, a purine base from a marine sponge, jaspis sp. *Journal of Natural Products, 57,* 837-838.

[63] Kamath, V. P.; Xue, J.; Juarez-Brambila, J. J. & Morris Jr., P. E. (2009). Alternative route towards the convergent synthesis of a human purine nucleoside phosphorylase inhibitor forodesineHCl. *Tetrahedron Letters, 50*, 5198-5200.

[64] Kitade, Y.; Mo-nguchi, Y.; Hirota, K. & Maki, Y. (1993). Synthesis of a ring-scission analogue of neplanocin A as a potential inhibitor of *S*-adenosylhomocysteine hydrolase. *Tetrahedron Letters, 34*, 6579-6580.

[65] Huang, W-S. ; Zhu, X. ; Wang, Y. ; Azam, M. ; Wen, M. ; Sundaramoorthi, R. ; Thomas, R. M. ; Liu, S. ; Banda, G. ; Lentini, S. P. ; Das, S. ; Xu, Q. ; Keats, J. ; Wang, F. ; Wardwell, S. ; Ning, Y. ; Snodgrass, J. T. ; Broudy, M. I. ; Russian, K. ; Daley, G. Q. ; Iuliucci, J. ; Dalgarno, D. C. ; Clackson, T. ; Sawyer , T. K. & Shakespeare, W. C. (2009). 9-(Arenethenyl)purines as dual Src/Abl kinase inhibitors targeting the inactive conformation: Design, synthesis, and biological evaluation. Journal of Medicinal Chemistry, 52, 4743–4756.

[66] Roberts, W. P.; Tate, M. E. & Kerr, A. (1977). Agrocin 84 is a 6-N-phosphoramidate of an adenine nucleotide analogue. *Nature, 265*, 379-381.

[67] Sudhadolnik, R. J. (1979). *Nucleosides as Biological Probes*. New York, John Wiley & Sons, 49-51.

[68] Katagiri, N.; Toyota, A.; Shiraishi, T.; Sato, H. & Kaneko, C. (1992). Synthesis of (1*R*, 4*S*, 5*R*)-9-(4, 5-bishydroxy-methylcyclopent-2-en-1-yl)-9*H*-adenine [(-)-BCA] and selective inhibition of human immune deficiency virus. *Tetrahedron Letters, 33,* 3507-3510.

[69] Huryn, D. M.; Sluboski, B. C.; Tam, S.Y.; Todaro, L. J. & Weigele, M. (1989). Synthesis of iso-ddA, member of a novel class of anti-HIV agents. *Tetrahedron Letters, 30,* 6259-6262.

[70] Ohrui, H. ; Kohgo, S. ; Kitano, K. ; Sakata, S. ; Kodama, E. ; Yoshimura, K. ; Matsuoka, M. ; Shigeta, S. & Mitsuya, H. (2000). Syntheses of 4‘-*C*-ethynyl-β-D-*arabino*- and 4‘-*C*-ethynyl-2‘-deoxy-β-D-*ribo*-pentofuranosylpyrimidines and purines and evaluation of their anti-HIV activity. Journal of Medicinal Chemistry, 43, 4516–4525.

[71] Chong, Y. ; Choo, H. ; Choi, Y. ; Mathew, J. ; Schinazi, R. F. & Chu, C. K. (2002). Stereoselective synthesis and antiviral activity of D-2', 3'-didehydro-2', 3'-dideoxy-2'-fluoro-4'-thionucleosides. Journal of Medicinal Chemistry, 45, 4888–4898.

[72] Matsubara S. (1980). Structure –activity relationships of cytokinins. *Phytochemistry, 19*, 2239-2253.

[73] Stuart, F. C. & Krikorian, A. D. (1971). Plants, chemicals and growth. Academic Press, New York.

[74] Watson, J. D.; Crick, F. H. C. (1953). General implications of the structure of deoxyribonucleic Acid, *171*, 964-967.

[75] Schlenk, F. (1987). The ancestry and adolescence of adenosine triphosphate. *Trends Biochemical Science, 12*, 367-368.

[76] Rall, T. W. & Sutherland, E. W. (1958). Formation of a cyclic adenine ribonucleotide by tissue particles. *The Journal of Biological Chemistry, 232*, 1065-1076.

[77] Schlenk, F. (1984). The dawn of nicotinamide coenzyme research. *Trends Biochemical Science, 9*, 286-288.

[78] Ghisla, S. & Massey, V. (1989). Mechanism of flavoprotein-catalyzed reactions. *181*, 1-17.

[79] Rosemeyer, H. (2004). The Chemodiversity of purine as a constituent of natural products. *Chemistry & Biodiversity, 1, 361*-401 and ref. cited therein.

[80] Lambert, J. B.; Marwood, J. F.; Davies, L. P. & Taylor, K. M. (1980). 1-methylisoguanosine: An orally active natural product with skeletal muscle and cardiovascular effects. *Life Science, 26*, 1069-1077.

[81] Fuhrman, F. A.; Fuhrman, G. J.; Kim Y.H.; Pavelka, L. A. & Mosher, H. S. (1980). Doridosine: a new hypotensive N-methylpurineriboside from the nudibranchanisodorisnobilis. *Science*, *207*, 193-195.

[82] Letham, D. S. Zeatin (1963). A factor inducing cell division isolated from Zea mays *Life Science*. 2, 569-573.

[83] Townsend, L. B. in "Handbook of biochemistry and molecular biology, Nucleic Acids' Vol.I, 3 rd edn., Ed. Fasman, G. D., Cleveland, CRC Press, 1975, 271.

[84] Hall, R. H.; Robins, M. J.; Stasiuk, L. & Thedford, R. (1966). Isolation of N6-(γ,γ-[dimethylallyl)adenosine from soluble ribonucleic acid. *Journal of American Chemical Society*, *88*, 2614-2615.

[85] Helgeson, J. P. & Leonard, N. J. (1966). Cytokinins: identification of compounds isolated from corynebacteriumfascians. Proceedings of the National Academy of Sciences of the United States of America *U. S. A.* 56, 60-63.

[86] Tokuda, S. & Kaneda, T. (1966). Journal of Japanese Society of Food and Nutrition.17, 297.

[87] Tokuda, S.; Kano, E. & Kaneda, T. (1972). *Journal of Japanese Society of Food and Nutrition,* 25, 609.

[88] Dimenna, W. S.; Piantadosi, C. & Lamb, G. R. (1978). Synthesis of potential hypolipidemic agents. Reaction of substituted phenyl 2, 3- epoxypropyl ethers with adenine, uracil and thymine. *Journal of Medicinal Chemistry, 21*, 1073-1076.

[89] Leonard, N. J. & Deyrup, J. A. (1962). The chemistry of Triacanthine.*Journal of American Chemical Society, 84*, 2148-2160.

[90] Patil, S. A.; Otter, B. A & Klein, R. S. (1994). 4-aza-7, 9-dideazaadenosine, a new cytotoxic synthetic C-nucleoside analogue of adenosine. *Tetrahedron Letters, 35*, 5339-5342.

[91] Kalla, R.V.; Elzein, E.; Perry, T.; Li, X.; Gimbel, A.; Yang, M.; Zeng, D. &Zablocki, J. (2008). Selective, high affinity A_{2B} adenosine receptor antagonists: N-1 monosubstituted 8-(pyrazol-4-yl) xanthenes. *Bioorganic and Medicinal Chemistry Letters, 18*, 1397-1401.

In: Bioactive Heterocycles
Editors: K. L. Ameta, R. P. Pawar and A. J. Domb

ISBN: 978-1-62257-636-4
© 2013 Nova Science Publishers, Inc.

Chapter 10

Bioactivities and Synthesis of Indoloquinoline Alkaloids: Cryptolepine, Isocryptolepine and Neocryptolepine

Prakash T. Parvatkar and Santosh G. Tilve
Department of Chemistry, Goa University, Taleigao Plateau, Goa, India

Abstract

Tetracyclic heteroaromatic compounds with linear or angularly fused indoloquinoline ring systems constitute an important structural moiety in natural products. The indoloquinoline alkaloids cryptolepine, isocryptolepine (cryptosanguino-lentine) and neocryptolepine (cryptotackieine) are three of the thirteen characterized alkaloids isolated from the roots of the West African shrub *Cryptolepis sanguinolenta.* They are important due to their wide spectrum of biological properties. These alkaloids possess an interesting anti-plasmodial activity against chloroquine-resistant *P. falciparum* and are known to intercalate in the DNA double helix resulting in dramatic changes in DNA conformation, thus inhibits the DNA replication and transcription. This chapter describes the brief biological activities and various important synthetic methods developed for the preparation of cryptolepine **1**, isocryptolepine **2** and neocryptolepine **3** during recent years.

Keywords: Alkaloid, cryptolepine, heteroaromatic, indoloquinoline, isocryptolepine, and neocryptolepine.

1. Introduction

In recent years, indoloquinoline alkaloids have received considerable attention due to their striking biological properties such a DNA intercalating [1] and anti-malarial properties

[2-4]. Malaria is one of the important infectious disease caused by the *Plasmodium* parasite (the most serious infections are due to *Plasmodium falciparum*). The roots of the West African plant *Cryptolepis sanguinolenta* [5-18] have long been used in folk medicine for the treatment of infectious diseases, amoebiasis, fever and malaria. Since 1974, a decoction of this plant is being used in the clinical therapy of rheumatism, urinary tract infections, malaria and other diseases [19 - 22]. This plant has been found to be a rich source of indoloquinoline alkaloids isolated through the efforts of several research groups [5-18].

Pyrrole ring of indole can be fused to pyridine ring of quinoline in four possible ways *i.e.* indolo[3,2-*b*]quinoline I, indolo[2,3-*b*]quinoline II, indolo[3,2-*c*]quinoline III and indolo[2,3-*c*]quinoline IV (Figure 1). Out of the four isomeric ring systems of indoloquinolines, only three I, II and III are found in natural alkaloids.

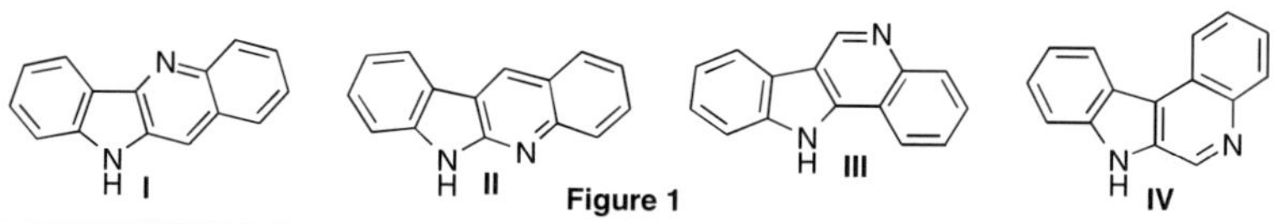

Figure 1

Cryptolepine **1**, isocryptolepine **2** and neocryptolepine **3** which are methyl derivatives of **I**, **III** and **II** are naturally occuring indoloquinolines isolated from the roots of the West African plant *Cryptolepis sanguinolenta* (Figure 2). A compound with indolo[2,3-*c*]quinoline **IV** ring system has not been found in nature.

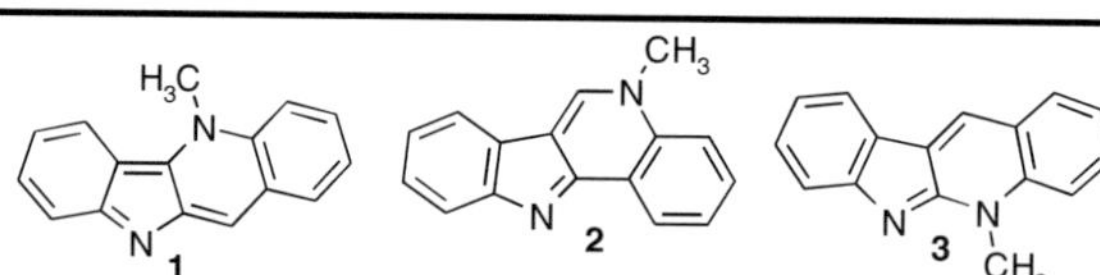

Figure 2

Cryptolepine 1 is a rare example of a natural product whose synthesis was reported prior to its isolation from nature. It was synthesized in 1906 by Fichter and Boehringer [23] for use as a possible dye while its isolation was reported in 1929 by Clinquart [24] from *Cryptolepis triangularis*. Gellert *et* al. [5] reported the isolation of cryptolepine from the roots of *Cryptolepis sanguinolenta* in 1951 and thereafter, several alkaloids have been isolated from this plant.

In 1995, another alkaloid was isolated from this plant by two independent research groups and named isocryptolepine by Pousset *et al.* [9] and cryptosanguinolentine by Sharaf *et al.* [25]. Isocryptolepine 2 is an angularly-fused alkaloid with indolo[3,2-*c*]quinoline ring system (III) whereas cryptolepine 1 is a linearly-fused alkaloid with indolo[3,2-*b*]quinoline ring system (I).

Subsequently in 1996, another linearly-fused alkaloid with indolo[2,3-*b*]quinoline ring system (II) was isolated from the same plant by two independent research teams and was named neocryptolepine by Pieter's group [8] and cryptotackieine by Schiff's group [25].

2. Biological Activities

Chemically, the tetracyclic heteroaromatic compounds 1, 2, & 3 are isomeric indoloquinolines. But more importantly, the two linearly-fused alkaloids cryptolepine 1 and neocryptolepine 3 and an angularly-fused alkaloid isocryptolepine 2 possesses an interesting anti-plasmodial activity [2-4, 26] against chloroquine-resistant *P. falciparum* and are known to intercalate in the DNA double helix resulting in dramatic changes in DNA conformation. Thus, they inhibit the DNA replication and transcription [1]. Cryptolepine has been used as a lead compound for the synthesis of anti-plasmodial agents [27–28]. Initially, neocryptolepine was reported to show an activity comparable to cryptolepine [2-3]. Recent studies have shown that it is seven times less active against the chloroquine-resistant *P. falciparum* Ghana-strain [29]. The strength and mode of binding of these alkaloids to DNA have been investigated by spectroscopy and X-ray analysis [30-31]. Cryptolepine binds 10-fold more tightly to DNA than other alkaloids and proves to be much more cytotoxic toward B16 melanoma cells [30]. It has also been reported that these compounds and their methyl derivatives display wide range of biological properties such as anti-muscarinic, antibacterial, anti-viral, anti-micotic, anti-hyperglycemic and cytotoxic properties *in vitro* and anti-tumor activity *in vivo* [18, 22, 32-35].

Due to the wide range of biological activities, these alkaloids have been targets of synthetic chemists in recent years.

3. Synthesis

3.1. Synthesis of Cryptolepine

One of the interesting methods describes the insertion of amino group at the electron rich 3-position of indole by the use of hypervalent iodine reagents. The key intermediate 1,3-bis-(2-nitrophenyl)propan-2-one **5** required for the synthesis of cryptolepine was readily prepared from 2-nitrophenyl acetic acid **4** by reactions with DCC in the presence of DMAP. The reduction of nitro groups with Fe powder followed by oxidative cyclization and subsequent *N*-methylation afforded the **1** (Scheme 1) [36].

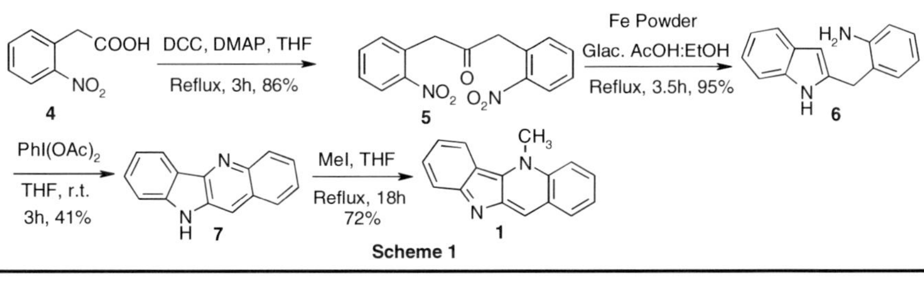

Scheme 1

Amido ketone 10 was prepared by directed lithiation of 8 followed by addition of *o*-nitrobenzaldehyde 9, subsequent oxidation of the resultant alcohol with MnO_2, reduction of nitro group using catalytic hydrogenation and *N*-benzoylation using benzoylchloride. The cyclized product 11 was obtained from 10 in 80% yield by initial 1,4-addition of amido anion followed by expulsion of the phenyl sulfonate. *N*-deprotection of 11 and subsequent reaction with $POCl_3$ followed by catalytic hydrogenolysis of the resultant chlorinated compound 12 afforded quindoline 7 in good yield (Scheme 2) [37].

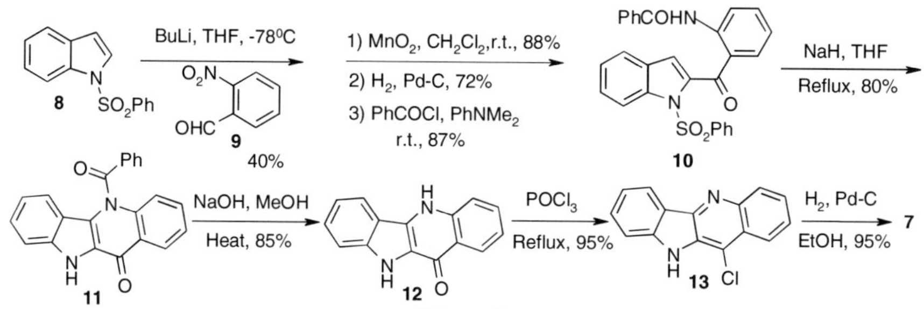

Scheme 2

Treatment of anthranilonitrilo derivative 14 with phenacyl bromide 15 in presence of K_2CO_3 gave the intermediate 16. Nucleophilic denitrocyclization [38] of 16 with NaH gave the required tetracyclic compound 12 which on treatment with PCl_5 afforded the corresponding chloro compound 13 in 70% yield [39] (Scheme 3).

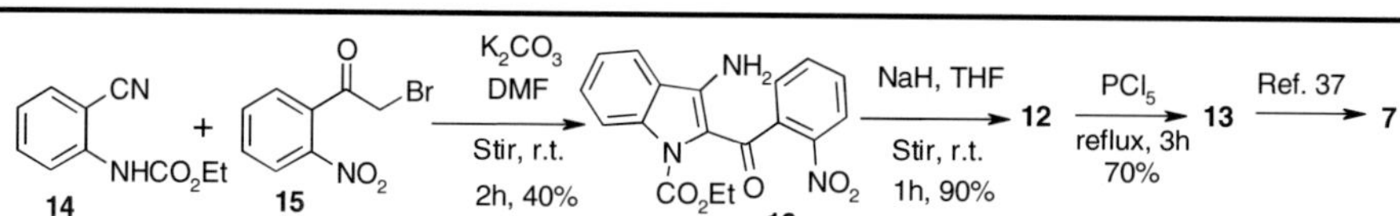

Scheme 3

A general method for the synthesis of various 2-substituted cryptolepines is described below which involves regioselective thermal cyclization and reductive cyclization using triethyl phosphite as the key steps (Scheme 4) [40].

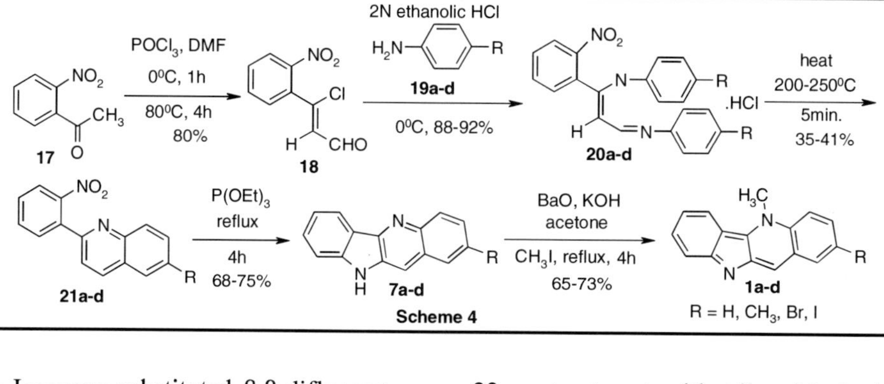

Scheme 4

o-Isocyano-substituted β,β-difluorostyrenes 22 on treatment with tributyltin hydride in the presence of a catalytic amount of AIBN and subsequent Pd-catalyzed coupling reaction with 23 afforded the 2,4-disustituted-3-fluoroquinolines 24 which, on cyclization followed by methylation furnished the 11-*n*-butyl and 11-isopropyl cryptolepines 1b-c (Scheme 5) [41].

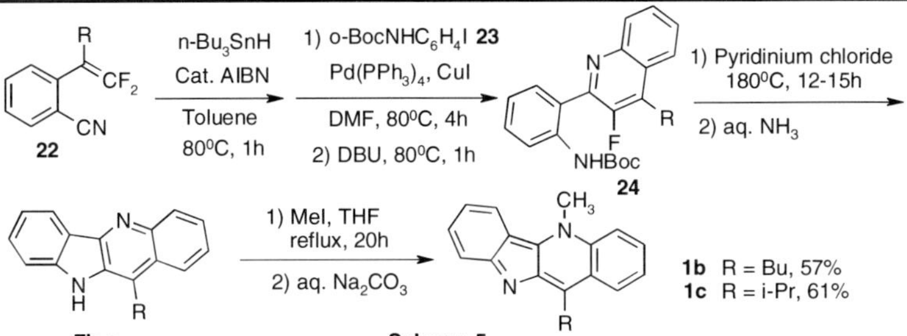

Scheme 5

Pd-catalyzed cross-coupling reaction between boronic acid 26 and 3-fluoro-2-iodoquinoline 25 using the Suzuki procedure [42 - 45] afforded the biaryl compound 27. The fluorine displacement on quinoline ring and cyclization of biaryl compound 27 in boiling pyridinium chloride [46] gave quindoline 7 in 83% yield (Scheme 6) [47].

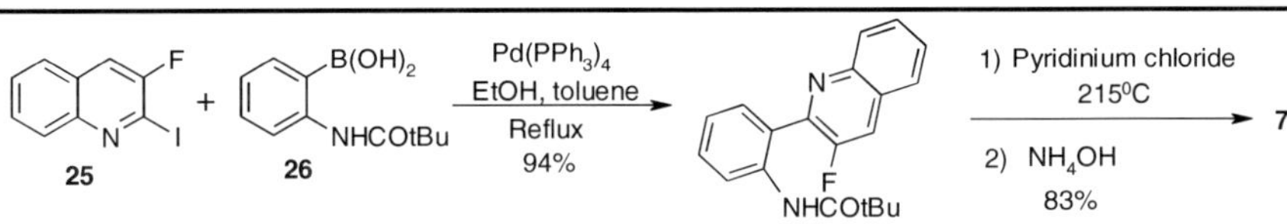

Scheme 6

Regioselective coupling reaction of 2,3-dibromoquinoline [48] 28 with 26 taking into consideration the fact that the α-heteroaryl halogen atom is more reactive than the β-halogen atom [49], gave *N*-pivaloyl-2-(2'-anilino)-3-bromoquinoline 29. Hydrolysis of 29 afforded the free amine 30 which underwent cyclization when heated at 200-220^{0}C in presence of pyridinium hydrochloride to give quindoline 7 (Scheme 7) [50].

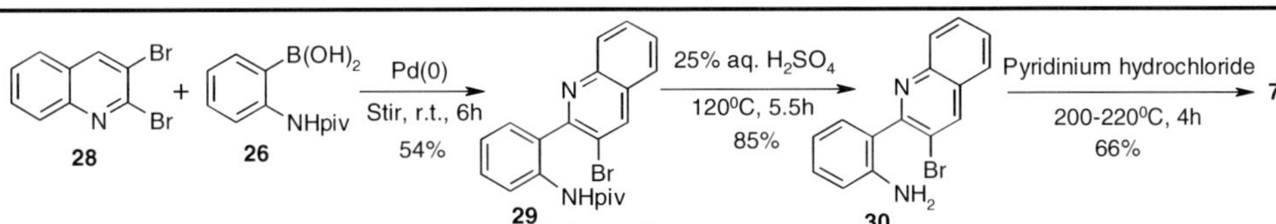

Scheme 7

Anilinoquinoline 32 was obtained by nucleophilic substitution of 3-bromoquinoline 31 with aniline 19 which was subjected to photochemical cyclization. Interestingly, both linearly-fused and angularly-fused products 7 and 33 were obtained. Methylation of 7 with dimethyl sulfate gave cryptolepine 1 (Scheme 8) [51].

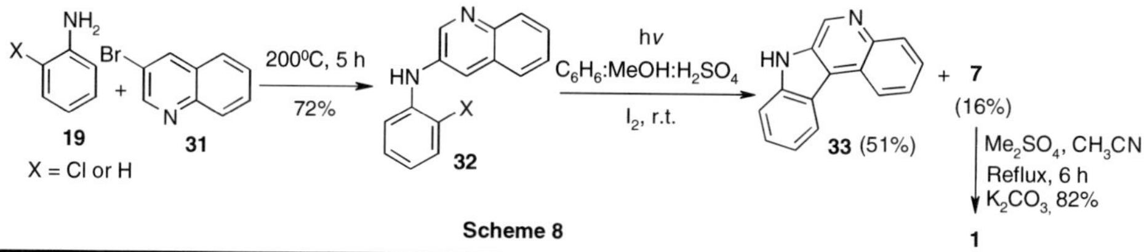

Scheme 8

N-Arylation of 3-aminoquinoline 34 with triphenylbismuth diacetate using metallic copper, followed by oxidative cyclization of the resultant anilinoquinoline 32 using palladium acetate furnished 10*H*-indolo[3,2-*b*]quinoline 7 (Scheme 9) [52].

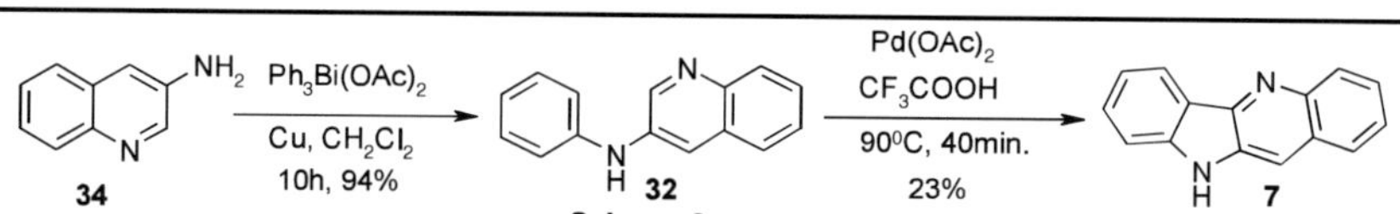

Scheme 9

3.2. Synthesis of Isocryptolepine (Cryptosanguinolentine)

The Stille coupling reaction of 2-tributylstannyl-*N*-protected indole 35 with 2-iodonitrobenzene 36 gave 2-(*o*-nitrophenyl)indole 37 which on reduction, *N*-formylation and *N*-methylation afforded the desired formamide 38. Final ring closure was achieved by refluxing compound 38 in ethanol in the presence of sulfuric acid to give isocryptolepine 2 (Scheme 10) [53].

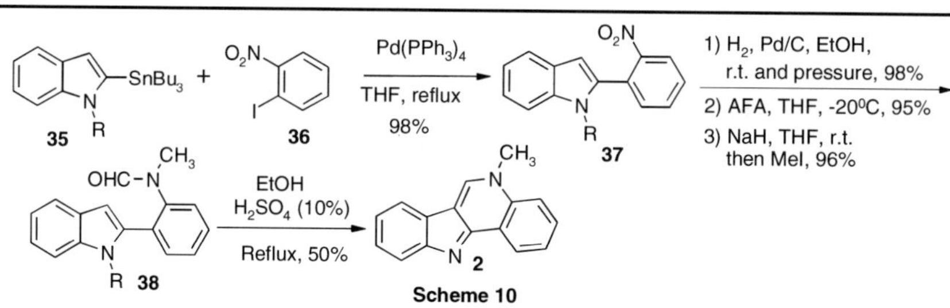

Scheme 10

Reaction of 39 with *N*-methyl aniline 40 in acetonitrile afforded acid 41 along with minor amounts of isomeric acid. The decarboxylative Heck-type cyclization of 41 was achieved using $Pd(OCOCF_3)_2$ and Ag_2CO_3 to give the compound 42 in 71% yield which on treatment with $LiAlH_4$ in hot dioxane furnished 2 in 98% yield (Scheme 11) [54].

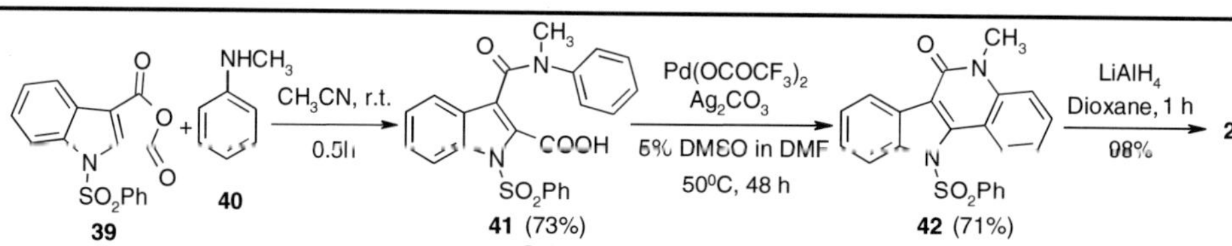

Scheme 11

Schiff's base 44, obtained by heating indole-3-carboxaldehyde 43 with aniline 19 in acetic acid, when irradiated at 253.7nm, underwent cyclization *via* initial photo-isomerization of the Schiff's base 44 from *E*- to *Z*-isomer followed by conrotatory ring closure and subsequent oxidation by iodine to give 11*H*-indolo[3,2-*c*]quinoline 45 (Scheme 12) [55].

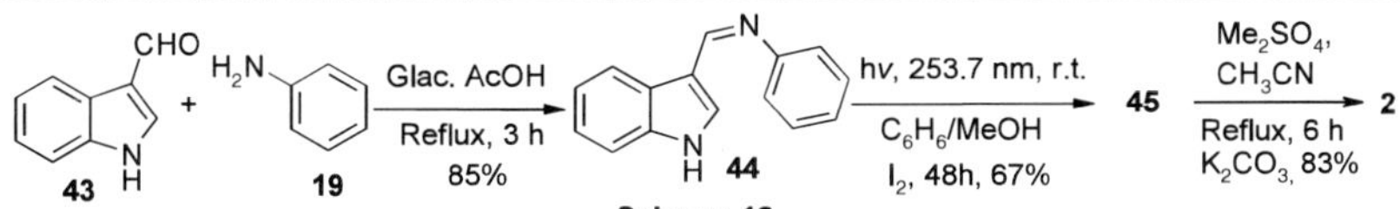

Scheme 12

Treatment of 1-phenylsulfonyl indole with LDA and reaction with cyclohexanone furnished the alcohol 46 which on deprotection and dehydration gave compound 47. Subsequent Vilsmeyer Haack formylation followed by oximination, intramolecular thermal electrocyclization, aromatization and *N*-methylation afforded 2 (Scheme 13) [56].

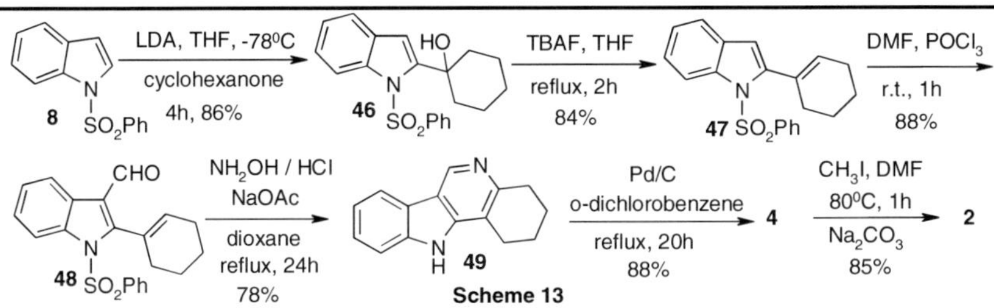

Scheme 13

The reaction of 3-bromoquinoline 31 with *N*-pivaloylaminophenyl boronic acid 26 in presence of Pd(0) catalyst afforded the desired biaryl compound 50 which, on hydrolysis with sulfuric acid, gave amine 51. The compound 51 was converted to azide 52 which, on nitrene insertion, gave exclusively the indolo[3,2-*c*]quinoline 45. Regioselective methylation on quinoline nitrogen using dimethyl sulfate yielded the isocryptolepine 2 (Scheme 14) [57].

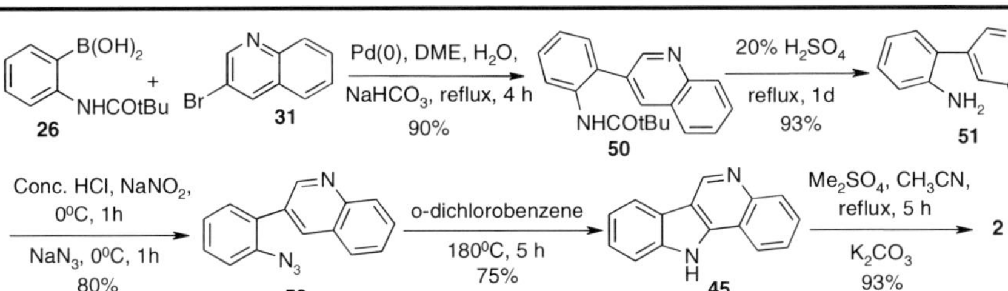

Scheme 14

Fischer indole reaction of 54 with 53 gave indoloquinoline 55 which exists predominantly in the hydroxy form 56 as confirmed by IR. The enol 56, when refluxed in $POCl_3$, afforded the corresponding chloride 57 which on catalytic hydrogenation yielded isocryptolepine 2 (Scheme 15) [58].

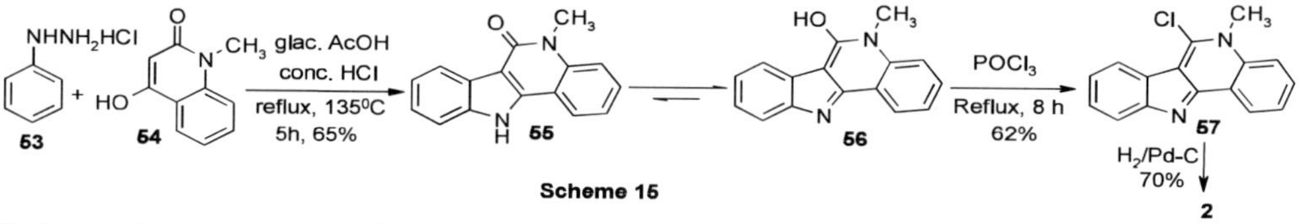

Scheme 15

The Pd-catalyzed 'amination-arylation' approach for the synthesis of isocryptolepine is described below (Scheme 16) [59]. This approach consists of two consecutive Pd-catalyzed reactions – a selective Buchwald-Hartwig [60 – 65] reaction of 2-chloroaniline 58 with 4-chloroquinoline 59 followed by an intramolecular arylation [66–68] of the resulting compound 60 to afford the 11*H*-indolo[3,2-*c*]quinoline 45.

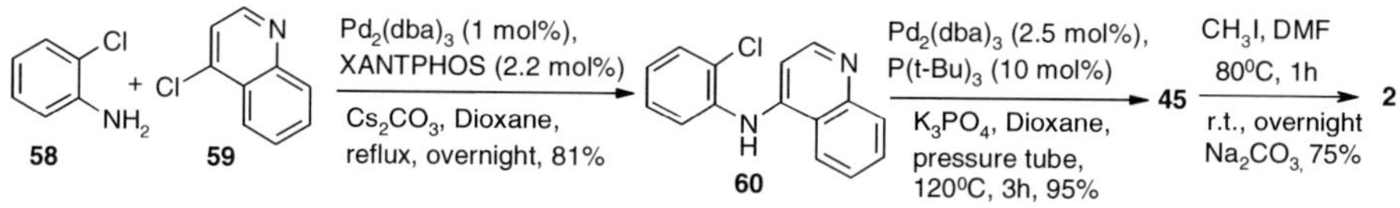

Scheme 16

Photocyclization of the anilinoquinolines 61, obtained by heating a mixture of 19 and 59 afforded 45. Regioselctive methylation of quinoline nitrogen of 45 yielded 2 (Scheme 17) [51].

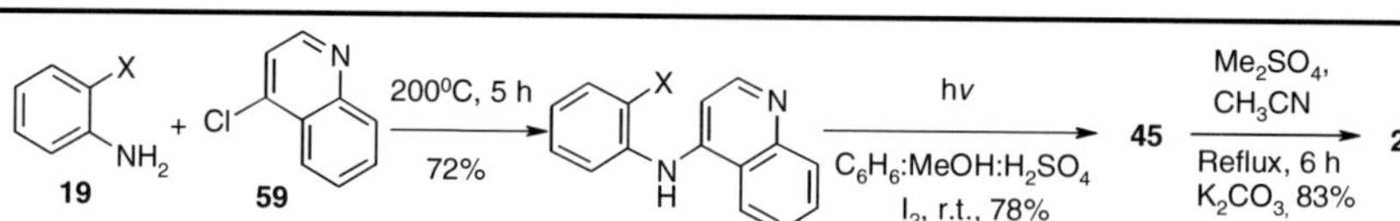

X = Cl, H, OH or OMe

Scheme 17

3.3. Synthesis of Neocryptolepine (Cryptotackieine)

Alkylation of indole with 2-nitrobenzyl bromide 63 yielded compound 64 which, on treatment with $SnCl_2 \cdot 2H_2O$, afforded 67 in 35% yield along with other two compounds 65 and 66 in 27% and 10%, respectively (Scheme 18) [69].

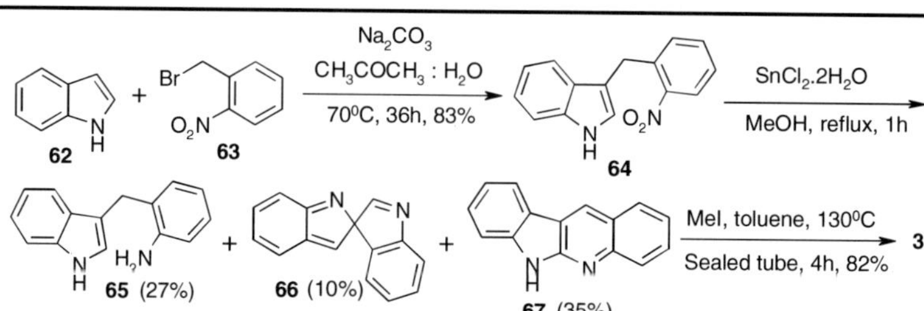

Scheme 18

Reaction of 68 with cyclohexanone 69 in presence of NaH underwent conjugate addition-elimination to give the corresponding adduct 70 which on heterocyclization with ammonium acetate yielded compound 71. Dethiomethylation of 71 with Raney Nickel and subsequent dehydrogenation with DDQ afforded 67 (Scheme 19) [70].

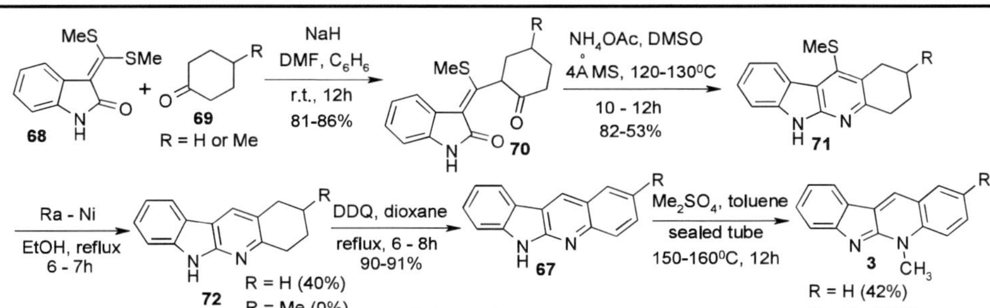

Scheme 19

A short method for the synthesis of neocryptolepine is achieved using Wittig reaction and one-pot reduction-cyclization-dehydration approach as the key steps (Scheme 20) [71].

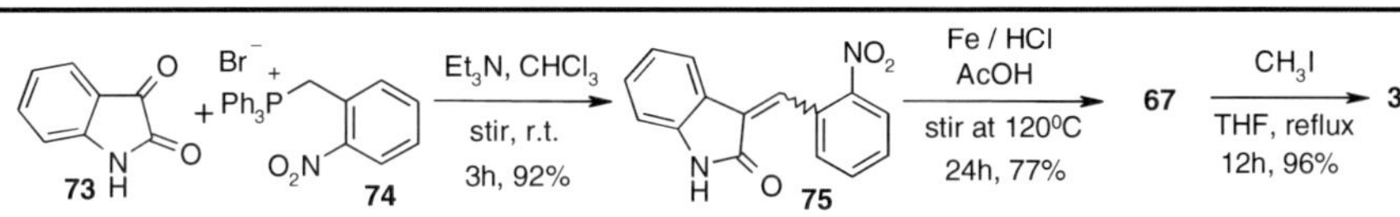

Scheme 20

One-pot synthesis of series of novel 6*H*-indolo[2,3-*b*]quinolines *via* sequential imination, nucleophilic addition and annulation is described using iodine as a catalyst (Scheme 21) [72].

Scheme 21

Heating the chloroindole derivative 76 with excess of *N*-methylaniline at reflux temperature gave the neocryptolepine 3 in one-pot (Scheme 22) [73].

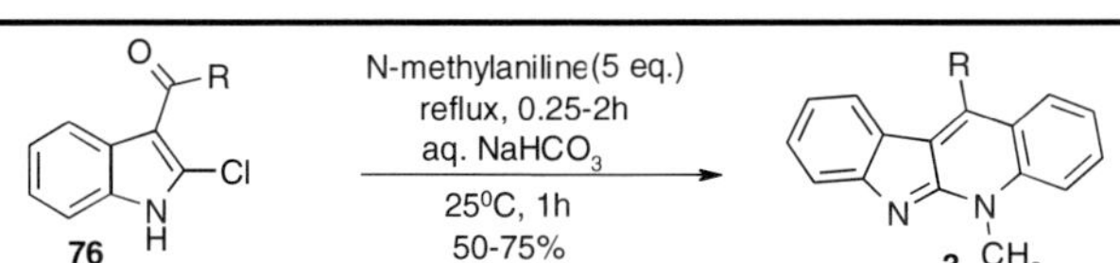

Scheme 22

The synthesis of neocryptolepine and their analogues is described starting from secondary amino benzaldehydes and indoles using p-TSA *via* indole annulation followed by oxidation in one-pot (Scheme 23) [74].

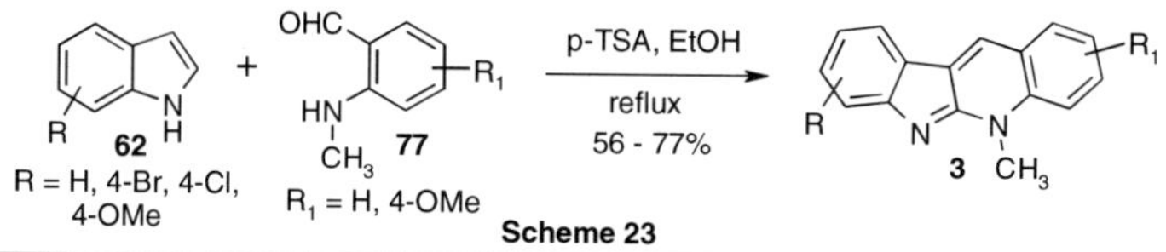

Scheme 23

Metal-free domino reaction of *N*-benzyl indole with 1-(2-tosylaminophenyl)ketone or 2-tosylaminobenzaldehyde using iodine afforded *N*-benzyl-indolo[2,3-*b*]quinolines. Debenzylation using $AlCl_3$ followed by regioselective methylation on quinoline nitrogen yielded cryptolepine and 11-methyl neocryptolepine in good overall yield (Scheme 24) [75].

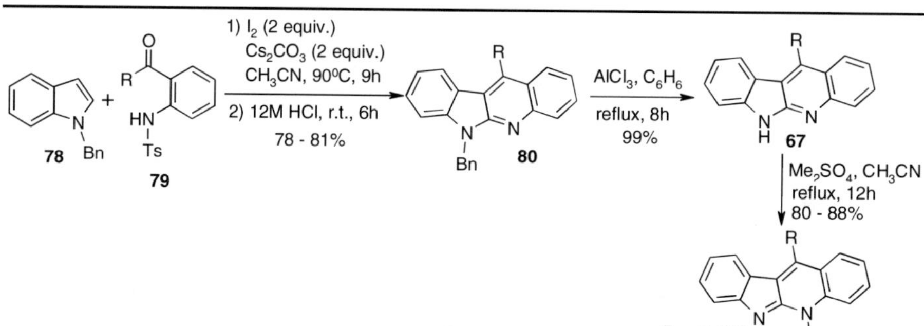

Scheme 24

3-Bromo-1*H*-2-quinoline 82 was prepared from 3-bromo-quinoline 31 *via* its *N*-oxide 81 which, on treatment with methyl iodide, gave *N*-methyl compound 83. Coupling reaction of 83 with 26 in the presence of Pd(0) catalyst afforded the biaryl compound 84. Hydrolysis of 84 with sulfuric acid followed by cyclization using $POCl_3$ furnished neocryptolepine 3 (Scheme 25) [57].

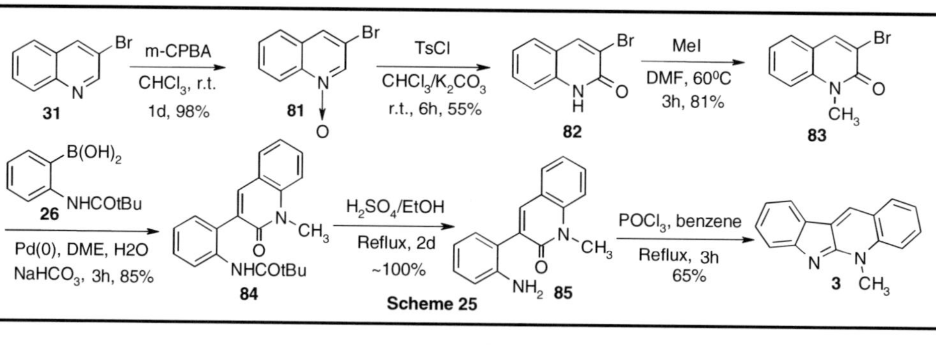

Scheme 25

The key intermediate 3-(2-aminophenyl)-quinolin-2-amine 89 was prepared by the condensation of *o*-aminobenzaldehyde with *o*-nitroaryl acetonitrile followed by reduction with Zn/AcOH. Treatment of 89 with $NaNO_2$/HCl followed by *N*-methylation furnished 3 (Scheme 26) [76].

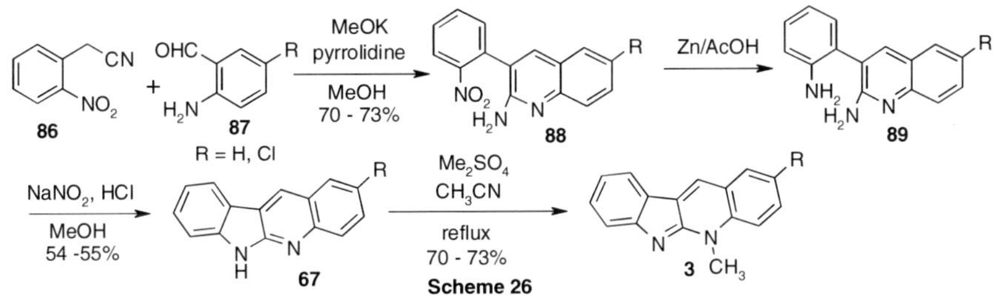

Scheme 26

Heating a mixture of 19 and 90 at 200^{0}C yielded the anilinoquinoline 91 which on photocyclization and subsequent methylation at quinoline nitrogen gave 3 (Scheme 27) [51].

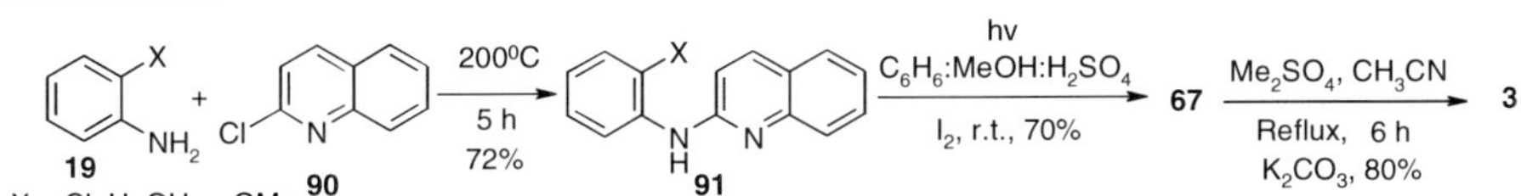

Scheme 27

Microwave irradiation of benzotriazoles 92 and 2-chloroquinoline 90 afforded the respective triazoles 93. The subsequent microwave irradiation of the resultant triazoles 93 in presence of acid gave the 6*H*-indolo[2,3-*b*]quinolines 67 (Scheme 28) [77].

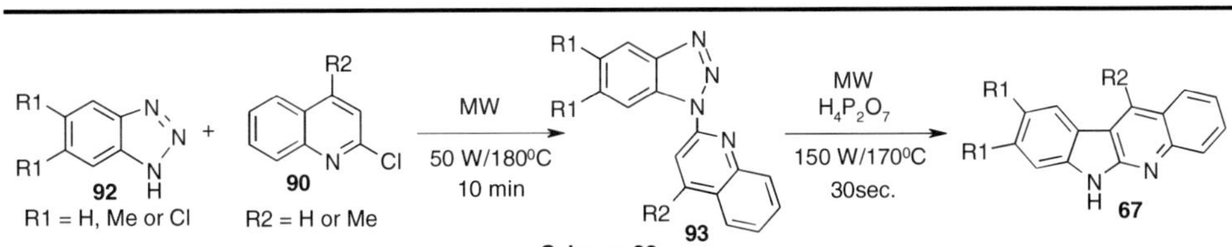

Scheme 28

2-Nitrophenyl acetic acid 4 on treatment with DCC in the presence of DMAP yielded the key intermediate 1,3-bis-(2-nitrophenyl)propan-2-one 5. Bromination of 5 followed by Favorskii rearrangement of the resultant bromo compound 94 and subsequent reduction-cyclization using Fe powder and *N*-methylation furnished neocryptolepine 3 (Scheme 29) [36].

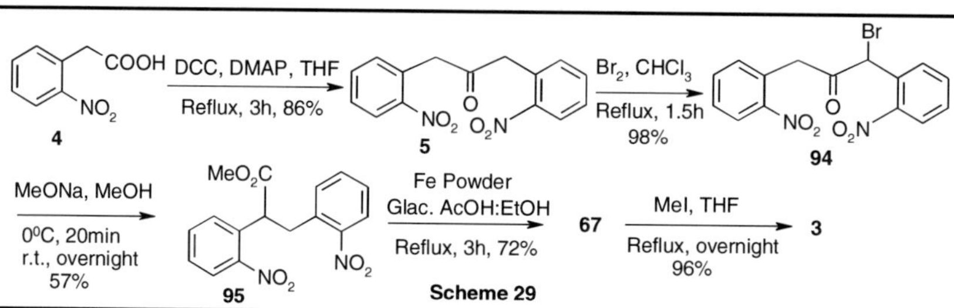

Scheme 29

Condensation of 2-nitrobenzaldehyde 9 with 2-nitrophenyl acetic acid 4 in refluxing acetic anhydride in the presence of Et_3N gave the α,β-unsaturated acid which on esterification afforded the ester 96 in good yield. Reduction with Fe powder furnished the 6*H*-indolo[2,3-*b*]quinoline 67 *via* double reduction-double cyclization reactions in one-pot (Scheme 30) [78].

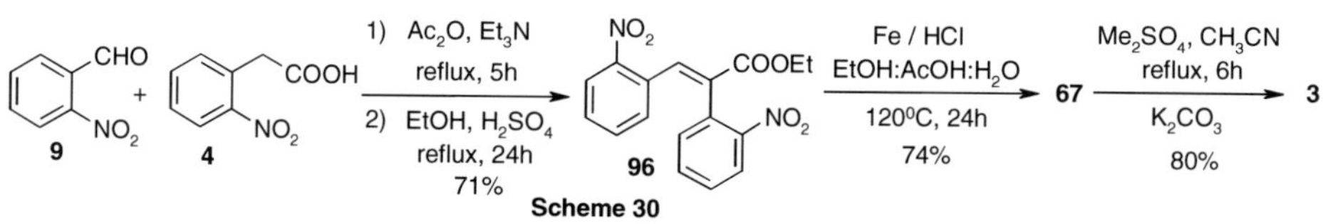

Scheme 30

Aza-Wittig reaction of the iminophosphorane 97 with phenyl isocyanate 98 yielded carbodiimide 99 and triphenylphosphine oxide which, without purification, was subjected to thermal treatment to give 67 and 2-anilinoquinoline 100 in 19% and 40% yield, respectively (Scheme 31) [79].

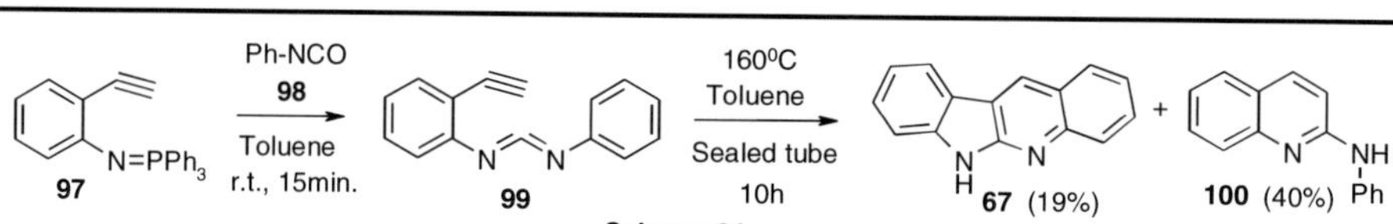

Scheme 31

Conclusion

Indoloquinoline alkaloids show the remarkable biological activities and constitute the lead scaffold for drug development. The synthesis of indoloquinoline alkaloids forms the important field of chemistry. Several synthetic strategies are now available which provides flexibility for introducing various substituents into the ring system. This chapter presents a collection of highly interesting and useful methods for the synthesis of different types of indoloquinoline alkaloids i.e. cryptolepine, isocryptolepine and neocryptolepine.

References

[1] Molina, A., Vaquero, J.J., Garcia-Navio, J.L., Alvarez-Builla, J., de Pascual-Teresa, B., Gago, F., Rodrigo, M.M., Ballesteros, M. Synthesis and DNA Binding Properties of γ-Carbolinium Derivatives and Benzologues. *J. Org. Chem.* 1996, *61*, 5587.

[2] Cimanga, K., De Bruyne, T., Pieters, L., Vlietinck, A.J., Turger, C.A. In Vitro and in Vivo Antiplasmodial Activity of Cryptolepine and Related Alkaloids from Cryptolepis sanguinolenta. *J. Nat. Prod.* 1997, *60*, 688.

[3] Paulo, A., Gomes, E.T., Steele, J., Warhurst, D.C., Houghton, P.J. Antiplasmodial Activity of Cryptolepis sanguinolenta Alkaloids from Leaves and Roots. *Planta Med.* 2000, *66*, 30.

[4] Miert, S.V., Hostyn, S., Maes, B.U.M., Cimanga, K., Brun, R., Kaiser, M., Matyus, P., Dommisse, R., Lemiere, G., Vlietinck, A., Pieters, L. Isoneocryptolepine, a Synthetic Indoloquinoline Alkaloid, as an Antiplasmodial Lead Compound. *J. Nat. Prod.* 2005, *68*, 674.

[5] Gellert, E., Hamet, R., Schlitter, E. Die Konstitution des Alkaloids Cryptolepin. *Helv. Chim. Acta* 1951, *34*, 642.

[6] Dwuma-Badu, D., Ayim, J.S.K., Fiagbe, N.Y.Y., Knapp, J.E., Schiff, P.L. Jr., Slatkin, D.J.J. Constituents of West African medicinal plants XX: Quindoline from Cryptolepis sanguinolenta. *Pharm. Sci.* 1978, *67*, 433.

[7] Ablordeppey, S.D., Hufford, C.D., Bourne, R.F., Dwama-Badu, D. 1H-NMR and 13C-NMR Assignments of Cryptolepine, A 3:4-Benz-δ-carboline Derivative Isolated from Cryptolepis sanguinolenta. *Planta Med.* 1990, *56*, 416.

[8] Cimanga, K., De Bruyne, T., Pieters, L., Claeys, M., Vlietinck, A. New alkaloids from *Cryptolepis sanguinolenta. Tetrahedron Lett.* 1996, *37*, 1703.

[9] Pousset, J.-L.; Martin, M.-T.; Jossang, A.; Bodo, B. Isocryptolepine from *Cryptolepis sanguinolenta. Phytochemistry* 1995, *39*, 735.

[10] Tackie, A.N., Sharaf, M.H.M., Schiff, P.L. Jr., Boye, G.L., Crouch, R.C., Martin, G.E. Assignment of the Proton and Carbon NMR Spectra of the Indoloquinoline Alkaloid Cryptolepine. *J. Heterocycl. Chem.* 1991, *28*, 1429.

[11] Spitzer, T.D., Crouch, R.C., Martin, G.E., Sharaf, M.H.M., Schiff, P.L. Jr., Tackie, A.N., Boye, G.L. Total Assignment of the Proton and Carbon NMR Spectra of the Alkaloid Quindoline - Utilization of HMQC-TOCSY to Indirectly Establish Protonated Carbon-Protonated Carbon Connectivities. *J. Heterocycl. Chem.* 1991, *28*, 2065.

[12] Tackie, A.N., Boye, G.L., Sharaf, M.H.M., Schiff, P.L. Jr., Crouch, R.C., Spitzer, T.D., Jhonson, R.L., Dunn, J., Minick, D., Martin, G.E. Cryptospirolepine, a Unique Spiro-non-acyclic Alkaloid Isolated from Cryptolepis sanguinolenta. *J. Nat. Prod.* 1993, *56*, 653.

[13] Crouch, R.C., Davis, A.O., Spitzer, T.,D., Martin, G.E., Sharaf, M.H.M., Schiff, P.L. Jr., Phoebe, C.H. Jr., Tackie, A.N. Elucidation of the Structure of Quindolinone, a Minor Alkaloid of *Cryptolepis sanguinolenta*: Submiligram ^{1}H- ^{13}C and ^{1}H-^{15}N Heteronuclear Shift Correlation Experiments Using Micro Inverse-Detection. *J. Heterocycl. Chem.* 1995, *32*, 1077.

[14] Paulo, A., Gomes, E.T., Hougton, P.J. New Alkaloids from Cryptolepis sanguinolenta. *J. Nat Prod.* 1995, *58*, 1485.

[15] Fort, D.M., Litvak, J., Chen, J.L., Lu, Q., Phuan, P.W., Cooper, R., Bierer, D.E. Isolation and Unambiguous Synthesis of Cryptolepinone: An Oxidation Artifact of Cryptolepine. *J. Nat. Prod.* 1998, *61*, 1528.

[16] Hadden, C.E., Sharaf, M.H.M., Guido, J.E., Robins, R.H., Tackie, A.N., Phoebe, C.H. Jr., Schiff, P.L., Jr., Martin, G.E. 11-Isopropylcryptolepine: A Novel Alkaloid Isolated from *Cryptolepis sanguinolenta* Characterized Using Submicro NMR Techniques. *J. Nat. Prod.* 1999, *62*, 238.

[17] Blinov, K., Elyashberg, M., Martirosian, E.R., Molodtsov, S.G., Williams, A.J.A.J., Tackie, A.N., Sharaf, M.H.M., Schiff, P.L. Jr., Crouch, R.C., Martin,G.E., Hadden, C.E., Guido, J.E., Mills, K.A. Quindolinocryptotackieine: the elucidation of a novel indoloquinoline alkaloid structure through the use of computer-assisted structure elucidation and 2D NMR. *Magn. Reson. Chem.* 2003, *41*, 577.

[18] Cimanga, K., DeBruyne, T., Lasure, A., Poel, B.V., Pieters, L., Claeys, M., Berghe, D.V., Vlietinck, A.J. *In Vitro* Biological Activities of Alkaloids from *Cryptolepis sanguinolenta*. *Planta Med.* 1996, *62*, 22.

[19] Bever, B.E.P.O. *Medicinal Plants in Tropical West Africa,* Cambridge University Press: Cambridge, 1986; *131*, 18.

[20] Boakye-Yiadom, K.Q.J. Antimicrobial Properties of Some West African Plants. *Crude Drug Res.* 1979, *17*, 78.

[21] Boye, G.L., Ampofo, O. *Proceeding of the First International Symposium on Cryptolepine*, Boakye-Yiadom, K.; Bamgbose, S. O. A. University of Science and Technology: Ghana, 1983; 37.

[22] Bierer, D.E., Fort, D.M., Mendez, C.D., Luo, J., Imbach, P.A., Dubenko, L.G., Jolad, S.D., Gerber, R.E., Litvak, J., Lu, Q., Zhang, P., Reed, M. J., Waldeck, N., Bruening, R.C., Noamesi, B.K., Hector, R.F., Carlson, T.J., King, S.R. Ethnobotanical-Directed Discovery of the Antihyperglycemic Properties of Cryptolepine: Its Isolation from *Cryptolepis sanguinolenta*, Synthesis, and *in Vitro* and *in Vivo* Activities. *J. Med. Chem.* 1998, *41*, 894.

[23] Fichter, F., Boehringer, R. Ueber Chindolin. (Over Quindoline). *Chem. Ber.* 1906, *39*, 3932.

[24] Clinquart, E. Sur la composition chimique de *Cryptolepis triangularies*. Plante congolaise. (On the Chemical Composition of *Cryptolepis triangularies*. Plants from the (Belgian) Congo). *Bull Acad. R. Med. Belg.* 1929, *9*, 627.

[25] Sharaf, M.H.M., Schiff, P.L. Jr., Tackie, A.N., Phoebe, C.H. Jr., Martin, G.E. Two New Indoloquinoline Alkaloids from *Cryptolepis sanguinolenta*: Cryptosanguinolentine and Cryptotackieine. *J. Heterocycl. Chem.* 1996, *33*, 239.

[26] Grellier, P., Ramiaramanana, L., Millerioux, V., Deharo, E., Schrevel, J., Frappier, F., Trigalo, F., Bodo, B., Pousset, J.-L. Antimalarial activity of Cryptolepine and Isocryptolepine, Alkaloids Isolated from *Cryptolepis sanguinolenta*. *Phytother. Res.* 1996, *10*, 317.

[27] Wright, C.W., Addae-Kyereme, J., Breen, A.G., Brown, J.E., Cox, M.F., Croft, S.L., Gokcek, Y., Kendrick, H., Phillips, R.M., Pollet, P.L. Synthesis and Evaluation of Cryptolepine Analogues for Their Potential as New Antimalarial Agents. *J. Med. Chem.* 2001, *44*, 3187.

[28] Arzel, E., Rocca, P., Grellier, P., Labaeid, M., Frappier, F., Gueritte, F., Gaspard, C., Marsais, F., Godard, A., Queguiner, G. New Synthesis of Benzo-δ- carbolines,

Cryptolepines, and Their Salts: *In Vitro* Cytotoxic, Antiplasmodial, and Antitrypanosomal Activities of δ-Carbolines, Benzo-δ-carbolines, and Cryptolepines. *J. Med. Chem.* 2001, *44*, 949.

[29] Jonckers, T.H.M., van Miert, S., Cimanga, K., Bailly, C., Colson, P., De Pauw-Gillet, M.-C., Van den Heuvel, H., Claeys, M., Dommisse, R., Lemiere, G.L.F., Vlietinck, A., Pieters, L. Synthesis, Cytotoxicity, and Antiplasmodial and Anti-trypanosomal Activity of New Neocryptolepine Derivatives. *J. Med. Chem.* 2002, *45*, 3497.

[30] Dassoneville, L., Bonjean, K., De Pauw-Gillet, M.C., Colson, P., Houssier, C., Quetin-Leclercq, J., Angenot, L., Bailly, C. Stimulation of Topoisomerase II-Mediated DNA Cleavage by Three DNA-Intercalating Plant Alkaloids: Cryptolepine, Matadine, and Serpentine. *Biochemistry* 1999, *38*, 7719.

[31] Lisgarten, J.N., Pous, J., Coll, M., Wright, C.W., Aymami, J. Crystallization and preliminary X- ray analysis of the antimalarial and cytotoxic alkaloid cryptolepine complexed with the DNA fragment d(CCTAGG)2. *Acta Crystallogr. D Biol. Crystallogr.* 2002, *58*, 312.

[32] Kirby, G.C., Paine, A., Warhurst, D.C., Noamese, B.K., Phillipson, J.D. *In vitro* and *in vivo* antimalarial activity of cryptolepine, a plant-derived Indoloquinoline. *Phytother. Res.* 1995, *9*, 359.

[33] Peczynska-Czoch, W., Pognan, F., Kaczmarek, L., Boratynski, J. Synthesis and Structure-Activity Relationship of Methyl-Substituted Indolo[2,3-*b*]quinolines: Novel Cytotoxic, DNA Topoisomerase II Inhibitors. *J. Med. Chem.* 1994, *37*, 3503.

[34] Cimanga, K., De Bruyne, T., Pieters, L., Totte, J., Tona, L., Kambu, K., Berghe, D.-V., Vlietinck, A.J. Antibacterial and antifungal activities of neocryptolepine, biscryptolepine and cryptoquindoline, alkaloids isolated from *Cryptolepis sanguinolenta*. *Phytomedicine* 1998, *5*, 209.

[35] Abblordeppey, S.Y., Fan, P., Clark, A.M., Nimrod, A. Probing the *N*-5 region of the indoloquinoline alkaloid, cryptolepine for anticryptococcal activity. *Bioorg. Med. Chem.* 1999, *7*, 343.

[36] Ho, T.-L., Jou, D.-G. Synthesis of Cryptolepine and Cryptoteckieine from a Common Intermediate. *Helv. Chim. Acta* 2002, *85*, 3823.

[37] Cooper, M.M., Lovell, J.M., Joule, J.A. Indole-β-nucleophilic substitution. Part 9 nitrogen nucleophiles. Syntheses of hydroxycryptolepine, cryptolepine, and quindoline. *Tetrahedron Lett.* 1996, *37*, 4283.

[38] Radl, S. Some Examples of Aromatic Nucleophilic Denitrocyclization Reactions. *Janssen Chimica Acta.* 1993, *11*, 12.

[39] Radl, S., Konvicka, P., Vachal, P. New Approach to the Synthesis of Benzofuro[3,2-b]quinolines, Benzothieno[3,2-b]quinolines and Indolo[3,2-b]quinolines. *J. Heterocycl. Chem.* 2000, *37*, 855.

[40] Dutta, B., Some, S., Ray, J.K. Thermal cyclization of 3-arylamino-3-(2-nitrophenyl)-propenal Schiff base hydrochlorides followed by triethyl phosphite mediated deoxygenation: a facile synthesis of quindolines. *Tetrahedron Lett.* 2006, *47*, 377.

[41] Mori, T., Ichikawa, J. Radical 6-endo-trig Cyclization of β,β-Difluoro-o-isocyanostyrenes: A Facile Synthesis of 3-Fluoroquinolines and Their Application to the Synthesis of 11-Alkylated Cryptolepines. *Synlett* 2007, 1169.

[42] Miyaura, N., Yanagi, T., Suzuki, A. The Palladium-Catalyzed Cross-Coupling Reaction of Phenylboronic Acid with Haloarenes in the Presence of Bases. *Synth. Commun.* 1981, 513.

[43] Godard, A., Rocca, P., Pomel, V., Thomas-dit-Dumont, L., Rovera, J.C., Thaburet, J.F., Marsais, F., Queguiner, G. Metalation in Connection with Cross-coupling Reactions. Coupling of Hindered Aryls for the Synthesis of 4-Phenylpyridines as Part of *Streptonigrin* and *Lavendamycin* Analogues. *J. Organomet. Chem.* 1996, *517*, 25.

[44] Alonso, F., Beletskaya, I.P., Yus, M. Non-conventional methodologies for transition-metal catalyzed carbon–carbon coupling: a critical overview. Part 2: The Suzuki reaction. *Tetrahedron* 2008, *64*, 3047.

[45] Bellina, F., Carpita, A., Rossi, R. Palladium Catalysts for the Suzuki Cross-Coupling Reaction: An Overview of Recent Advances. *Synthesis* 2004, 2419.

[46] Rocca, P., Marsais, F., Godard, A., Queguiner, G. Connection between metalation and cross-coupling strategies. A new convergent route to azacar-bazoles. *Tetrahedron* 1993, *1*, 49.

[47] Arzel, E., Rocca, P., Marsais, F., Godard, A., Queguiner, G. First halogendance reaction in quinoline series: application to a new synthesis of quindoline. *Tetrahedron Lett.* 1998, *39*, 6465.

[48] Buurman, H. Reactivity of 2,3-, 2,4-, and 3,4-Dibromoquinoline Towards Potassium Amide in Liquid Ammonia. *Recl. Trav. Chim. Pays-Bas* 1973, *304*, 305.

[49] Ford, A., Sinn, E., Woodward, S. Exploitation of differential reactivity of the carbon–chlorine bonds in 1,3-dichloroisoquinoline. Routes to new N,N-chelate ligands and 1,3-disubstituted isoquinolines. *J. Chem. Soc., Perkin Trans. 1* 1997, 927.

[50] Csanyi, D., Timari, G., Hajos, G. An alternative synthesis of quindoline and one of its closely related derivatives. *Synth. Commun.* 1999, *29*, 3959.

[51] Dhanabal, T., Sangeetha, R., Mohan, P.S. Heteroatom directed photoannulation: synthesis of indoloquinoline alkaloids: cryptolepine, cryptotackieine, cryptosanguinolentine, and their methyl derivatives. *Tetrahedron* 2006, *62*, 6258.

[52] Fan, P., Ablordeppey, S.Y. An Alternative Synthesis of 10*H*-Indolo[3,2-*b*]quinoline and its Selective *N*-Alkylation. *J. Heterocycl. Chem.* 1997, *34*, 1789.

[53] Murray, P. E.; Mills, K.; Joule, J. A. Synthesis of Isocryptolepine. *J. Chem. Res. (S)* 1998, 377.

[54] Miki, Y., Kuromatsu, M., Miyatake, H., Hamamoto, H. Synthesis of benzo-γ-carboline alkaloid cryptosanginolentine by reaction of indole-2,3-dicarboxylic anhydrides with anilines. *Tetrahedron Lett.* 2007, *48*, 9093.

[55] Kumar, R.N., Suresh, T., Mohan, P.S.A photochemical route to synthesize cryptosanguinolentine. *Tetrahedron Lett.* 2002, *43*, 3327.

[56] Hingane, D.G., Kusurkar, R.S. An efficient new route towards biologically active isocryptolepine and γ-carboline derivatives using an intramolecular thermal electrocyclization strategy. *Tetrahedron Lett.* 2011, *52*, 3686.

[57] Timari, G., Soos, T., Hajos, G. A Convenient Synthesis of Two New Indoloquinoline Alkaloids. *Synlett* 1997, 1067.

[58] Dhanabal, T., Sangeetha, R., Mohan, P.S. Fischer indole synthesis of the indoloquinoline alkaloid: cryptosanguinolentine. *Tetrahedron Lett.* 2005, *46*, 4509.

[59] Jonckers, T.H.M., Maes, B.U.W., Lemiere, G.L.F., Rombouts, G., Pieters, L., Haemers, A., Dommisse, R.A. Synthesis of Isocryptolepine via a Pd-Catalyzed 'Amination-Arylation' Approach. *Synlett* 2003, 615.
[60] Frost, C.G., Mendonca, P. Recent developments in aromatic heteroatom coupling reactions. *J. Chem. Soc., Perkin Trans. 1* 1998, 2615.
[61] Muci, A.R., Buchwald, S.L. Practical Palladium Catalysts for C-N and C-O Bond Formation. *Top. Curr. Chem.* 2002, *219*, 131.
[62] Buchwald, S.L., Mauger, C., Mignani, G., Scholz, U. Industrial-Scale Palladium-Catalyzed Coupling of Aryl Halides and Amines –A Personal Account. *Adv. Synth. Catal.* 2006, *348*, 23.
[63] Hartwig, J.F. Discovery and Understanding of Transition-Metal-Catalyzed Aromatic Substitution Reactions. *Synlett* 2006, 1283.
[64] Joucla, L., Djakovitch, L. Transition Metal-Catalyzed, Direct and Site-Selective N1-, C2- or C3- Arylation of the Indole Nucleus: 20 Years of Improvements. *Adv. Synth. Catal.* 2009, *351*, 673.
[65] Kienle, M., Dubbaka, S.R., Brade, K., Knochel, P. Modern Amination Reactions. *Eur. J. Org. Chem.* 2007, 4166.
[66] Ames, D.E., Bull, D. Some reactions of 3-halogenocinnolines catalyzed by palladium compounds. *Tetrahedron* 1982, *38*, 383.
[67] Iwaki, T., Yasuhara, A., Sakamoto, T. Novel synthetic strategy of carbolines via palladium-catalyzed amination and arylation reaction. *J. Chem. Soc., Perkin Trans. 1* 1999, 1505.
[68] Bedford, R.B., Cazin, C.S.J.A novel catalytic one-pot synthesis of carbazoles via consecutive amination and C–H activation. *Chem. Commun.* 2002, 2310.
[69] Sharma, S., Kundu, B. Unprecedented $SnCl_2 \cdot 2H_2O$-mediated intramolecular cyclization of nitroarenes via C–N bond formation: a new entry to the synthesis of cryptotackieine and related skeletons. *Tetrahedron Lett.* 2008, *49*, 7062.
[70] Sundaram, G.S.M., Venkatesh, C., Syam Kumar, U.K., Ila, H., Junjappa, H. A Concise Formal Synthesis of Alkaloid Cryptotackiene and Substituted 6H-Indolo[2,3-b]quinolines. *J. Org. Chem.* 2004, *69*, 5760.
[71] Parvatkar, P.T., Tilve, S.G. An effcient synthesis of indoloquinoline alkaloid—neocryptolepine (cryptotackieine). *Tetrahedron Lett.* 2011, *52*, 6594.
[72] Parvatkar, P.T., Parameswaran, P.S., Tilve, S.G. An Expeditious I_2-Catalyzed Entry into 6H- Indolo[2,3-b]quinoline System of Cryptotackieine. *J. Org. Chem.* 2009, *74*, 8369.
[73] Engqvist, R., Bergman, J. An improved synthesis of neocryptolepine. *Org. Prep. Proced. Int.* 2004, *36*, 386.
[74] Vecchione, M.K., Sun, A.X., Seidel, D. Divergent reactions of indoles with aminobenzaldehydes: indole ring-opening *vs.* annulation and facile synthesis of neocryptolepine. *Chem. Sci.* 2011, *2*, 2178.
[75] Ali, S., Li, Y.-X., Anwar, S., Yang, F., Chen, Z.-S., Liang, Y.M. One-Pot Access to Indolo[2,3-b]quinolines by Electrophile-Triggered Cross-Amination /Friedel–Crafts Alkylation of Indoles with 1-(2-Tosylaminophenyl)ketones. *J. Org. Chem.* 2012, *77*, 424.

[76] Haddadin, M.J., Bou Zerdan, R.M., Kurth, M.J., Fettinger, J.C. Efficient Syntheses of the Unknown Quinolino[2,3-*c*]cinnolines; Synthesis of Neocryptolepines. *Org. Lett.* 2010, *12*, 5502.

[77] Vera-Luque, P., Alajarin, R., Alvarez-Builla, J., Vaquero, J.J. An Improved Synthesis of α-Carbolines under Microwave Irradiation. *Org. Lett.* 2006, *8*, 415.

[78] Parvatkar, P.T., Parameswaran, P.S., Tilve, S.G. Double reductive cyclization: a facile synthesis of the indoloquinoline alkaloid cryptotackieine. *Tetrahedron Lett.* 2007, *48*, 7870.

[79] Alajarin, M., Molina, P., Vidal, A. Formal Total Synthesis of the Alkaloid Cryptotackieine (Neocryptolepine). *J. Nat. Prod.* 1997, *60*, 747.

In: Bioactive Heterocycles
Editors: K. L. Ameta, R. P. Pawar and A. J. Domb
ISBN: 978-1-62257-636-4
© 2013 Nova Science Publishers, Inc.

Chapter 11

Synthesis and Biological Evaluation of Flavones

Swapnil R. Sarda[1], Sunil R. Mirgane[1], Wamanrao N. Jadhav[2], Sunil K. Wasmatkar[3], Satish U. Deshmukh[3], Sanjay B. Ubale[3] and Abraham J. Domb[4*]

[1] Department of Chemistry, J.E.S. College, Jalna, India
[2] Organic Chemistry Synthesis Laboratory, Dnyanopasak College, Parbhani, India
[3] Department of Chemistry, Deogiri College, Aurangabad, India
[4] School of Pharmacy, Department of Medicine, The Hebrew University of Jerusalem, Jerusalem, Israel

Abstract

Flavonoids are becoming very popular because of their many health-promoting effects. Different compounds comprises the flavonoids; among them, flavones is a group of oxygen-containing heterocycles. Members of the flavone class have been associated with a wide variety of biological activities and may be useful in the treatment of certain diseases. This chapter explains the biological significance as well as synthesis and isolation of flavones.

1. Introduction

Flavonoids belong to the group of natural substances with different phenolic structures found in fruit, vegetables, grains, bark, roots, stems, flowers, tea, wine etc. [1]. These natural products were known for their beneficial health effects. More than 4000 varieties of flavonoids have been identified, which are responsible for the attractive colors of flowers, fruit and leaves [2]. Since flavonoids are naturally present in fruits, vegetables and tea, they

[*] Corresponding Email: avid@ekmd.huji.ac.il

are an integral part of the human diet [3]. Flavonoids are water soluble polyphenolic molecules, having a wide variety of phytochemicals present in the human diet. Indgested flavonoids are absorbed by the gastrointestinal tract and have been found in body fluids and tissues. Flavonoid intake may reduce the risk of several age-related chronic diseases [4].

Flavonoids are commonly known for their antioxidant activity. This effect is due to the physiological activity of flavonoids in the reduction of oxidative stress, inhibiting low-density lipoproteins (LDL) oxidation, platelet aggregation, acting as vasodilators in blood vessels, promoting fibrinolysis, acting as immunomodulators, anti-inflammatory agents and anti-tumor agents. The widespread distribution of flavonoids, their variety and relatively low toxicity as compared to other active compounds of plant mean that many animals, including humans, ingest significant quantities of it in their diet. Some epidemiologic studies reported a cardioprotective role of flavonoids against coronary heart disease [5] i.e. flavonoids may reduce mortality from coronary heart disease. Epidemiological studies have illustrated that heart diseases are inversely related to flavonoid intake. A survey of the literature provides information that flavonoids containing methylenedioxy group ($-O-CH_2-O-$) widely occur in natural plant pigments. Flavonoids have received much attention over the past decade because of their variety of potential beneficial effects.

Dietary flavonoids are probably ingested at sufficient levels to have significant pharmaceutical effects. Most of the flavonoids are very potent antioxidants [6] because they have chelate metal ions, scavenge oxygen free radicals and prevent the oxidation of low density lipoprotein (LDL) [7]. Literature survy revales that flavonoids prevent the oxidation of low-density lipoprotein, thereby reducing the risk for the development of atherosclerosis.

The most important dietary sources of flavonoids are fruits, tea and soybean. Green and black tea contains about 25% percent flavonoids. Other important sources of flavonoids are apple (quercetin) and citrus fruits (rutin and hesperidin). Flavonoids have antioxidant activities, and some of the activities attributed to flavonoids include: anti-allergic, anti-cancer, antioxidant, anti-inflammatory and anti-viral activities.

Red wine contains high levels of flavonoids, mainly quercetin and rutin. Many studies have confirmed that one or two glasses of red wine daily can protect against heart disease. Tea flavonoids have many health benefits. Tea flavonoids reduce the oxidation of low-density lipoprotein; lower the blood levels of cholesterol and triglycerides. Soy flavonoids (isoflavones) can also reduce blood cholesterol and can help to prevent osteoporosis. Soy flavonoids are also used to ease menopausal symptoms.

Other known biological properties of flavonoids include, anti-allergic activity [8], and modulation of monooxygenase activity [9]. Flavonoids can pass the blood–brain barrier as they were found to accumulate in rat brains [10]. Certain flavonoids inhibit the cellular excretion of some drugs, a process mediated by P-glycoprotein [11].

The flavonoids consist of 6 major subgroups: chalcone, flavone, flavonol, flavanone, anthocyanins and isoflavonoids. Together with carotenes, flavanoids are also responsible for the coloring of fruits, vegetables and herbs. Many types of compounds comprise the flavonoids; one of the most abundant being the flavones, a group of oxygen containing heterocycles with 2-phenylbenzopyrone ring system. Members of the flavone class are associated with a wide variety of biological activities [12] and used in the treatment of certain diseases.

The flavones (2-phenyl-chromen-4-one or 2-phenyl-4*H*-1-benzopyran-4-ones) are widely distributed in vascular plants [13] and possess biological activities such as antioxidant, inhibition of HIV-1 proteinase and anti-cancer agent [14]. Flavones occupy a special place in the realm of natural and synthetic organic chemistry owing to their useful biological activities [15] such as antibacterial, antifungal and used in the treatment of other various diseases [16]. Substituted flavones like amino flavones have been studied as tyrosine kinase inhibitors and anti-mitotic agents [17]. These compounds showed biological activities [18] like anti-viral, anti-hepatotoxic, anti-inflammatory etc.

2. Biological Significance

2.1. Flavones as Anti-cancer Agent

The naturally occurring aurone and flavones [19] isolated from *Uvaria hamiltonii* act as anti-cancer agents (**1**). Xing Zheng *et al.* [20] synthesized a series of chrysin derivatives and tested in *vitro* anti-cancer activities against SGC-7901 and HT-29 cells. Non-prenylated rotenoids, potent breast cancer resistance protein inhibitors have been reported [21]. 7,8-hydroxy-flavone showed anti-proliferative [22] activity against MCF-7 human breast cancer cells. Synthetic flavonoids were screened for their effect on cell proliferation and apoptosis in a human colonic cell line (**2**). [23]

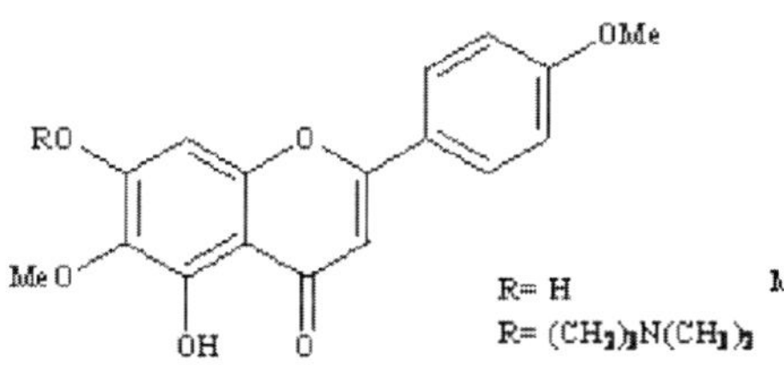

(2)

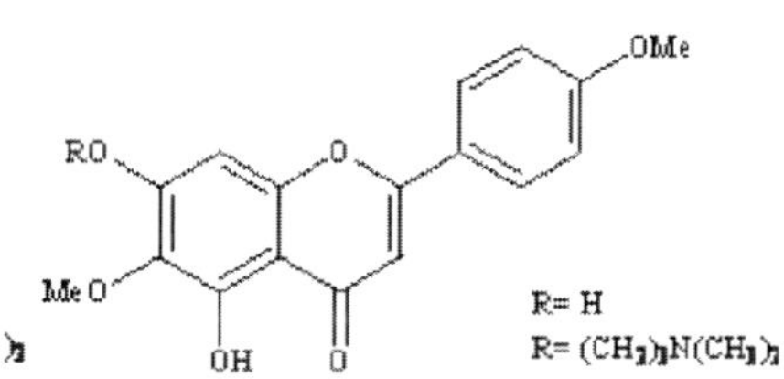

(2)

In vitro biological evaluations of novel 7-dialkylaminoalkyl cytotoxic pectolinarigenin derivatives against human cancer cell lines have been reported [24]. The anti-proliferative activity [25] of several flavonoids isolated from *Linaria reflexa* and their derivatives were evaluated *in vitro* by the SRB assay against the tumor cell lines. Flavone desmosdumotin B (**3**) and its analogs [26] were evaluated as *in vitro* inhibitors of human cancer cell growth. Among them, trihydroxyacetophenone derivatives showed good activity.

Radix Scutellariae dried root of the medicinal plant *Scutellariae baicalensis Georgi*, exhibits a variety of therapeutic effects and has a long history of the application in traditional formulations as well as in modern herbal medications. Flavones in *Radix Scutellariae* existing in the forms of aglycons and glycosides show great potential in the treatment of inflammation, cancers and virus-related diseases [27].

Sutradhar *et al.* [28] isolated 5,7-dihydroxy-3-isoprenyl flavone and 5-hydroxy-3-isoprenyl flavones and reported for analgesic and anti-inflammatory activities. Polymethoxyflavonoids exist in citrus plants showing anti-carcinogenic and anti-inflammatory activities [29]. 5-carboxymethyl-4',7-dihydroxyflavone were isolated from the herb *Selaginella moellendorffii Hieron* and was reported as anti-HBV and cytotoxic activity [30].

Maiti *et al.* [31] synthesized zapotin (5,6,2‘,6‘-tetramethoxyflavone) (4), component of the edible fruit *Casimiroa edulis,* induced both cell differentiation and apoptosis with cultured human promyelocytic leukemia cells (HL-60 cells), act as a potential cancer chemopreventive agent.

Flavone acetic acid (FAA) showed tumor selectivity in experimental animal models. 3-Amino derivative of flavone 8-acetic acid is a potent anti-cancer agent [32]. A series of apigenin-based flavonoid dimers, linked by poly-ethylene glycol chains of various lengths, have been synthesized and modulate the drug chemosensitivity and retention in breast and leukemic MDR cells [33]. Aitken *et al.* [34] reported anti-tumor activity of flavone-8-acetic acid (FAA) (5) with a 6-methyl substituent.

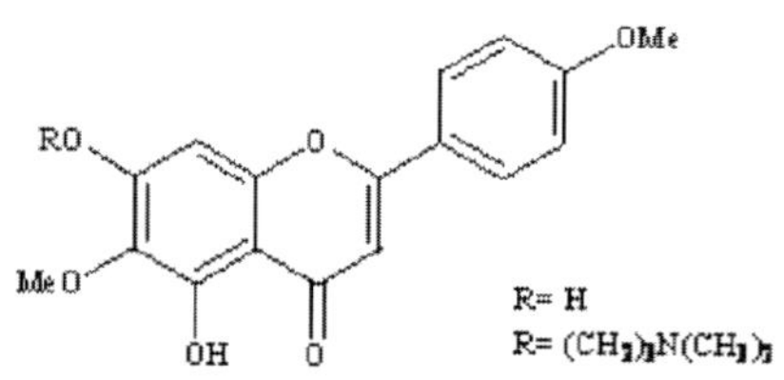

(2)

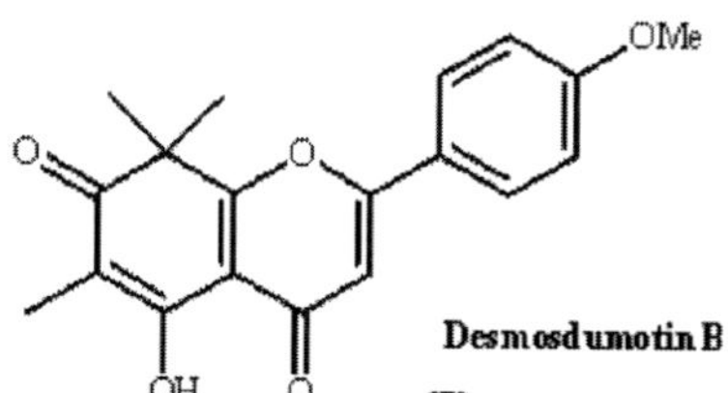

(3)

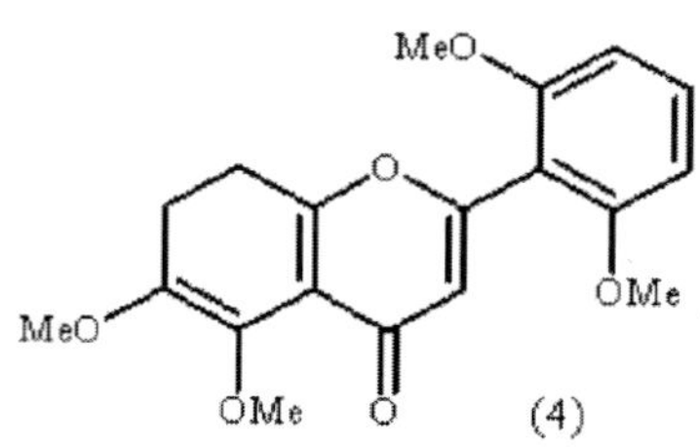

(4)

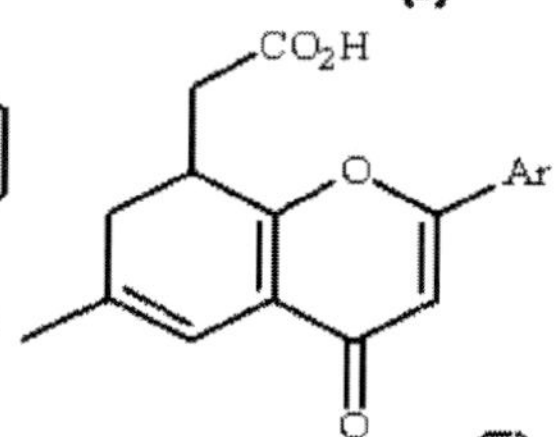

(5)

2.2. Flavones as Anti-Inflammatory Agents

5,7,2'-trihydroxy-6,4'-dimethoxyflavone and 5,2',4'-trihydroxy-6,7,8-trimethoxy flavone isolated from aerial parts of *Tamarix dioica*, along with hexacosyl-pcoumarate, gardenins A, B, C and E, nevadensin A and apigenin. Anti-viral activity and anti-invasive activity of Gardening B against solid tumors were reported [35].

In vitro anti-inflammatory activity of 3-*o*-methyl-flavones (6) isolated from *Siegesbeckia glabrescens* has been studied [36]. Several reports showed anti-inflammatory activity of (7, 8) isolated flavones [37]

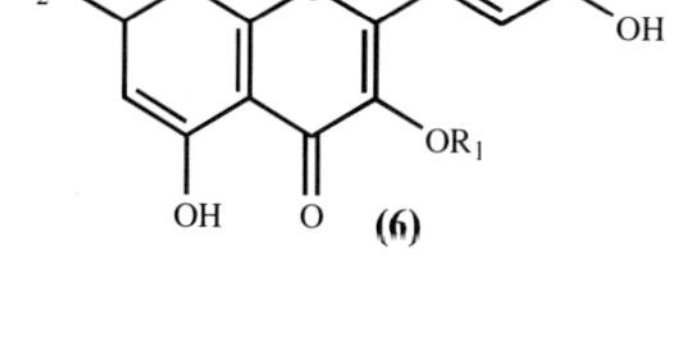

(6)

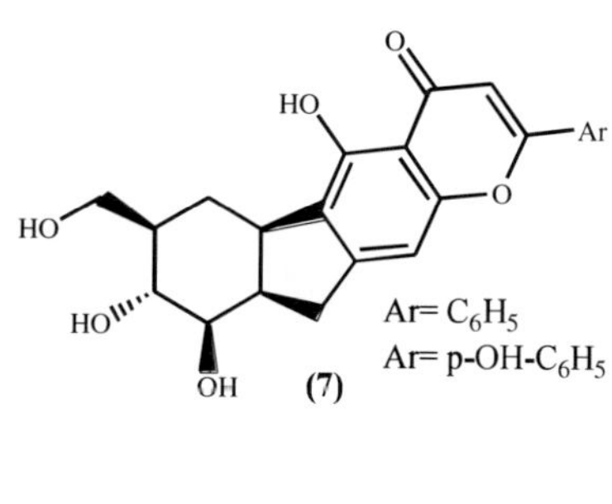

(7)

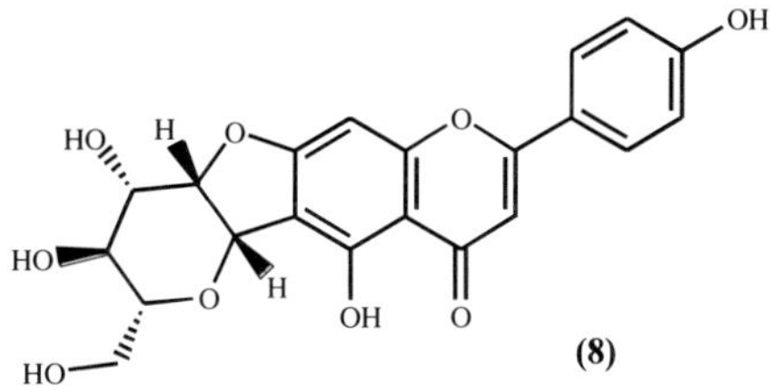

(8)

Bioactive flavones eupatilin and jaceosidin found in the medicinal herbs of the genus *Artemisia* exhibit various antioxidant, anti-inflammatory, anti-allergic, and anti-tumor activities. The inhibitory activities of eupatilin and jaceosidin on seven major human cytochrome P 450 enzymes in human liver microsomes were reported by Ji, Hye Young *et al.* [38].

2.3. Flavones as Anti-HIV Agents

Veljko *et al.* [39] explained that flavonoid compounds are an important natural source of anti-retrovirals for AIDS therapy due to their significant anti-HIV-1 activity and low toxicity. The *o*-((3-hydroxyethyl) derivatives of diosmin and rutin, i.e. Hidrosmin and troxerutin respectively, are widely used for the treatment of vascular disorders. Hydroxylated flavones inhibit retroviral reverse transcriptases, including HIV reverse transcriptases [40] as well as inhibit protein tyrosine kinases [41]. In fact a naturally occurring hydroxylated flavone and several related compounds have recently shown to inhibit HIV-induced synctium formation [42]. In addition, these substances have been found to possess anti-cancer and chemo-preventative activity.

2.4. Flavones as Antimalarial Agents

A series of flavonoid (9) derivatives containing a piperazinyl chain showed antiplasmodial (antimalarial) activity [43].

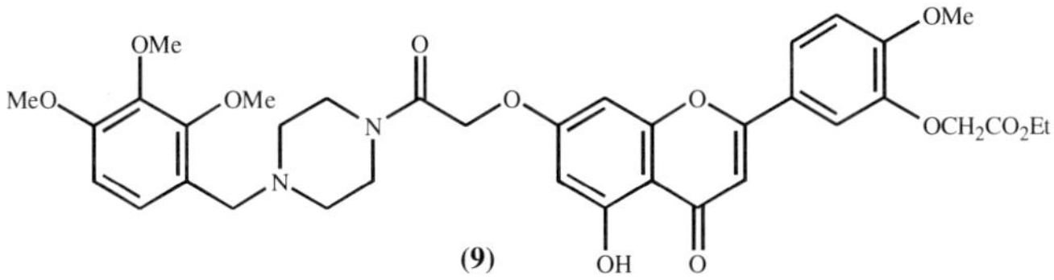

(9)

Biflavonoids i.e. Ginkgetin-sialic acid conjugates (10, 11) showed a significant survival effect in the influenza-virus-infected mice [44].

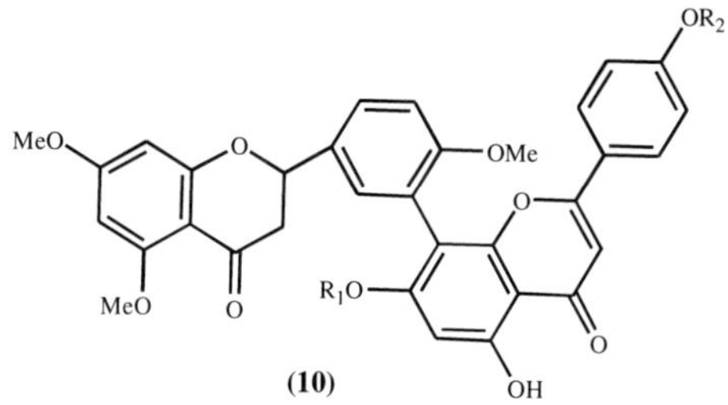

(10)

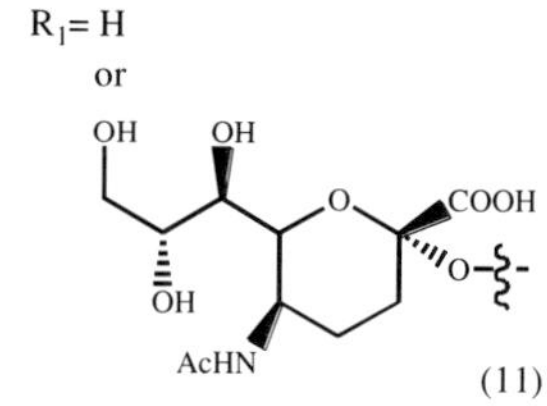

(11)

2.5. Flavones as Antibacterial & Antifungal Agents

Diwakar *et al.* [45] synthesized a series of 3-((*Z*)-2-(4-nitrophenyl)-2-(1*H*-tetrazol-5-yl) vinyl)-4*H*-chromen-4-ones assayed in *vitro* for antibacterial activities against gram-negative and gram-positive bacteria. A series of 8-aminomethylated derivatives were prepared and subjected for α-glucosidase inhibition activities [46]. Dauzonne *et al.* [47] screened *in vitro* cytotoxity of a series of 3-amino flavones and concluded that 4'-methoxy group is important for their activity. Series of chrysin [48] analogues (5,7-dihydroxy flavone) were prepared and screened for antibacterial activity against a panel of susceptible and resistant gram-positive and gram-negative organisms.

Flavone i.e. glycoside (12) was isolated from the leaves of *Vitex negundo* [49] and evaluated for antifungal and antibacterial activity against *Trichophyton mentagrophytes* and *Cryptococcus neoformans* at MIC. Some flavones also exhibit hypotensive, hypothermic, anti-allergic and anti-platelet activity [50]. Parmar *et al.* [51] synthesized number of flavones and screened for bactericidal, fungicidal, anti-yeast and anti-viral properties. A novel series of nitrogen-containing flavonoids [52] (13, 14) were showed to be potent cyclin-dependent kinases (CDKs) inhibitors.

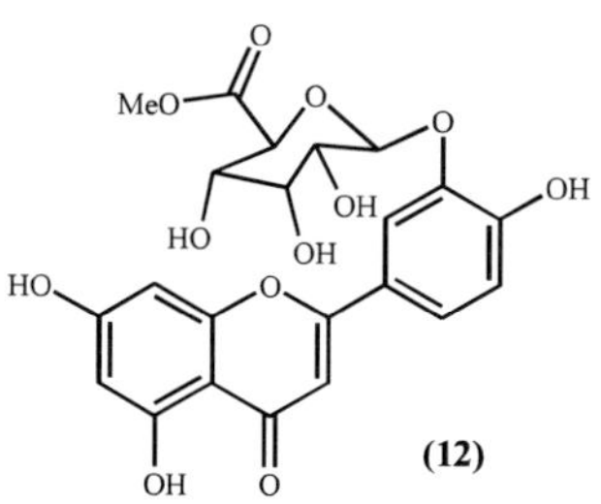

(12)

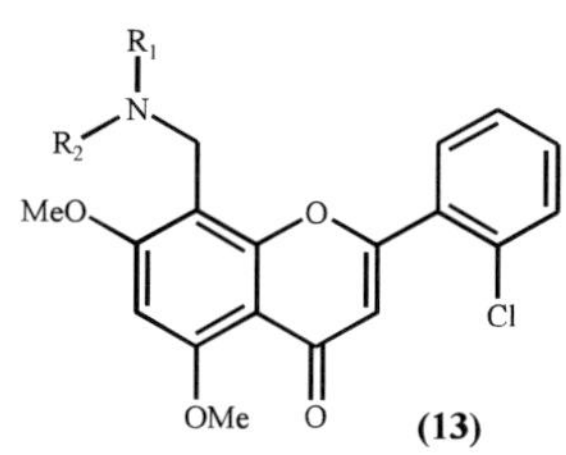

(13)

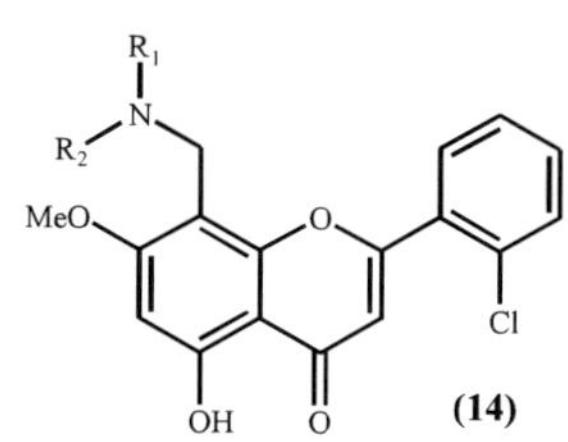

(14)

2.6. Flavones as Neuroprotective & Anti-tuberculosis Agent

Narender *et al.* [53] described anti-dyslipidemic activity of furano-flavonoids isolated from *Indigofera tinctoria.* A series of C–C biflavones [54] was investigated inhibitory activity against phospholipase A_2 (PLA2). Kang *et al.* [55] reported neuroprotective effects of naturally occurring biflavonoids on oxidative stress-induced and amyloid β peptide-induced cell death in neuronal cells. Larget *et al.* [56] have synthesized certain new substituted flavones and reported as efficient neuroprotective agents. Cardiovascular activity of these flavones were described [57].

2,3-Diaryl benzopyrans and analogs (15,16,17) reported as cycloxygenase-2 inhibitors were subjected to quantitative structure–activity relationship (QSAR) analysis and design new class of potent selective COX-2 inhibitors [58].

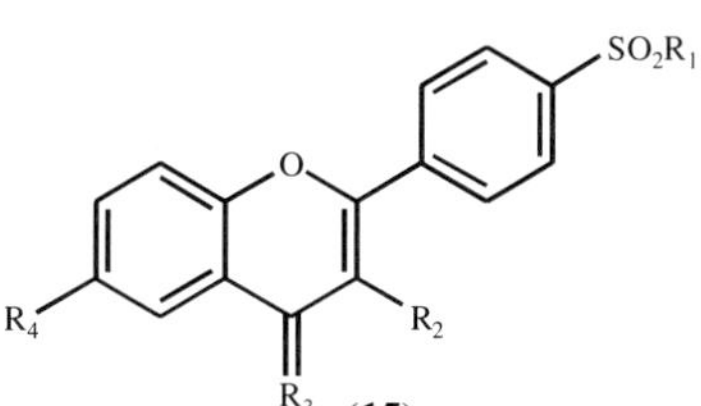

(15)

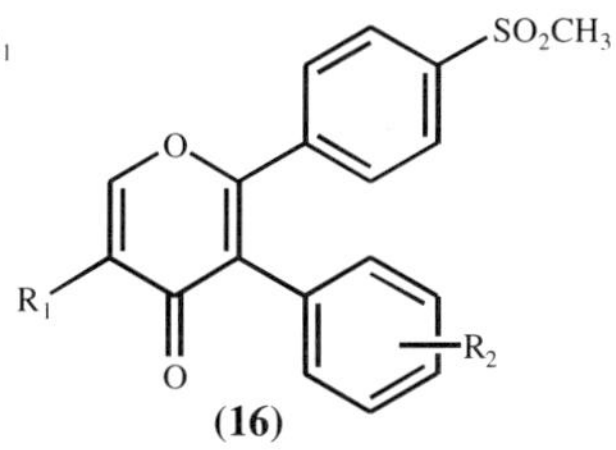

(16)

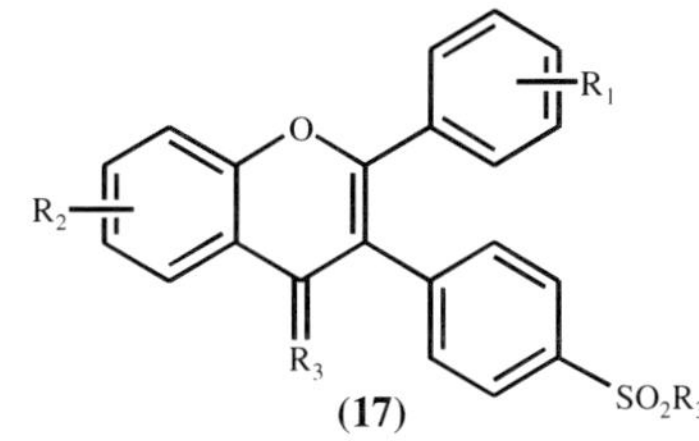

(17)

Dao *et al.* [59] found that chrysin derivatives with 3′,4′-dichloro substituent (18) exhibited good inhibitory activity of prostaglandin production. Lee *et al.* [60] found increased anti-*P*-glycoprotein activity of baicalein by the alkylation of ring in flavones (19). Antiproliferative activity [61] of several natural and synthetic flavonoids was evaluated in *vitro* against a cell line derived from human cervical carcinoma (WISH cells). The screening results for a series of naturally occurring and synthetic biflavonoids (20) against *Mycobacterium tuberculosis* [62] were reported. 5,4'-Dihydroxy-3,6,7,8,3'-pentamethoxy-flavone and 5,4',5'-trihydroxy-3,6,7,4-tetramethoxyflavone were isolated [63] from *G. resinosa* leaves and stems, were reported as *in vitro* antineoplastic activity in nasofaringeal carcinoma.

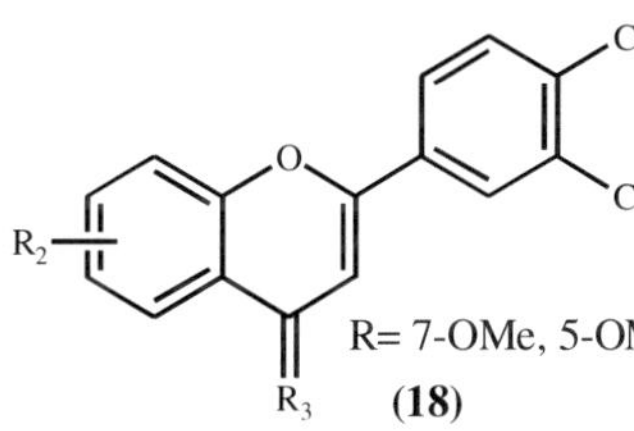

R= 7-OMe, 5-OMe, 7-OH

(18)

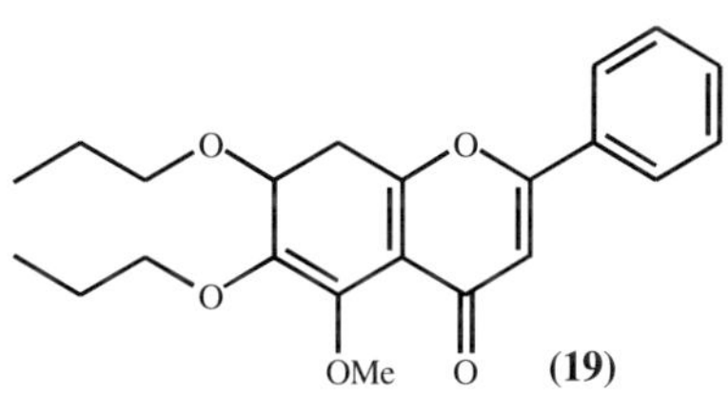

(19)

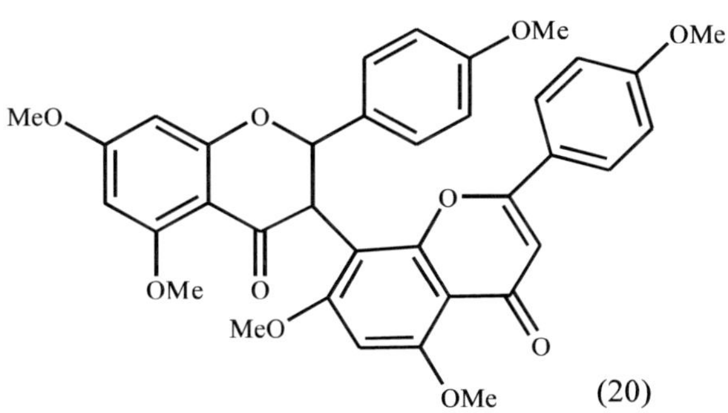

(20)

Shin *et al.* [64] synthesized a series of chrysin derivatives and tested *in vivo* against the diabetes mellitus. Chrysin itself inhibited insulin release by 40-60%. Robustaflavone, (21) a naturally occurring biflavanoid isolated from the seed kernel extract of *rhus succedanea*, was found to be a potent in *vitro* inhibitor of hepatitis B [65]. Halogenated or nitrated flavone compounds (22, 23) were found to be potent central benzodiazepine receptor (BDZ-R) ligands [66].

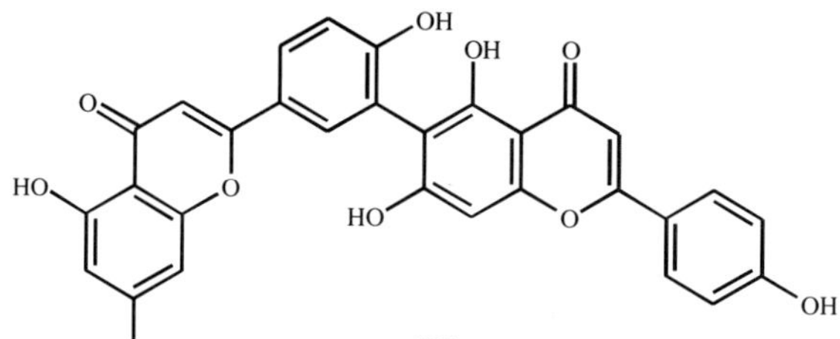

(21)

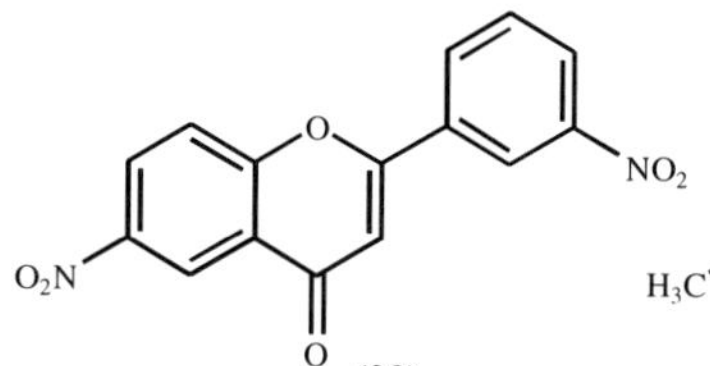

(22)

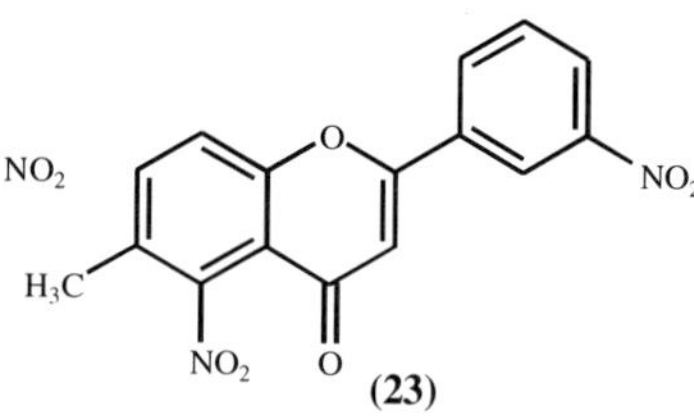

(23)

2.7. Flavones as Anti-convulsant Agents

Bioactive flavones isolated from *Scutellaria baicalensis* (baicalin, baicalein and oroxylin A) and reported convulsion-related activities [67]. Baicalin showed anti-convulsant effects in *vitro* studies. Anti-convulsive effects of baicalein were mediated by the benzodiazepine (BZD) binding site of GABAA receptor. The 5, 7-dihydroxyl groups present in the flavones played a key role in inducing convulsion-related activities.

2.8. Other Activities of Flavones

Several flavone moiety compounds showed gastroprotective activity, antioxidants and were also used as metal ion-chelators. Colombo *et al.* described the characterization of flavones from sugarcane *(Saccharum officinarum*) juice by high-performance liquid chromatography [68]. Hydroxymethoxyflavones was extracted from peels of *Citrus sudachi Hort. ex Shirai*. A mixture of sudachitin (4',5,7-trihydroxy-3',6,8-trimethoxyflavone) and demethoxysudachitin (4',5,7-trihydroxy- 6,8-dimethoxyflavone) were obtained [69]. Apigenin-7-O-b-D-glucoside and luteolin-7-O-b-D-glucoside bioactive flavones [70] were isolated from *atricaria chamomilla L.*

Li, Chenrui *et al.* [71] reported major bioactive flavones in proprietary traditional Chinese medicine products and reference the herb of *Radix Scutellariae*. Wei, Yun *et al.* isolated bioactive flavonoids from *Eupatorium adenophorum Spreng* using stepwise elution by high-speed countercurrent chromatography [72].

Zhang *et al.* [73] reported the potential of oral and pulmonary nanocrystals to enhance the bioavailability of baicalein, a bioactive flavonoid isolated from the root of *Scutellaria baicalensis Georgi*. The metabolites of baicalein in human plasma were reported, which is one of the main bioactive flavones in *Scutellaria baicalensis Georgi*. Possible metabolic pathways of baicalein in rat serum were proposed by Guo *et al.*[74].

3. Methods of Synthesis

Various methods have been reported for the synthesis of flavones [75], including Baker-Venkatraman rearrangement, Allan-Robinson synthesis, Von-Konstanecki method, Wittig reaction from 2-hydroxychalcones etc. several catalyst were used for this purpose such as molecular iodine, $FeCl_3$, SeO_2, Koser's reagent, clay, H_2SO_4, Palladium-Thiourea, Wells-Dawson heteropolyacid, NaOAc/AcOH, $Br_2/CHCl_3$, EtOH/HCl etc. However, many of these methods suffered from harsh reaction condition, toxic reagents, strong acidic or basic conditions, prolonged reaction time, poor yield and low selectivity. A large number of alternative procedures have been developed in the past few years. Although several modifications have been made to counter these problems, there is still a need to develop better methods for synthesis of such potent flavones.

The interest in the biological properties of flavones has resulted in intense synthetic efforts towards the synthesis of various flavones. Flavones have been synthesis from A) 2-hydroxychalcones, B) flavanones, C) Baker-Venkatraman rearrangement, D) Wittig reaction, E) Von-Konstanecki method, F) Allan-Robinson synthesis, etc.

3.1. From Chalcones

Flavones have been synthesized by oxidative cyclization of 2-hydroxychalcones using various reagents under different conditions. The synthesis of flavones starting from 2-hydroxychalcones accomplished by various research groups is recorded as follows:

3.1.1. Using Molecular Iodine

Agrawal *et al.* [76] synthesized flavones from 2-hydroxychalcone, a small amount of sulfuric acid and a catalytic amount of iodine in dimethylsulphoxide, at 100 ^{0}C for 40 min. Flavones were synthesized [77] from 2-hydroxy-chalcones by their oxidative cyclization using iodine in DMSO. Singh *et al.* [78] reported synthesis of flavones from corresponding chalcones using DMSO and iodine.

Synthesis of 2-benzoyl-3-methyl-7-(substituted aryl)-5-furo [3,2-g] benzopyran-5-ones [79] was carried out from orthohydroxy chalcones in DMSO and iodine.

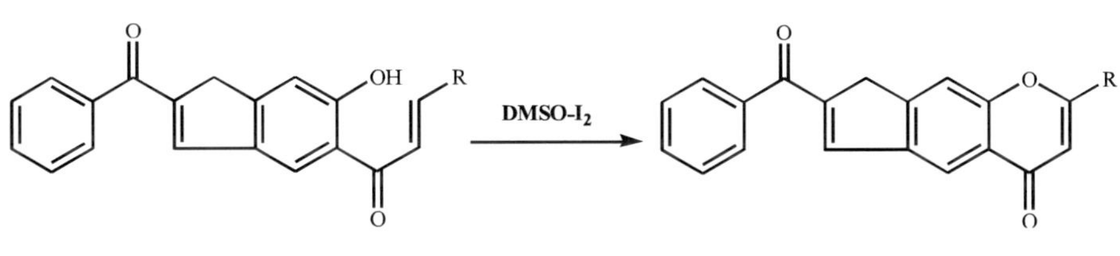

Synthesis of flavones were reported by cyclodehydrogenation of 2-hydroxychalcone in the presence of iodosobenzene diacetate/KOH in methanol [80]. Lokhande *et al.* [81] reported synthesis of flavones by deprotection of *o*-allyl protecting groups under oxidative cyclization of 2-allyloxychalcones.

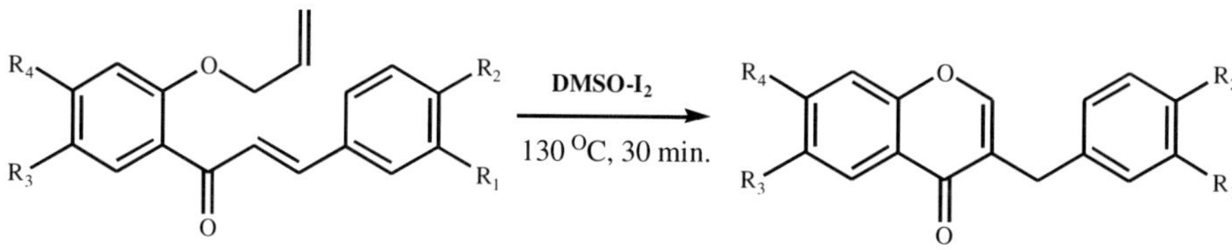

3.1.2. Using 2,3-Dichloro-5,6-Dicyano-l,4-Benzoqiunone (DDQ)

Imafuku *et al.* [82] synthesized flavones from 2-hydroxy-chalcones using 2,3-Dichloro-5,6-dicyano-l,4-benzoqiunone(DDQ). Hossain *et al.* [83] synthesized naturally occurring flavone *(genkwanin)* from respective chalcone using DDQ.

3.1.3. Using Diphenyl Sulphide

Flavones were synthesized from 2-hydroxy-chalcones using diphenyl sulphide. Amzad *et al.* [84] reported synthesis of flavones using DMSO/I_2, diphenyl sulphide or DDQ,

Furanoflavones (lanceolatin-B) [85] were synthesized from β-resacetophenone *via* chalcone precursors followed by the treatment of DMSO/I_2 or diphenyl sulphide or DDQ.

3.1.4. Using Seleniumdioxide

Flavones were also synthesized by oxidative cyclization of 2-Hydroxychalcone using selenium dioxide in DMSO [86]. Earlier flavones were synthesized from 2-hydorxychalcones using seleniumdioxide in isoamylalcohol [87]. Dawane *et al.* [88] synthesized 2-(9-anthryl) chromones by the oxidative cyclization of l-(2-hydroxyphenyl)-3-(9-anthryl)-2-propen-1-one using SeO_2 in isoamyl-alcohol.

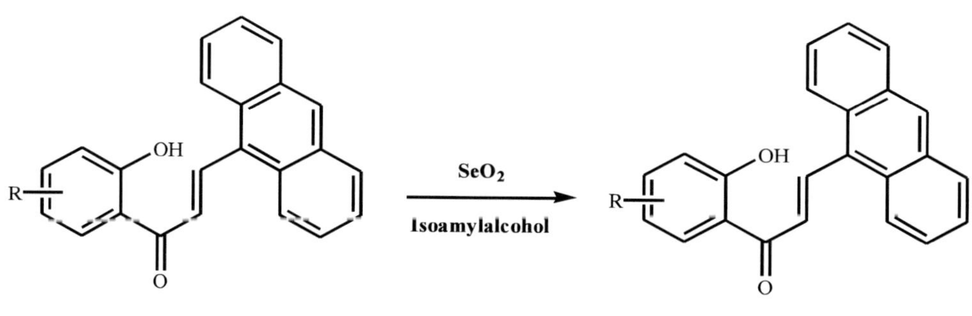

3.1.5. Bromination of Chalcones

Bromination of chalcones, followed by treatment of resulting dibromides with KOH (Emilewicz-Von Kostanecki) gave the flavone-6-carboxylic acids as on heating 2-hydroxychalconedibromide in alkali, flavones [89] were obtained. Synthesis of flavones by dehydrobromination followed by cyclization of corresponding dibromo-2'-hydroxychalcones in pyridine or acetic acid and potassium acetate were reported [90].

A variety of flavones were synthesized from 2-acetoxychalcones [91] by bromination using *n*-tetrabutylammonium tribromide (TBATB) in the presence of $CaCO_3$ in CH_2Cl_2–MeOH (5:2) at 0–5 ^{O}C followed by dehydrobromination and finally, cyclization by treating with 0.1 M sodium methoxide solution.

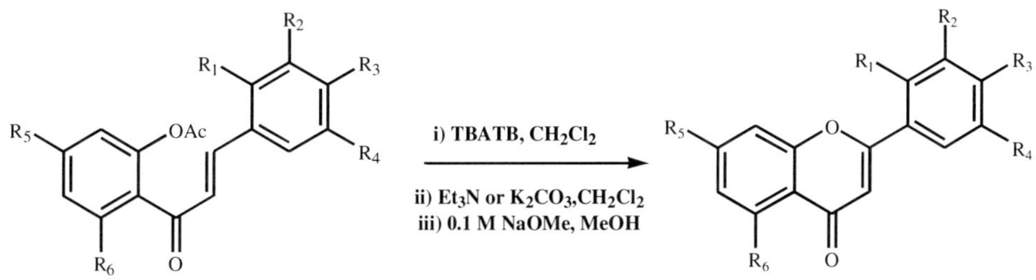

6,8-Dibromoflavones and 8-bromoflavones were synthesized by bromination [92] with a combination of vanadium pentoxide, hydrogen peroxide and ammonium bromide in dichloromethane-water at 0–5 °C, followed by dehydrobromination using 0.2 M ethanolic KOH solution at ice-bath temperature.

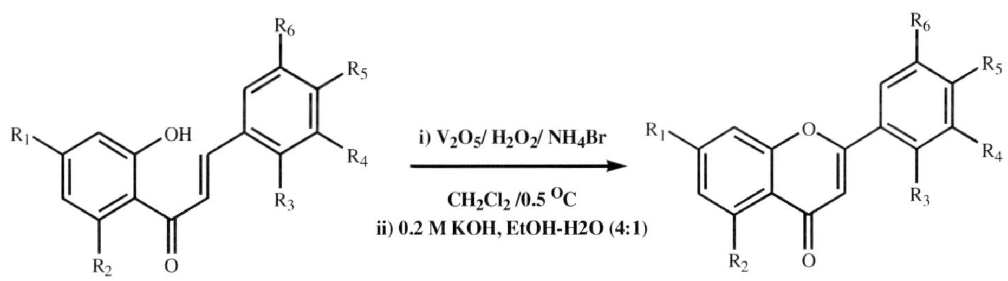

Meyer *et al.* [93] described the conversion of **α-**azidochalcone [**α**-azido-4-(benzyloxy)-4-methyl-2-(*p*-tosyloxy)-chalcone] to flavones.

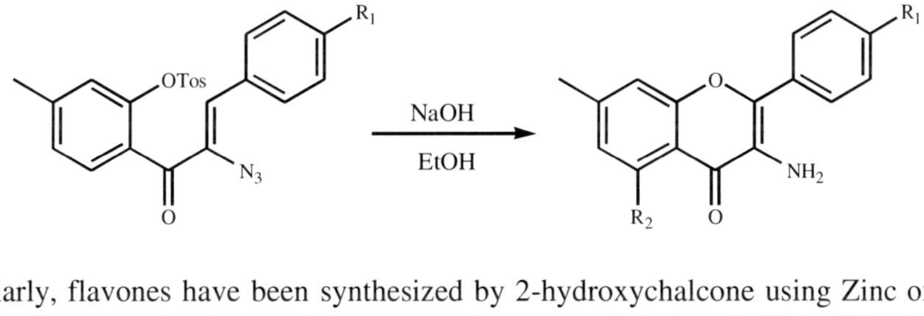

Similarly, flavones have been synthesized by 2-hydroxychalcone using Zinc oxide [94], Nikel Peroxide [95], Sodium Periodate [96], Palladous Chloride ($PdCl_2$) [97], Oxalic acid

[98], Lithium Chloropaladite (LCP) [99], Dicyanoanthracene (DCA) [100] and Potassium-ferricyanide [101], etc.

The acetylation of the dihydrochalcone afforded the acetylated product which, furthering the rearrangement in basic conditions and subsequent cyclization in acidic medium, afforded the corresponding flavone [102]. Mokle, S. S. *et al.* [103] synthesized bioactive flavones by oxidative cyclization of chalcones. Ferrocenyl-substituted aurones, flavones, and flavonols were synthesized by Tiwari, Keshri Nath *et al.* [104].

3.2. From Flavanones

Singh *et al.* [105] synthesized flavones by the dehydrogenation of flavanones with thallium (III) acetate in acetic acid.

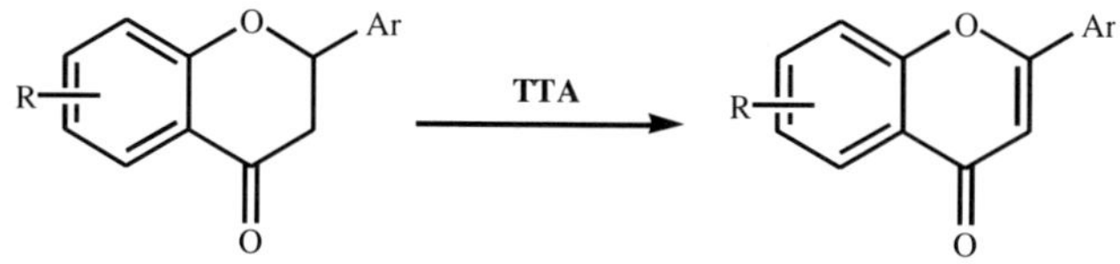

Dehydrogenation of flavanones to flavones were reported by using manganese (III) acetate in acetic acid and perchloric acid [106], nickel peroxide in benzene [107], DDQ [2,3-Dichloro-5,6-dicyano-l,4-benzoqiunone] [108], DMSO-I_2 [109], CAN and H_2O_2 [110] etc.

3.3. Baker-Venkatraman Rearrangement

The Baker-Venkataraman rearrangement is a reaction of 2-acetoxyacetophenones with base to form 1,3-diketones. This rearrangement reaction proceeds *via* enolate formation followed by acyl transfer often used to synthesize chromones and flavones. This reaction is named after the two scientists, W. Baker and Krishnaswamy Venkataraman.

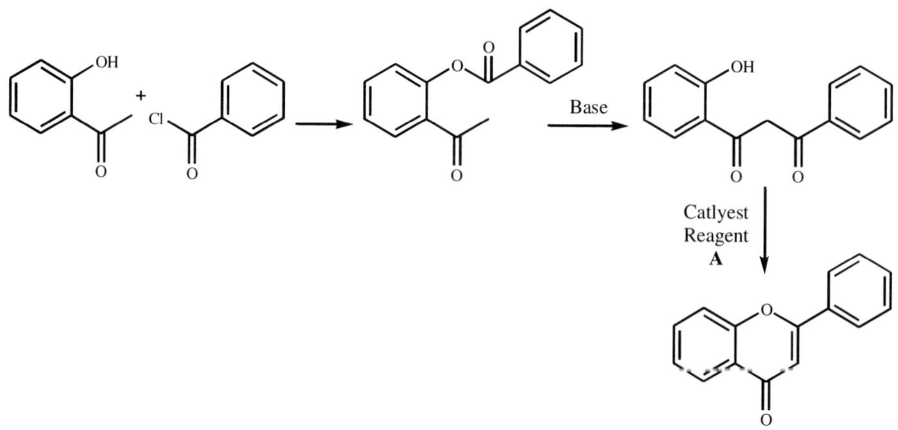

For the synthesis of flavones *via* Baker-Venkataraman reaction [111], 2-hydroxy acetophenone was converted to ester, furthering the rearrangement by intramolecular Claisen condensation in the presence of base afforded 1,3-diketone, which was then dehydrative cyclization to flavones by treatment with acid or various reagents.

The known catalyst or reagent (A) used for the cyclization of 1,3-diketone to flavones:

Sulphuric acid	Ref-[112]
Sulphuric acid - Acetic acid	Ref-[113]
Acetic acid-hydrochloric acid	Ref-[114]
Acetic acid-hydrobromic acid	Ref-[115]
Acetic acid-sodium acetate	Ref-[116]
Phase transfer catalyst	Ref-[117]

Ethylene glycol	Ref-[118]
p-toluene sulphonic acid	Ref-[119]
Potassium carbonate	Ref-[120]
Iodine	Ref-[121]
Organolithium reagents	Ref-[122]
Indium (III) chloride	Ref-[123]

However, these simplified procedures have certain limitations. Substituted dimethoxy-acetophenones failed to undergo conversion into the corresponding dibenzoyl-methanes. Some researchers have found that the conventional Baker-Venkataraman approach was not suitable for synthesizing large amounts of flavones since low yields and product isolation complications. Several methods were reported to overcome these problems. Daniel *et al.* [124] reported silica-supported ($H_2P_2W_{18}O_{62}.24H_2O$) Wells-Dawson acid as reusable, heterogeneous catalysts for the synthesis of flavones and chromones by cyclization of 1-(2-hydroxyphenyl)-3-aryl-1,3-propanediones.

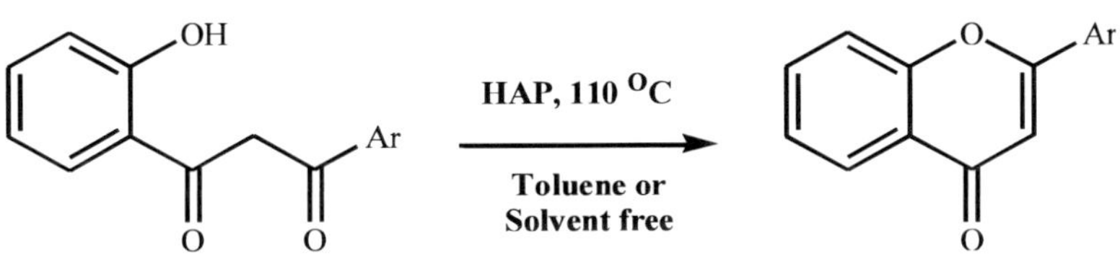

A microwave synthesis of flavones and chromones via the cyclization of 1-(2-hydroxyaryl)-3-aryl-1,3-propanedione using cupper chloride ($CuCl_2$) is described [125]. Solid-state [126] dehydrative cyclization of *o*-hydroxydibenzoylmethanes on a clay surface using microwaves has been reported.

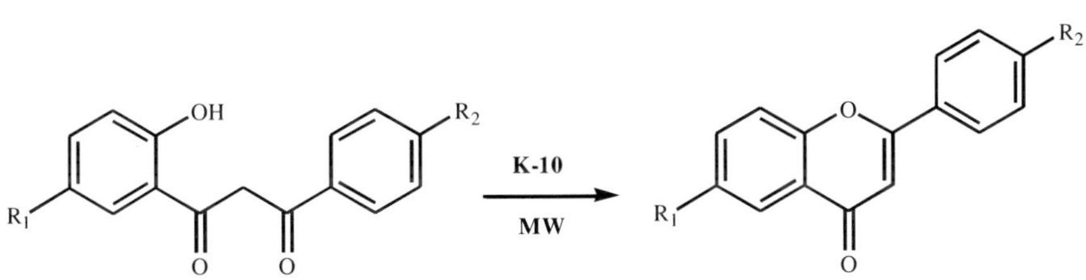

Flavones [127] were synthesized in one pot reaction by the condensation of 2,6-dihydroxyacetophenone with an alkoyl chloride in the presence of K_2CO_3. Lee *et al.* [128] synthesized flavones from 2-methoxybenzoic acids.

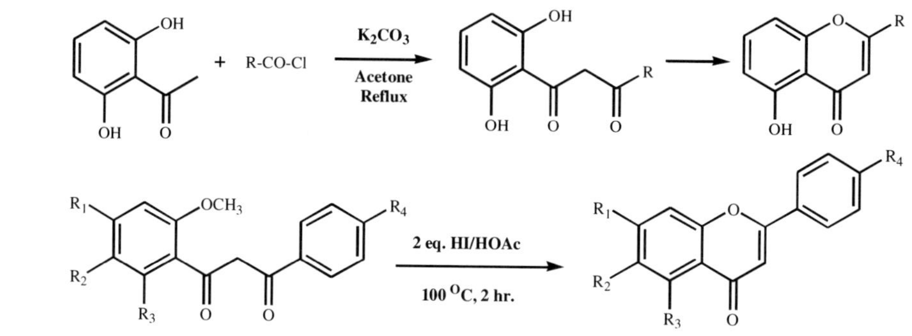

Lijun Tang *et al.* [129] described the synthesis of 6-amino-7-hydroxyflavone. 1-(2,4-dihydroxy-5-nitrophenyl)-3-phenyl-1,3-propanedione transformed into 7-hydroxy-6-nitroflavone using $AcOH/H_2SO_4$, followed by reduction to afford the 6-amino-7-hydroxyflavone.

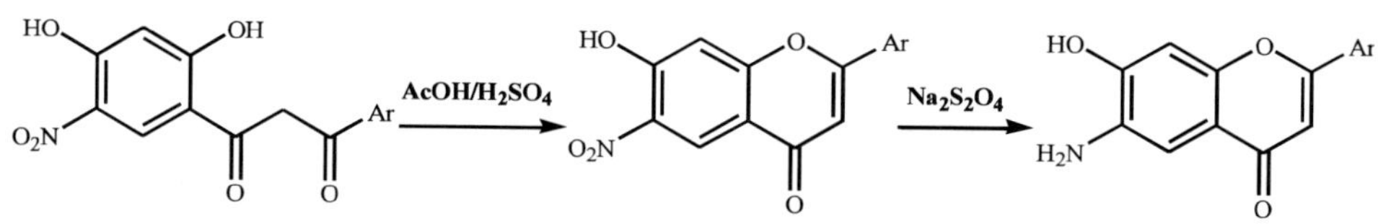

Ganguly *et al.* [130] modified the Baker-Venkataraman reaction. Hydroxy acetophenone were heated with acyl chloride in the presence of 1,8-diazabicyclo [5.4.0]undec-7-ene (DBU) and pyridine to obtain 3-acyl flavones.

3.4. Wittig Reaction

Kumar *et al.* [131] synthesized 4*H*-chromen-4-ones *via* intramolecular Wittig reaction.

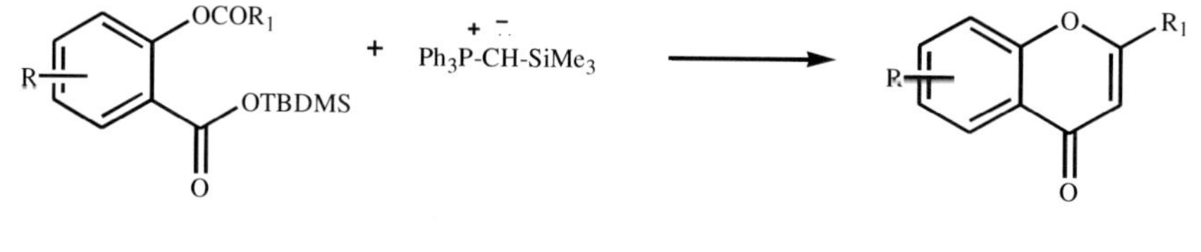

Hercovet *et al.* [132] reported Wittig reaction between an *o*-hydroxy-phenacylidene (triphenyl) phosphorane and a substituted benzoyl chloride to prepare flavones. Phosphorane in boiling toluene reacted with acid chloride in the presence of pyridine to give an unstable intermediate, which spontaneously undergoes intramolecular olefimation to give the flavone [133].

Floc'h *et al.* [134] described intramolecular Wittig reaction. Reaction of acetophenone dibenzoate with bromine followed by triphenylphosphine affords the triphenyl phosphonium salt, which on treatment with sodium carbonate followed by hydrolysis with sodium hydroxide, gave flavones. Some dihydroxy flavones [135] were prepared by acylation, intramolecular olefimation and saponification of the two remaining ester groups.

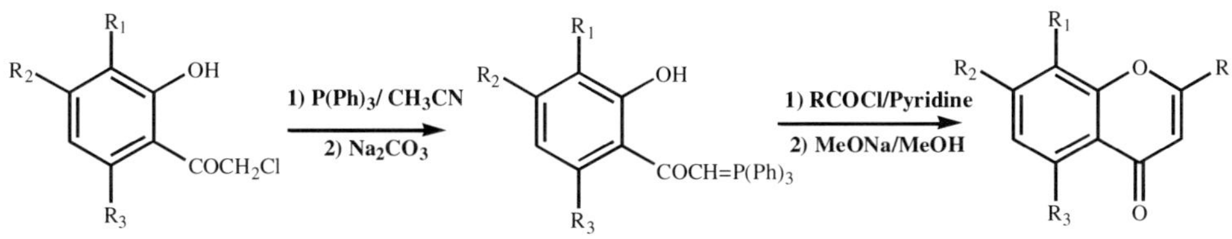

Flavones have been synthesized by carbonylative annulations [136] of *o*-iodophenol acetates and acetylenes mediated by palladium-thiourea complex in the presence of base at 40 ^{O}C under a balloon pressure of carbon monoxide.

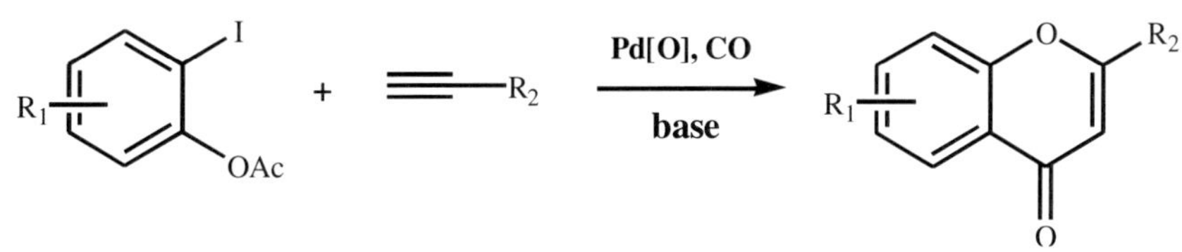

Nagarathnam *et al.* [137] reported methyl salicylate LiMHDS catalyzed condensation of methyl 2-{(*t*-butyldimethylsilyl)-oxyl}-benzoate with acetophenones followed by treatment of intermediates with H_2SO_4 in acetic acid at 95-100 ^{O}C gave flavones.

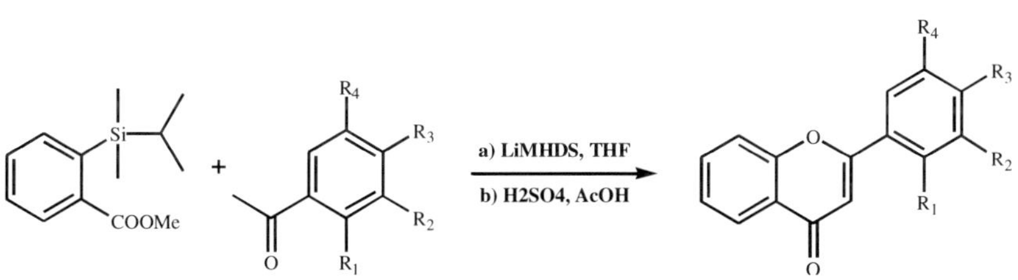

Phenylpropynoic [138] acid on condensation with phenol by using N,N'-dicyclohexyl carbodiimide (DCC), pyridine as solvent yields *o*-hydroxyaryl phenyl ethynyl ketones, which on cyclization carried out using refluxing suspension of K_2CO_3 in acetone, gave flavones as the major product.

3.5. The Von-Konstanecki Method

The method involves a reaction of *o*-methoxybenzoate and acetophenone in the presence of sodium metal followed by treatment of acid and by elimination of water in order to form the flavones [139].

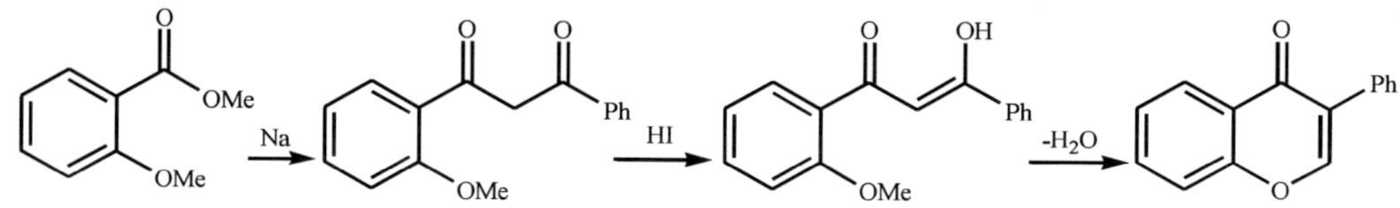

3.6. Allan-Robinson Synthesis

Synthesis of flavones or isoflavones has been reported by condensing *o*-hydroxyaryl ketones with anhydrides of aromatic acids and their sodium salts [140].

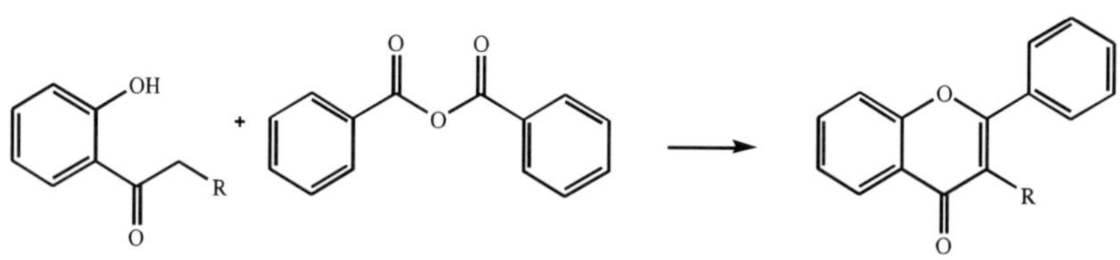

3.7. Other Methods

Paquette *et al.* [`141] reported salicyaldehyde reaction with 1-morpholino-l -phenyl ethylene in benzene to form an adduct, It was further oxidized to flavone with chromium trioxide-pyridine. The direct conversion [142] of *o*-hydroxyacetophenone to flavone occurs on reaction with benzaldehydes in perchloric acid in the presence of orthoformates. 2-Hydroxy-2-methoxyacetophenones [143] was condensed with 4-(benzyloxy) benzoic acid anhydride in presence of the potassium salt of 4-(benzyloxy) benzoic acid as base. These corresponding benzyl ethers were synthesized by an Allan-Robinson reaction. The acid hydrolysis of which offered the flavones.

Synthesis of new substituted 3-(l*H*-imadazol-1-yl) flavones with a montmorillonite-KSF catalyst has been reported [144]. Synthesis of *o*-tolyl chromones [145] from propiophenones using phase transfer catalysis has been reported.

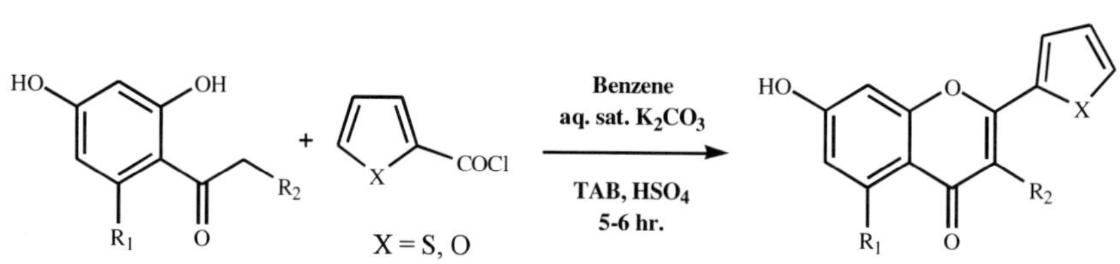

Organic reactions under solvent-free conditions are advantageous because of enhanced selectivity and efficiency, ease of manipulation, and because of toxic and volatile solvents are avoided. Solvent-free approaches involve mechano-chemical mixing (grinding), ionic liquid, microwave (MW) irradiation of neat reactants (undiluted), or catalysis by the surfaces of inexpensive and recyclable mineral supports, such as alumina, silica, clay, or 'doped' surfaces. Solvent-free microwave-assisted chemical reactions in combination with solid supported reagents were used to carry out a reaction in shorter time as compare to other conventional reaction methods. The use of solvent-free conditions with heterogeneous catalysts represents one of the more powerful green chemical procedures, Microwave irradiation, as an unconventional energy source, has been used for a variety of organic transformations wherein chemical reactions are accelerated because of selective absorption of microwave energy by polar molecules; non-polar molecules are inert to the microwave irradiation.

Recently flavones were synthesized by using I_2-Al_2O_3 under microwave irradiation using 2-hydroxy chalcone [146]. This method has several advantages such as shorter reaction time, excellent yield of products and a solvent-free reaction.

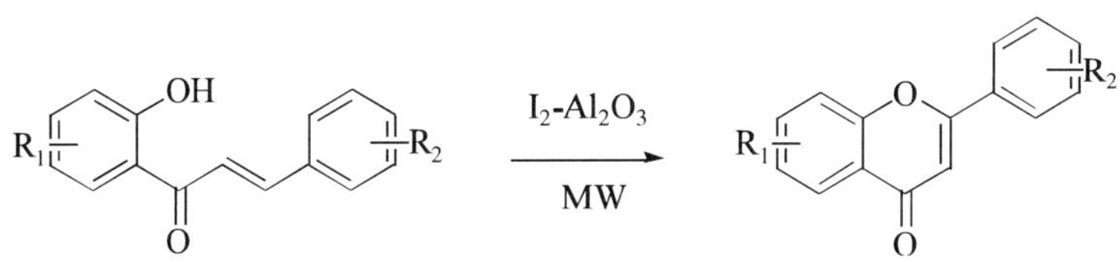

Flavones were also synthesized by using ethyl ammonium nitrate ionic liquid $[EtNH_3]NO_3$ as well as n-butyl-3-methyl-imidazolium tetrafluoroborate ionic liquid $[bmim]BF_4$. Flavones synthesized *via* Baker-Venkataraman reaction, using ethyl ammonium nitrate ionic liquid $[EtNH_3]NO_3$ under microwave irradiation [147]. The conversion of 1,3-diaryl diketones to flavones in the presence of ionic liquid $[EtNH_3]NO_3$ under microwave irradiation render these method superior as clean, simple and ecofriendly.

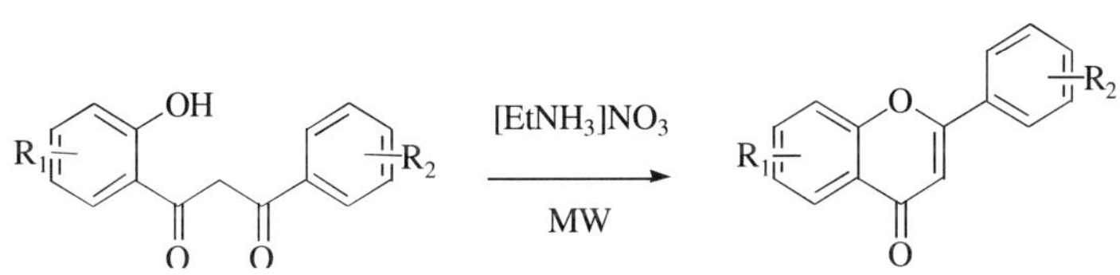

Similarly, flavones synthesized from 1,3-diketones *via* Baker-Venkataraman reaction using n-butyl-3-methyl-imidazolium tetrafluoroborate ionic liquid $[bmim]BF_4$ at 100 ^{O}C. The dehydrative cyclization of 1,3-(diaryl) diketones to flavones in ionic liquids $[bmim]BF_4$ has been reported [148]. This simplified procedure finds wide utility in organic synthesis due to the presence of ionic liquid without any additional catalyst or hazardous solvent.

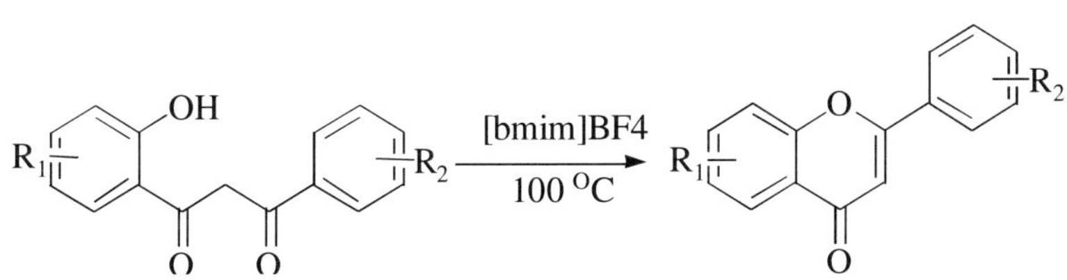

Characteristic Test for Flavones:

Flavones do not give violet coloration with $FeCl_3$ solution and pink coloration with cone. H_2SO_4 and Wilson Test [149] were negative.

References

[1] Middleton, E.J. *Adv. Exp. Med. Biol.* 1998, **439**, 175-182.

[2] a)Groot, D., Rauen, H.U. *Fundam. Clin Pharmacology*, 1998, **12,** 249-255. b) Tanaka, T., Iinuma, M., Yuki, K., Fujii, Y., Mizuno, M. 1992, *Phytochem.* 31, 993.

[3] Kuhnau, J. *World Rev. Natr. Diet.* 1976, 24, 117-191. b) Middleton, E. *J. Ann. Allergy* 1988, 61, 63.

[4] Brigitte, A.G. Paul, E.M., Jeffrey, B.B. *J. of Medicinal Food.* 2005, 8(3), 281-290.

[5] Hertog, M.G.L., Feskens, E.J.M., Hollman, P.C.H., Katan, M.B., Kromhout, D. *Lancet*, 1993, 342, 1007-1011.

[6] a) Heim, K.E., Tagliaferro, A.R., Bobilya, D.J. *J. Nutr. Biochem.* 2002, 13, 572-584. b) Harborne, J.B., Williams, C.A. *Nat. Prod. Rep.* 2001, 18, 310.

[7] Hollman, P.C.H., Van Trijp, J.M.P., Buysman, M.N.C.P., v.d. Gaag, M.S., Mengelers, M.J.B., de Vries, J.H.M., Katan, M.B. *FEBS Lett.* 1997, 418, 152.

[8] Bear, W.L., Teel, R.W. *Anticancer Res.* 2000, 20, 3609-3614.

[9] a) Lasker, J.M., Huang, M.T., Conney, A.H. *J. Pharmacol. Exp. Ther.* 1984, 229, 162-170. b) Moon, J.Y., Lee, D.W., Park, K.H. *Xenobiotica*, 1998, 28, 117-126.

[10] Datla, K.P., Hristidou, M., Widmer, W.W., Rooprai, H.K., Dexter, D.T., *NeuroReport*, 2001, 12, 3871-3875.

[11] a) Takanaga, H., Ohnishi, A., Yamada, S., Matsuo, H., Morimoto, S., Shoyama, Y., Ohtani, H., Sawada, Y. *J. Pharmacol. Exp. Ther.* 2000, 293, 230-236. b) Choi, C.H., Sun, K.H., An, C.S., Yoo, J.C., Hahm, K.S., Lee, I.H., Sohng, J.K., Kim, Y.C. *Biochem. Biophys. Res. Commun.* 2002, 295, 832-840.

[12] a) Welton, A.F., Tobias, L.D., Fiedler-Nagy, C., Andereon, W., Hope, W., Me ern, K., Coffey, J.W. In *Plant Flavonoids in Biology and Medicine;* Cdy, V., Middleton, E., Jr., Harborne, J.B., Eda., Alan R. Lisa, Inc.: New York, l986, p 231. (b) Selway, J.W.T. In *Plant Flauonoida in Biology and Medicine;* Cody, V., Middleton, E., Jr., Harborne, J.B. Eda., Alan, R. Lisa, Inc.: New York, 1986; p 621. (c) Brandi, M.L. *Bone Miner.* 1992, *19 (suppl.),* 53. (d) Hagiwara, M., Inoue, S., T d a , T., Nunoki, K., Ito, M., Hidaka, H. *Biochem. Pharmacol.* 1988,37,2987.

[13] Harborne, J.B., Williams, C.A. *Nat. Prod. Rep.* 2001, 18, 310.

[14] Gee, J.M., Johnson, I.T. *J. Curr. Med. Chem.* 2001, 8, 1245-1255.

[15] Cabrera, M., Simoens, M., Falchi, G., Lavaggi, M.L., Piro, O.E., Castellano, E.E., Vidal, A., Azqueta, A., Monge, A., Ceráin, A.L., Sagrera, G., Seoane, G., Hugo Cerecetto, H., González, M. *Bioorg. Med. Chem.* 2007, 15(10), 3356-3367.

[16] a) Nijveldt, R.J., Nood, E., Hoorn, D., Boelens, P.G., Norren, K., Leeuwen, P. *Am. J. Clin. Nutr.* 2001, 74, 418. b) Heim, K.E., Tagliaferro, A.R., Bobilya, D.J. *J. Nutr. Biochem.* 2002, 13, 572.

[17] Havsteen, *Biochem. Pharmacol*, 1983, 32, 1141.

[18] a) Vrysein, R., Everaert, L., VanHoof, L.M., Vlietinck, A., Vanden, B.D.A., Boeye, A. *Antiviral Res.* 1987, 7, 35. b) Hjelle, J.J., Brzeznicka, E.A.A., Klassen, C.D. *J. Pharmacol. Expt. Ther.* 1986, 236, 526.

[19] Lawrence, N.J., Rennison, D., McGown, A.T., Hadfield, J.A. *Bioorg. Med. Chem. Lett.* 2003, 13(21), 3759-3763.

[20] Zheng, X., Meng, W.D., Xu, Y.Y., Cao, J.G., Qing, F.L. *Bioorg. Med. Chem. Lett.* 2003, 13(5), 881-884.
[21] Ahmed-Belkacem, A., Macalou, S., Borrelli, F., Capasso, R., Fattorusso, E., Taglialatela-Scafati, O., Pietro, A.D. *J. Med. Chem.* 2007, 50 (8), 1933-1938.
[22] Pouget, C., Lauthier, F., Simon, A., Fagnere, C., Basly, J., Delage, C., Chulia, A. *Bioorg. Med. Chem. Lett.* 2001, 11(24), 3095-3097.
[23] Daskiewicz, J., Depeint, F., Viornery, L., Bayet, C., Comte-Sarrazin, G., Comte, G., Gee, J.M., Johnson, I.T., Ndjoko, K., Hostettmann, K., Barron, D. *J. Med. Chem.* 2005, 48(8), 2790-2804.
[24] Bonesi, M., Tundis, R., Deguin, B., Loizzo, M.R., Menichini, F., Tillequin, F., Menichini, F. *Bioorg. Med. Chem. Lett.* 2008, 18(20), 5431-5434.
[25] Tundis, R., Deguin, B., Loizzo, M.R., Bonesi, M., Statti, G.A., Tillequin, F., Menichini, F. *Bioorg. Med. Chem. Lett.* 2005, 15(21), 4757-4760.
[26] Nakagawa-Goto, K., Bastow, K.F., Wu, J., Tokuda, H., Lee, K. *Bioorg. Med. Chem. Lett.* 2005, 15(12), 3016.
[27] Li, Chenrui, Lin, Ge, Zuo, Zhong. *Biopharmaceutics & Drug Disposition*, 2011, 32(8), 427-445.
[28] Sutradhar, Ranajit K., Rahman, A.K.M. Matior, Ahmad, Mesbah U., Bachar, Sitesh, C. *Phytochemistry Lett.* 2008, 1(4), 179-182.
[29] Wang, Q., Wu, Z., Liu, L., Zou, L., Luo, M., Sumbayev, V.V., Youji, H. 2010, 30(11), 1682-1688.
[30] Cao, Y., Tan, Nin-Hua., Chen, Ji-Jun., Zeng, G., Ma, Y., Wu, Y., Yan, H., Yang, J., Lu, L-F., Wang, Q. *Fitoterapia*, 2010, 81(4), 253-258.
[31] Maiti, A., Cuendet, M., Kondratyuk, T., Croy, V.L., Pezzuto, J.M., Cushman, M. *J. Med. Chem.*, 2007, 50(2), 350-355.
[32] Ribby, M.C., Double, J.A.; *Anticancer Drugs,* 1993, 4, 3-17.
[33] Chan, K., Zhao, Y., Burkett, B.A., Wong, I.L.K., Chow, L.M.C., Chan, T.H. *J. Med. Chem.* 2006, 49 (23), 6742-6759.
[34] Aitken, R.A., Bibby, M.C., Double, J.A., Phillips, R.M., Sharma, S. *Bioorg. Med. Chem. Lett.* 1994, 4(19), 2313-2316.
[35] Parmar, V.S., Bisht, K.S., Sharma, S.K., Jain, R., Taneja, P., Singh, S., Simonsen, O. *Boll, Per M. Phytochemistry*,1994, 36(2), 507-11.
[36] Kim, J.Y., Lim, H.J., Ryu, J. *Bioorg. Med. Chem. Lett.* 2008, 18(4), 1511-1514.
[37] a) Ballesteros, J.F., Sanz, M.J., Ubeda, A., Miranda, M.A., Iborra, S., Paya, M., Alcaraz, M.J. *J. Med. Chem.* 1995, 38(14), 2794-2797. b) Nakatsuka, T., Tomimori, Y., Fukuda, Y., Nukaya, H. *Bioorg. Med. Chem. Lett.* 2004, 14(12), 3201-3203. c) Furuta, T.; Kimura, T.; Kondo, S.; Mihara, H.; Wakimoto, T.; Nukaya, H.; Tsuji, K.; Tanaka, K. *Tetrahedron,* 2004, 60(42), 9375-9379.
[38] Ji, Hye Young; Kim, Sung Yeon; Kim, Dong Kyun; Jeong, Ji Hyun; Lee, Hye Suk, *Molecules*, 2010, 15, 6466-6475.
[39] Veljkovic, V.; Mouscadet, J.; Veljkovic, N.; Glisic, S.; Debyser, Z.; *Bioorg. Med. Chem. Lett.* 2007, 17(5), 1226-1232.
[40] a) Inouye, Y.; Yamaguchi, K.; Take, Y.; Nakamura, S. *J. Antibiotics*, 1989, 42, 1523. b) Nakane, H.; Ono, K. *Biochemistry,* 1990, 29, 2841.

[41] a) Geahlen, R.L.; Koonchanok, N.M.; McLaughlin, J.L.; Pratt, D.E.; *J. Nat. Product*, 1989, 52, 982. b) Hagiwara, M.; Inouc, S.; Tanaka, T.; Nunoki, K.; Hidaka, I.H. *Biochem. Pharmacol*. 1988, 37, 2987.

[42] Hatano, T.; Yasuhara, T.; Mujamoto, K.; Okuda, T. *Chem. Pharm. Bull*, 1988, 36, 2286.

[43] Auffret, G.; Labaied, M.; Frappier, F.; Rasoanaivo, P.; Grellier, P.; Lewin, G. *Bioorg. Med. Chem. Lett*. 2007, 17(4), 959-963.

[44] Miki, K.; Nagai, T.; Suzuki, K.; Tsujimura, R.; Koyama, K.; Kinoshita, K.; Furuhata, K.; Yamada, H.; Takahashi, K. *Bioorg. Med. Chem. Lett*. 2007, 17(3), 772-775.

[45] Diwakar, S.D.; Bhagwat, S.S.; Shingare, M.S.; Gill, C.H. *Bioorg. Med. Chem. Lett*. 2008, 18(16), 4678-4681.

[46] Babu, T.H.; Subbarao, V.R.; Tiwari, A.K.; Babu, K., Srinivas, P.V.; Ali, A.Z.; Rao, J.M. *Bioorg. Med. Chem. Lett*. 2008, 18(5), 1659-1662.

[47] Dauzonne, D., Folleas, B., Martinez, L., Chabot, G.G. *Eur. J. Med. Chem*. 1997, 32, 71-82.

[48] Suresh Babu, K., Hari Babu, T., Srinivas, P.V., Kishore, K.H., Murthy, U.S.N., Rao, J.M. *Bioorg. Med. Chem. Lett*. 2006, 16(1), 221-224.

[49] Sathiamoorthy, B., Gupta, P., Kumar, M., Chaturvedi, A.K., Shukla, P.K., Maurya, R. *Bioorg. Med. Chem. Lett*. 2007, 17(1), 239-242.

[50] Mazzci, M., Balli, A., Roma, G., Braccio, M.D., Leoncini, G., Buzzi, E., Maresca, M. *Eur. J. Med. Chem*. 1988, 23, 237.

[51] Parmar, V.S., Bisht, K.S., Jain, R., Singh, S., Sharma, S.K., Gupta, S., Malhotra, S., Tyagi, O.D., Vardhan, A., Pati, N.H. *Indian J. Chem*. 1996, 35B, 220-232.

[52] Liu, T., Xu, Z., He, Q., Chen, Y.;,Yang, B., Hu, Y. *Bioorg. Med. Chem. Lett*. 2007, 17(1), 278-281.

[53] Narender, T., Khaliq, T., Puri, A., Chander, R. *Bioorg. Med. Chem. Lett*. 2006, *16(13)*, 3411-3414.

[54] Chen, J., Chang, H.C., Kim, H.P., Park, H. *Bioorg. Med. Chem. Lett*. 2006, 16(9), 2373-2375.

[55] Kang, S.S., Lee, J.Y., Choi, Y.K., Song, S.S., Kim, J.S., Jeon, S.J., Han, Y.N., Son, K.H., Han, B.H. *Bioorg. Med. Chem. Lett*. 2005, 15(15), 3588-3591.

[56] Larget, R., Lockhart, B., Renard, P., Largeron, M. *Bioorg. Med. Chem. Lett*. 2000, 10, 835-838.

[57] a) Frédérique, A.A., Acker, V., Hageman, J.A., Guido, R.M., Haenen, M., Wim, J.F., van der Vijgh, V., Aalt Bast, A., Wiro, M.P., Menge, B. *J. Med. Chem*. 2000, 43 (20), 3752-3760. b) Yue-ling, Z., Chi, W.T., Long, Y., Ling, C.Y., Ming, T.C., Cherng, C. *Chin. Pharm. J*. (Taipei), 1999, 51(1), 49-63.

[58] Prasanna, S., Manivannan, E., Chaturvedi, S.C. *Bioorg. Med. Chem. Lett*. 2004, 14(15), 4005-4011. b) Joo, Y.H., Kim, J.K., Kang, S., Noh, M., Ha, J., Choi, J.K., Lim, K. M., Chung, S. *Bioorg. Med. Chem. Lett*. 2004, 14(9), 2195-2198. c) Joo, Y.H., Kim, J.K., Kang, S., Noh, M., Ha, J., Choi, J.K., Lim, K.M., Lee, C.H., Chung, S. *Bioorg. Med. Chem. Lett*. 2003, 13(3), 413-417.

[59] Dao, T.T., Yeon Sook Chi, Y.S., Jeongsoo Kim, J., Hyun Pyo Kim, H.P., Sanghee Kim, S., Haeil Park, H. *Bioorg. Med. Chem. Lett*. 2004, 14(5), 1165-1167.

[60] Lee, Y., Yeo, H., Liu, S., Jiang, Z., Savizky, R.M., Austin, D.J., Cheng, Y. *J. Med. Chem*. 2004, 47(22), 5555-5566.

[61] Blank, V.C., Poli, C., Marder, M., Roguin, L.P. *Bioorg. Med. Chem. Lett.* 2004, 14(1), 133-136.

[62] Lin, Y., Michael, T., Flavin, M.T., Cassidy, C.S., Mar, A., Chen, F. *Bioorg. Med. Chem. Lett.* 2001, 11(16), 2101-2104.

[63] Bittner, M.L., Silva, M.J., Vargas, J.C., Watson, W.H. *Boletin de la Sociedad Chilena de Quimica*, 1982, 27(2), 291-292.

[64] Shin, J., Kim, K., Kim, M., Jeong, J., Kim, B. *Bioorg. Med. Chem. Lett.* 1999, 9(6), 869-874.

[65] Lin, Y., Zembower, D.E., Flavin, M.T., Schure, R.M., Anderson, H.M., Korba, B.E., Chen, F. *Bioorg. Med. Chem. Lett.* 1999, 7(17), 2325-2328.

[66] a) Marde, M., Zinczuk, J., Colombo, M.I., Wasowski, C., Viola, H., Wolfman, C., Medina, J.H., Rúveda, E.A., Paladini, A.C. *Bioorg. Med. Chem. Lett.* 1997, 7(15), 2003-2008. b) Huang, X., Liu, T., Gu, J., Luo, X., Ji, R,, Cao, Y., Xue, H., Wong, T., Bing, L., Wong, B.L., Pei, G., Jiang, H., Chen, K. *J. Med. Chem.* 2001, 44 (12), 1883-1891.

[67] Yoon, S.Y., dela P., Ike, C., Shin, C.Y., Son, K.H., Lee, Y., Soo, R.J.H., Cheong, J.H., Ko, K.H., *European J. Pharmacology*, 2011, 659(2-3), 155-160.

[68] Colombo, R., Yariwake, J.H., Queiroz, E.F., Ndjoko, K., Hostettmann, K. *J. Braz. Chem. Soc.* 2009, 20(9), 1574-1579.

[69] Tsukayama, Masao; Sasaki, Takahiro; Yamamoto, Kanji; Kawamura, Yasuhiko; Ichikawa, Ryouichi, *Nippon Shokuhin Kagaku Kogaku Kaishi,* 2010, 57(10), 427-433.

[70] Yang, Yansong; Pan, Langsheng, *Yingyong Huagong*, 2008, 37(6), 697-698.

[71] Li, Chenrui, Zhou, Limin, Lin, Ge, Zuo, Zhong, *J. Pharmaceutical and Biomedical Analysis*, 2009, 50(3), 298-306.

[72] Wei, Yun, Zhang, Kai, Zhang, Guoliang, Ito, Yoichiro. *J. Liquid Chromatography & Related Technologies,* 2011, 34(20), 2505-2515.

[73] Zhang, Jianjun, Lv, Huixia, Jiang, Kun, Gao, Yuan. *International J. Pharmaceutics* 2011, 420(1), 180-188.

[74] a) Guo, Xiao-Yu, Yang, Lin, Chen, Ying, Wang, Qi-Fang, Sun, Qi-Shi, Che, Yi-Xin, Che, Qing-Ming. *J. Asian Natural Products Research,* 2011, 13(9), 861-868. b) Yang, Lin, Guo, Xiao-Yu, Wang, Qi-Fang, Chen, Ying, Che, Yi-Xin, Che, Qing-Ming, J. *Chinese Pharm.l Sci.s,* 2011, 20(3), 275-281.

[75] Das, S., Rosazza, J.P.N.*J.* Nat. Prod. 2006, *69* (3), 499-508.

[76] Agrawal, N.N., Soni, P.A. *Indian J. Chem.* 2005, 44B, 2601-2603.

[77] a) Doshi, A.G., Soni, P.A., Ghiya, B.G. *Indian J. Chem.* 1986, 25B, 759. b) Ankhiwala, M.D. *J. Inst. Chem.* 1995, 67(2), 58-61. c) Ankhiwala, M.D. *J. Inst. Chem.,* 1995, 67(4), 121-122.

[78] Singh, O.V., Khanna, S.M.S., Garg, C.P., Kapoor, R.P. *Indian J. Chem.*1993, 32B, 1241-1248.

[79] Rao, P.S., Reddy, K.V.V., Ashok, D. *Indian J. Chem.* 2000, 393, 557.

[80] Gyoergy, L., Katalin, G., Sandor, A., Gabor, B. *Leibigs Ann.* 1995, 9, 1711-1715.

[81] Lokhande, P.D., Sakate, S.S., Taksande, K.N., Navghare, B. *Tetrahedron Lett.* 2005, 46(9), 1573-1574.

[82] Imafuku, K., Honda, M., McOmie, J. *Synthesis,* 1987, 199. b) Hossain, M.A., Islam, A. *Indian J. Chem.* 1993, 32B, 1058-1060.

[83] Hossain, M.A. *Indian J. Chem.* 1997, *36B,* 927-928. c) Hossain, M.A. *Indian J. Chem.* 1999, 38B, 590-592.

[84] a) Hossain, M.A., Ismail, Z. *Indian J. Chem.* 2005, *44B*, 413-415. b) Sayed, A. *J. Chem. Sci.* 2004, 116(6), 325–331.

[85] Sayed, A. *Acta Chim. Slov.* 2004, 51, 447-452.

[86] Makrandi, J. K., Seema, P. *Chem. & Industry,* 1989, 607-608.

[87] Mahal, H.S., Rai, H.S., Venkataraman, K. *J. Chem. Soc.* 1935, 866.

[88] Dawane, B.S., Vibhute, Y.B., Chavan, V.P., Mane, A.S., Shingare, M.S. *Korean J. Med. Chem.* 2000, 10(2),78-83. d) Subbanwad, G.R., Vibhute, Y.B. *J. Ind. Chem.* Soc. 1991, 68, 570-571.

[89] Pfister, J.R., Wymann, W.E., Schuler, M.E., Roszkowski, A.P. *J, Med. Chem.* 1980, 23, 335-338.

[90] Dawane, B.S., Vibhute, Y.B., Chavan, V.P., Mane, A.S., Shingare, M.S. *Korean J. Med. Chem.* 2000, 10(2), 78-83.

[91] Bose, G., Mondal, E., Khan, A.T., Bordoloi, M.J. *Tetrahedron Lett.* 2001, 42(50), 8907-8909.

[92] Khan, A.T., Goswami, P. *Tetrahedron Lett.* 2005, 46(30), 4937-4940.

[93] Meyer, N.D., Haemers, A., Mishra, L., Pandey, H.K., Pieters, L.A.C., Berghe, D.A.V., Vlientinck, A. J. *J. Med. Chem.* 1991, 34, 736-746.

[94] Saravanamurugan, S., Palanichamy, M., Arabindoo, B., Murugesan, V. *Catal. Commun.* 2005, 6(6), 399-403.

[95] Malik, U.K., Saha, M.M., Malik, A.K. *Indian. J. Chem.* 1989, 28B, 970-972.

[96] Hans, N., Grover, S.K. *Synth. Commun.* 1993, 23(7), 1021-1023.

[97] Artur, M.S.S., Jose, A.S.C. *Bull. Liaison-Group Polyphenols*, 1992, 16(1), 157.

[98] Zambare, A.S., Sangshetti, J.N., Kokare, N.D., Shinde, D.B. *Chinese Chem. Lett.* 2009, 20(2), 171-174.

[99] Kasahare, A., Izumi, T., Ooshima, M. *Bull. Chem. Soc. Jpn.* 1974, 47, 2226.

[100] Pandey, G., Krishna, A., Kumarswamy, G. *Tetrahedron Lett.* 1987, 28, 4615.

[101] Krishnamurthi, M., Parthasarthi, M. *J. Indian J. Chem.* 1980, 19B, 816.

[102] Costantino, L., Rastelli, G., Gamberini, M.C., Vinson, J.A., Bose, P., Iannone, A., Staffieri, M., Antolini, L., Corso, A.D., Mura, U., Albasini, A. *J. Med. Chem.* 1999, 42, 1881-1893.

[103] Mokle, S.S., Sayyed, M.A., Vibhute, A.Y., Khansole, S.V., Nalwar, Y.S., Vibhute, Y.B. *Research J. Pharmacy and Technology*, 2009), 2(4), 846-849.

[104] Tiwari, K.N., Monserrat, J.P., de Montigny, F., Jaouen, G., Rager, M.N., Hillard, E. *Organometallics,* 2011, 30(20), 5424-5432.

[105] Singh, O.V., Kapoor, R.P. *Tetrahedron Lett.* 1990, 31, 1459.

[106] Singh, O.V., Muthukrishnan, M., Raj, G. *Synth. Commun.* 2005, 35(20), 2723.

[107] Malik, U. K., Saha, M.M., Malik, A.K. *Indian J. Chem.* 1989, 28B, 910-972.

[108] a) Bhat, V.S., Joshi, V.S. Indian J. Chem. 1991, 30 B, 976-977. b) Shankar, C. G.; Mallaiah, B.V., Srimannarayana, G. Synthesis, 1983, 310. c) Krishnamurthi, M.., Parthasarthi, J. *Indian J. Chem.*, 1980, 19B, 816.

[109] Fatma, W., Iqbal, J., Manchanda, V., Shaida, W.A., Rehman, W.*J. Chem. Res.(S)*, 1984, 298, *J. Chem. Res.* 1984, (M), 2656. b) Doshi, Y.A.C., Soni, P.A., Ghiya, B.J. *Indian J. Chem.* 1986, 25B, 259.

[110] Chakrabarty, K., Chawla, H.M., Suresh, V.V. *Indian J. Chem.* 1992, 31B, 464. b) Chawla, H.M., Sharma, S.K. *Bull. Chem. Soc. Fr.* 1990, 127, 656.
[111] Baker, W.J. *J. Chem. Soc.* 1933, 1381-1389. b) Mahal, H.S., Venkataraman, K. *ibid, J. Chem. Soc.* 1934, 1767-1769. c) Diana, C.G.P., Silva, A.M.S., Jose, A.S.C. *New J. Chem.* 2000, 24, 85È92.
[112] Lee, J.I., Son, H.S., Park, H. *Bull. Korean Chem. Soc.* 2004, 25(12), 1945-1946.
[113] a) Fougerousse, A., Gonzalez, E., Bouillard, R. *J. Org. Chem.* 2000, 65, 583-586. b) Ismail, K.A., Aziem, T.A. E. *Eur. J. Med. Chem.* 2001, 36, 243-253.
[114] Damle, S.V., Muley, P.R. *Indian. J. Hetro. Chem.*1999, 9, 81-86. b) Gadaginamath, G.S, Kavali, R.K. *Indian J. Chem.* 1999, 38B, 178-182. c) Mazumdar, A.K.D., Karmakar, P.K., Das, S.C, Saha, S.C., Banerji, K.D. *J. Indian Chem. Soc.* 1992, 69, 880.
[115] Virkar, V.V., Wheeler, T. S. *ibid,* 1939, 1679-1681.
[116] a) Goker, H., Siizen, S., Kus, C., Boykin, D.W. *Hetro. Commun.* 2000, 6(5), 385. b) Kumar, P.E., Prasad, K.J.R. *Indian J. Chem.* 1999, 38B, 1277-1279.
[117] Jain, P.K., Makrandi, J.K., Grover, S.K. *Synthesis,* 1982, 221. b) Saxena, S., Makrandi, J.K., Grover, S.K. *Synthesis,* 1985, 697.
[118] Palkar, R.B., Master, H.E. *bid. J. Hetro. Chem.* 1997, 7, 25-30.
[119] Jain, P.K., Makrandi, J.K., Grover, S.K. *Curr. Scince.* 1981, 50, 857-858.
[120] Gupta, V.N., Seshadri, T.R. *J. Sci. Indust. Res (India),* 1957, 16B, 116-119.
[121] a) Makrandi, J.K., Kumari, V. *Chem. & Industry,* 1988, 630. b) Paquette, L.A. Stucki, H. *J. Org. Chem.* 1966, 31, 1232-1235.
[122] Banerji, A., Goomer, N.C. *Synthesis,* 1980, 874. b) Cushman, M., Nagarathnam, D. *Tetrahedron Lett.* 1990, 31, 6497.
[123] *Lee, Y.R., Kang, K.Y. Lett. Org. Chem.* 2007, 4(6).
[124] Daniel, O.B., Gustavo, P.R., Jorge, L.J., Juan, C.A., Graciela, T.B., Horacio, J.T. *Arkivoc*, 2008, xi, 123-130.
[125] Kabalka, G.W., Mereddy, A.R. *Tetrahedron Lett.* 2005, 46(37), 6315-6317.
[126] Varma, R.S., Saini, R.K., Kumar, D. *J. Chem. Res.* 1998, *(S)*, 348-349.
[127] Okombi, S., Schmidt, J., Mariotte, A., Perrier, E., Boumendjel, A. *Chem. Pharm. Bull.* 2005, 53(11), 1460-1462.
[128] Lee, J., Son, H.S., Jung, M.G. *Bull. Korean Chem. Soc.* 2005, 26(9), 1461.
[129] Tang, L., Zhang, S., Yang, J., Gao, W., Cui, J., Zhuang, T. *Molecules,* 2004, 9, 842.
[130] Ganguly, A.K., Kaur, S., Mahata, P.K., Biswas, D., Pramanik, B.N., Chan, T. M. *Tetrahedron Lett.* 2005, 46, 4119-4121.
[131] Kumar, P., Bodas, M.P. Org. Lett., 2000, *2* (24), 3821-3823.
[132] Hercovert, A., Corre, L.M. *Synthesis,* 1982, 597.
[133] Orjales, A., Berisa, A., Cires, L.A., *Indian J. Chem.* 1994, 33B, 27-31.
[134] Floc'h, Y.L., Lefeuvre, M. *Tetrahedron Lett.* 1986, 27(45), 2751.
[135] Floc'h, Y.L., Lefeuvre, M. *Tetrahedron Lett.* 1986, 27(45), 5503-5504.
[136] a) Miao, H., Yang, Z. *Org. Lett.* 2000, 2(12), 1765-68. b) Kalinin, V.N., Shostakovsky, M.V., Ponomaryov, A.B. *Tetrahedron Lett.* 1990, 31, 4073.
[137] Nagarathnam, D., Cushman, M. *Tetrahedron*, 1991, 47(28), 5071-5076.
[138] a) Garcia, H., Iborra, S., Primo, J., Miranda, M.A. *J. Org. Chem.* 1986, *51*, 4432-4436. b) Ares, J.J. *J. Med. Chem.* 1995, 38, 4937-4943.
[139] Nakanishi, K. *Natural Products Chemistry*. 2, Tokyo: Kodansha Ltd. 225-228.

[140] a) Allan, J., Robinson, R. *J. Chem. Soc.* 1924, 125, 2192. b) Horie, T., Kawamura, Y., Tsukayama, M., Yoshizaki, S. *Chem. Pharm. Bull.* 1989, 32, 1216. c) Twardhan, S.A., Gupta, A.S. *J. Chem. Res. (S)* 1984, 395. d) Dutta, P.K., Bagchi, D., Pakrashi, S.C. *Indian J. Chem.* 1982, 21B, 1037. d) Wagner, H., Maurer, I., Farkas, L., Strelisky, J. *Tetrahedron,* 1977, 33, 1405.

[141] Paquette, L.A., Stucki, H. *Org. Chem.* 1966, 31, 1232-1235.

[142] Oganesyan, E.T., Vasilenko, Y.K., Khachatryan, M.M., Psychen, A.I. *Khim. Farm. Zh,* 1989, *23,* 1353. *Chem. Abstr.* 1990, 122, 171980.

[143] Boumendjel, A. F., Bois, C., Beney, A.M., Mariotte, G., Couseil, A.D. *Bioorg. Med. Chem. Lett.* 2001, 11, 75-77.

[144] Uchil, V.R.; Joshi, V. *Indian J. Hetro. Chem.* 2001, 10, 161-166.

[145] Kumari, T.U., Krupadanam, G.L.D., Srimannarayana, G. *Indian. J. Chem.* 1998, 37B, 847-850.

[146] Swapnil R. Sarda, Wamanrao N. Jadhav, Rajendra P. Pawar, *International J. Chem. Tech. Res.* 2009, 1(3), 539-543.

[147] Swapnil R. Sarda Mohsin Y. Pathan, Vijaykumar V. Paike, Pandurang R. Pachmase, Wamanrao N. Jadhav, Rajendra P. Pawar. *Arkivoc*, 2006, xvi, 43-48.

[148] Bhosale, R.S., Sarda, S.R., Giram, R.P., Raut, D.S., Parve, S.P., Ardhapure, S.S.,Pawar, R. P. *J. of the Iranian Chemical Society***,** 2009, 6(3), 519-522.

[149] Wilson, C.W. *J. Am, Chem. Soc.* 1938, 2303.

In: Bioactive Heterocycles
Editors: K. L. Ameta, R. P. Pawar and A. J. Domb

ISBN: 978-1-62257-636-4
© 2013 Nova Science Publishers, Inc.

Chapter 12

Bioactivity and Synthesis of Pyrazoline Motifs

Sunil U. Tekale[1*], Vivekanand B. Jadhav[1], Rupali L. Magar[2], Chabubai S. Patil[2], Rajita D. Ingle[2], Saroj R. Bembalkar[2] and Yeshwant B. Vibhute[3]

[*1] Department of Chemistry, Shri Muktanand College, Gangapur, MS, India
[2] Department of Chemistry, Deogiri College, Aurangabad, MS, India
[3] Department of Chemistry, Yeshwant Mahavidyalaya, Nanded, MS, India

Abstract

Pyrazolines constitute an important class of heterocyclic compounds that are associated with immense pharmacological significance. The present chapter describes the biological potential and various synthetic methods for pyrazoline heterocycle containing organic compounds. Different biologically active heterocyclic compounds possessing pyrazolines in their structures are discussed in detail for their utility as antidepressant, cytotoxic, anticancer, anticonvolusant and much other therapeutic potential.

1. Introduction

Pyrazole (**1a**) is a simple, five membered aromatic heterocyclic compound consisting of three carbon atoms and two nitrogen atoms in adjacent positions. The partially reduced form of pyrazole is called pyrazoline **(1b),** which exists in tautomeric forms (**1c, 1d**). These pyrazoline compounds possess interesting pharmacological properties and hence they are used as raw materials in drug synthesis.

[*] Corresponding Email: tekale.sunil@rediffmail.com

pyrazole a | pyrazoline b | Tautomeric forms of pyrazoline c d

(1)

These play a crucial role in heterocyclic chemistry and are extensively used as synthons in organic synthesis.

Pyrazolines are important, 5-membered, electron rich nitrogen heterocyclic compounds possessing diverse biological and pharmacological activities. Numerous pyrazoline derivatives have been found to possess considerable biological activities that stimulated the research activity in this field. Organic compounds containing pyrazoline heterocyclic moiety as the basic skeleton have been proved to possess remarkable biological activities, such as being antibacterial, antifungal, antimicrobial, antipyretic' and antiinflammatory agents. They also possess some selective receptor activity.

Diversely substituted pyrazoline motifs and their derivatives with different functional groups in their skeletons are important biological agents; hence research activity is directed towards this class of organic compounds. QSAR study of these compounds helps in designing their structures. In particular, they are widely used as antitumor, antiviral, antiparasitic, anti-tubercular and insecticidal agents and exhibit potent selective activities such as nitric oxide synthase (NOS) inhibitor and cannabinoid CB1 receptor antagonist activity. The pharmaceutical importance of these compounds lies in the fact that they can be effectively utilized as antinociceptive, antidepressant, antiamoebic, tranquillizing, immune-suppressive, anti-arthirtic and antimycobacterial agents. Some of these compounds also have anti-inflammatory, antidiabetic and anaesthetic properties.

2-Pyrazolines are found in several pharmacologically potent molecules, such as indoxacarb (insecticide), azolid/ tandearil (anti-inflammatory) and anturane (uricosuric). The minor structural variation of these molecules provides diversified activities, which are useful for the development of new therapeutic agents. Furthermore, pyrazolines play a crucial role in the development of heterocyclic chemistry and organic synthesis.

The present chapter provides a compilation overview of biological activities along with the recent methods of synthesis of these pyrazoline motifs.

2. Biological Significance of Pyrazolines and their Derivatives

Different biologically active heterocyclic compounds possessing pyrazolines as their ring motifs can be categorized into following types:

- Antimicrobial pyrazolines
  - Antibacterial and antifungal pyrazolines
  - Antiviral pyrazolines
- Antidepressant pyrazolines

- Cytotoxic and anticancer pyrazolines
- Anticonvolusant pyrazolines
- Anti-inflammatory and analgesic pyrazolines
- Antiamoebic pyrazolines
- Enzyme inhibitors/modulator pyrazolines
- Anti-tubercular and antimycobacterial pyrazolines
- Anti-malerial pyrazolines
- Miscellaneous activities of pyrazolines

2.1. Antimicrobial (Anti-bacterial, Antifungal and Anti-Viral) Pyrazolines

The different micro-organisms responsible for causing several infectious diseases in both human and plants include various bacteria, fungi and viruses. Bacteria are prokaryotic microorganisms, typically few micrometers in length.

Some species of bacteria occurring in nature are pathogenic and cause infectious diseases such as cholera, syphilis, anthrax, leprosy etc. Commonly, antibiotics are used to treat bacterial infections. Bacteria are often maligned as the causes of human and animal disease (like this one, *Leptospira*, which causes serious disease in livestock). Similarly, some species of fungi are also responsible for causing certain diseases. An antibacterial is a compound or substance that kills or slows down the growth of bacteria whereas antifungal agents are the compounds used against various types of fungi.

Microbial infections may produce pain and inflammation in the regions of affected areas.

An enormous amount of research is being undertaken to investigate bio-activities associated with pyrazolines. Various reports are available in literature. The common microorganisms evaluated for the biological activity studies of pyrazolines involve gram positive and gram negative bacteria, such as *Escherichia coli, Staphylococcus aureus*, *Salmonella typhimurium*, *Bacillus cereus* and *Streptococcus faecalis* and fungi like *Aspergillus niger, Aspergillus flavus, Fusarium oxysporium, Trichoderma species, Fusarium monaliforme* etc.

The standard drugs employed are ampicillin, ciprofloxacin, streptomycin, tetracycline etc. The usually adapted methods for activity evaluation are either disc diffusion or agar method.

2.1.1. Antibacterial and Antifungal Pyrazolines

Ozdemir A. et al. [1] reported on some 1-(4-aryl-2-thiazolyl)-3-(2-thienyl)-5-aryl-2-pyrazolines **(2)** and evaluated their antimicrobial activity against *Escherichia coli, Staphylococcus aureus, Salmonella typhimurium, Bacillus cereus, Streptococcus faecalis, Aeromonas hydrophila, Candida albicans* and *Candida glabrata*. The compounds were observed to show promising activity.

Bakr F. Abdel-Wahab **[2]** synthesized 1-(benzofuran-2-yl)-4-nitro-3-arylbutan-1-ones **(3)** and 3-(benzofuran-2-yl)-4,5-dihydro-5-aryl-1-[4-(aryl)-1,3-thiazol-2-yl]-1*H*-pyrazoles and evaluated their antibacterial and antifungal activities at 100 μg concentration.

Some of the compounds showed excellent antimicrobial activities, better than the control drugs.

The pyrazolines (4) synthesized under solvent free conditions by Zeba N. Siddiqui [3] were tested for antimicrobial activity by the disk diffusion method. The compounds were shown to be potential agents having antibacterial and antifungal activities as good as standard antibiotics like ciprofloxacin, griseofulvin and fluconazole.

3-aryl-5-(pyridin-3-yl)-1-thiocarbamoyl-2-pyrazolines (5) were prepared by M. Shekarchi et al. [4] and reported their antibacterial activities against *Escherichia coli, Staphylococcus aureus, Staphylococcus epidermidis, Pseudomonas aeruginosa, Bacillus subtilis* and *Micrococcus luteus*. The antifungal activity of these compounds against *Candida albicans* and *Candida globrata* were found to be inactive. These compounds showed good antibacterial activity. An important conclusion of the work was the activity of these compounds against M. tuberculosis.

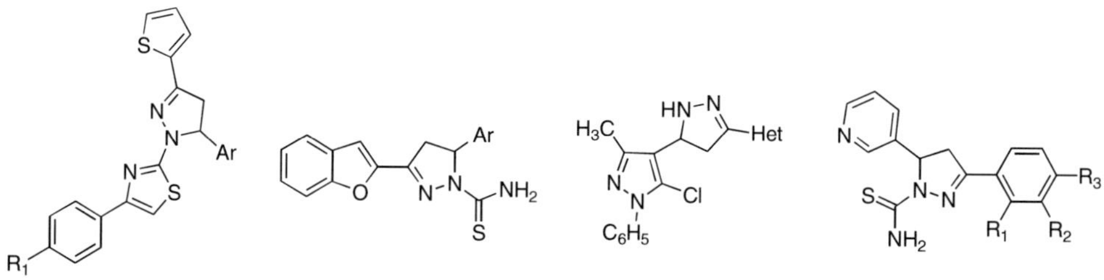

(2) (3) (4) (5)

Davood Azarifar and Maseud Shaebanzadeh [5] studied the synthesis and antimicrobial evaluation of some 3, 5-dinaphthyl substituted 2-pyrazolines (8). The compounds were tested on several bacterial strains including *Escherichia coli, Staphylococcus aureus, Klebsiella pneumoniae, Proteus mirabilis* etc. The compounds containing chloro, hydroxo and dimethylamino - $N(CH_3)_2$ group as substituents on the naphthalene rings were highly potential as antimicrobial agents. The substantial antimicrobial activity of carboxamido derivative was attributed to the presence of $-CONH_2$.

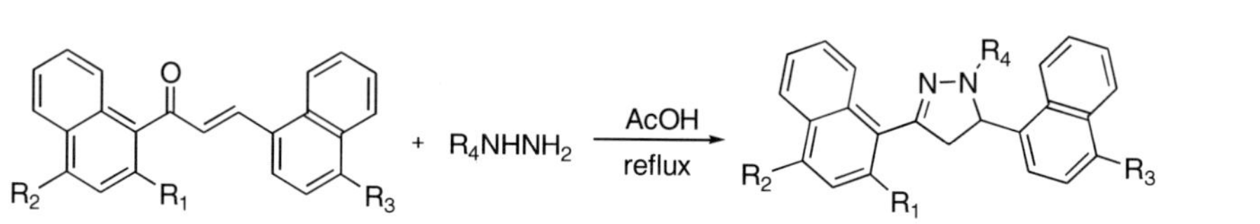

(6) (7) (8)

R_1= H, OH; R_2= H, Cl, Me
R_4= H, Ph, $CONH_2$

Dhananiaya Nauduri and G. B. S. Reddy [6] reported some antibacterial and antifungal pyrazolines **(14)**. The antibacterial study was compared with ciprofloxacin and the antifungal activity of the compounds was compared with naftifine as the standard. The 4', 4"-NMe_2, 4', 4"-NO_2, 4', 4"-OMe substituents on the phenyl or styryl ring showed the highest activity in the case of both the gram positive and gram negative bacteria.

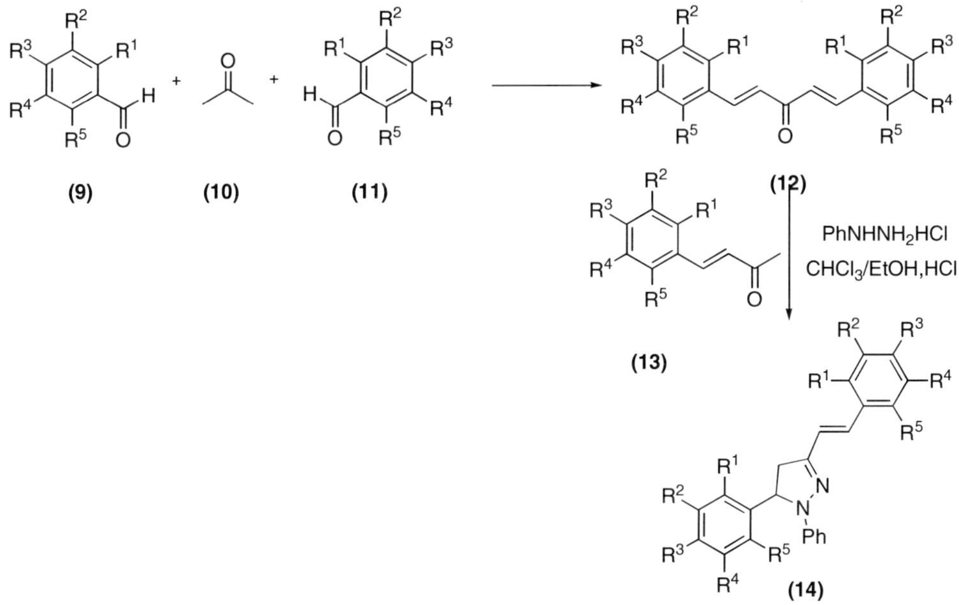

(9) (10) (11) (12) (13) (14)

Karen L. Stirrett et al. [7] **synthesized compounds (15) having structures similar to siderophores and demonstrated their antimicrobial activity against *Mycobacterium tuber-culosis* and *Yersinia pestis* under iron-limiting conditions.**

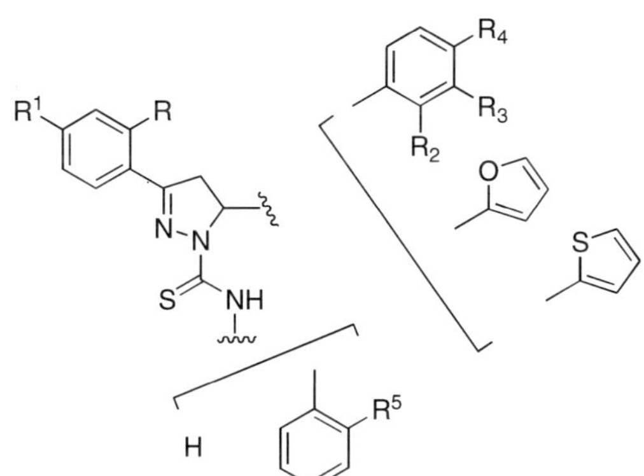

(15)

The chalcones obtained from 4-thiomethyl benzaldehyde and various ketones were cyclized to form various pyrazolines (19), (20) and evaluated for antibacterial activity against *Staphylococcus aureus, pseudomonas aeruginosa, E. coli* etc with the standard ciprofloxacin by Mithun Ashok and B. Shivaram Holla [8]. Many of the compounds were found to be highly active agents against these bacteria. The compounds 1-acetyl-3-phenyl-5-(4-methylthiophenyl)-4,5-dihydropyrazole and 1-acetyl-3-(4-chlorophenyl)-5-(4-methylthio-phenyl)-4,5-hihydropyrazole showed the highest activity.

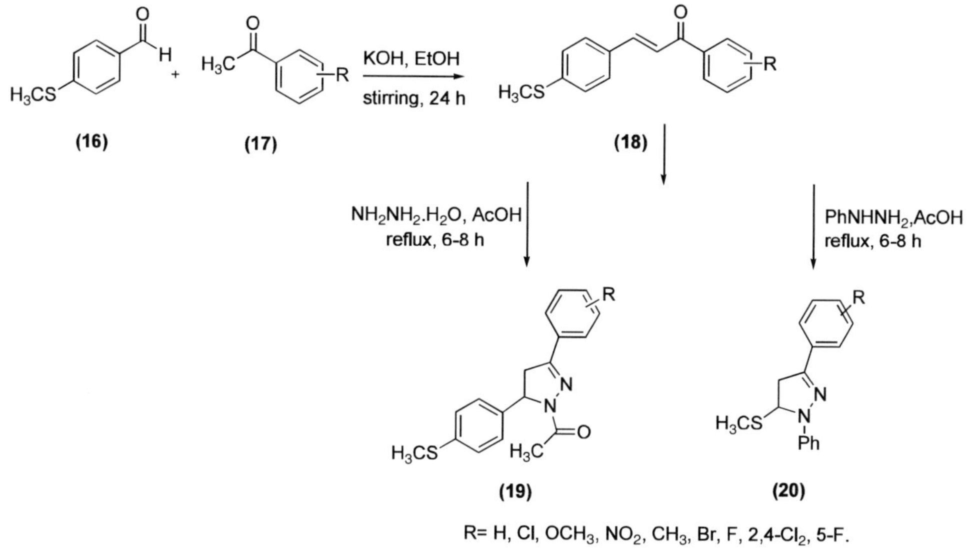

A. Solankee [9] prepared 2-(4'- chlorophenylamino)-4-(4'-fluorophenylamino)-6-[4'-{1"-phenyl-5"-(phenyl / substituted phenyl)-2"-pyrazolin-3"-yl}phenylamino]-s triazine **(29),** which were screened for in vitro antibacterial activity using gram positive and gram negative bacteria. The bacteria used were *S. aureus, B. subtilis*, *E. coli* and *S. Paratyphi-B* using the agar diffusion method. The activity results were compared with antibiotic ciprofloxacin. Almost all compounds exhibited promising results.

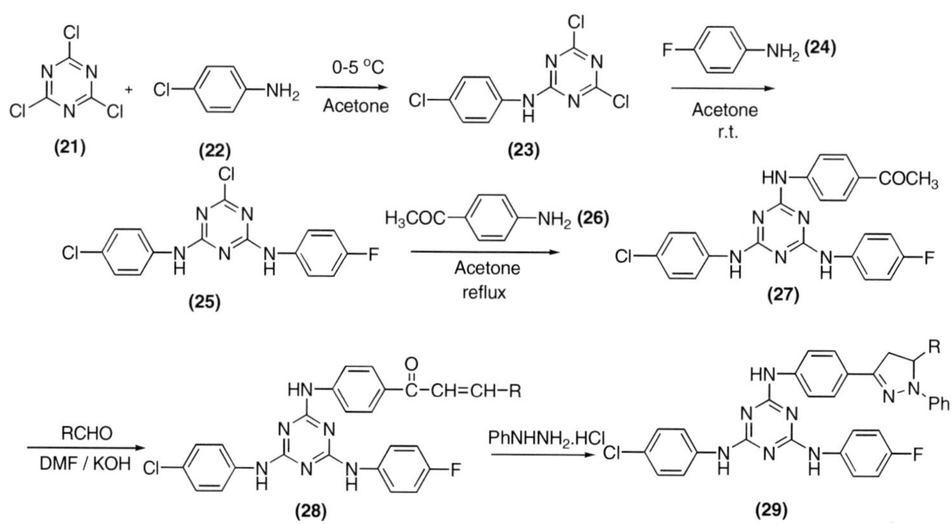

S. S. Mokle et al. [10] obtained 2-pyrazolines containing quinoline moiety using triethanolamine as the reaction solvent **(31)** and evaluated their antibacterial nature. Many of the compounds exhibited potent activity. The antimicrobial activities of the synthesized compounds were determined by agar well diffusion method against *Bacillus subtilis, E. coli, Ervinia carotovara* and *Xanthomanas citri.* Two of the compounds showed better activity when tested on *Bacillus subtilis* and *Ervinia carotovara* than the standard ampicillin drug and the remaining compounds showed moderate to good antibacterial activity.

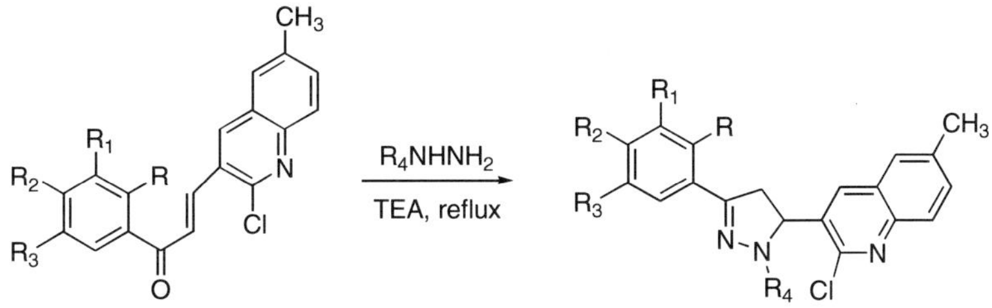

(30) **(31)**

B. Jayashankara and K. M. Lokanatha Rai [11] synthesized imidazole pyrazolines **(36)** by 1, 3-dipolar cycloaddition on phenyl hydrazones **(35)** and evaluated them for their antimicrobial studies using the disc diffusion method and microdilution method against different microbes such as bacteria - *Escherichia coli, Xanthomonas oryzae, Bacillus substilis, Pseudomonas fluorescens, Xanthomonas campestris pvs* - and fungi, namely, *Aspergillus niger, Aspergillus flavus, Fusarium oxysporium, Trichoderma species, Fusarium* and *monaliforme.* The microbial evaluation was carried out using streptomycin and tetracycline as the reference drugs. The compounds showed good activity.

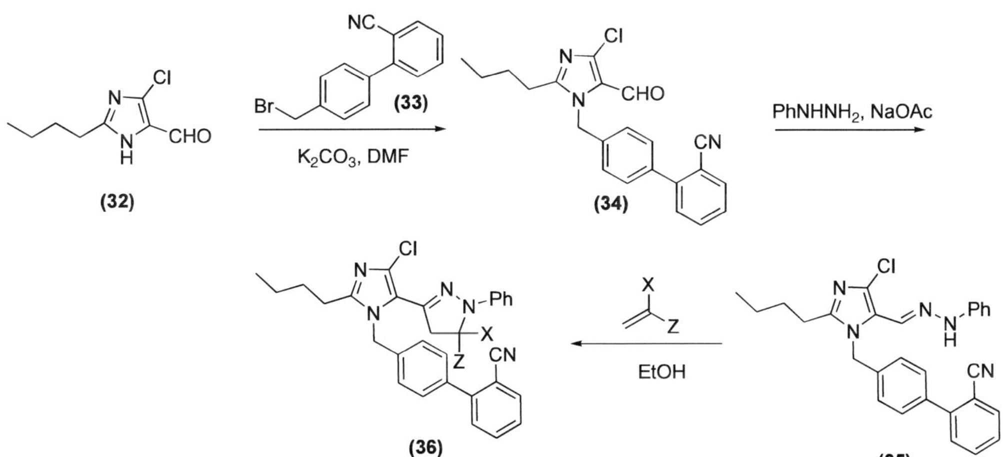

X = H, CH_3

Z = Ph, CH_2CN, CH_2Cl, CH_2Br, CH_2OH, COOMe, COOEt

Bromo-benzothiazolo pyrazolines **(38)** were synthesized by Bharat Kumar [12] and tested for their microbial activity against pathogenic bacteria *S. aureus, B. subtilis, E. coli* and *S. typhi* using streptomycin, chloramphenicol and penicillin as the standard drugs. Among the compounds tested, most of them were found to be active against all bacteria except gram negative *E. coli.*

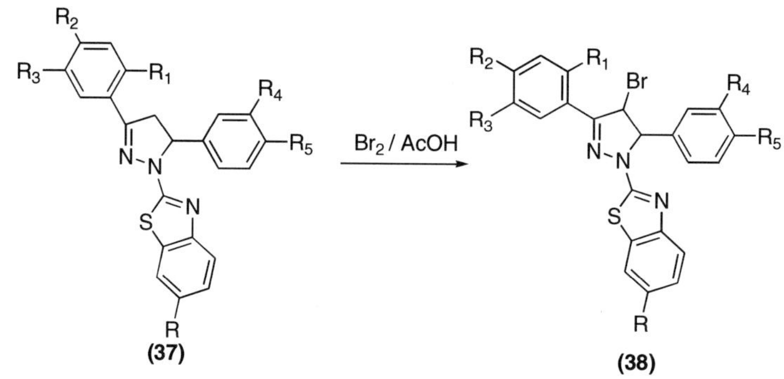

R = CH_3, OCH_3, R_1 = H, OH R_2 = H, Cl, CH_3,
R_3 = H, Cl, R_4 = H, OCH_3 R_5 = H, OCH_3

Sagar K. Mishra [13] reported the synthesis of pyrazolinones (45) which were checked for activity against bacterial strains such as *Staphylococcus aereus, Enterococcus faecalis* and gram negative bacteria (*Escherichia coli, Pseudomonas aeruginosa, Klebsiella pneumonia* and *Proteus mirabilis*) , fungi including *Aspergillus Niger, Candida krusei, Candida albicans* and *Candida tropicalis*. The 2-pyrazolin-4-one derivatives with 4-nitro- and 4-chlorophenyl substitutions at 5-position of pyrazoline ring are most active. 4-Bromo-2-pyrazoline derivatives and their 4-keto analogues with phenyl and 4- chlorophenyl substitutions at 5-position of pyrazoline ring and 2-pyrazolin-4-one derivative with 4- bromophenyl substitutions at 5-position of pyrazoline ring exhibited significant antibacterial activity against all. The antifungal activity study revealed the highest activity for 4-bromo-2-pyrazoline derivative and its 4-keto analogue with phenyl substitution at 5-position of pyrazoline ring against *A. niger* at 1000 mg/disc.

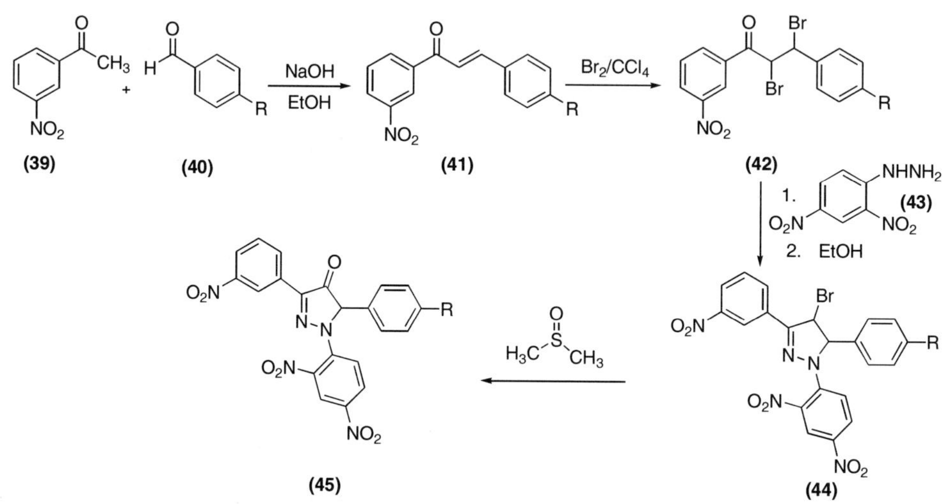

(R = H, NO_2, Cl, Br, OCH_3)

P. S. Patel [14] synthesized acetyl-5-(substituted phenyl)-{3-[4-(2-methyl-4-benzylidene-5- oxo-imidazol-1-yl)]phenyl}-4,5-dihydropyrazolines (50) and anti-microbial activities were tested by Cup-Borer method using standard drugs like penicillin, kanamycine, baycor 25 w.p and amphotericin B. The compounds showed moderate to good activity against these microorganisms.

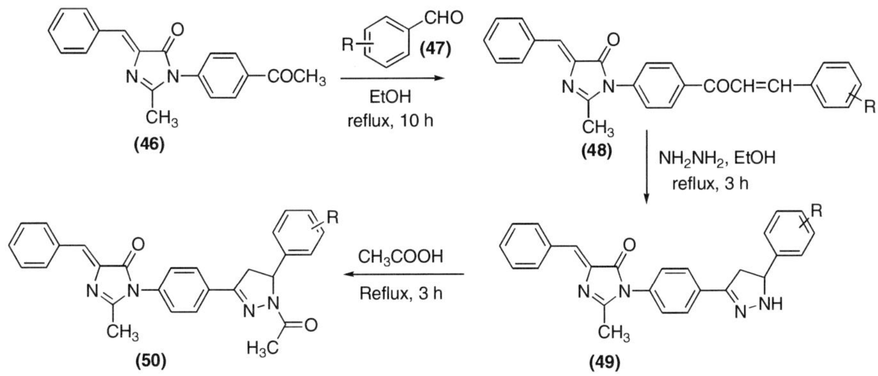

R= 4-Cl, 2-Cl, 3,4-$(OCH_3)_2$, 2-NO_2, 2-OH, etc

B. S. Dawane [15] synthesized 1-(4-(4′-chlorophenyl)-2-thiazolyl)-3-aryl-5-(2-butyl-4-chloro-1H-imidazol-5yl)-2-pyrazolines (57) in PEG as the reaction medium and reported their antimicrobial evaluation against *Escherichia coli, Salmonella typhi, Staphylococcus aureus, Bacillus subtilis, Aspergillus niger, Trichoderma viridae, Penicillium chrysogenum, Fusarium moniliforme* and *Candida albicans*. A majority of the compounds showed potent antibacterial and antifungal activity.

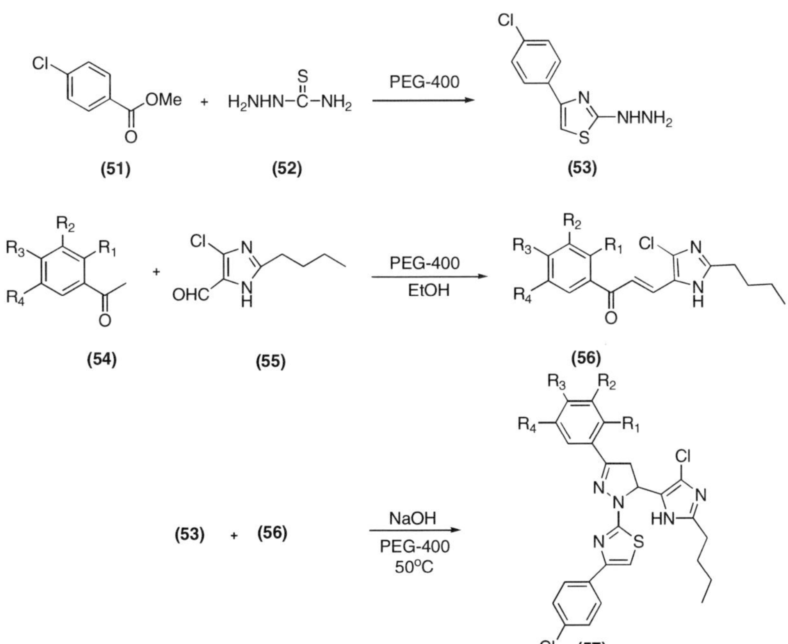

Shankaraiah G. Konda [16] reported the synthesis and antibacterial studies of pyrazolines containing naphthalene moiety (61). The antibacterial activity of these compounds was monitored by the agar well diffusion method against gram-positive *Bacillus subtilis* and

Staphylococcus aureus and gram-negative *Escherichia coli* and *Salmonella typhi* bacteria species using antibiotic penicillin (25 μg/mL) as the reference antibacterial substance for comparison purpose. All the cited compounds exhibited moderate to good activity against these bacterial stains.

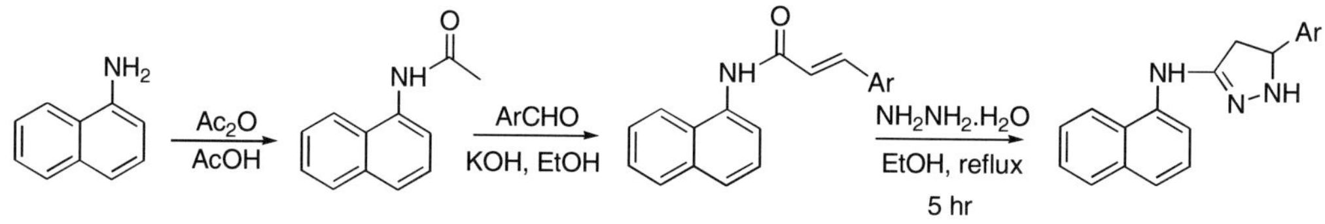

(58) **(59)** **(60)** **(61)**

Shah N. N. [17] reported piperazine containing pyrazolines (63) possessing moderate to good antimicrobial activity against several bacterial and fungal species using the cup plate method. The greater broad spectrum of biological and medicinal utilities of these compounds was suggested because of the presence of piperazine ring, chloro, bromo and iodo types of substituents.

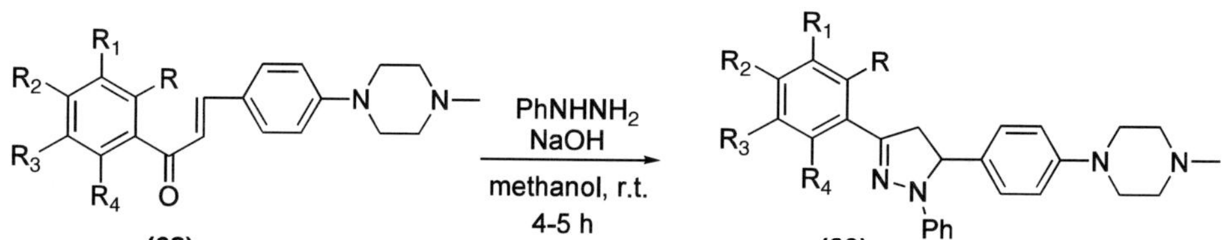

(62) **(63)**

2.1.2. Antiviral Pyrazolines

John R. Goodell [18] reported pyrazolines (64) which were identified as viable leads and subjected to additional evaluation using an authentic viral titer reduction assay employing an epidemic strain of WNV. The compounds were further evaluated in an *Anti-West Nile* virus activity studies.

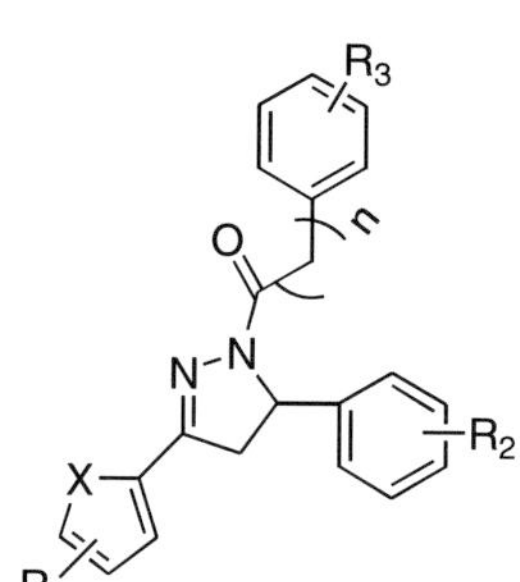

(64)

K. Sujatha [19] reported the CAN catalyzed synthesis of pyrazolines (66) in water as the reaction medium and investigated their antiviral activity against *peste des petits ruminant* virus (PPRV) with the standard drug ribavirin.

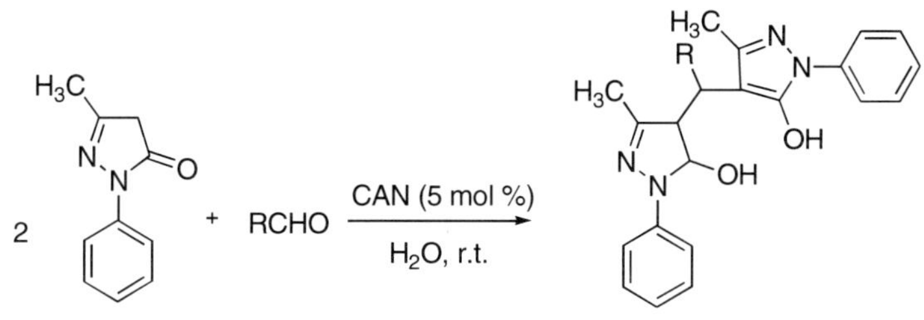

(65) (66)

2.2. Antidepressant Pyrazolines

An antidepressant is a psychiatric medication used to treat and cure mood disorders such as major depression, dysthymia, anxiety disorders etc. In simple words, *antidepressants* are the medicines that are used against mental depression. The term antidepressant is applied to any compound which can improve a clinically depressed mood. Antidepressants have sometimes been used to treat snoring and migraines also. Drugs including the monoamine oxidase inhibitors (MAOIs), tricyclic antidepressants (TCAs) etc are available as anti-depressants.

Zuhal Ozdemir [20] reported the antidepressant and anticonvulsant activity of the 3-(2-furyl)-pyrazolines. The antidepressant studies of these compounds were evaluated on albino mice by Porsolt's behavioral despair method. Compounds (67) were found to be highly promising for their anticonvulsant activities.

M. C. Cardia [21] reported the synthesis and inhibitory activity of *N*1-propanoyl-3,5-diphenyl-4,5-dihydro-(1*H*)-pyrazole (68) derivatives against MAO. A majority of the compounds showed inhibitory activity with micromolar values and MAO-A selectivity.

1-Thiocarbamoyl-3,5-diaryl-4,5-dihydro-(1H)- pyrazole (69) derivatives were synthesized by Franco Chimenti [22] et al. and were proven to be active against the A and B isoforms of monoamine oxidase (MAO).

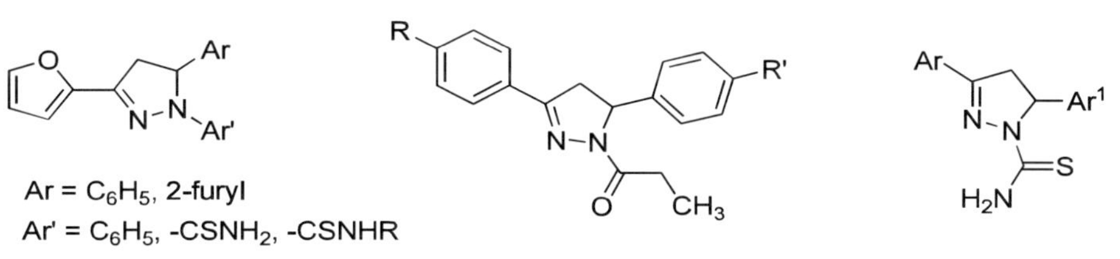

(67) (68) (69)

The antidepressant activity of 1,3,5-triphenyl-2-pyrazolines and 3-(2″-hydroxy naphthalen-1″-yl)-1, 5-diphenyl-2-pyrazolines was reported by Y. R. Prasad et al. [23] (70), (71). 1,3,5-triphenyl-2-pyrazolines were synthesized by reacting 1,3-diphenyl-2-propene-1-one with phenyl hydrazine hydrochloride. Other five new 3-(2″-hydroxy naphthalen-1″-yl)-1,5-diphenyl-2-pyrazolines were also synthesized by reacting 1-(2′-hydroxynaphthyl)-3-phenyl-2-propene-1-one with phenyl hydrazine hydrochloride. The ‘Porsolt behavioural despair test’ was employed on Swiss-Webster mice method to study the antidepressant activity of these compounds. Studies have revealed that compounds with electron-releasing groups like dimethyl amino, methoxy and hydroxyl substituents on both the aromatic rings at positions 3 and 5 of pyrazolines considerably enhanced the antidepressant activity as compared with the pyrazolines having no substituents on the phenyl rings.

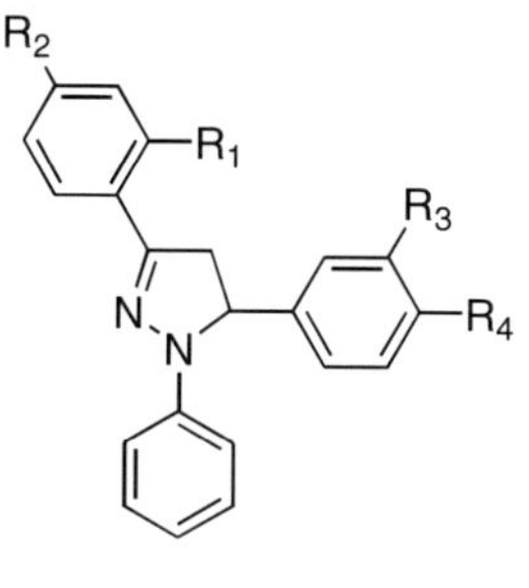

(70)

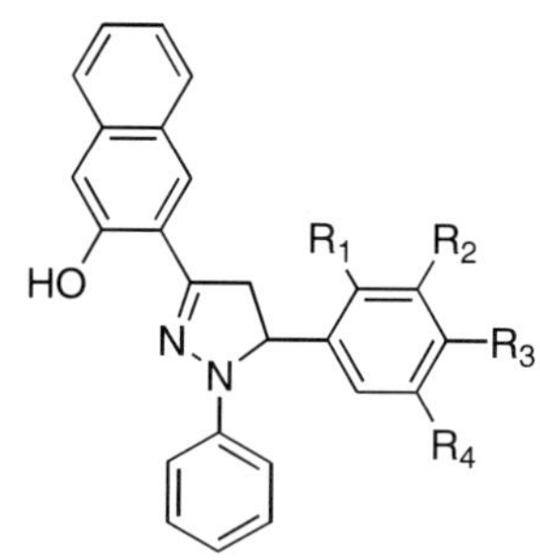

(71)

2.3. Cytotoxic and Anticancer Pyrazolines

Cells are the building blocks of living world. Cytotoxicity is the quality of being toxic to cells. Researchers are trying to develop new cytotoxic compounds so that such compounds can function as therapeutic agents for different targets, such as rapidly dividing cancer cells.

Cancer (also referred as malignant tumor or malignant neoplasm) is the uncontrolled growth of abnormal cells in body. Cancer is caused when the growth of cells in the body is out of control and cells divide rapidly or when the organ cells forget to die. Cancer can affect people irrespective of age. Currently cancer is becoming a striking disease all over the world. It is the second leading cause of death in the world (the first being cardiovascular diseases).

The literature survey reveals that in 2007 cancer caused about 13% of all human deaths worldwide (7.9 million). Hence, it is essential to develop new potent anticancer compounds for the survival of mankind.

Chemotherapy (cancer chemotherapy) is the treatment of cancer with an antineoplastic drug. Anticancer drugs normally act directly against abnormal proteins in cancer cells. Diversely substituted pyrazolines may function as promising therapeutic agents for cancer tissues.

M. Gordaliza et al. [24] reported some fused ring pyrazolines (72) as the cytotoxic compounds during culture cells studies of P-388 murine leukemia, A-549 lung carcinoma and HT-29 colon carcinoma.

Fedele Manna [25] synthesized some pyrazolines (73) and evaluated them for anticancer activity and for their ability to inhibit P-glycoprotein-mediated multidrug resistance.

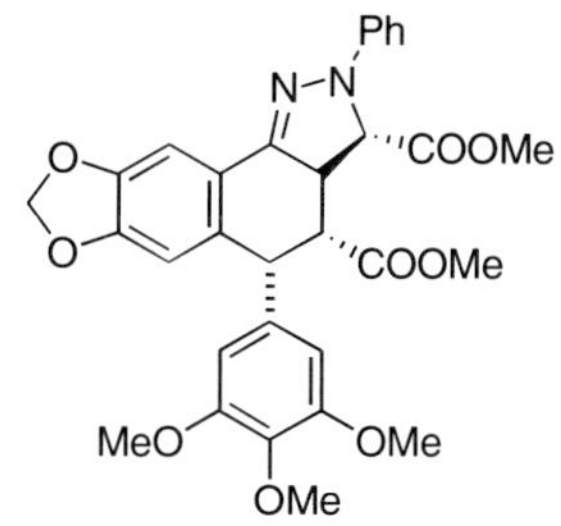

(72)

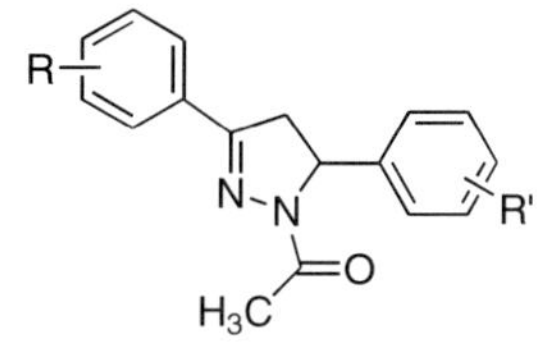

(73)

Amnerkar [26] studied the anticancer activity of some aminobenzothiazoles on cancer cell lines from ovarian cancer, renal cancer, lung cancer, leukemia etc. Among the five compounds evaluated for anticancer activity studies, the compounds (74) and (75) were highly effective.

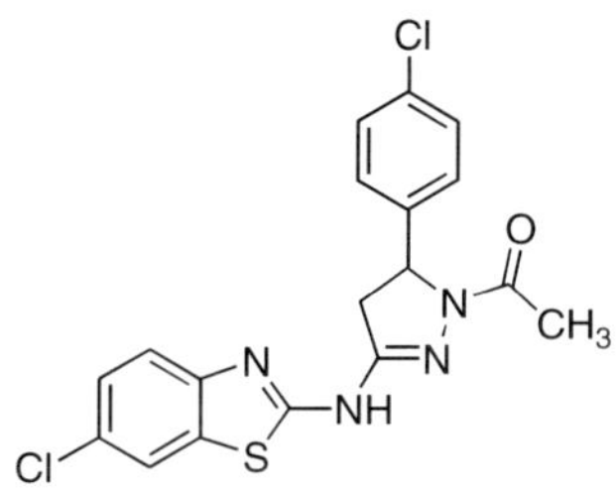

(74)

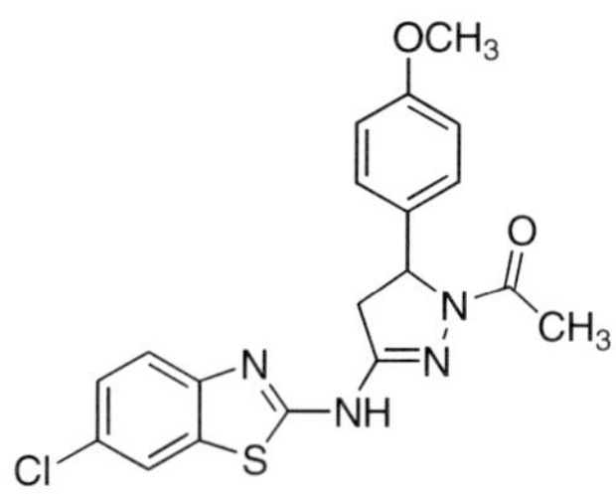

(75)

D. Havrylyuk [27] synthesized some thiazolone containing compounds (76), (77) and evaluated their anticancer activity. Most of them displayed the *in vitro* anticancer activity on leukemia, melanoma, lung, colon, CNS, ovarian, renal, prostate and breast cancer cell lines. The most active compound (Ar = 4-OH-3, 5-$(OMe)_2$-C_6H_2; R = 2-OH) exhibited selective influence on colon cancer cell lines.

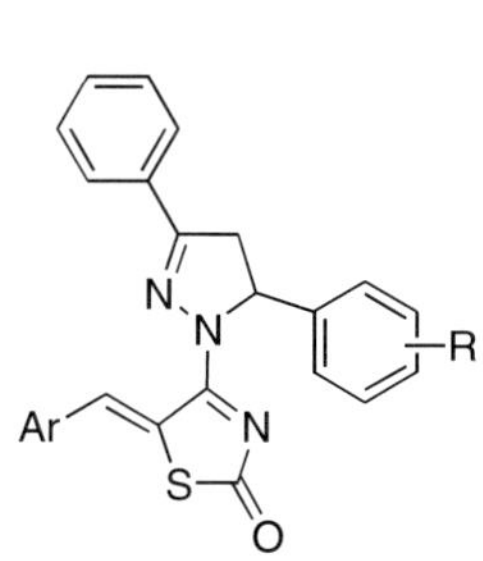

(76)

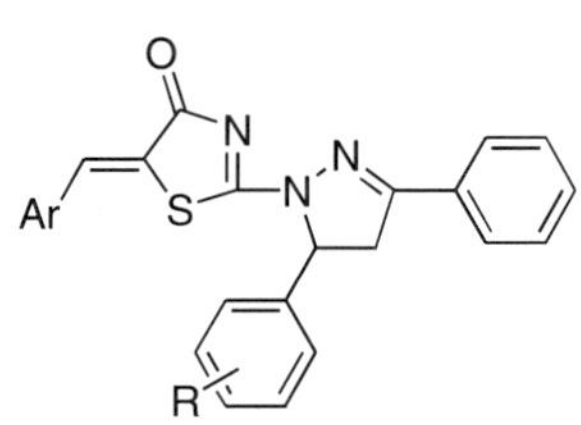

(77)

Sherif A. F. Rostom [28] synthesized sulfonamide, sulfonylurea (-thiourea) pharmacophores and some derived thiazole ring systems substituted indeno [1, 2-*c*] pyrazol (in) es (78), (79) and evaluated for their antitumor activity. Most of the compounds exhibited promising activity against subpanel tumor cell lines. Compound 3, 4-(3-(4-chlorophenyl)-4*H*-indeno [1, 2-*c*] pyrazol- 2-yl)-benzenesulfonamide proved to be the most active.

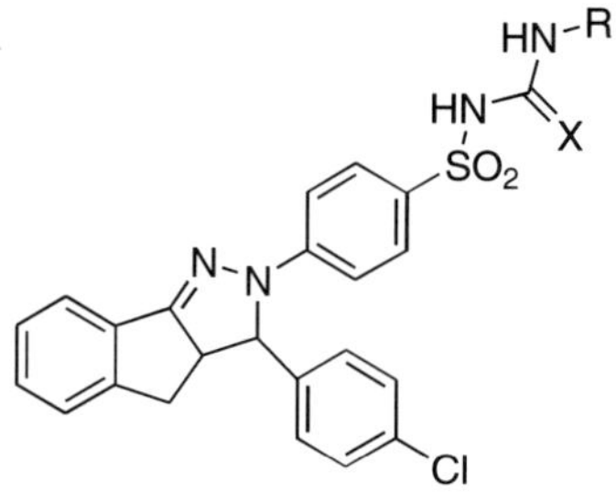

(78)

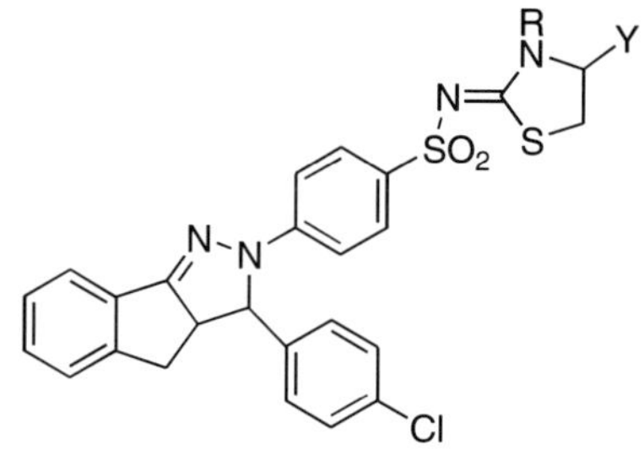

(79)

Pyrazolines bearing indolyl moiety in their structures were synthesized and tested for their antitumor activity by Magdy A. H. Zahran [29]. Chalcones (82) were obtained from the Claisen–Schmidt condensation with acetophenone derivatives (81) under microwave irradiation, which underwent cyclization reaction with phenylhydrazine to afford the corresponding indolylpyrazoline analogues (83). The synthesized compounds were evaluated against the human hepatocellular carcinoma cell line (Hep-G2). Some compounds showed high inhibition against the human hepatocellular carcinoma cell line (Hep-G2).

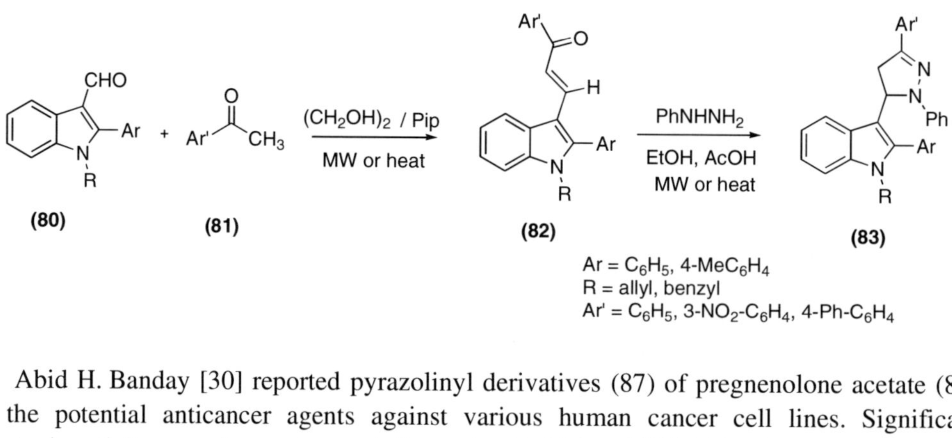

(80) **(81)** **(82)** **(83)**

Ar = C_6H_5, 4-MeC_6H_4
R = allyl, benzyl
Ar' = C_6H_5, 3-NO_2-C_6H_4, 4-Ph-C_6H_4

Abid H. Banday [30] reported pyrazolinyl derivatives (87) of pregnenolone acetate (84) as the potential anticancer agents against various human cancer cell lines. Significant cytotoxic activity was observed, especially against HT-29 and HCT-15 cell lines.

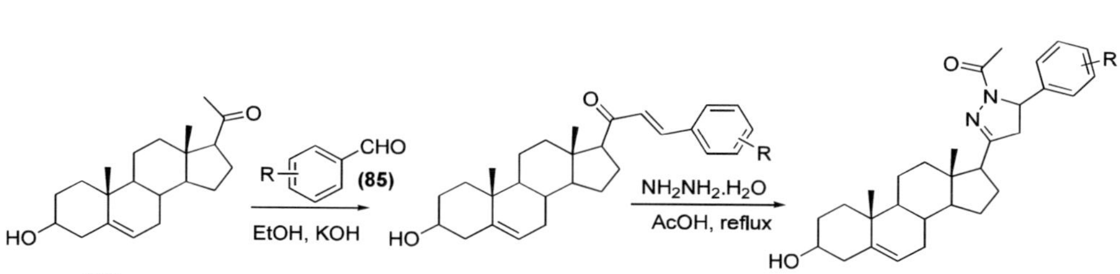

(84) **(86)** **(87)**

2.4. Anticonvolusant Pyrazolines

Anticonvulsants are the pharmaceuticals used in the treatment of epileptic seizures. Hence more accurately anticonvulsants are called antiepileptic drugs (AEDs), sometimes referred to as antiseizure drugs. The activity of these compounds as the potent pharmacophore is due to ability of the two-electron donor system to bind with the receptor site; hence, research continues for the development and synthesis of novel pyrazolines as therapeutic agents.

Z. Ozdemir et al. [31] reported the 3-(2-thienyl) pyrazolines (92), (93) and (94) as antidepressants and anticonvulsants. The antidepressant activities were investigated by Porsolt's behavioral despair test (forced swimming) on albino mice and compared with tranylcypromine. Anticonvulsant activities were determined by maximal electroshock seizure (MES) and subcutaneous pentylenetetrazole (metrazol) (scMet.) tests. Neurotoxicities were determined by therotarod toxicity test on albino mice. N-phenylthiocarbamoyl at 1-position of the pyrazoline ring exhibited remarkable antidepressant and anticonvulsant activity.

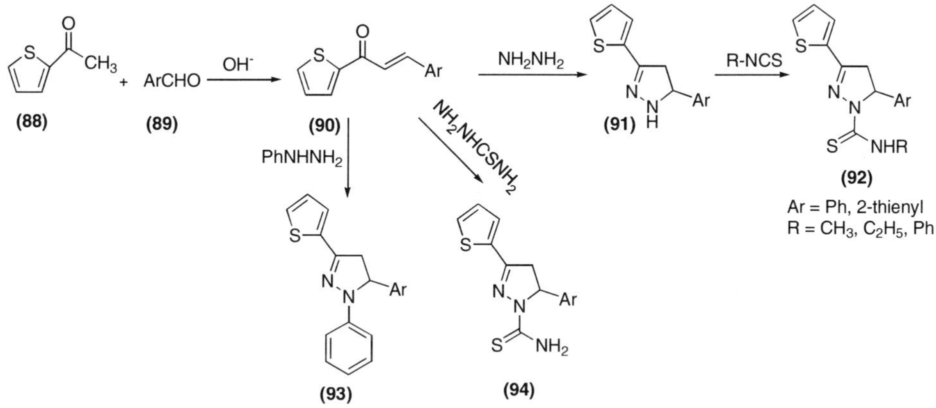

The pyrazoline-1-carboxamide derivatives (98) reported by A. A. Siddiqui [32] were studied for neurotoxicity in albino mice. Their results showed that all of the tested compounds were protective against MES at 100-300 mg/kg dose levels. The compounds with electron withdrawing groups were found to be more active as anticonvulsants than those with electron releasing groups.

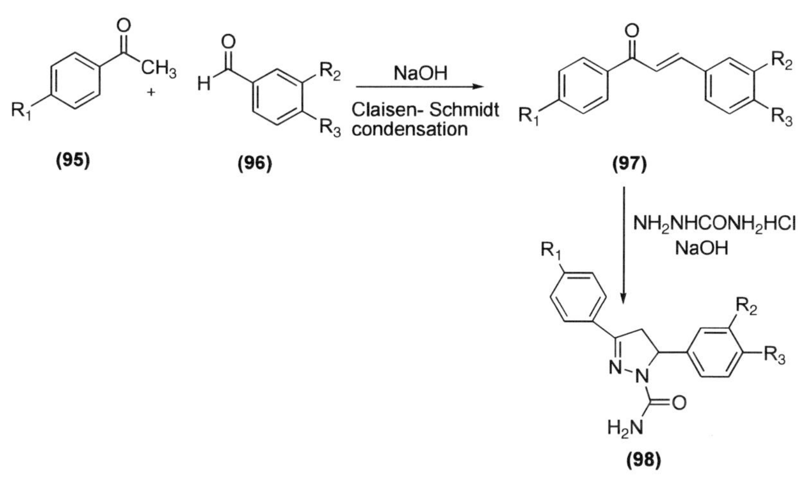

2.5. Anti-inflammatory and Analgesic Pyrazolines:

Anti-inflammatory action refers to the property of a substance or treatment that reduces inflammation. Anti-inflammatory drugs constitute about half of analgesics, which kill pain by reducing and affecting the central nervous system.

The cyclooxygenase enzyme inhibited by non-steroidal anti-inflammatory drugs (NSAIDs) has 2 different forms: COX1 and COX2. The adverse effects of NSAIDs were caused by blocking the COX1 (constitutive) enzyme, with the analgesic effects mediated by the COX2 (inducible) enzyme. The COX2 inhibitors were thus developed to inhibit only the COX2 enzyme (traditional NSAIDs block both versions in general). These drugs (such as rofecoxib and celecoxib) are equally effective analgesics when compared with NSAIDs, but cause less gastrointestinal hemorrhage in particular.

An analgesic (pain killer) is a drug used to relieve pain. Analgesic drugs act in various ways on the peripheral and central nervous systems. Analgesics are selective central nervous system depressants used to relieve pain. The term analgesic means "without pain."

Analgesics may relieve pain by preventing the release of acetylcholine. Enkephalin molecules are released from a nerve cell and bind to analgesic receptor sites on the nerve cell sending the impulse. The binding of enkephalin or morphine-like drugs changes the shape of the nerve sending the impulse to prevent the cell from releasing acetylcholine. As a result, the pain impulse cannot be transmitted and the brain does not receive pain.

Phenazone (99) also named as phenazon/antipyrine or analgesine is a pyrazoline ring containing commercial analgesic and antipyretic drugs.

M. V. R. Reddy [33] reported the anti-inflammatory activity of some 1-(4-sulfamylphenyl)-3-trifluoromethyl-5-indolyl pyrazolines (100) and observed them to be the cyclooxygenase-2 (COX-2) and lipoxygenase (LOX) inhibitors.

Barsoum [34] studied bis (1-acyl-2-pyrazolines) (101) as the potential compounds with anti-inflammatory and molluscicidal properties. Anti-inflammatory and ulcerogenic activities of the prepared pyrazolines were evaluated with the standard drug (indomethacin). Many of the tested compounds showed remarkable anti-inflammatory properties with an ulcerogenic liability, lower than that of the standard used drug. The molluscicidal activity of the compounds was tested against Biomphalaria alexandrina snails.

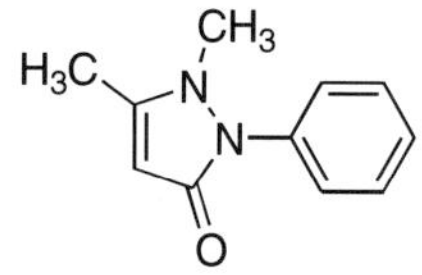

(99)

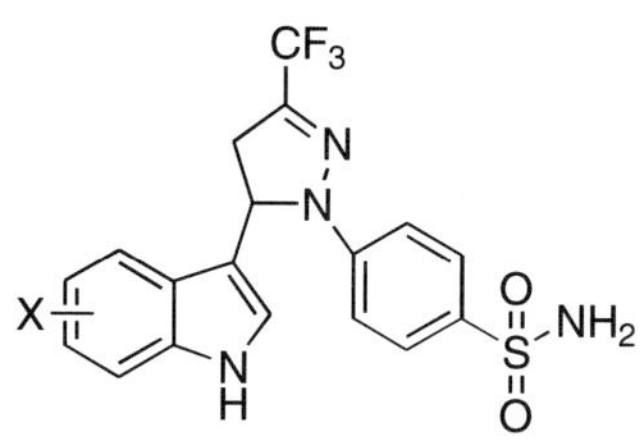

(100)

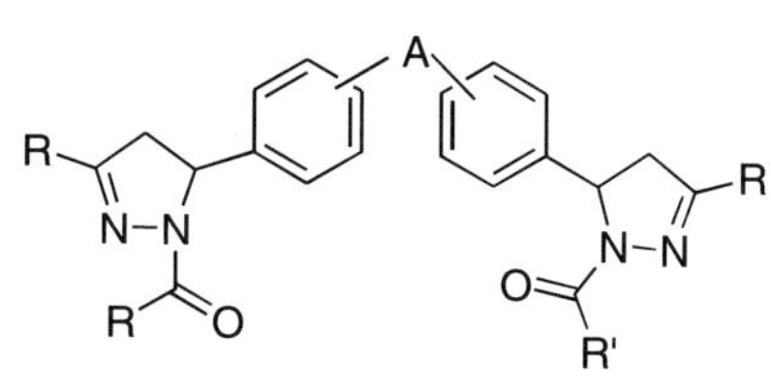

(101)

Suresh Khode [35] synthesized 5-(substituted) aryl-3-(3-coumarinyl)-1-phenyl-2-pyrazolines (102) and demonstrated them to be active anti-inflammatory and analgesic agents at the concentrations of 200 mg/kg b.w. The compounds exhibited significant anti-inflammatory activity in models of acute inflammation of rat edema paw.

Among the various acridinyl pyrazolines synthesized by **T. Chandra [36]**, the compound1-(2′,4′-Chloroacridine-9′-yl)-3-(5′-pyridine-4-yl)-(1,3,4-oxadiazol-2-yl-thiomethyl)-pyrazole-5-one (103) was reported to be the most potent anti-inflammatory and analgesic agent.

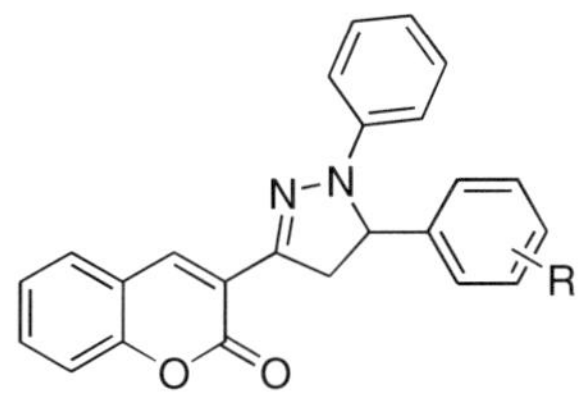

(102)

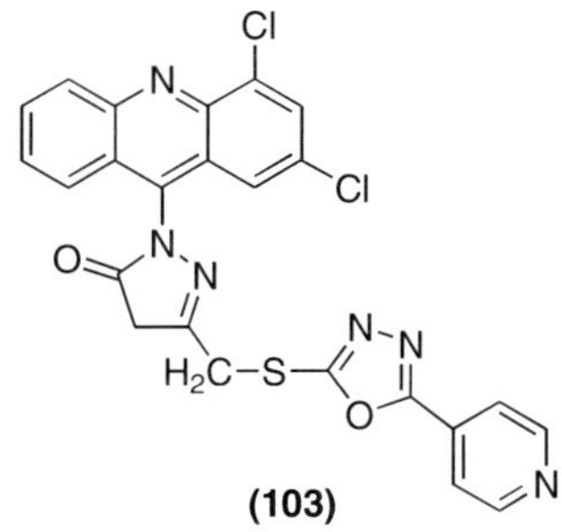

(103)

Mohammad Amir [37] studied the anti-inflammatory and analgesic activity of 3-(4-biphenyl)-5-substituted phenyl-2-pyrazolines and 1-benzoyl-3-(4-biphenyl)-5-substituted phenyl-2-pyrazolines (104), (105). The results were compared with the standard drug flurbiprofen.

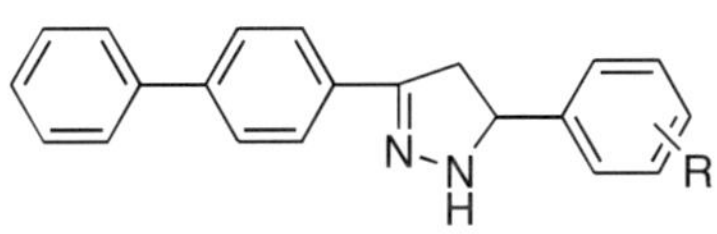

(104)

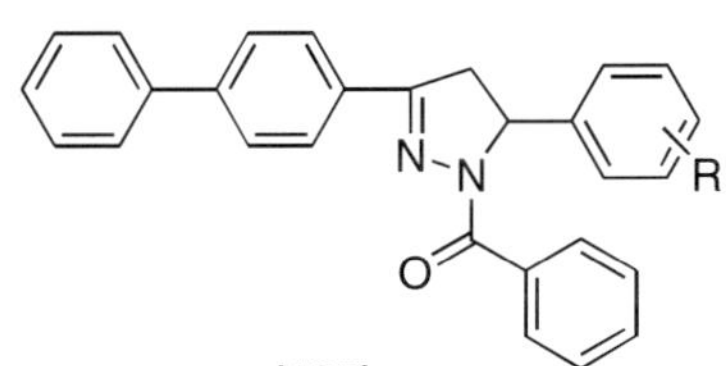

(105)

R. S. Joshi et al. [38], reported 3, 2-(4, 5-dihydro-5-(4-morphilinophenyl)-1H-pyarazol-3-yl) phenols and its N-phenylpyrazol-1-carbothioamides (106), (107) as the potent analgesic and anti-inflammatory compounds, as compared to the standard drug diclofenac.

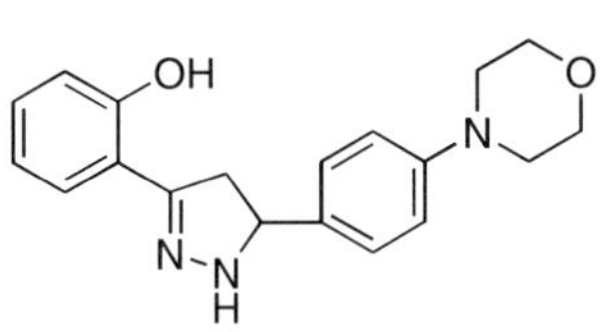

(106)

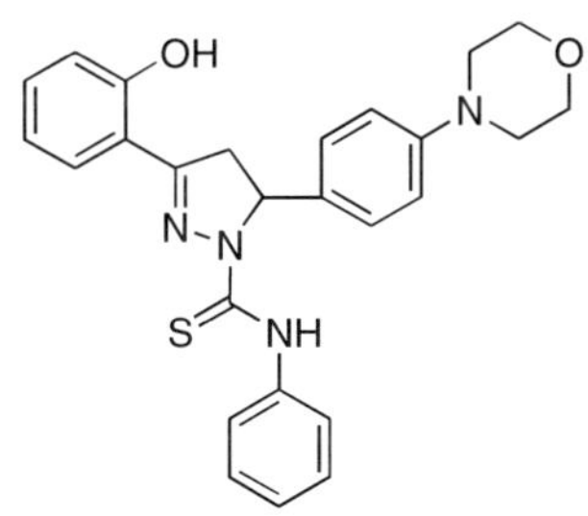

(107)

The synthesis and anti-inflammatory activity of pyrazolines (111) bearing thiophene ring was carried out by B. Ramesh and T. Sumana [39], using aceclofenac as the reference drug. The synthesized compounds showed anti-inflammatory activity in the range of 50-80 %; whereas the standard drug showed 80-85 % inhibition in paw edema. The results of anti-inflammatory activity indicated that some compounds showed maximum activity.

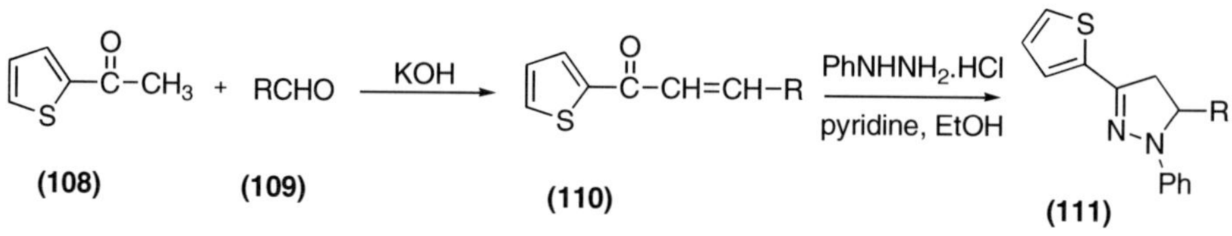

R = 2-thienyl, 4-(NMe_2)C_6H_4, 4-Cl-C_6H_4, 4-F-C_6H_4, 2,4-$Cl_4C_6H_3$

Pyrazolines (115) bearing benzene sulfonamide moiety were reported as potential anti-inflammatory and anti-cancer agents by Sameena Bano [40]. The compounds were synthesized by reacting chalcones (112)/flavanones (113) with 4-hydrazinon-benzenesulfonamide hydrochloride (114) and their anti-inflammatory and anti-cancer activities were evaluated. Some of the reported compounds exhibited good anti-inflammatory activity comparable to the reference drug celecoxib.

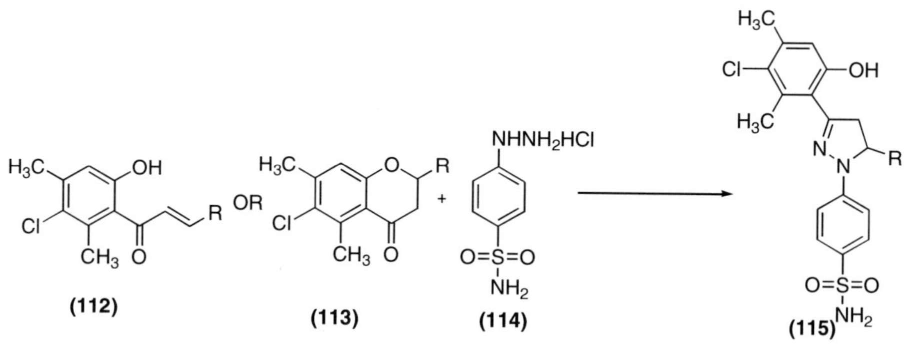

1-acetyl/propyl-3-aryl-5-(5-chloro-3-methyl-1-phenyl-1H-pyrazol-4-yl)-2-pyrazolines (118) were obtained by K. S. Girisha [41] by the reaction of propenones, hydrazine and acetic/propionic acid and evaluated for analgesic and anti-inflammatory activity. The results were compared with pentazocin and diclofinac sodium respectively as the standard drugs.

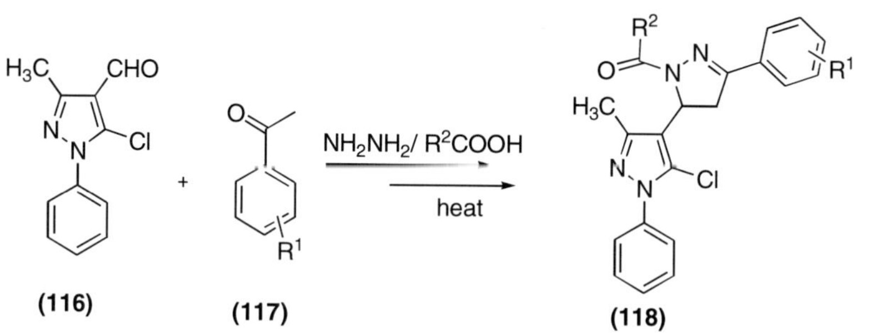

I. G. Rathish et al. [42] reported benzene sulfonamide containing pyrazolines (121) and monitored their anti-inflammatory activity. Some of the compounds were more potent than the reference drug celecoxib.

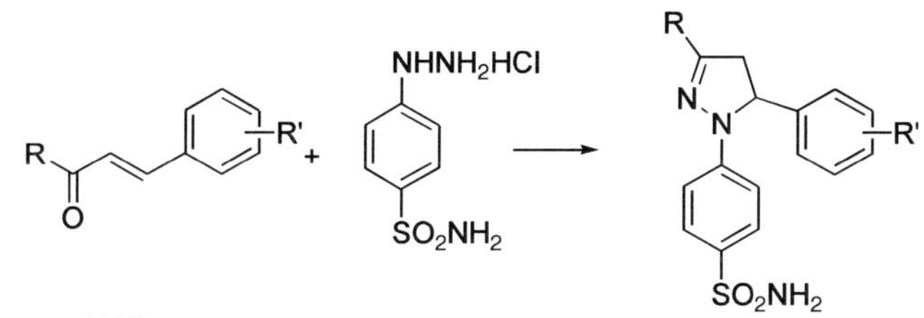

(119) **(120)** **(121)**

Pankaj Malhotra [43] carried out microwave assisted synthesis of some 3, 5-diaryl substituted –2-pyrazolines **(125)** and evaluated their anti-inflammatory activity. The reports revealed that the compounds showed an appreciable amount of activity. Among these, the compounds with (R = H, R_1 = furan); (R = H, R_1 = methoxy phenyl); (R = -H, R_1 = thiophene); and (R = -H, R_1 = p-amino phenyl) (R = -H, R_1 = trimethoxy phenyl) (R = -H, R_1 = 3- methoxy 2-hydroxy phenyl) (R = -OH, R_1 = 2-hydroxy phenyl) showed potency almost equivalent to the standard drug.

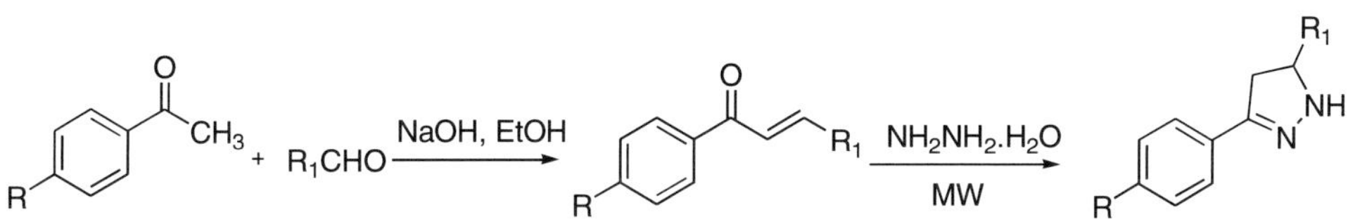

(122) **(123)** **(124)** **(125)**

2.6. Anti-amoebic Pyrazolines

Amoebiasis, or amebiasis, refers to the infection caused by the *amoeba entamoeba histolytica. E. histolytica* infections occur in tissue of the intestine and / or liver. The drugs used in the treatment of areas of infection are referred to as amoebicides. It is estimated that about 40 to 50 million people get infected with *E. histolytica*. As a resultof which there is a need for the development of amoebicides.

M. Abid and A. Azam [44] reported the 1-*N*-substituted cyclized pyrazoline analogues of thiosemicarbazones and demonstrated their antiamoebic activities. The compounds were obtained from the cyclization between Mannich bases and thiosemicarbazides as the precursors. The activities were tested on the *HM1*:*1MSS* strain of *Entamoeba histolytica* using the microdilution method. It was observed that 3-chloro and 3-bromo substituents on the phenyl ring at position 3 of the pyrazoline ring enhanced the antiamoebic activity as compared to unsubstituted phenyl ring (126).

Asha Budakoti [45] studied the synthesis and in vitro antiamoebic activity of new Pd (II) complexes with 1-N-substituted thiocarbamoyl-3,5-diphenyl-2-pyrazolines (127) using microdilution method against the HM1:IMSS strain of *Entameoba histolytica*, and the results were compared with metronidazole as the standard drug. Some compounds exhibited promising activity compared to other ligands and complexes.

A. R. Bhat etal [46] synthesized bis-pyrazolines (128) and evaluated their antiamoebic activity as inhibitors of growth of *Entamoeba histolytica* by the microdilution method against the *HM1*: *IMSS* strain of *Entamoeba histolytica* and the results were compared with the standard drug, metronidazole.

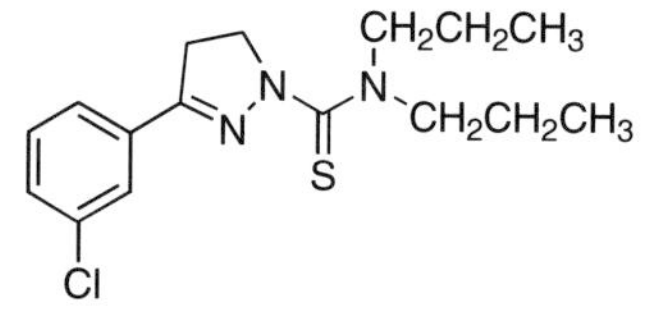

(126)

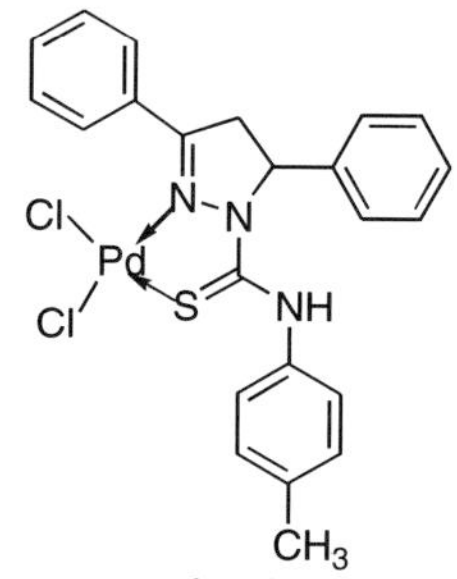

(127)

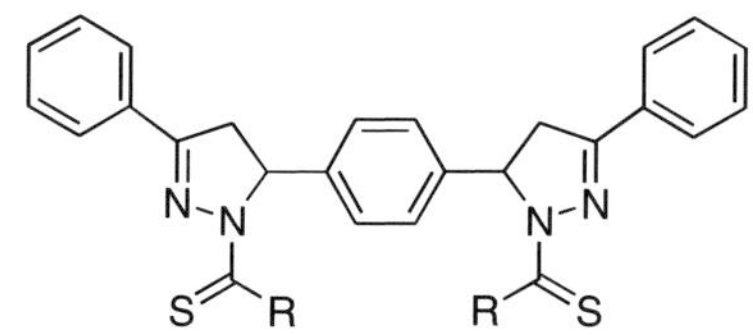

(128)

The synthesis and antiamoebic activity of some 3-(3-bromo phenyl)-5-phenyl-1-(thiazolo [4,5-b] quinoxaline-2-yl)-2-pyrazolines (129) by the microdilution method against HMI:IMSS strain of *Entamoeba histolytica* is also reported in literature [47]. The IC_{50} values were compared with the standard drug metronidazole.

M. Abid et al. [48] reported the synthesis and antiamoebic activity of new 1-*N*-substituted thiocarbamoyl-3-phenyl-2-pyrazolines (130) by the cyclization between Mannich bases and thiosemicarbazides.

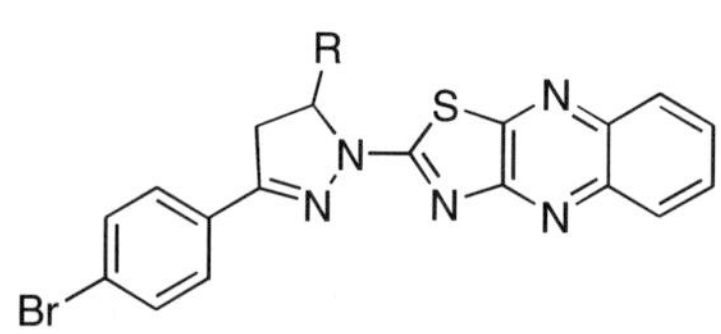

(129)

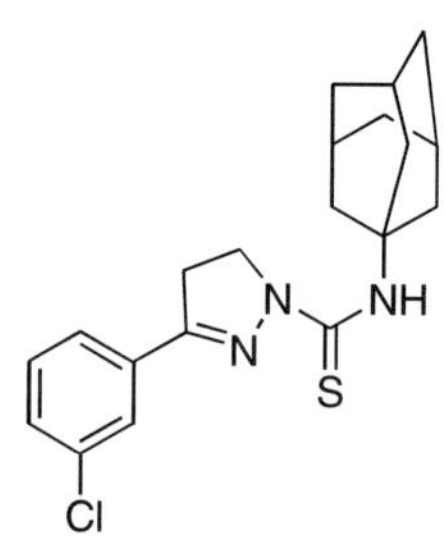

(130)

M. Abid and A. Azam [49] reported the synthesis of 1-(thiazolo [4, 5-b] quinoxaline-2-yl)-3-phenyl-2-pyrazolines from the cyclization between Mannich bases and unsubstituted thiosemicarbazide and studied their antiamoebic activity. The studies were carried out on the HM1: IMSS strain of *Entamoeba histolytica* parasite using the microdilution method. The compound (131) was observed to be a better inhibitor of Entamoeba histolytica compared to the standard drug metronidazole.

Asha Budakoti and coworkers [50] reported the synthesis and antiamoebic activity of palladium (II) complexes of 1-N-substituted thiocarbamoyl-3, 5-diphenyl-2-pyrazoline derivatives. The complex (132) was significantly potent against the HM1: IMSS strain of *Entamoeba histolytica* when compared with the standard drug metronidazole.

Faisal Hayat et al. [51] synthesized pyrazoline derivatives bearing quinoline moiety (133) and evaluated their antiamoebic activity and cytotoxicity. The antiamoebic activity was tested against the HM1: IMSS strain of *Entamoeba histolytica*. Some of the synthesized compounds were found to be active on the human breast cancer MCF-7 cell line.

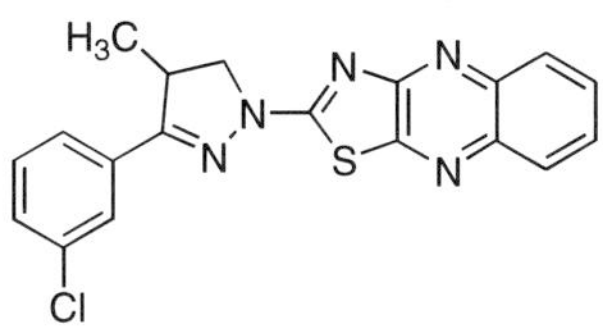

(131)

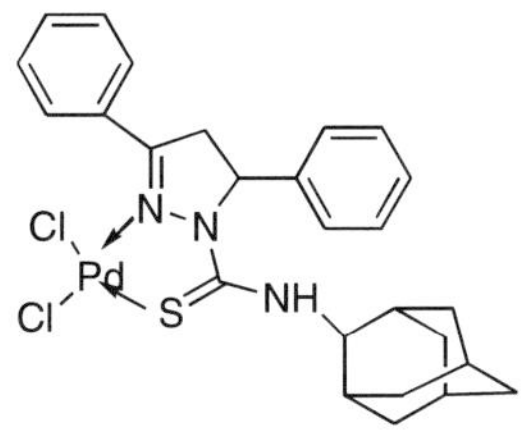

(132)

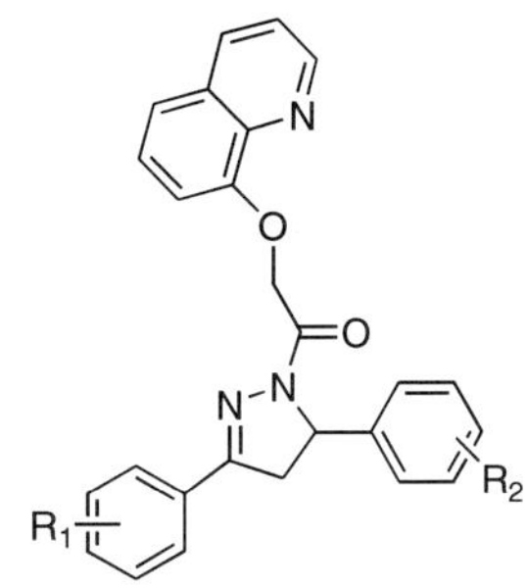

(133)

2.7. Enzyme Inhibitors/Modulator Pyrazolines:

An enzyme inhibitor is a molecule that binds to enzymes and decreases their activity. Since blocking an enzyme's activity can kill a pathogen, many drugs are enzyme inhibitors. Thus their discovery and improvement is an active area of research in biochemistry and pharmacology.

X. Zhang [52] reported pyrazolines bearing amide linkage in their structures as the potent tissue-selective androgen receptor modulators (SARMs). The molecules possessing strong electron-withdrawing groups at R^1 and R^2 positions and a small group at the R^5 and R^6 position were found to exhibit potent AR antagonist activity. Further the (*S*)-isomers were more potent AR agonist activity than the corresponding (*R*)-isomer (134). The compounds have significant antiandrogenic effects on the prostate in intact rats with endogenous testosterone.

Fedele Manna [53] reported the synthesis of some pyrazolines (135) and carried out an investigation of their affinity to bind with P-glycoprotein. Some of these compounds have the ability to inhibit P-glycoprotein-mediated multidrug resistance by directly binding to a purified protein domain containing an ATP-binding site and a modulator interacting region. These were also evaluated for their anticancer activity.

A. J. Roecker [54] synthesized and reported the 1, 4-diaryl-4, 5-dihydropyrazoles (136) as potent inhibitors of the mitotic kinesin KSP.

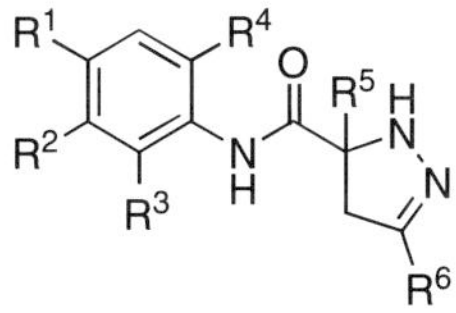

(134)

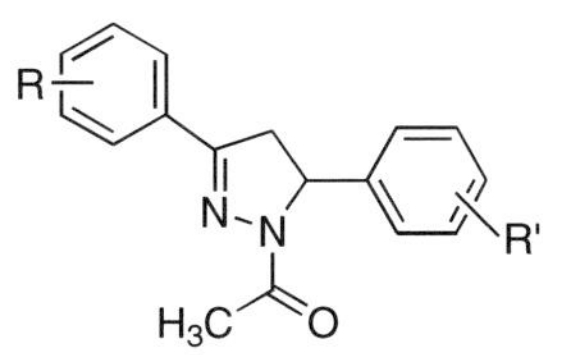

(135)

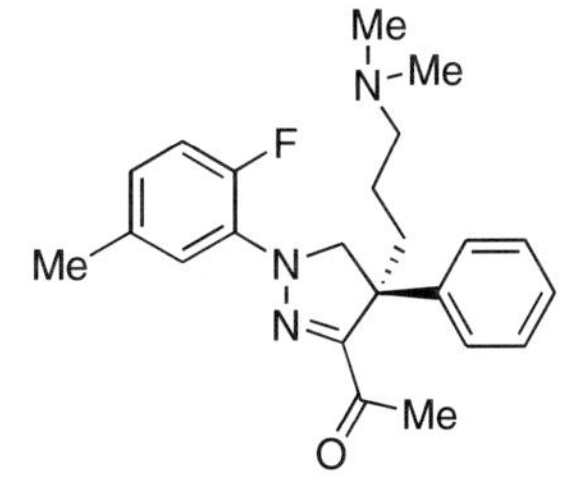

(136)

David G. Jones et al. [55] reported **b** non-steroidal mimetics (137) of mifepristone and progesterone. exhibited functional antagonism of the progesterone receptor.

Abd El-Galil E. Amr [56] synthesized pyrazolines from 3β-hydroxyandrostan-17-one as the starting material and evaluated the antiandrogenic activity of 3-substituted androstano [17, 16-*c*]-5′-aryl-pyrazoline and their derivatives (138). The antiandrogenic activity of the compounds was compared with cyproterone as the positive control. Some of the compounds exhibited better antiandrogenic activity than the reference drug.

Tae-Sook Jeong et al. [57] reported the 3-(3,5-di-tert-butyl-4-hydroxyphenyl)-5-(multi-substituted 4-hydroxyphenyl)-2-pyrazolines (139), which act as inhibitors of acyl-CoA: cholesterol acyltransferase in human.

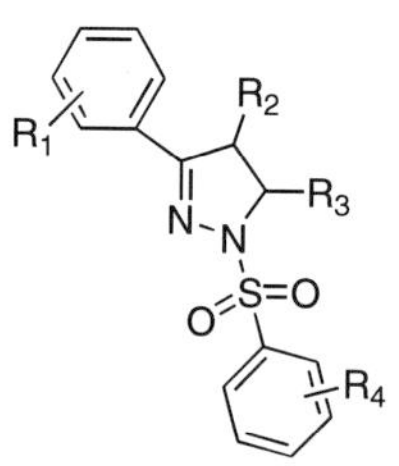

(137)

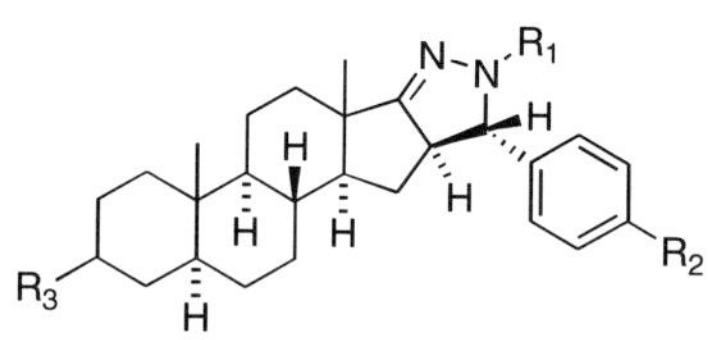

(138)

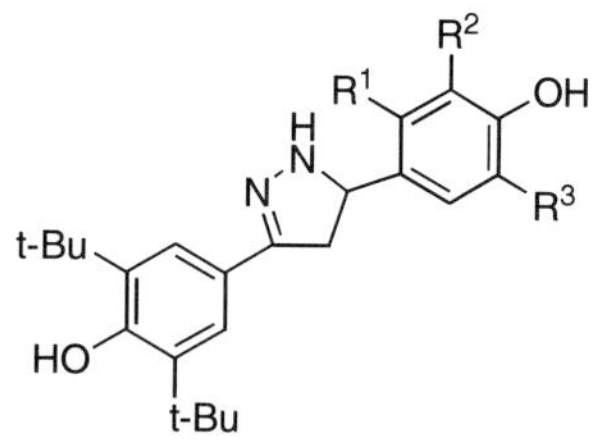

(139)

M. Camacho et al. [58] reported that the 4, 5-dihydro-1H-pyrazole derivatives (140) have inhibitory activity for the nNOS activity in the brains of rats.

M. Dora Carrion [59] reported some benzoylpyrazolines (141) using 1, 3-dipolar-cycloaddition reaction through a series of reactions. The compounds were shown to be the inhibitors of nitric oxide synthase (NOS), showing better affinity towards the inducible NOS isoform.

N1-thiocarbamoyl-3, 5-di (hetero) aryl-4, 5-dihydro-(1H)-pyrazoles (142) were reported as the inhibitors of the A and B isoforms of human monoamine oxidase (hMAO) enzyme by F. Chimenti et al. [60]. Some of these compounds had selective inhibitory activity against hMAO-B in the micromolar range. The most active compound of the series is the compound N1-thiocarbamoyl-3-(fur-2′-yl)-5-(4′-fluoro-phenyl)-4, 5-dihydro-(1H)-pyrazole.

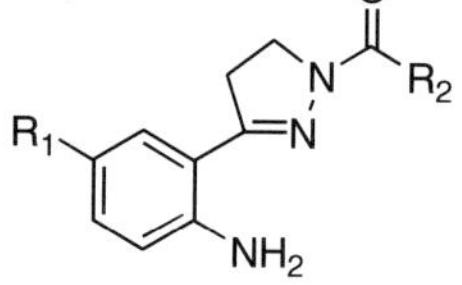

(140)

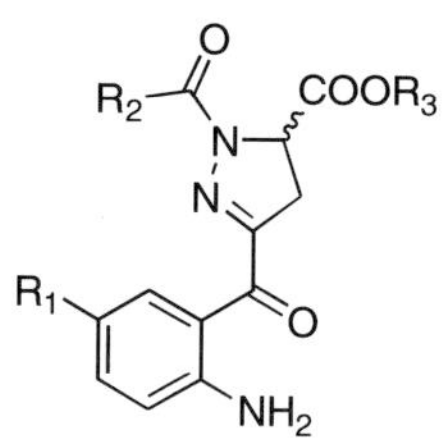

(141)

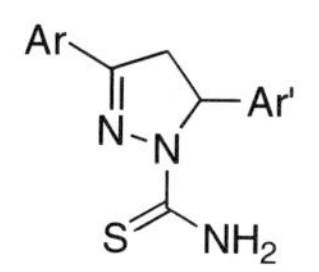

(142)

The prenylated pyrazolines (143) were reported as the MAO-B inhibitors which were tested on human monoamine oxidase-A and -B isoforms of the enzyme [61]. It was found that substituents like benzyloxy or chlorine atoms show excellent interaction with the active site of hMAO-B.

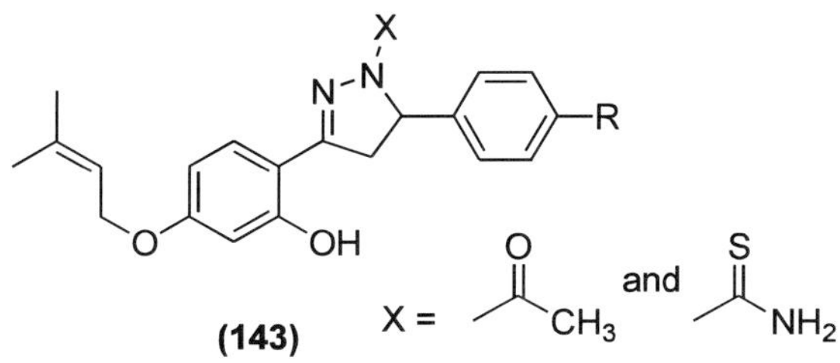

(143)

2.8. Anti-tubercular and Anti-mycobacterial Pyrazolines

The word "tuberculosis" (MTB or TB) is used to describe the cases of the lethal, infectious disease caused by the various strains of mycobacteria, usually *Mycobacterium tuberculosis*. Tuberculosis generally attacks the lungs; it can also affect other parts of body. It is spread through the air when people who have an active MTB infection cough, sneeze or otherwise transmit their saliva through the air. One-third of the world's population is thought to be infected with M. tuberculosis, and new infections occur at a rate of about one per second. More than 50 % of the seriously infected patients of TB die due to this chronic disease. Hence, it becomes essential to develop new antitubercular drugs.

Imidazole containing pyrazolines was reported on by D. Zampieri [62]. These 1-(3, 5-diaryl-4, 5-dihydro-1H-pyrazol-4-yl)-1H-imidazole derivatives (144) were synthesized and tested for in vitro antifungal and antimycobacterial activities. These showed excellent antifungal activity against a clinical strain of Candida albicans and an interesting antitubercular activity against *Mycobacterium tuberculosis* $H_{37}Rv$ reference strain.

M. Shaharyar [63] synthesized different N^1-nicotinoyl-3-(4′-hydroxy-3′-methyl phenyl)-5-[(sub) phenyl]-2-pyrazolines (145) and studied their *vitro* antimycobacterial activity against *Mycobacterium tuberculosis* and INH-resistant *M. tuberculosis* (INHR-MTB) using the agar dilution method. Among the synthesized compounds, the N^1-nicotinyl-3-(4′-hydroxy-3′-methyl phenyl)-5-(1″-chlorophenyl)-2-pyrazoline was found to be the most active agent against MTB and INHR-MTB, with the minimum inhibitory concentration of 0.26 μM.

M. J. Ahsan [64] synthesized some fluorine containing 3-substituted-N-aryl-6, 7-dimethoxy-3a,4-dihydro-3H-indeno[1,2-c]pyrazole-2-carboxamide pyrazolines and studied their antimycobacterial activity by the two-fold serial dilution technique. All the newly synthesized compounds showed moderate to high inhibitory activities against *Mycobacterium tuberculosis* $H_{37}Rv$ and INH resistant M. tuberculosis. The compound N,3-bis(4-fluorophenyl)-6,7-dimethoxy-3a,4-dihydro-3H-indeno[1,2-c]pyrazole-2-carboxamide (146) was found to be the most promising against M. tuberculosis $H_{37}Rv$ and isoniazid resistant M. tuberculosis with the minimum inhibitory concentration of 0.78 μM.

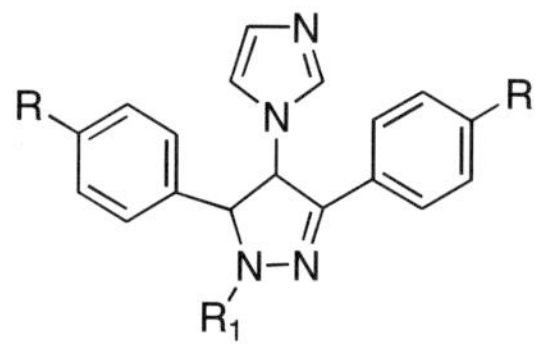

(144)

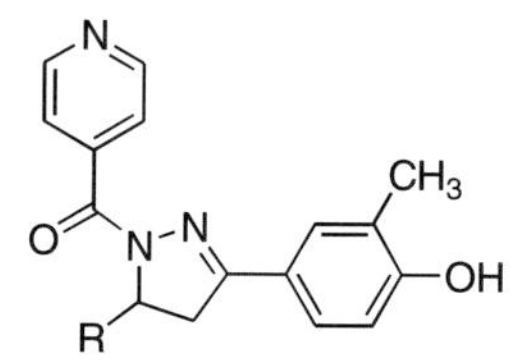

(145)

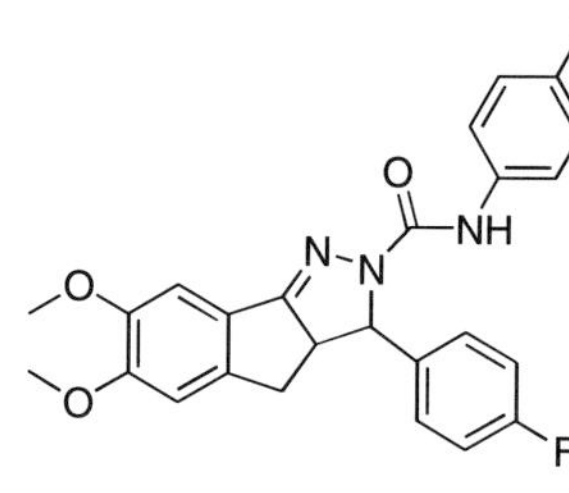

(146)

M. J. Ahsan et al. [65] reported some antitubercular 3a,4-dihydro-3H-indeno[1,2-c]pyrazole-2-carboxamide/carbothioamides (147). The compounds showed moderate to high inhibitory activities against *Mycobacterium tuberculosis $H_{37}Rv$* and INH resistant M. tuberculosis. The compound 3-(4-fluorophenyl)-6,7-dimethoxy-3a,4-dihydro-3H-indeno[1,2-c]pyrazole-2-carbothioamide (148) was found to be the most active against M. tuberculosis $H_{37}Rv$ and isoniazid resistant M. tuberculosis with minimum inhibitory concentrations of 3.12 μM and 6.25 μM respectively.

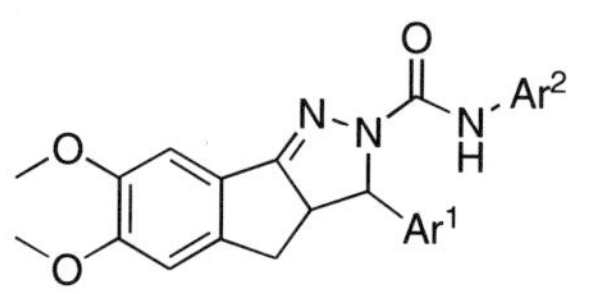

(147)

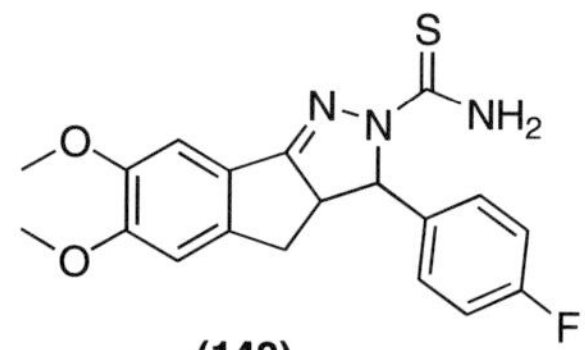

(148)

Mohammad Shaharyar [66] reported some phenoxy acetic acid derivatives as anti-mycobacterial agents. The 2-(4-formyl-2-methoxyphenoxy) acetic acid, when condensed with various ketones in methanolic KOH solution, yielded the corresponding chalcones, which reacted with appropriate acid hydrazide in glacial acetic acid and led to the formation of phenoxy acetic acid derivatives (149), (150). The synthesized compounds were evaluated for their anti-mycobacterial activities against *Mycobacterium tuberculosis* $H_{37}Rv$. The majority of the compounds were found to possess promising activity against the *mycobacterium*.

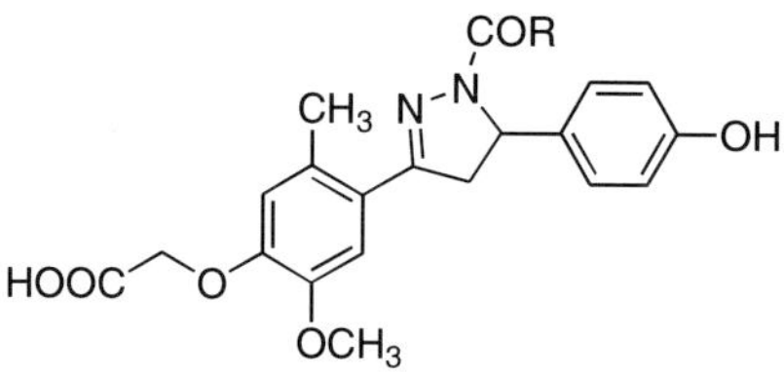

(149)

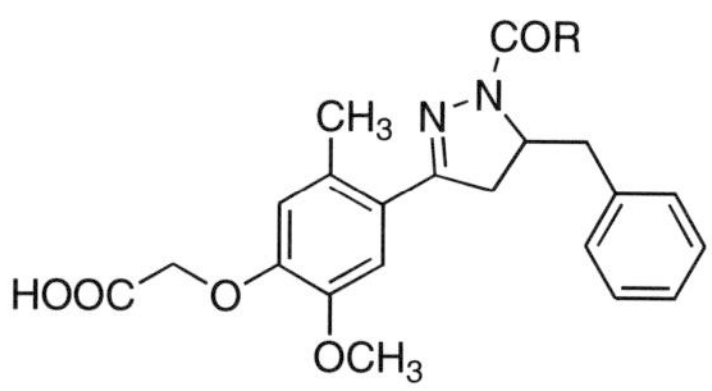

(150)

Several pyrazolines bearing thioamide linkage viz 5-(-4-(substituted) phenyl)-3-(4-hydroxy-3-methylphenyl)-4,5-dihydro-1H-1-pyrazolyl-2-toluidino methane thione and 5-(substituted) phenyl-3-(4-hydroxy-3-methylphenyl)-4,5-dihydro-1H-1-pyrazolyl-2-methoxy-anilino methane thione (151), (152) were prepared by Mohamed Ashraf Ali [67] and their anti-tubercular activities were evaluated. The compounds were synthesized by the condensation reaction between hydrazine hydrate and chalcones, followed by treatment with appropriate aryl isothiocyanate, which yielded N-substituted pyrazoline derivatives. The anti-tubercular activity was tested against *Mycobacterium tuberculosis* H37Rv using the BACTEC 460 radiometric system. Among the derived compounds; anilino-3-(4-hydroxy-3-methylphenyl)-5-(2, 6-dichlorophenyl)-4, 5-dihydro-1H-1-pyrazolylmethanethione was found to be a more active agent against *M. tuberculosis H37Rv*, with the minimum inhibitory concentration of 0.0034 μM.

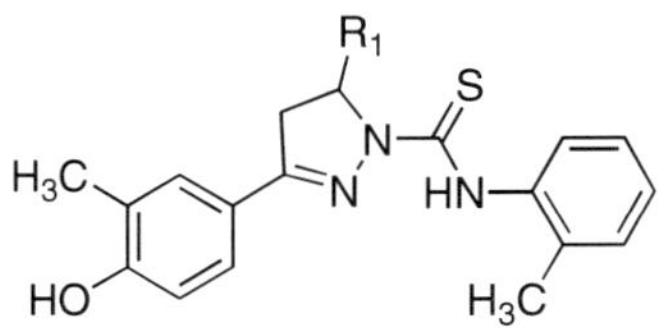

(151)

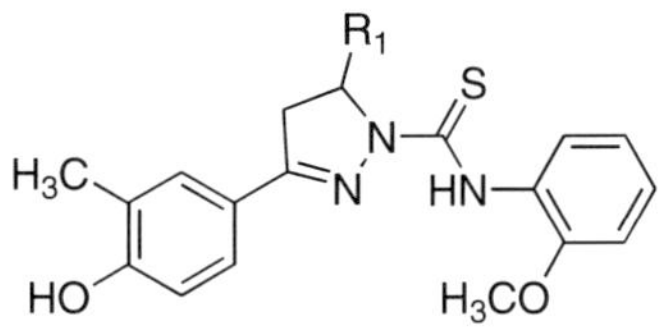

(152)

Mohamed Jawed Ahsan [68] evaluated the antimycobacterial activity of carboxamide pyrazolines (153) against Mycobacterium tuberculosis $H_{37}Rv$ and multi-drug resistant *M. tuberculosis* (MDR-TB). 3-(4-fluorophenyl)-N-(4-chlorophenyl)-6,7-dimethoxy-3a,4-dihydro-3H-indeno [1,2-c] pyrazole-2-carboxamide (154) was found to be the most promising compound against *M. tuberculosis.*

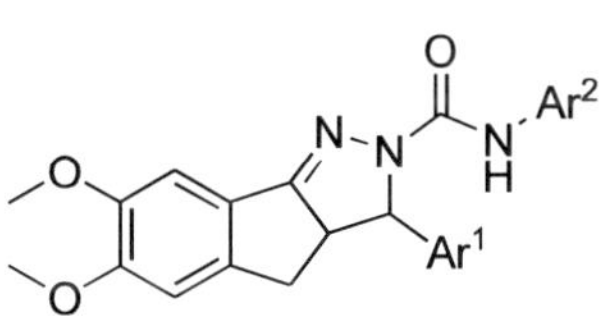

(153)

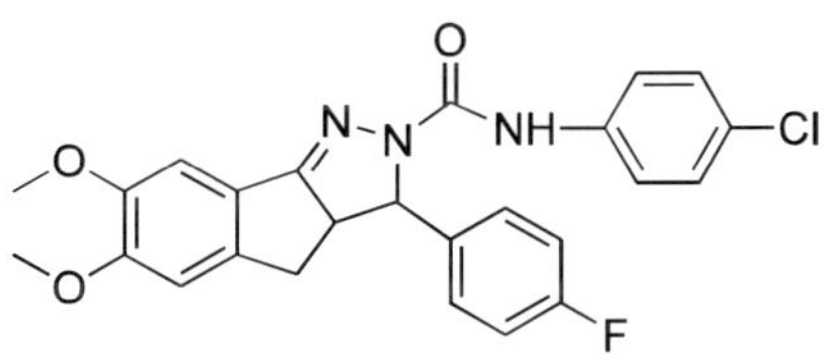

(154)

2.9. Anti-malerial Pyrazolines

1, 3, 5-trisubstituted pyrazolines (155) were reported as antimalerial agents by B. N. Acharya [69] with good efficiency against chloroquine sensitive (MRC-02) and chloroquine resistant (RKL9) strains of plasmodium falciparum. The antimalerial activity was related by studying their β-hematin formation inhibition ability. Some of the compounds reported showed more potential antimalerial activity than chloroquines against P. falciparum stain.

The study of synthesis and ability of 3-aryl-4, 5-dihydropyrazole-1-carbothioamides (156) to inhibit monoamine oxidase was studied by E. Maccioni [70]. Almost all the compounds possess high selectivity for the B isoform of the enzyme, irrespective of the nature of the substituents on the heterocyclic ring. The inhibitory activity of the compounds was measured against human recombinant MAO isoforms, which were expressed in baculovirus infected BTI insect cells using docking experiments. The 3-(4-methoxyphenyl) substituted derivative was observed to be active among the compounds evaluated for the activity studies.

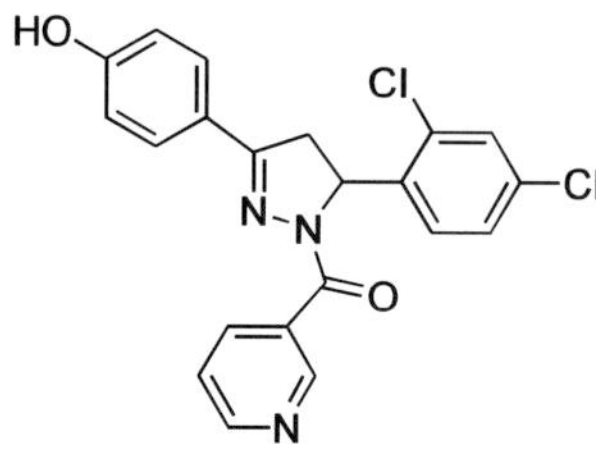

(155)

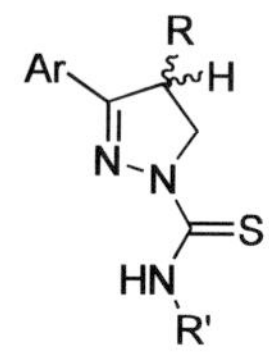

R = H, CH_3
R' = H, CH_3

(156)

2.10. Miscellaneous Activities of Pyrazolines

The hypotensive activity of pyazolines was monitored by Turan et al. [71]. 1-(4-arylthiazol-2-yl)-3, 5-diaryl-2-pyrazoline derivatives (157) were investigated for their hypotensive activity by the tail-cuff method using clonidine as the reference drug. All the reported compounds were found to have appreciable hypotensive activities.

Z. A. Kaplancikli [72] reported the antinociceptive activities of some benzoxazole/benzimidazole–pyrazolines (158). The compounds showed remarkable antinociceptive activities in both hot plate and acetic acid-induced writhing tests.

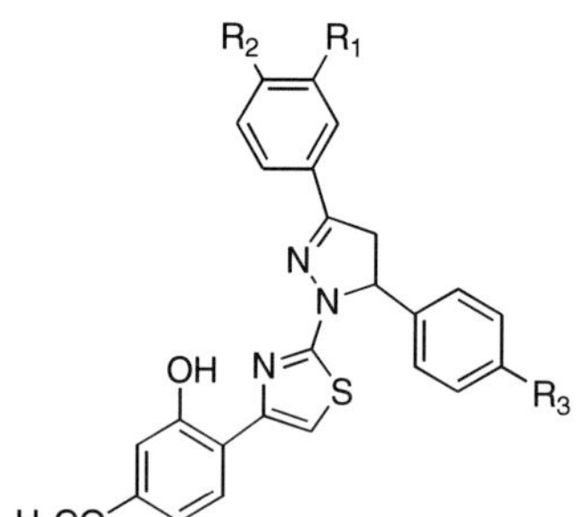

(157)

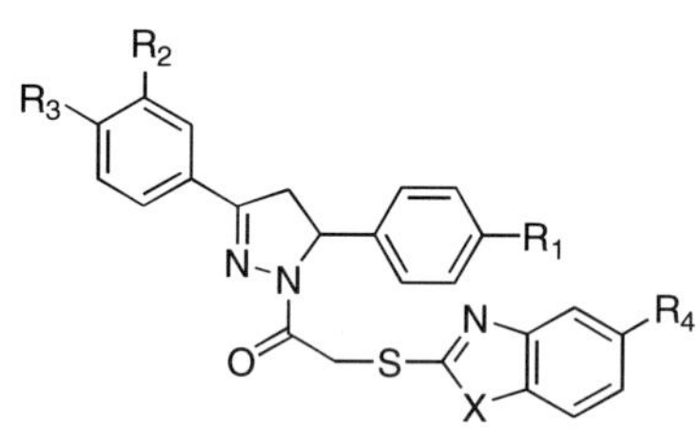

X = O, NH
R_1= H, OCH_3
R_2 =H, CH_3
R_3 = H, CH_3
R_4 = H, Cl

(158)

Ahn et al. [73] reported some novel cyano-pyrazolines (159) as potent anti-diabetic agents. These have the ability to inhibit dipeptidyl peptidase. Furthermore the (S)-configuration was found to be more effective.

The antioxidant properties of pyrazolines were evaluated by Babu et al. [74]. The studies were carried out at the concentrations of 1000, 500, 250, 100, 50, 25 and 10 $mg.mL^{-1}$ against the standard drug, ascorbic acid. Some of the compounds (160) have good activity, comparable with the standard ascorbic acid.

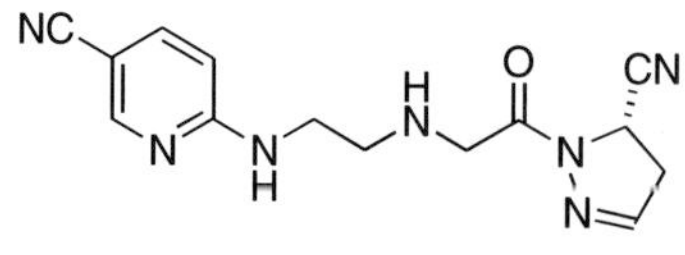

(159)

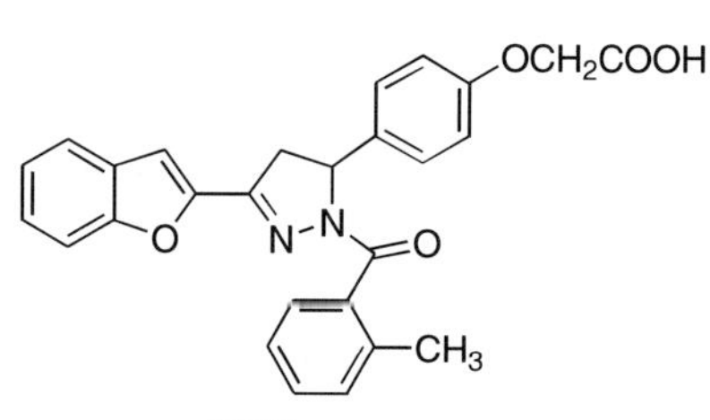

(160)

Silver K. S. [75] synthesized pyrazolines (161), (162) and studied the insecticidal properties of these compounds on the basis of pharmacological, electrophysiological and toxicological properties.

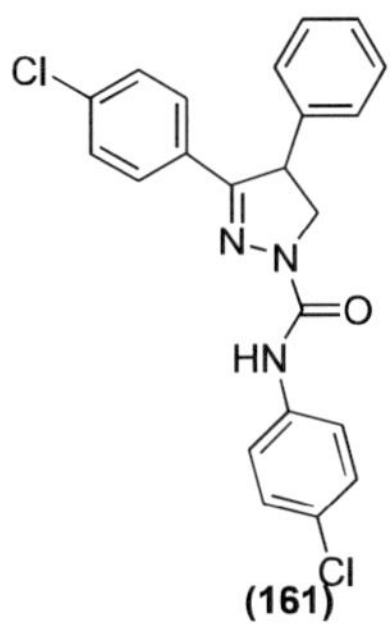

(161)

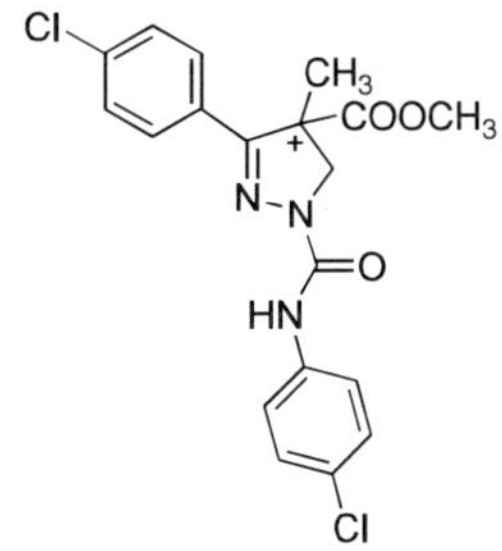

(162)

Marlie Johnson et al. [76] synthesized some acetyl- and non- acylated pyrazoline derivatives of combretastatin-A4 and evaluated their activity against cell lines. Compound (163) was the most active compound in the series. From the molecular modeling studies, these compounds have twisted conformation.

The compound (164) was shown by G. Holan [77] to possess insecticidal and anti-acetylcholinesterase activity.

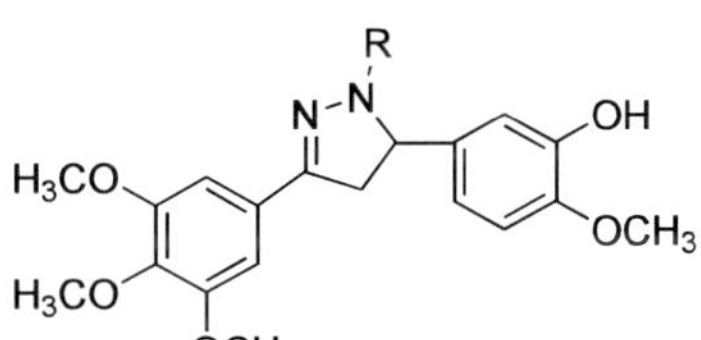

R = H, Ac

(163)

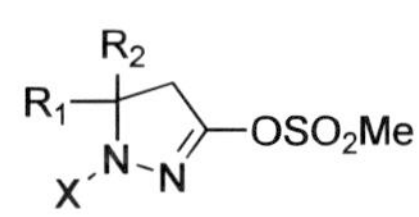

(164)

Some pyrazolines show excellent luminescent properties. Z. Y. Lu [78] reported the synthesis and application of fluorescent pyrazolines (169). Three different pyrazolines were synthesized that showed strong blue fluorescent properties, which can be used as blue organic electroluminescence materials (OELMs).

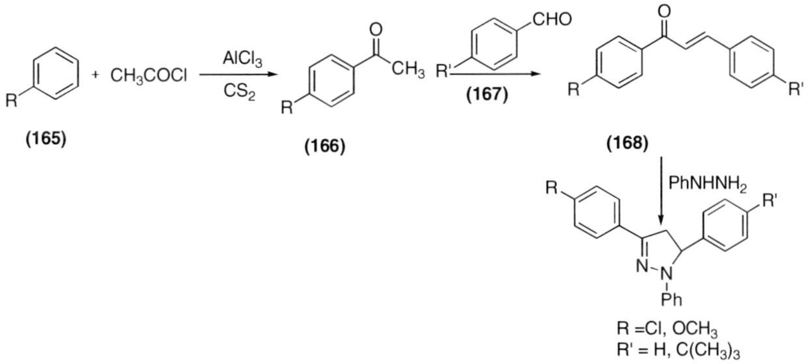

R =Cl, OCH_3
R' = H, $C(CH_3)_3$

(169)

3. Synthesis of Pyrazolines

Pyrazolines are usually synthesized by the cyclization reaction between chalcones and hydrazines under acidic or basic conditions. The most simple, straightforward and classical synthesis of these compounds involves the base-catalyzed Claisen-Schmidt condensation reaction of aromatic ketones and aldehydes to give α, β - unsaturated ketones (chalcones) which are then cyclized in the presence of suitable cyclizing agent with hydrazines to afford the corresponding 2-pyrazolines. Synthetic protocols employing microwave irradiations are well reported in literature. Here is a survey of some methods for the synthesis of pyrazolines.

B. Shivaram Holla [79] synthesized pyrazolines (173) by the condensation reaction between chalcones (172) and hydrazines using amberlyst-15 catalyst. The catalyst was non-toxic, stable, heterogeneous and recyclable.

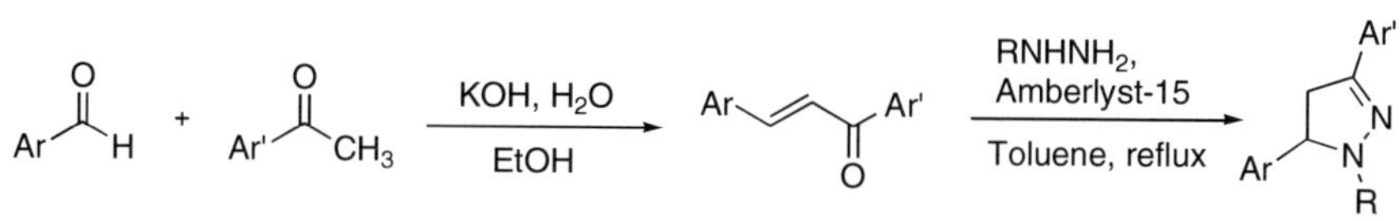

(170) (171) (172) (173)

Ji-Tai Li [80] synthesized 1, 3, 5-triaryl-2-pyrazolines (176) by the reaction between chalcones (174) and phenyl hydrazine hydrochloride (175) in an aqueous medium under ultrasound irradiation in the presence of sodium actetate in acetic acid. This method affords a high yield of products in a shorter duration (1.5-2 h) under mild reaction conditions.

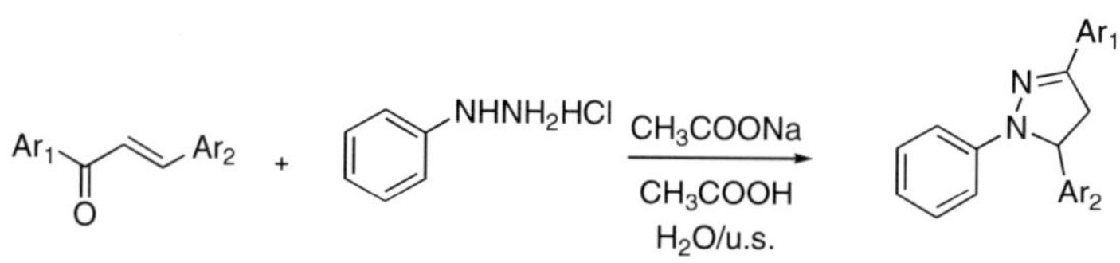

(174) (175) (176)

Vimesh M. Patel and Kishor R. Desai [81] suggested microwave-promoted, solvent-free and eco-friendly synthesis of fluorine-containing 1-thiocarbamoyl-3, 5-diphenyl-2-pyrazolines (180) over potassium carbonate in better yields.

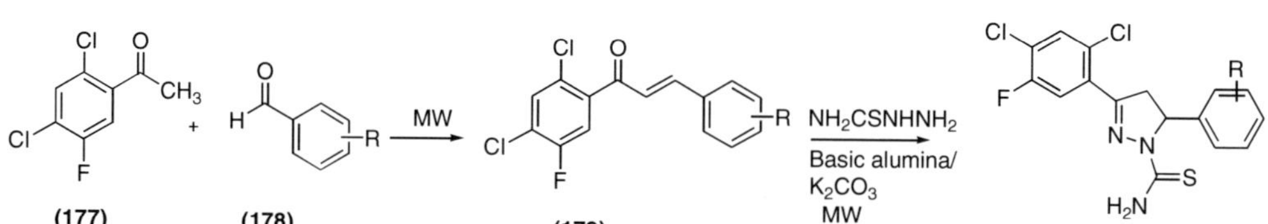

(177) (178) (179) (180)

R =H, 2-NO$_2$, 3-NO$_2$, 2-Cl, 4-Cl, 4-NMe$_2$, etc

Karolin Alex [82] reported the zinc triflate catalyzed one pot method for the regioselective synthesis of pyrazolines (184). In this method, hydrazines were treated with 3-

butynol (182) in the presence of a catalytic amount of zinc triflate to give pyrazoline derivatives.

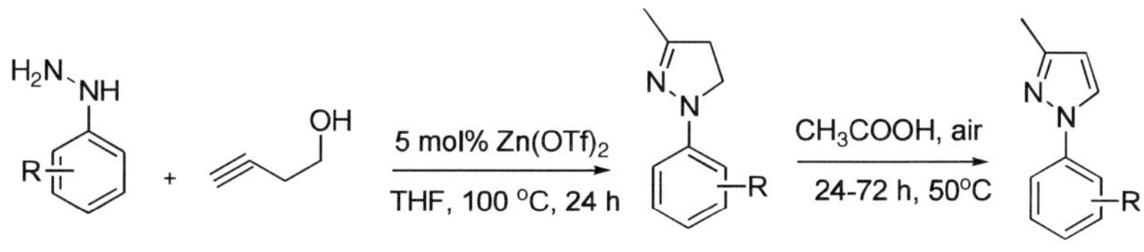

(181) (182) (183) (184)

Lars O. Lungs and John Nielsen [83] developed a method for the solid phase synthesis of 3-amino-2-pyrazolines (187) via the conjugate addition of hydrazines to α, β-unsaturated nitriles and then by cyclization to afford the corresponding 3-amino-2-pyrazolines.

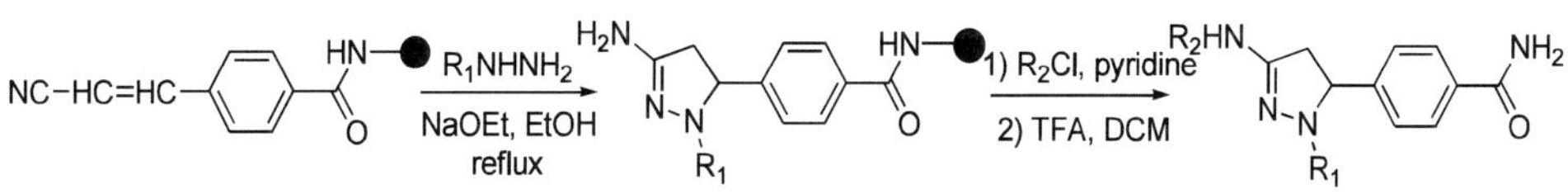

(185) (186) (187)

R_1= aryl, heteroaryl
R_2= acetyl, p-tosyl or 4-OMe-benzoyl

Y. Ju and R. S. Varma [84] synthesized pyrazolines (190) from the condensation reaction between hydrazines and alkyl dihalides in the presence of potassium carbonate and NaOH in aqueous medium under microwave irradiation in excellent yield.

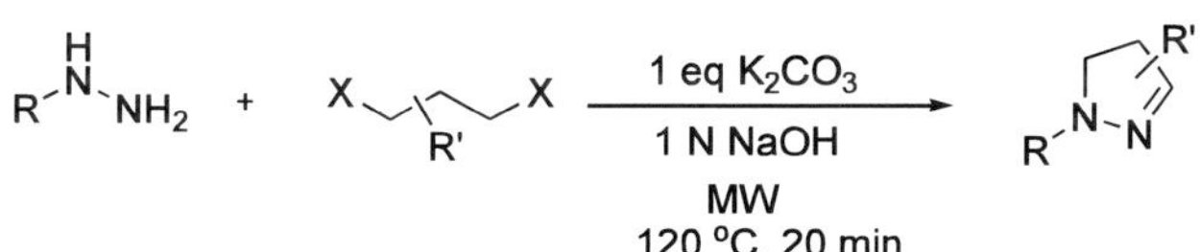

(188) (189) (190)

R = Ar, alkyl
X = Br, Cl , OTs

S. L. Cui et al. [85] reported that aziridines on treatment with DEAD undergoes ring expansion to afford the corresponding pyrazolines (193).

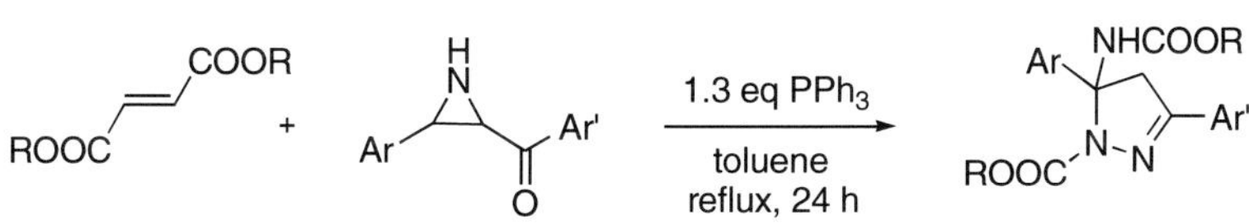

(191) (192) (193)

A catalyst free green protocol for the synthesis of pyrazolines (196) under solvent free conditions was reported by Daniela Vuluga [86]. The reaction involves 1, 3-dipolar addition of alkynes to α-diazocarbonyl compounds. The reaction is simple and requires no work up or purification techiques after completion.

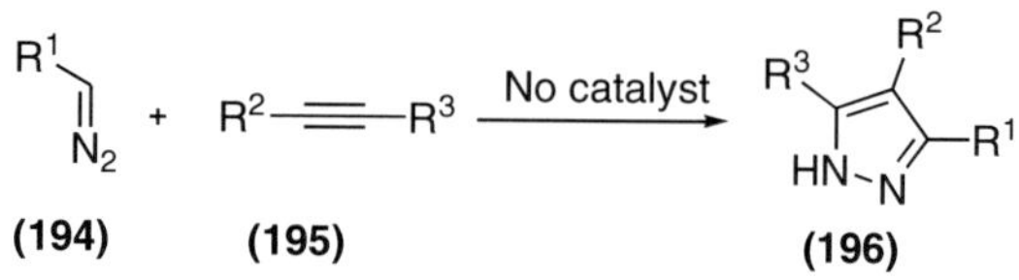

Conclusion

Variously substituted pyrazoline skeleton containing heterocyclic compounds exhibit remarkable biological and pharmacological activities. Research is under investigation to develop new therapeutic agents based upon pyrazoline ring. This chapter will really help researchers in this area for the development of novel bioactive compounds possessing pyrazoline motifs.

References

[1] Ozdemir, A.; Turan, Z. G.; Kaplancıklı, Z.A.; Revial G. & Guven, K. (2007). Synthesis and antimicrobial activity of 1-(4-aryl-2-thiazolyl)-3-(2-thienyl)-5-aryl-2-pyrazoline derivatives. *European Journal of Medicinal Chemistry*, *42*, 403-409.

[2] Abdel-Wahab, B. F.; Abdel-Aziz, H. A. & Ahmed, E. M. (2009). Synthesis and antimicrobial evaluation of 1-(benzofuran-2-yl)-4-nitro-3-arylbutan-1-ones and 3-(benzofuran-2-yl)-4,5-dihydro-5-aryl-1-[4-(aryl)-1,3-thiazol-2-yl]-1*H*-pyrazoles. *European Journal of Medicinal Chemistry, 44*, 2632-2635.

[3] Siddiqui, Z. N.; Musthafa, T. N. M.; Ahmed, A. & Khan, A. U. (2011). Thermal solvent-free synthesis of novel pyrazolyl chalcones and pyrazolines as potential antimicrobial agents. *Biorganic and Medicinal Chemistry Letters, 21,* 2860-2865.

[4] Shekarchi, M.; Pirali-Hamedani, M.; Navidpour, L.; Adib, N. & Shafiee, A. (2008). Synthesis, antibacterial and antifungal activities of 3-aryl-5-(pyridin-3-yl)-4, 5-dihydropyrazole-1-carbothioamide derivatives. *Journal of Iranian Chemical Society, 5*, 150-158.

[5] Azarifar, D. & Shaebanzadeh, M. (2002). Synthesis and characterization of new 3, 5-dinaphthyl substituted 2-pyrazolines and study of their antimicrobial activity. *Molecules*, *7*, 885-895.

[6] Nauduri, D. & Reddy, G. B. S. (1998). Antibacetrials and antimycotics: Part 1: Synthesis and activity of 2-pyrazoline derivatives. *Chemical Pharmaceutical Bulleitin, 46*, 1254-1260.

[7] Stirrett, K. L.; Ferreras, J. A.; Jayaprakash, V.; Sinha, B. N.; Ren, T. & Quadri, L. E. N. (2008). Small molecules with structural similarities to siderophores as novel antimicrobials against *mycobacterium tuberculosis* and *yersinia pestis*. *Bioorganic & Medicinal Chemistry Letters*, *18,* 2662-2668.

[8] Ashok, M. & Holla, B. S. (2006). Convenient synthesis of some 3, 5-arylated 2-pyrazolines carrying 4-methylthiophenyl moiety and evaluation of their antimicrobial activity. *Journal of Pharmacology and Toxicology, 1*, 464-470.

[9] Solankee, A.; Solankee, S. & Patel, G. (2008). Synthesis and antibacterial evaluation of some novel isoxazole and pyrazoline derivatives. *Rasayan Journal of Chemistry, 1*, 581-585.

[10] Mokle, S. S.; Vibhute, A. Y.; Khansole, S. V.; Zangade, S. B. & Vibhute, Y. B. (2010). Synthesis, characterization and antibacterial activity of some new 2-pyrazolines using triethanolamine as reaction solvent. *Research Journal of Pharmaceutical, Biological and Chemical Sciences, 1*, 631-638.

[11] Jayashankara, B. & Rai, K. M. L. (2008). Synthesis and antimicrobial studies of new series of pyrazoline bearing *bis*-heterocycles *via* 1, 3-dipolar cycloaddition reactions. *E-Journal of Chemistry, 5*, 309-315.

[12] Kumar, B.; Pathak, V.; Rani, S.; Kant, R. & Tewari, I. C. (2009). Synthesis and antimicrobial activity of some bromo-benzothiazolo pyrazolines. *International Journal of Microbiology Research, 1*, 20-22.

[13] Mishra, S. K. ; Sahoo, S. ; Panda, P. K. ; Mishra, S. R. ; Mohanta, R. K. ; Ellaiah , P. & Panda, C. S. (2007). Synthesis and antimicrobial activity of some 1-(2, 4-dinitrophenyl)-3-(3-nitrophenyl)-5-(4-substituted phenyl)-4-bromo-2-pyrazolines and 1-(2,4-dinitrophenyl)-3-(3-nitrophenyl)-5-(4-substituted phenyl)-2-pyrazolin-4-ones. *Acta Poloniae Pharmaceutica n Drug Research, 64*, 359-364.

[14] Patel, P. S.; Shah, R. A.; Trivedi, D. K. & Vyas, P. J. (2010). Synthesis, characterization and antimicrobial activities of 1-acetyl-5-(substituted phenyl)-{3-[4-(2-methyl-4-benzylidene-5-oxo-imidazol-1-yl)] phenyl}-4, 5-dihydropyrazole. *Der Pharma Chemica, 2*, 168-172.

[15] Dawane, B. S.; Konda, S. G.; Mandawad, G. G. & Shaikh, B. M. (2010). Poly (ethylene glycol) (PEG-400) as an alternative reaction solvent for the synthesis of some new 1-(4-(4′-chlorophenyl)-2-thiazolyl)-3-aryl-5-(2-butyl-4-chloro-1*H*-imidazol-5yl)-2-pyrazolines and their *in vitro* antimicrobial evaluation. *European Journal of Medicinal Chemistry, 45,* 387-392.

[16] Konda, S. G.; Valekar, A. H.; Lomate, S. T.; Lokhande, P. D. & Dawane, B. S. (2010). Synthesis and antibacterial studies of some new pyrazoline and isoxazoline derivatives *Journal of Chemical and Pharmaceutical Research, 2*, 1-6.

[17] Shah, N. N.; Ziauddin, H. M.; Zameer, M.; Hingole, S. S. & Baseer, M. A. (2010). An introduction of some novel pyrazolines to synthetic chemistry and antimicrobial studies. *Journal of Chemical and Pharmacutical Research, 2*, 441-445.

[18] Goodell, J. R.; Puig-Basagoiti, F.; Forshey, B. M.; Shi, P-Y. & Ferguson, D. M. (2006). Identification of compounds with *anti-west Nile virus* activity. Journal of Medicinal Chemistry, *49*, 2127–2137.

[19] Sujatha, K. ; Shanthi, G. ; Selvam, N. P. ; Manoharan, S. ; Perumal, P. T. & Rajendran, M. (2009). Synthesis and antiviral activity of 4,4′-(arylmethylene)bis(1*H*-pyrazol-5-ols) against *peste des petits ruminant virus* (PPRV). *Bioorganic & Medicinal Chemistry Letters, 19*, 4501-4503.

[20] Ozdemir, Z.; Kandilci, H. B.; Gumusel, B.; Calis, U. & Bilgin, A. A. (2007). Synthesis and studies on antidepressant and anticonvulsant activities of some 3-(2-furyl)-pyrazoline derivatives. *European Journal of Medicinal Chemistry, 42,* 373-379.

[21] Chimenti, F.; Fioravanti, R.; Bolasco, A.; Manna, F.; Chimenti, P.; Secci, D.; Rossi, F.; Turini, P.; Ortuso, F.; Alcaro, S. & Cardia, M. C. (2008). Synthesis, molecular modeling studies and selective inhibitory activity against MAO of *N*1-propanoyl-3, 5-diphenyl-4, 5-dihydro-(1*H*)-pyrazole derivatives. *European Journal of Medicinal Chemistry, 43,* 2262-2267.

[22] Chimenti, F.; Maccioni, E.; Secci, D.; Bolasco, A.; Chimenti, P.; Granese, A.; Befani, O.; Turini, P.; Alcaro, S.; Ortuso, F.; Cirilli, R.; Torre, F. L.; Cardia, M. C. & Distinto, S. (2005). Synthesis, molecular modeling studies and selective inhibitory activity against monoamine oxidase of 1-thiocarbamoyl-3, 5-diaryl-4, 5-dihydro-(1*H*) - pyrazole derivatives. Journal of Medicinal Chemistry, *48*, 7113–7122.

[23] Prasad, Y. R. ; Rao, A. L. ; Prasoona, L. ; Murali, L. & Kumar, P. R. (2005). Synthesis and antidepressant activity of some 1, 3, 5-triphenyl-2-pyrazolines and 3-(2″-hydroxy naphthalen-1″-yl)-1, 5-diphenyl-2-pyrazolines. *Bioorganic & Medicinal Chemistry Letters*, *15*, 5030-5034.

[24] Gordaliza, M. ; del Corral, J. M. M. ; Castro, M. A. ; Lopez-Vazquez, M. L. ; Feliciano, A. S. ; Garcia-Gravalos, M. D. & Carpy, A. (1995). Synthesis and evaluation of pyrazolignans: A new class of cytotoxic agents. *Bioorganic & Medicinal Chemistry, 3,* 1203-1210.

[25] Manna, F. ; Chimenti, F. ; Fioravanti, R. ; Bolasco, A. ; Secci, D. ; Chimenti, P. ; Ferlini, C. & Scambia, G. (2005). Synthesis of some pyrazole derivatives and preliminary investigation of their affinity binding to P-glycoprotein. *Bioorganic & Medicinal Chemistry Letters, 15*, 4632-4635.

[26] Amnerkar, N. D. & Bhusari, K. P. (2010). Preliminary anticancer activity of some prop-2-eneamido, thiazole and 1-acetyl-pyrazolin derivatives of aminobenzothiazoles. *Digest Journal of Nanomaterials and Biostructures*, *5*, 177-184.

[27] Havrylyuk, D. ; Zimenkovsky, B. ; Vasylenko, O. ; Zaprutko, L. ; Gzella, A. & Lesyk, R. (2009). Synthesis of novel thiazolone-based compounds containing pyrazoline moiety and evaluation of their anticancer activity. *European Journal of Medicinal Chemistry, 44*, 1396-1404.

[28] Rostom, S. A. F. (2006). Synthesis and in vitro antitumor evaluation of some indeno [1, 2-*c*] pyrazolines substituted with sulfonamide, sulfonylurea (-thiourea) pharmacophores and some derived thiazole ring systems. *Bioorganic and Medicinal Chemistry, 14,* 6475-6485.

[29] Zahran, M. A. H.; Salama, H. F.; Abdin; Y. G. & Gamal-Eldeen, A. M. (2010). Efficient microwave irradiation enhanced stereoselective synthesis and antitumor activity of indolylchalcones and their pyrazoline analogues. *Journal of Chemical Science, 122,* 587–595.

[30] Banday, A. H. ; Mir, B. P. ; Lone, I. H. ; Suri, K. A. & Kumar, H. M. S. (2010). Studies on novel D-ring substituted steroidal pyrazolines as potential anticancer agents. *Steroids, 75*, 805-809.

[31] Ozdemir, Z.; Kandilci, H. B.; Gumusel, B.; Calis, U. & Bilgin, A. A. (2008). Synthesis and studies on antidepressant and anticonvulsant activities of some 3-(2-thienyl) pyrazoline derivatives. *Archieves of Pharmaceutical and Chemical Life Sciences, 341,* 701–707.

[32] Siddiqui, A. A.; Rahman, M. A.; Shaharyar, M. & Mishra, R. (2010). Synthesis and anticonvulsant activity of some substituted 3, 5-diphenyl-2-pyrazoline-1-carboxamide derivatives. *Chemical Sciences Journal, 8*, 1-10.

[33] Reddy, M. V. R. ; Billa, V. K. ; Pallela, V. R. ; Mallireddigari, M. R. ; Boominathan, R. ; Gabriel, J. L. ; Reddy , E. P. & Reddy, M. V. R. (2008). Design, synthesis, and biological evaluation of 1-(4-sulfamylphenyl)-3-trifluoromethyl-5-indolyl pyrazolines as cyclooxygenase-2 (COX-2) and lipoxygenase (LOX) inhibitors. *Bioorganic & Medicinal Chemistry, 16,* 3907-3916.

[34] Barsoum, F. F.; Hosni, H. M. & Girgis, A. S. (2006). Novel bis(1-acyl-2-pyrazolines) of potential anti-inflammatory and molluscicidal properties, *Bioorganic & Medicinal Chemistry*, *14*, 3929-3937.

[35] Khode, S.; Maddi, V.; Aragade, P.; Palkar, M.; Ronad, P. K.; Mamledesai, S.; Thippeswamy, A. H. M. & Satyanarayana, D. (2009). Synthesis and pharmacological evaluation of a novel series of 5-(substituted) aryl-3-(3-coumarinyl)-1-phenyl-2-pyrazolines as novel anti-inflammatory and analgesic agents. *Journal of Medicinal Chemistry*, *44*, 1682-1688.

[36] Chandra, T.; Garg, N.; Lata, S. Saxena, K. K. & Kumar, A. (2010). Synthesis of substituted acridinyl pyrazoline derivatives and their evaluation for anti-inflammatory activity. *European Journal of Medicinal Chemistry*, *45*, 1772-1776.

[37] Amir, M.; Kumar, H. & Khan, S. A. (2008). Synthesis and pharmacological evaluation of pyrazoline derivatives as new anti-inflammatory and analgesic agents. *Bioorganic & Medicinal Chemistry Letters, 18*, 918-922.

[38] Joshi, R. S. ; Mandhane, P. G.; Diwakar, S. D. ; Dabhade, S. K. & Gill, C. H. (2010). Synthesis, analgesic and anti-inflammatory activities of some novel pyrazolines derivatives. *Bioorganic & Medicinal Chemistry Letters, 20*, 3721-3725.

[39] Ramesh, B. & Sumana, T. (2010). Synthesis and anti-inflammatory activity of pyrazolines. *E-Journal of Chemistry, 7*, 514-516.

[40] Bano, S.; Javed, K.; Ahmad, S.; Rathish, I. G.; Singh, S. & Alam, M. S. (2011). Synthesis and biological evaluation of some new 2-pyrazolines bearing benzene sulfonamide moiety as potential anti-inflammatory and anti-cancer agents. *European Journal of Medicinal Chemistry, 46,* 5763-5768.

[41] Girisha, K. S.; Kalluraya, B.; Narayana, V. & Padmashree. (2010). Synthesis and pharmacological study of 1-acetyl/propyl-3-aryl-5-(5-chloro-3-methyl-1-phenyl-1*H*-pyrazol-4-yl)-2-pyrazoline. *European Journal of Medicinal Chemistry*, *45*, 4640-4644.

[42] Rathish, I.G.; Javed, K.; Ahmad, S.; Bano , S.; Alam, M. S.; Pillai, K. K; Singh, S.& Bagchi, V. (2009). Synthesis and anti-inflammatory activity of some new 1, 3, 5-trisubstituted pyrazolines bearing benzene sulfonamide. *Bioorganic & Medicinal Chemistry Letters*, *19*, 255-258.

[43] Malhotra, P.; Pattan, S. & Nikalje, P. (2010). Microwave assisted synthesis and anti-inflammatory activity of 3, 5-diaryl substituted–2-pyrazolines. *International Journal of Pharmacy and Pharmaceutical Sciences*, *2*, 21-26.

[44] Abid, M. & Azam, A. (2005). Synthesis and antiamoebic activities of 1-*N*-substituted cyclised pyrazoline analogues of thiosemicarbazones. *Bioorganic & Medicinal Chemistry, 13,* 2213-2220.

[45] Budakoti, A.; Abid, M. & Azam, A. (2007). Syntheses, characterization and in vitro antiamoebic activity of new Pd (II) complexes with 1-*N*-substituted thiocarbamoyl-3, 5-

diphenyl-2-pyrazoline derivatives. *European Journal of Medicinal Chemistry, 42*, 544-551.

[46] Bhat, A. R.; Athar, F. & Azam, A. (2009). Bis-pyrazolines: Synthesis, characterization and antiamoebic activity as inhibitors of growth of Entamoeba histolytica. *European Journal of Medicinal Chemistry, 44,* 426-431.

[47] Budakoti, A.; Bhat, A. R.; Athar, F. & Azam, A. (2008). Syntheses and evaluation of 3-(3-bromo phenyl)-5-phenyl-1-(thiazolo[4,5-*b*] quinoxaline-2-yl)-2-pyrazoline derivatives. *European Journal of Medicinal Chemistry, 43*, 1749-1757.

[48] Abid, M.; Bhat, A. R.; Athar, F. & Azam, A. (2009). Synthesis, spectral studies and antiamoebic activity of new 1-*N*-substituted thiocarbamoyl-3-phenyl-2-pyrazolines. *European Journal of Medicinal Chemistry, 44*, 417-425.

[49] Abid, M. & Azam, A. (2006). Synthesis, characterization and antiamoebic activity of 1-(thiazolo [4, 5-*b*] quinoxaline-2-yl)-3-phenyl-2-pyrazoline derivatives. *Bioorganic and Medicinal Chemistry Letters, 16*, 2812-2816.

[50] Budakoti, A.; Abid, M. & Azam, A. (2006). Synthesis and antiamoebic activity of new 1-*N*-substituted thiocarbamoyl-3, 5-diphenyl-2-pyrazoline derivatives and their Pd (II) complexes. *European Journal of Medicinal Chemistry, 41*, 63-70.

[51] Hayat, F.; Salahuddin, A. ; Umar, S. & Azam, A. (2010) Synthesis, characterization, antiamoebic activity and cytotoxicity of novel series of pyrazoline derivatives bearing quinoline tail. *European Journal of Medicinal Chemistry, 45*, 4669-4675.

[52] Zhang, X.; Li, X. ; Allan, G. F.; Sbriscia, T.; Linton, O.; Lundeen, S. G. & Sui, Z. (2007). Design, Synthesis and in vivo SAR of a novel series of pyrazolines as potent selective androgen receptor modulators. Journal of Medicinal Chemistry, *50*, 3857–3869.

[53] Manna, F.; Chimenti, F.; Fioravanti, R.; Bolasco, A.; Secci, D.; Chimenti, P. Ferlini, C. & Scambia, G. (2005). Synthesis of some pyrazole derivatives and preliminary investigation of their affinity binding to P-glycoprotein. *Bioorganic & Medicinal Chemistry Letters, 15*, 4632-4635.

[54] Roecker , A. J.; Coleman, P. J.; Mercer, S. P.; Schreier, J. D.; Buser, C. A.; Walsh, E. S.; Hamilton, K.; Lobell, R. B.; Tao, W.; Diehl, R. E.; South, V. J.; Davide, J. P.; Kohl, N. E.; Yan, Y., Kuo, L. C.; Li, C.; Fernandez-Metzler, C.; Mahan, E. A.; Prueksaritanont, T. & Hartman, G. D. (2007). Kinesin spindle protein (KSP) inhibitors. Part 8: Design and synthesis of 1, 4-diaryl-4, 5-dihydropyrazoles as potent inhibitors of the mitotic kinesin KSP. *Bioorganic & Medicinal Chemistry Letters, 17*, 5677-5682.

[55] Jones, D. G.; Liang, X.; Stewart, E. L.; Noe, R. A. ; Kallander, L. S.; Madauss, K P.; Williams, S. P.; Thompson, S. K.; Gray, D. W. & Hoekstra, W. J. (2005). Discovery of non-steroidal mifepristone mimetics: Pyrazoline-based PR antagonists. *Bioorganic & Medicinal Chemistry Letters, 15*, 3203-3206.

[56] Amr, A. E-G. E.; Abdel-Latif, N. A. & Abdalla, M. M. (2006). Synthesis and antiandrogenic activity of some new 3-substituted androstano [17, 16-*c*]-5′-aryl-pyrazoline and their derivatives. *Bioorganic & Medicinal Chemistry, 14,* 373-384.

[57] Jeong, T-S.; Ki, K. S.; An, S-J. ; Cho, K-H. ; Lee, S. & Lee, W. S. (2004). Novel 3, 5-diaryl pyrazolines as human acyl-CoA: cholesterol acyltransferase inhibitors. *Bioorganic & Medicinal Chemistry Letters, 14,* 2715-2717.

[58] Camacho, M. E.; Leon, J.; Entrena, A.; Velasco, G.; Carrion, M. D.; Escames, G.; Vivo, A. ; Acuna-Castroviejo, D.; Gallo, M. A. & Espinosa, A. (2004). 4, 5-Dihydro-

1*H*-pyrazole derivatives with inhibitory nNOS activity in rat brain: Synthesis and structure-activity relationships. Journal of Medicinal Chemistry, *47*, 5641–5650.

[59] Carrion, M. D.; Camacho, M. E.; Leon, J.; Escames, G.; Tapias, V.; Acuna-Castroviejo, D.; Gallo, M. A. & Espinosa, A. (2004). Synthesis and iNOS/nNOS inhibitory activities of new benzoylpyrazoline derivatives. *Tetrahedron, 60*, 4051-4069.

[60] Chimenti, F. ; Carradori , S. ; Secci, D. ; Bolasco, A.; Bizzarri, B. ; Chimenti, P.; Granese, A.; Yanez, M. & Orallo, F. (2010). Synthesis and inhibitory activity against human monoamine oxidase of N1-thiocarbamoyl-3,5-di(hetero)aryl-4,5-dihydro-(1*H*)-pyrazole derivatives. *European Journal of Medicinal Chemistry, 45*, 800-804.

[61] Fioravanti, R.; Bolasco, A.; Manna, F.; Rossi, F.; Orallo, F.; Yanez, M.; Vitali, A.; Ortuso, F. & Alcaro, S. (2010). Synthesis and molecular modelling studies of prenylated pyrazolines as MAO-B inhibitors. *Bioorganic & Medicinal Chemistry Letters, 20,* 6479-6482.

[62] Zampieri, D. ; Mamolo, M. G.; Laurini, E.; Scialino, G.; Banfi, E. & Vio, L. (2008). Antifungal and antimycobacterial activity of 1-(3, 5-diaryl-4, 5-dihydro-1*H*-pyrazol-4-yl)-1*H*-imidazole derivatives. *Bioorganic & Medicinal Chemistry, 16,* 4516-4522.

[63] Shaharyar, M.; Siddiqui, A. A.; Ali, M. A.; Sriram, D. & Yogeeswari, P. (2006). Synthesis and in vitro antimycobacterial activity of N^1-nicotinoyl-3-(4′-hydroxy-3′-methyl phenyl)-5-[(sub) phenyl]-2-pyrazolines. *Bioorganic & Medicinal Chemistry Letters, 16*, 3947-3949.

[64] Ahsan , M. J.; Samy, J. G.; Dutt, K. R.; Agrawal, U. K.; Yadav, B. S.; Vyas, S.; Kaur, R. & Yadav, G. (2011). Design, synthesis and antimycobacterial evaluation of novel 3-substituted-*N*-aryl-6, 7-dimethoxy-3*a*,4-dihydro-3*H*-indeno[1,2-*c*]pyrazole-2-carboxamide analogues. *Bioorganic & Medicinal Chemistry Letters, 21,* 4451-4453.

[65] Ahsan, M. J.; Samy, J. G. ; Soni, S.; Jain, N. ; Kumar, L. ; Sharma, L. K.; Yadav, H.; Saini, L.; Kalyansing, R. G.; Devenda, N. S.; Prasad, R. & Jain, C. B. (2011). Discovery of novel antitubercular 3a,4-dihydro-3*H*-indeno[1,2-*c*]pyrazole-2-carboxamide/carbothioamide analogues. *Bioorganic & Medicinal Chemistry Letters, 21,* 5259-5261.

[66] Shaharyar, M.; Siddiqui, A. A. & Ali, M. A. Synthesis and evaluation of phenoxy acetic acid derivatives as anti-mycobacterial agents. (2006). *Bioorganic & Medicinal Chemistry Letters, 16,* 4571-4574.

[67] Ali, M. A.; Shaharyar, M. & Siddiqui, A. A. (2007). Synthesis, structural activity relationship and anti-tubercular activity of novel pyrazoline derivatives. *European Journal of Medicinal Chemistry, 42*, 268-275.

[68] Ahsan, M. J.; Samy, J. G.; Khalilullah, H.; Bakht, M. A. & Hassan, M. Z. (2011). Synthesis and antimycobacterial evaluation of 3a,4-dihydro-3*H*-indeno [1,2-c] pyrazole-2-carboxamide analogues. *European Journal of Medicinal Chemistry, 46,* 5694-5697.

[69] Acharya, B. N.; Saraswat, D. ; Tiwari, M. ; Shrivastava, A. K. ; Ghorpade, R. ; Bapna, S. & Kaushik, M. P. (2010). Synthesis and antimalarial evaluation of 1, 3, 5-trisubstituted pyrazolines. *European Journal of Medicinal Chemistry, 45,* 430-438.

[70] Maccioni, E.; Alcaro, S.; Orallo, F.; Cardia, M. C.; Distinto, S.; Costa, G.; Yanez, M. ; Sanna, M. L.; Vigo, S. ; Meleddu, R. & Secci, D. (2010). Synthesis of new 3-aryl-4,5-dihydropyrazole-1-carbothioamide derivatives. An investigation on their ability to inhibit monoamine oxidase. *European Journal of Medicinal Chemistry, 45*, 4490-4498.

[71] Turan, Z. G.; Chevallet, P.; Kilic, F. S. & Erolc, K. (2000). Synthesis of some thiazolyl-pyrazoline derivatives and preliminary investigation of their hypotensive activity. *European Journal of Medicinal Chemistry, 35*, 635-641.

[72] Kaplancikli, Z. A.; Turan-Zitouni, G.; Ozdemir, A.; Can, O. D. & Chevallet, P. (2009). Synthesis and antinociceptive activities of some pyrazoline derivatives. *European Journal of Medicinal Chemistry, 44,* 2606-2610.

[73] Ahn , J. H. ; Kim, H-M.; Jung, S. H.; Kang, S. K.; Kim, K. R.; Rhee, S. D.; Yang, S-D.; Cheon, H. G. & Kim, S. S. (2004). Synthesis and DP-IV inhibition of cyano-pyrazoline derivatives as potent anti-diabetic agents. *Bioorganic & Medicinal Chemistry Letters, 17*, 4461-4465.

[74] Babu, V. H.; Sridevi, C.; Joseph, A. & Srinivasan, K. K. (2007). Synthesis and biological evaluation of some novel pyrazolines. *Indian Journal of Pharmaceutical Sciences*, *69*, 470-473.

[75] Silver, K. S. & Soderlund, D. M. (2005). Action of pyrazoline-type insecticides at neuronal target sites. *Pesticide Biochemistry and Physiology, 81*, 136-143.

[76] Johnson, M.; Younglove, B. ; Lee, L.; Blanc, R. L.; Holt, H.; Hills, P.; Mackay, H. ; Brown, T. ; Mooberry, S. L. & Lee, M. (2007). Design, synthesis, and biological testing of pyrazoline derivatives of combretastatin-A4. *Bioorganic & Medicinal Chemistry Letters, 17,* 5897-5901.

[77] Holan, G.; Virgona, C. T.; Watson, K. G. & Finkelstein, B. L. (1996). Synthesis, insecticidal and anti-acetylcholinesterase activity of a new class of heterocyclic methanesulfonates. *Bioorganic & Medicinal Chemistry Letters, 6,* 77-80.

[78] Lu, Z. Y.; Zhu, W. G.; Jiang, Q. & Xie, M. G. (1999). Synthesis of pyrazolines used as blue organic electroluminescence materials. *Chinese Chemical Letters, 10*, 679–682.

[79] Holla, B. S.; Mahalinga, M.; Poojari, B.; Ashok, M. & Akberali, P. M. (2006). Synthesis of pyrazolines promoted by Amberlyst-15 catalyst. *Indian Journal of Chemistry*, *45B*, 568-571.

[80] Li, J-T.; Zhang, X-H. & Lin, Z-P. (2007). An improved synthesis of 1, 3, 5-triaryl-2-pyrazolines in acetic acid aqueous solution under ultrasound irradiation. *Beilstein Journal of Organic Chemistry, 3*, 1-4.

[81] Patel, V. M. & Desai, K. R. (2004). Eco-friendly synthesis of pyrazoline derivatives over potassium carbonate. *Arkivoc*, *i,* 123-129.

[82] Alex, K.; Tillack, A.; Schwarz, N. & M. Beller. (2008). Zinc-Catalyzed Synthesis of pyrazolines and pyrazoles via hydrohydrazination. *Organic Letters*, *10*, 2377-2379.

[83] Lungs, L. O. & Nielsen, J. (1998). Solid-phase synthesis of 3-amino-2-pyrazolines. *Tetrahedron Letters, 39,* 5845-5848.

[84] Ju, Y. & Varma, R. S. (2006). Aqueous N-Heterocyclization of Primary amines and hydrazines with dihalides: Microwave-assisted syntheses of *N*-azacycloalkanes, isoindole, pyrazole, pyrazolidine, phthalazine derivatives. *Journal of Organic Chemistry*, *71*, 135-141.

[85] Cui, S-L.; Wang, J. & Wang, Y-G. (2008). Facile access to pyrazolines via domino reaction of the Huisgen zwitterions with aziridines. *Organic Letters*, *10*, 13-16.

[86] Vuluga, D.; Legros, J.; Crousse, B. & Bonnet-Delpon, D. (2009). Synthesis of pyrazoles through catalyst-free cycloaddition of diazo compounds to alkynes. *Green Chemistry, 11*, 156-159.

Index

A

B

C

D

E

N

O

P

Q

R

S

T

U

V

W

X

Y

Z